SECOND EDITION

INTRODUCTION TO
STATISTICS
CONCEPTS AND APPLICATIONS

SECOND EDITION

INTRODUCTION TO
STATISTICS
CONCEPTS AND APPLICATIONS

David R. Anderson
University of Cincinnati

Dennis J. Sweeney
University of Cincinnati

Thomas A. Williams
Rochester Institute of Technology

WEST PUBLISHING COMPANY
St. Paul   New York   Los Angeles   San Francisco

Copyediting: Linda Thompson
Composition: Graphic World
Text design: Janet Bollow
Cover design: Imagesmythe
Cover painting: *Figure 1,* Jasper Johns, 1969. © Jasper Johns/VAGA New York 1990.

COPYRIGHT © 1986, 1991 by WEST PUBLISHING COMPANY
50 West Kellogg Boulevard
P.O. Box 64526
St. Paul, MN 55164-0526

LIBRARY OF CONGRESS CATALOGING-IN-PUBLICATION DATA

Anderson, David Ray, 1941—
    Introduction to statistics : concepts and applications / David R.
Anderson, Dennis J. Sweeney, Thomas A. Williams. —2nd ed.
        p.   cm.
    Includes index.
    ISBN 0-314-56669-4
    1. Statistics.   I. Sweeney, Dennis J.   II. Williams, Thomas
Arthur, 1944- .   III. Title.
QA276.12.A44   1991
519.5—dc20   90-19481
                                                                    CIP

Contents

CHAPTER 3

Descriptive Statistics II:
Measures of Location and Dispersion   56

STATISTICS IN THE NEWS   57

CHAPTER 4

Introduction to Probability 98

STATISTICS IN THE NEWS 99

CHAPTER 5

Random Variables and Discrete Probability Distributions 139

STATISTICS IN THE NEWS 140

CHAPTER 6   Continuous Probability Distributions   180

CHAPTER 7   Sampling and Sampling Distributions   215

CHAPTER 8

Inferences About a Population Mean   252

STATISTICS IN THE NEWS   253

CHAPTER 9   **Inferences About a Population Proportion**   303

CHAPTER 10   **Inferences About Means and Proportions with Two Populations**   328

CHAPTER 11   ## Inferences About Population Variances   366

CHAPTER 12   ## Experimental Design and the Analysis of Variance   391

CHAPTER 16

Tests of Goodness of Fit and Independence 604

STATISTICS IN THE NEWS 605

CHAPTER 17

Nonparametric Methods 622

STATISTICS IN THE NEWS 623

List of Statistics in the News

Preface

The purpose of this book is to provide a comprehensive treatment of introductory statistics for students from a wide variety of academic backgrounds. The text is applications oriented, and has been written with the needs of the nonmathematician in mind. The mathematical prerequisite is a course in college algebra.

Methodology and Applications Integrated

We have taken care to provide a sound methodological development. Throughout the text we have utilized notation that is generally accepted for the topic being covered. Thus, students will find that the text provides good preparation for the study of more advanced statistical material. A bibliography that should prove useful as a guide to further study has been included as an appendix.

Applications of the statistical methodology are an integral part of the organization and presentation of the material. Each chapter begins by motivating the student with a general interest news article that demonstrates a use of the statistical procedures that will be introduced in the chapter. Statistical techniques are then introduced using examples where the techniques have been successfully applied. The discussion and development of each technique is centered around an application setting, with the statistical results providing information helpful in solving the underlying problem.

Changes in the Second Edition

The first edition of this text was entitled *Statistics: Concepts and Applications;* the title of this edition has been changed to *Introduction to Statistics: Concepts and Applications.* We feel that this title better reflects the level of the material. In making modifications for this new edition we have maintained the presentation style of the previous edition. However, this edition includes a much more complete and detailed topical coverage than the first edition. For example, we have added to the material on exploratory data analysis, included sections on the Poisson, the hypergeometric, and exponential probability distributions, expanded the treatment of experimental design, and have added a new chapter on model building in regression analysis. The more significant changes in the second edition are summarized below.

Notes and Comments

At the end of many sections we have provided "notes and comments" designed to give the student additional insights about the statistical methodology presented in the section. The notes and comments include warnings and/or limitations about the methodology, recommendations for application, brief descriptions of additional technical considerations, and so on. It is hoped that this feature will expand the student's understanding of statistics and the student's ability to use the material.

Examples and Problems Based on Real Data

Sources such as *U.S.A. Today, Psychology Today, New England Journal of Medicine,* and *The Wall Street Journal* have been used to provide real data and real case studies with demonstrate the use of statistics in a variety of disciplines including science, sociology, psychology, health care, government and public policy, and business and economics. In some instances, data from the above sources enable the student to study methodology and solve problems based on real data. The use of real data means that the student not only learns about the statistical methodology but also learns about an application and the types of studies encountered in practice.

Descriptive Statistics

The three chapters on descriptive statistics from the first edition have been combined into two chapters. We now include material on the use of the standard deviation in the computation and interpretation of z-scores. The empirical rule has been added to Chapter 3 to further demonstrate the use of the standard deviation. Outliers and procedures for identifying outliers are now covered. In addition, the exploratory data analysis material on box plots and fences has been expanded. The presentation of frequency distributions has been simplified.

Probability and Probability Distribution

The Poisson and hypergeometric probability distributions have been added to Chapter 5. In Chapter 6 we have also included a section on the exponential probability distribution.

Sampling and Sampling Methods

Chapter 7 now includes a more thorough discussion of the important properties of point estimators. In addition, the section on other sampling methods such as stratified sampling and cluster sampling has been expanded.

Hypothesis Testing

Throughout the text more emphasis is placed on the proper formulation of hypotheses and reaching the proper conclusion when the hypothesis test leads to "Reject H_0" and "Do Not Reject H_0." The appropriate uses and limitations of hypothesis testing in research studies, in testing assumptions, and in decision making are discussed. The use and interpretation of p-values has been added and is used throughout the text.

Experimental Design

The presentation of experimental design in Chapter 12 has been revised and expanded. A new section on factorial experiments has been included, and in each section we have simplified the presentation by reducing the emphasis on computational formulas. The detailed formulas have been removed from the text presentation and are available in the chapter appendix.

New Chapter on Model Building in Regression Analysis

The presentation of regression analysis has been expanded from two chapters. Simple linear regression and correlation is covered in Chapter 13. Multiple linear regression is covered in Chapter 14. A Chapter 15 on regression analysis and model building has been added. The new chapter presents the general linear model and describes variable selection procedures such as stepwise regression and backwards elimination. Additional uses of residual analysis to identify outliers, influential observations and autocorrelation are also covered.

Microcomputer Software

Available to adopters is a revised and expanded version of *The Data Analyst*, an IBM-compatible microcomputer software package developed by the author. *The Data Analyst* version 2.0 can be used to solve problems and analyze data using descriptive statistics, confidence intervals for means and proportions of one and two populations, analysis of variance, simple and multiple linear regression, and more. Data sets from the text as well as additional data sets are available on a special data diskette accompanying *The Data Analyst*.

New Problems

Over 150 new problems have been added to this edition. As discussed earlier, many of these problems are based on real data and real case studies.

Features Retained in the Second Edition

We appreciate the acceptance and positive response to the first edition of this text. Accordingly, in making the above modifications to the text we have maintained many of the features of the previous edition. The more significant first edition features that are also included in this second edition are summarized below.

Learning Objectives

Each chapter begins with a statement of learning objectives under the heading of "What You Will Learn in This Chapter." This list contains the concepts the student will be expected to master and should help guide the student's study of the material.

Statistics in the News

Each chapter opens with a general interest news article that demonstrates a use of the statistical procedures that will be introduced in the chapter. These "Statistics in the News" applications are based on actual news articles appearing in magazines and newspapers such as *Time, USA Today, The New York Times,* and so on. The "Statistics in the News" is a condensed version of the original article and specifically focuses on the use of statistics as reported in the publication. Topical selections include the cost of a college education, marriage statistics, pay differentials for men and women, and professional sports; these selections have been made in an attempt to capture the student's interest and indicate that the statistical procedures that he or she is about to learn have some interesting applications.

Chapter Pedagogy

Each chapter introduces statistical methodology in the context of examples which demonstrate the use of the methodology in a wide variety of general interest applications. Problems are provided after each section to enable the student to check his or her progress. Answers to the even-numbered problems are provided at the back of the book. Each chapter concludes with a summary that reviews the key concepts and topics that have been introduced in the chapter. A glossary follows with review definitions of the statistical terms found in the chapter, and a key formula section itemizes the important equations that the student should know how to apply. A review quiz is then included to reinforce the key concepts presented. Supplementary exercises, based on the material throughout the chapter, are presented to provide additional opportunities to practice applying the methodology presented. Where appropriate, computer printouts are provided; these printouts demonstrate how computer packages can be used to provide the statistical computation and summaries.

Review Quizzes

The review quizzes consist of true-false and multiple choice questions. Each review quiz provides the student with an opportunity to evaluate his or her progress after the chapter material has been covered. Answers for the review quiz questions are included at the back of the book.

Computer Exercises

Eleven chapters conclude with an exercise which has been designed to be solved using a statistical computer package. These exercises, which are available for assignment at the option of the instructor, contain larger data sets than would normally be processed with hand calculations. Students may treat the exercises as cases in which the computer results and their personal evaluation and judgment may combine to provide the desired solution and/or recommendation.

Ancillaries

Accompanying the text is a complete package of support materials. A solutions manual, prepared by the authors, contains complete solutions to all exercises in the text. A test bank contains multiple-choice questions followed by problems, and WESTTEST Computerized Testing also accompanies the package. Transparency masters display worked-out demonstration problems, and a student workbook (prepared by Meredith Many and Charlotte Lewis, University of New Orleans) provides chapter summaries and glossaries, formula references, and additional practice problems with solutions. The Data Analyst microcomputer software package, developed by the authors, is available free to adopters, and students may purchase the software shrinkwrapped with the text for a small additional charge. In addition, the MYSTAT statistical software package may also be purchased shrinkwrapped with the text for a small additional charge.

Acknowledgments

We owe a debt to many of our colleagues and friends for their helpful comments and suggestions during the development of this manuscript. Among these are Paul B. Berger, Ben P. Bockstege, John M. Burns, Louis J. Cote, Robert H. Cranford, Henry Crouch, Carl Cuneo, Shirley Dowdy, David Gillman, Penelope Greene, Terry H. Hughes, Robert L. Lacher, Stanley M. Lukawecki, David R. Lund, Daniel E. McNamara, Robert Mee, Jeff Mock, Alex Papadopoulos, Charles D. Reinauer, Franklin D. Rich, Arnold L. Schroeder, A.K. Shah, W. Robert Stephenson, Bill Stines, Barbara Treadwell, David L. Turner, and Vasant B. Waikar.

We are also indebted to our editor, Mary C. Schiller, for her counsel and support during the preparation of this text. We would like to express our appreciation to Tad Bornhoft, our production editor, for the outstanding job he performed in pulling all the pieces together to produce this book. Finally, a sincere thank you to everyone else at West Publishing Company that played a part in helping us complete this project.

David R. Anderson
Dennis J. Sweeney
Thomas A. Williams

1 Introduction

Contents

The Trend Toward More Education

For most of us, having a good education is one of the keys to personal success. Today it is generally true that individuals with more education have more career opportunities and a greater potential for obtaining higher-paying jobs. But what is the trend in education today? Are more individuals going to college? Are workers in the United States becoming better educated?

U. S. Department of Labor statistics show that the amount of education received by people in the work force is increasing. In 1978, 20% of workers were college graduates. In 1988, the percentage of workers who were college graduates had increased to 25%. In 1978, 36% of the work force had at least attended college. In 1988, the percentage of workers who had at least attended college had increased to 45%. The changes taking place between 1978 and 1988 clearly show a trend toward workers having more education.

With this trend extended into the future, it is anticipated that in the 1990s over one-half of the workers will have attended college and as many as

Graduation is a big day in the life of many.

one worker in three will be a college graduate. If you will be seeking employment and a good-paying job in the future, education is a key to success. To compete effectively with others in the work force, more education is necessary.

Based on "More Workers Have More Education," *USA Today,* December 9, 1988.

The next time you read a newspaper, look for statistical information such as the following:

46% of the people surveyed believe that the president is doing a good job in foreign affairs.

The average selling price of a new house is **$91,500.**

The unemployment rate is **5.0%.**

The average starting salary for a new college graduate is **$26,712.**

New car sales are up **2.4%** over last year.

The numerical facts in the preceding examples (46%, $91,500, 5.0%, $26,712, 2.4%) are commonly referred to as *statistics*. Statistics such as these communicate information that enables us better to understand the subject being discussed. For example, instead of reporting that prices of new houses are "high," the newspaper article has specifically informed the reader that the average price of a new house is $91,500. Similarly, instead of reporting that new car sales are up "slightly" over last year, the newspaper article has informed the reader that the amount of the increase in new car sales is 2.4%. The chapter-opening Statistics in the News article on the trend in worker education used the statistics of 36% of 1978 workers with at least some college education and 45% of 1988 workers with at least some college education to conclude that there is a trend toward more college education.

In common or everyday usage, the term statistics simply refers to numerical facts, or data, that communicate information. However, the field, or subject, of statistics involves much more than simply numerical facts. Thus, a broader definition of the term statistics is required. The field or subject of statistics about which you will be learning in this text is defined as follows:

> **Definition of Statistics**
>
> Statistics is the science of collecting, analyzing, presenting and interpreting data.

In this chapter, we provide an introduction to the fundamental concepts of the field of statistics. We begin by defining the terms *population* and *sample*. Then we discuss statistical methods for summarizing a set of data, the objective being to present the data in a more convenient and easily interpreted form. Next we discuss the process of statistical inference and its role in helping us draw conclusions about a population based on information obtained from a sample. Finally, the chapter concludes by describing several actual studies from a diversity of fields in which the methods of statistics have been successfully applied.

1.1   The Population and the Sample

Consider the situation in which a political party would like to estimate the percentage of registered voters who favor the party's presidential candidate. How could the party obtain such an estimate? First of all, there are approximately 100 million registered voters in the United States. The collection of all registered voters forms the *population* for the study. In general, the population is defined as follows:

> **Population**
>
> A population is the set of all elements of interest in a particular study.

In theory, every individual in the population of registered voters could be contacted and asked if he or she preferred the party's candidate. Such a collection of data from every element in a population is called a *census*.

We can see that attempting to take a census of 100 million registered voters is impractical both in terms of time and cost. Instead, let us suppose that the political party will select a subset of 1500 registered voters believed to be representative of the population of 100 million registered voters. This subset of voters is referred to as a *sample*. In general, a sample is defined as follows:

> **Sample**
>
> A sample is a subset of the population.

Suppose that the sample of 1500 registered voters showed that 600 voters favor the party's candidate. Expressing the sample result as a percentage, we find that (600/1500) × 100% = 40% of the registered voters in the sample favor the candidate. The sample statistic of 40% would be used as an estimate of the percentage of all registered voters (the population) who favor the candidate.

Many situations have characteristics similar to those in the *preceding* illustration. Often there exists a large group (individuals, voters, households, products, customers, and so on) about which information is being sought. Because of time, cost, or other considerations, information is collected only from a small portion, or subset, of the group. As defined earlier, the larger group of items in the study is called the population. The smaller group of items actually contacted is called the sample.

1.2 Data and Data Summarization

When we want to learn about the characteristics of a particular population, we generally begin by selecting a sample of elements from the population. Then, through the process of *observation*, we record facts and figures about the sample elements. These facts and figures are referred to as *data*. Data collected through this process provide the raw material for statistical analysis.

Whenever we collect data for a particular study, we will be interested in summarizing the data in order to present it in a more convenient or more easily interpreted form. For example, as Figure 1.1 shows, unsummarized data is unstructured, and the volume of data available can overwhelm the user. However, with the aid of statistical procedures for data summarization, we see that the condensed, or summarized, data are easier for the user to interpret. Graphical, tabular, and numerical methods for summarizing data are referred to as *descriptive statistics*. Methods of descriptive statistics are presented in Chapters 2 and 3.

An Example: The Cost of Attending College

Suppose that we are interested in learning about the cost of attending an accredited college or university in the United States. It is generally understood that the cost of attending college increases each year, but just what is the annual cost of attending college? The *1988 Information Please Almanac* lists over 1700 accredited senior colleges and universities in the United States. Let us use this list of colleges and universities as the population and attempt to learn more about the annual room and board cost associated with attending accredited colleges or universities in the United States.

We might attempt to observe or collect data on the annual room and board cost for all the colleges and universities in the population. However, suppose we feel the population is too large for a census. The data for a sample of 50 colleges and universities have been collected and are shown in Table 1.1. By studying this data, we can learn about the annual room and board cost for each of the 50 colleges and universities in the sample. However, it is difficult to make any general statements about the costs by looking at the data in this form. In order to provide more useful information about annual room and

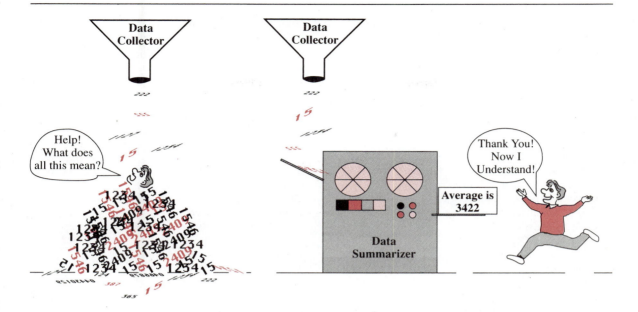

Before Data Summarization   After Data Summarization

FIGURE 1.1   The Process of Data Summarization

board cost, the data have been summarized in Table 1.2. A graphical summary of the data is shown in Figure 1.2.

If the data summarization step is of value, you should find the summaries in Table 1.2 and Figure 1.2 helpful. For example, referring to these summaries, we can observe that 28 of the 50 colleges and universities in the sample (56%) have annual room and board costs between $2,500 to $3,499, inclusive. Four colleges and universities (8%) are in the categories of $4,000–$4,499 and $4,500–$4,999. Three colleges and universities (6%) are in the lowest-cost category, with a cost less than $2,000.

Instead of using a tabular or graphical procedure, another way of summarizing data is to provide a single numerical measure that is representative of the data. For example, we could add the 50 data values in Table 1.1 together and divide by 50 to compute the average annual room and board cost for the sample. Doing so provides a value of $2,999. The value $2,999 is easily interpreted as the average annual room and board cost for the sample of 50 colleges and universities used in this study.

Perhaps with these descriptive summaries of the data you would be ready to conclude that annual room and board cost runs in the neighborhood of $3,000. Costs over $4,000 per year or under $2,000 per year can be found at some schools, but such annual room and board costs are relatively uncommon. Thus, through the statistical processes of sampling (collecting data from sampled elements) and summarization, we have provided information that provides insight and understanding about the annual room and board cost associated with attending college.

TABLE 1.1	Annual Room and Board Cost for a Sample of 50 U.S. Colleges and Universities

University of Alabama—Birmingham	$2,898	Kansas State University	2,286
Alcorn State University	1,650	Lynchburg College	3,400
Amherst College	3,600	Marshall University	2,932
Appalachian State University	1,650	McNeese State University	1,650
Arizona State University	2,500	University of Miami	4,080
Baylor University	3,336	Minnesota Bible College	3,800
Bennington College	3,140	Morgan State University	3,220
Boise State University	2,415	Mount Holyoke College	3,475
Brigham Young University	2,500	University of Nebraska	2,170
University of California—Los Angeles	2,950	State University of New York—Buffalo	3,080
Case Western Reserve University	4,000		
Catholic University of America	3,500	University of North Carolina	3,055
Central Michigan University	2,616	Ohio Wesleyan University	3,661
Clemson University	2,150	Oregon State University	2,445
University of Colorado	2,962	Pepperdine University	4,535
Drake University	3,130	University of Pittsburgh	2,930
Eastern Kentucky University	2,150	Providence College	3,700
Emory University	3,612	St. Mary of the Woods College	2,805
Fairleigh Dickinson University	3,955	University of South Carolina	2,500
University of Florida	3,240	University of South Florida	2,520
Furman University	3,328	Syracuse University	4,430
Georgia Institute of Technology	3,390	University of Texas	3,200
Grand Valley State College	2,750	Washington and Lee University	3,308
University of Illinois	3,370	University of Washington	2,590
Indiana State University	2,312	William Penn College	2,180
		University of Wisconsin	2,900

Source: "Accredited United States Senior Colleges and Universities," *The 1988 Information Please Almanac,* 41st Ed. Boston: Houghton Mifflin, 1988.

TABLE 1.2	Tabular Summary of the Annual Room and Board Cost for a Sample of 50 U.S. Colleges and Universities

ANNUAL ROOM AND BOARD COST	NUMBER OF SCHOOLS	PERCENT
$1,500–1,999	3	6
$2,000–2,499	8	16
$2,500–2,999	14	28
$3,000–3,499	14	28
$3,500–3,999	7	14
$4,000–4,499	3	6
$4,500–4,999	1	2
Total	50	100

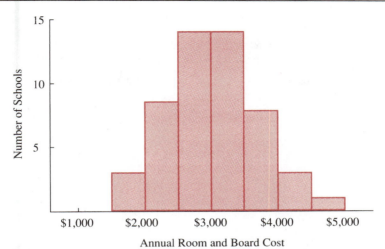

FIGURE 1.2 Histogram of the Annual Room and Board Cost for a Sample of 50 U.S. Colleges and Universities

1.3 Statistical Inference and Probability

Much of the value of statistics is that it provides methods for using sample data to learn about characteristics of a population. This process, referred to as *statistical inference*, is defined as follows:

> **Statistical Inference**
>
> Statistical inference is the process of using information contained in a sample to make estimates or test claims about the characteristics of a population.

It is usually through statistical inference that we learn about a particular population. In many cases, statistical inference forms the basis for drawing conclusions or making decisions.

As an illustration of statistical inference, let us return to the sample data on annual room and board cost shown in Table 1.1. In the previous section we found that the average annual room and board cost for the sample of 50 colleges and universities was $2,999. If we use this value as an *estimate* of the average annual room and board cost for the population of over 1700 accredited colleges and universities in the United States, we are making a statistical inference. This statistical inference process is illustrated in Figure 1.3.

Whenever we make an inference about a population based on information contained in a sample, we have to recognize that since the sample consists of only a portion or subset of the population, the sample results will not be exactly the same as if the entire population had been used to obtain the data. Hence, it is desirable to provide an indication of the precision of the sample results in terms of estimating the population characteristics. This is where probability plays an important role in statistical inference.

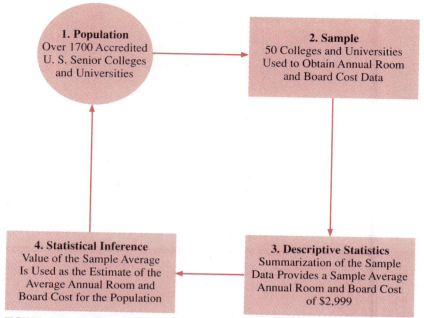

FIGURE 1.3 The Process of Statistical Inference for Estimating the Average Annual Room and Board Cost at U.S. Colleges and Universities

Whenever statisticians make statements about the quality or precision of sample estimates, a measure of uncertainty based on the error due to sampling is included. For instance, using the sample data in Table 1.1, a statistician might state that the estimate of the average annual room and board cost for the population is $2,999 with a margin of error of plus or minus $200. With the help of probability theory, the statistician can state how likely, or probable, it is that the sample result is within the stated margin of error from the actual population average. In the above example, the appropriate statement might be that there is a .95 probability that the estimated average cost of $2,999 is within $200 of the true population average. By applying probability concepts to the analysis of data in a sample, we will learn how to provide estimates of the characteristics of a population and how to provide probability-based statements about the quality or precision of the estimates. The procedures and formulas for determining the precision of sample estimates are presented in Chapters 7 and 8.

1.4   Some Illustrative Applications

In this section we present some recent studies that illustrate how statistics are used in practice.

MEDICINE

Researchers at Cornell University studied the effect of tight neckties on the flow of blood to the head and the possible decrease in the brain's ability to respond to visual information.

Results from a sample of businessmen yielded an estimate that 67% of businessmen wear their ties too tight and that 12% of businessmen wear ties tight enough to interfere with blood flow and brain function. (*Medical Self Care*, July–August 1988).

EDUCATION

Do males and females score differently on the mathematical and verbal portions of the Scholastic Aptitude Test (SAT)? Historically male examinees score an average of about 50 points higher than female examinees on the mathematics portion. Using samples of test scores for men and women who scored high on the mathematics portion, researchers at the Educational Testing Service found differences in performance on the verbal portion. The women tended to score higher than the men. (*Journal of Educational Measurement*, Spring 1987).

BUSINESS

Where do people place their investment dollars? A recent sample conducted by *The Wall Street Journal* showed that most investors' dollars were in the stock market. Responses to the question of where portfolio dollars were placed showed 42% of total investment dollars in the stock market, with an estimated 91% of all investors owning at least some stocks. Real estate was the second-largest investment dollar category (19%), with approximately 47% of all investors owning at least some real estate. Money-market funds, bank accounts, stock mutual funds and municipal bonds were listed as other, less used investment alternatives. (*The Wall Street Journal*, December 2, 1988)

SCIENCE

Scientists have developed a new window glass that will darken whenever the temperature outside is hot or whenever the sunlight is too bright. The window glass will lighten up again when outdoor conditions change. By using computers to sense weather conditions and control the degree of window darkening, offices on the sunny side of an office building can have darkened or shaded windows. Offices on the dark side of the building can have clear windows. Laboratory tests indicate that office buildings can save up to 25% on heating and cooling costs and up to 50% on lighting costs by using the "smart window glass." Expect to see this type of glass on automobile sunroofs in the near future. (*Popular Science*, December 1987)

PSYCHOLOGY

Raymond Moody has published the books entitled *Life After Life* and *The Light Beyond*. In these books he describes the remarkable mental journeys of people who have come close to death but have not died. Some near-death experiences involved "out-of-body" experiences and other common experiences such as traveling through a tunnel. A Gallup poll found that 15% of American adults polled reported some form of a near-death experience. (*Psychology Today*, September 1988)

SOCIOLOGY

People are not getting married as early as they used to. Statistics compiled by *American Demographics* magazine show that the median age at first marriage for men is now 25.9 years, whereas the median age at first marriage for women is now 23.6 years. Thirty years ago, these ages were 22.5 years for men and 20.1 years for women. These data lead to the inference that both men and women are delaying the decision to get married. (*The Wall Street Journal*, October 22, 1988)

TRAVEL AND TOURISM

In many areas of the world, travel and tourism can be considered the number one industry. This is certainly true for island areas such as Hawaii. The Hawaii Visitors Bureau monitors the number of visitors and uses sample information to determine how much money is spent and where it is spent. In 1988, statistical data provided by the Bureau led to the estimate that visitors coming to Hawaii from the mainland spend an average of $102 per day per person. Lodging accounted for approximately 37% of the daily expenditure, whereas travel accounted for approximately 11%. The per-person daily expenditures were expected to increase 9 to 12% in 1989. (*St. Petersburg Times*, December 11, 1988)

STATISTICS AND MICROCOMPUTERS

The growing use of the microcomputer is making statistical software available to more and more people. People were asked in what area they employ microcomputer statistical analysis as part of a sample survey sponsored by *Info World*. The primary area of usage was in education, with 28.8% of the survey respondents indicating use in this field. Other areas of statistics and microcomputer usage occurred in the fields of economic trend analysis, engineering, marketing research, personnel administration and medicine. (*Info World*, September 19, 1988)

1.5   *The Data Analyst*—A Microcomputer Statistical Software Package

Microcomputer software packages are making many statistical techniques easier to use. *The Data Analyst* is a statistical software package that has been developed to accompany this text. It is designed for solving problems in the text as well as small-scale problems that may be encountered in practice.

    The Data Analyst consists of modules, or programs, that enable the user to perform statistical analysis in the following topical areas:

- Descriptive statistics
- Interval estimation for one population
- Interval estimation for two populations
- Analysis of variance
- Regression and correlation

    The use of *The Data Analyst* with this text is optional. The software package itself contains instructions for installing the system on your microcomputer; the remainder of this section provides an introductory description of the key features of the software package.

Top Level Menu

The Data Analyst is a menu-driven system; that is, users communicate with the system by selecting an option from a list provided. The user can direct the package to select a statistical routine, accept a data set, and display the statistical results. The first menu that appears on the screen is the "Top Level Menu." The choices on this menu provide access to the corresponding statistical modules or programs that can be used to analyze a data set. As mentioned earlier, the options for statistical analysis include descriptive statistics, interval estimation, analysis of variance, regression and correlation analysis. Once the desired option has been selected, the user can select the desired data set and the statistical results will be displayed on the screen.

The type of information provided and its interpretation varies with the statistical module selected. By reading the corresponding chapter in the text, you should be able to interpret the output information. After the solution and other output information have been displayed on the screen, the user will be provided with the option to have the results sent to a printer.

The advantage of *The Data Analyst's* menu system is that users do not have to learn a special command language to use the system; effort can be focused on learning how the computer package can aid in the analysis of data.

Data Menu

The "Data Selection Menu" enables the user to select a data set for statistical analysis. The choices available from the menu are as follows:

1. Create a New Data Set
2. Retrieve a Previously Saved Data Set
3. Continue with the Current Data Set
4. Delete a Previously Saved Data Set
5. Return to the Top Level Menu

Saving, Retrieving, and/or Deleting Data Sets

The Data Analyst allows the user to save data sets for future use on the disk drive that is specified by the user when the software package is initially started. When this option is selected, instructions for naming the data set will appear on the screen. The data set will then be saved automatically, using the name specified. An option is also provided to save the data as a Minitab file so it can be read by the Minitab software package.

When reentering *The Data Analyst* at a later date, the user may select the retrieve option from the "Data Selection Menu" in order to recall a previously saved data set. When the data set is no longer needed, the delete option can be used to erase the data set from the data disk.

Further Advice about Data Input

When using *The Data Analyst,* you may find the following data input suggestions helpful.

1. Do not enter commas (,) with your input data. For example, to enter the numerical value of 104,000, simply type the six digits 104000.
2. Do not enter the dollar sign ($) for profit or cost data. For example, a cost of $20.00 should be entered as 20.
3. Do not enter the percent sign (%) if percentage input is requested. For a percentage of 25%, simply enter 25. Do not enter 25% or 0.25.
4. If the computer did not interpret your input correctly (for example, you tried to input a comma), the message "Redo from start" may appear. This message refers to the

input question or prompt to which you are currently responding. The message means to respond to the same question or prompt again.

5. For data values containing the digit zero, be sure to enter the numeric 0 rather than the letter O.

6. Occasionally data may be recorded with fractional values such as 1/4, 2/3, 5/6, and so on. The data input for the computer must be in decimal form. The fraction of 1/4 can be entered as .25. However, the fractions 2/3 and 5/6 have repeating decimal forms. In cases such as these we recommend the convention of rounding to five places. Thus, the corresponding decimal values of .66667 and .83333 should be entered.

7. Finally, we recommend that in general you attempt to scale extremely large input data so that smaller numbers may be input and operated on by the computer. For example, 2,500,000 may be scaled to 2.5 with the understanding that the data used in the problem reflect millions.

SUMMARY

Although the term *statistics* as used in everyday language refers to numerical facts or data, the field of statistics requires a broader definition. The field of statistics is the science of collecting, analyzing, presenting and interpreting data. Two components of most statistical studies are the population and the sample.

The field of descriptive statistics is concerned with utilizing tabular, graphical and numerical methods to summarize data. Statistical inference is the process of making estimates or testing claims about a population based on the data available in a sample. Probability plays an important role in statistical inference by enabling the statistician to make statements concerning the precision of a sample result.

Because the methods of statistics have been successfully applied in so many diverse fields of study, nearly every college student is required to take a course in statistics. Overviews of several actual statistical studies were presented in Section 1.4. We concluded the chapter with an introduction to *The Data Analyst* statistical software package.

GLOSSARY

Population   The set of all elements (individuals, households, products, customers, and so on) of interest in a particular study.

Sample   A subset of the population.

Data   The facts and figures collected in a statistical study.

Census   A collection of data from every element in a population.

Descriptive Statistics   Tabular, graphical and numerical methods of data summarization for the purpose of better presentation and interpretation.

Statistical Inference   The process of using data obtained from a sample to make estimates or test claims about the characteristics of a population.

TRUE/FALSE

1. A population is the set of all elements of interest in a particular study.

2. A sample may be larger than the population.

3. Methods for data summarization are referred to as prescriptive statistics.

4. The use of probability enables the statistician to make statements concerning the precision of statistical inferences.

MULTIPLE CHOICE

5. When data are collected for only a subset of the elements of interest, we are using a
   a. population
   b. sample
   c. statistical inference
   d. summary

6. Descriptive statistics is that branch of statistics concerned with
   a. arriving at a conclusion for the population based upon sample information
   b. the summarization and presentation of data
   c. statistical inference

7. Statistical inference is that branch of statistics concerned with
   a. arriving at a conclusion about the population based upon sample information
   b. the summarization and presentation of data

8. In a recent study based upon an inspection of 100 homes in Central City, 60 homes were found to violate one or more city codes. Based upon this information, the city manager released a statement that 60% of all Central City's homes were in violation of city codes. The manager's statement is an example of
   a. descriptive statistics
   b. statistical inference

9. Refer to question 8. The manager's statement that 60% of all Central City's homes are in violation of city codes is
   a. exactly correct
   b. only an approximation, since it is based upon sample information
   c. very misleading, since it is based upon a study of only 100 homes

10. The statement "Based on sample data, we expect to sell $50,000 worth of snow removal equipment this winter" is an example of
   a. descriptive statistics
   b. statistical inference

SUPPLEMENTARY EXERCISES

1. Discuss the difference between the everyday concept of *statistics* as numerical facts or data and the concept of *statistics* as a field of knowledge and study.

2. A 1988 *Newsweek*/Gallup poll investigated whether or not adults preferred to stay at home or go out as their favorite way of spending time in the evening. The poll of 1500 adults

concluded that the majority of adults (70%) indicated that "staying at home with family" was the favorite evening activity.
a. What is the population of interest in this study?
b. What was the size of the sample used?
c. Where was a descriptive statistic used in this study?
d. Describe the process of statistical inference in this study.

3. In a recent study of causes of death in males 60 years of age and older, a sample of 120 men indicated that 48 had died due to some form of heart disease.
a. Develop a descriptive statistic that can be used as an estimate of the proportion of males 60 years of age or older who die from some form of heart disease.
b. Discuss the role of statistical inference in this type of medical research.

4. The 25th Annual Report on Shoplifting in Supermarkets (*Commercial Service Systems, Inc.*, 1987) used 391 supermarkets in Southern California to compile the following statistics on supermarket shoplifters caught in the act:

Most items stolen were valued at between $1 and $5.
Most shoplifters were male (56%).
The most common age of shoplifters was under 30 (51%).

Answer the following questions, assuming that the purpose of the study was to present statistical data on the national trend and impact of shoplifting in supermarkets.
a. Did this study use a census or a sample?
b. Cite two descriptive statistics used.
c. What warning would you issue if the results of the study were to be used to make statistical inference about the national trend and impact of shoplifting in supermarkets?

5. Select a recent copy of the newspaper *USA Today*.
a. Note four examples of statistical information.
b. For each example, indicate the descriptive statistics used and discuss any statistical inferences made.

6. A 7-year medical research study (*Journal of the American Medical Association,* December 1984) reported that women whose mothers took the drug DES during pregnancy were *twice* as likely to develop tissue abnormalities that might lead to cancer as women whose mothers did not take the drug.
a. This study involved the comparison of two populations. What were the populations involved?
b. For the population of women whose mothers took the drug DES during pregnancy, a sample of 3980 women showed 63 developed tissue abnormalities that might lead to cancer. Provide a descriptive statistic estimating the number of women out of 1000 in this population who have the tissue abnormalities.
c. For the population of women whose mothers did not take the drug DES during pregnancy, what is the estimate of the number of women out of 1000 who would be expected to have the tissue abnormalities?
d. Medical studies of diseases and disease occurrence often use a relatively large sample (3980 as noted in part b). Why is this done?

7. A firm is interested in testing the advertising effectiveness of a new television commercial. As part of the test, the commercial is shown on a 6:00 P.M. local news program in Denver, Colorado. Two days later a market research firm conducts a telephone survey to obtain information on recall rates (percentage of viewers who recall seeing the commercial) and impressions of the commercial.
a. What is the population for this study?
b. What is the sample for this study?

8. The quality control department of a large manufacturing firm is responsible for maintaining product specifications for a variety of production line operations.

    a. List some of the information that the quality control department might want in order to determine whether or not product specifications are being met.

    b. Why would the firm be interested in sampling concepts from the area of statistics?

9. Comment on the problem of misleading statistical data that may result in each of the following situations.

    a. In order to estimate the support for a particular political candidate, a pollster visits a major shopping center from 10:00 A.M. to 3:00 P.M. and interviews shoppers.

    b. In order to determine the favorite type of vacation for families in the United States, a study by a Florida tourism promoter reports interviews with 50 out-of-state families visiting Orlando, Florida.

    c. A door-to-door interviewer is instructed to open each interview with the words, "Hello, I am conducting a survey for H&G Soap Products Company. Do you like using H&G products?"

10. A sample of midterm grades for five students in Chemistry 121 showed the following results: 72, 65, 82, 90, 76. A total of 180 students were enrolled in the course. Which of the following statements are correct and which should be challenged as being too generalized?

    a. The average midterm grade for the sample of five students is 77.

    b. The average midterm grade for all 180 students is 77.

    c. An estimate of the average midterm grade for all 180 students is 77.

    d. More than half of all students who take this exam will score between 70 and 85.

    e. If five other students are included in the sample, their grades will be between 65 and 90.

11. Refer to the study on the expenditures for a Hawaiian vacation as presented in Section 1.4 (*St. Petersburg Times,* December 11, 1988). Assume two people are planning a 10-day trip to Hawaii and that the round trip airfare is $595 per person.

    a. Develop an estimate of the total amount of money that they will spend on the Hawaiian vacation. Is it possible that they could spend more than the amount you have estimated? Explain.

    b. Estimate how much they will spend on lodging. Estimate how much they will spend on transportation while in Hawaii.

    c. What is the population used in the study presented in Section 1.4?

    d. Describe the statistical inference process for this example.

12. Refer to the study on age at time of marriage presented in Section 1.4 (*The Wall Street Journal,* October 22, 1988). Answer the following questions.

    a. Define the population for this study.

    b. Was the statistical information based on a sample or a census? Explain.

    c. Define the descriptive statistics used.

    d. Describe the statistical inference process as it relates to this study.

    e. Use the statistical results provided in the example to estimate how much longer men are now waiting to marry compared with 30 years ago.

    f. Are women waiting less, about the same, or longer than men? Use descriptive statistics to support your answer.

13. What makes us happy? A survey conducted by California economist Harry Beiderman used a sample of 673 people to investigate this question (*Psychology Today,* November 1988). The top three responses were family and friends (215 people), hobbies and entertainment (135 people), and accomplishments (87 people).

    a. Develop descriptive statistics that indicate the percentage of people who selected each of the top three responses.

b. Other factors mentioned by people in the survey included nature, health, food, money, surprises, and sex. What percentage of people in the survey selected factors such as these as making them happy?

14. In their book *The Case Against the SAT,* authors James Crouse and Dale Trusheim of the University of Delaware claim that "despite the SAT's (Scholastic Aptitude Test) ability to predict educational success, the SAT is unnecessary." A sample of 2800 high school seniors who had taken the SAT was used in one part of their study. College admissions decisions were made two ways:

- The students were admitted based solely on their high school class rank.
- The students were admitted based solely on their SAT score rank.

The study reported that the two methods provided the *same admission decisions* for 2,338 students.

a. What is the population in this study?

b. What is the sample in this study?

c. Compute the descriptive statistic that indicates SAT scores lead to the same admission decision as using high school class rank.

d. Considering the fact that high school students throughout the United States pay a combined total of $20 to $22 million dollars to take the SAT each year, do you feel the statistical information provided supports the author's contention that the SATs are unnecessary? Explain.

2

Descriptive Statistics I: Tabular and Graphical Methods

What You Will Learn in This Chapter:

- The nominal, ordinal, interval and ratio measurement scales

- The distinction between qualitative and quantitative data

- How to construct and interpret summarization procedures for qualitative data such as frequency distributions, percent frequency distributions, bar graphs and pie charts

- How to construct and interpret summarization procedures for quantitative data such as stem-and-leaf displays, frequency distributions, percent frequency distributions, histograms, frequency polygons and ogives

- How to construct and interpret summarization procedures for bivariate data

Contents

Soft Drinks: America's Favorite Beverage

What is your favorite beverage? If you are a typical American, recent statistics show your favorite beverage should be a soft drink. An average American drinks approximately 10 to 12 cans or bottles of soft drinks per week, resulting in a consumption of 46.4 gallons per year. The graphical summary in the accompanying figure shows how annual soft-drink consumption per person compares with the annual consumption of bottled water, fruit juice, milk and beer. As you can see, soft drinks are clearly number one.

Soft-drink lovers spend over $25 billion per year. Competition for a share of this market is intense; new products are proliferating and much money is being spent on advertising campaigns. Including diet and caffeine-free soft drinks, consumers may now choose from approximately 250 brands. Coca-Cola and Pepsi continue to be the big names in the industry, with their brands accounting for approximately 90% of the market.

The future promises a continued growth, with manufacturers introducing even more new products in an attempt to capture a larger share of the market. However, it is doubtful that the soft-drink market can continue growing at its current rate. By one estimate, if the growth rate of the past 30 years continues over the next 30 years, by the year 2020 every American will down 1900 bottles of soda annually. That is more than 5 soft drinks per day.

Based on "A Healthy Helping," *Psychology Today,* October 1988.

Fifteen little leaguers claim soft drinks are number 1.

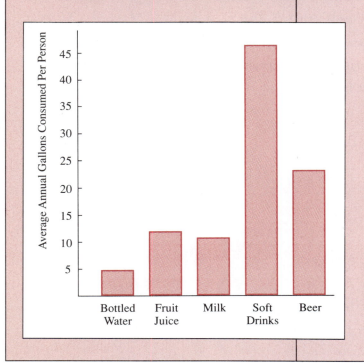

The purpose of this chapter is to introduce tabular and graphical procedures that are commonly used to summarize data. We begin with a definition of data and measurement. Data are shown to be generated by four measurement scales: nominal, ordinal, interval, and ratio. The major portion of this chapter is devoted to presenting tabular and graphical methods that are helpful in summarizing the various types of data. The ultimate purpose of these summarization procedures is to present data in a more easily interpreted form.

2.1 Data and Scales of Measurement

Data are the facts and figures that are collected, analyzed, presented, and interpreted in a statistical study. All the data collected in a particular study are referred to as the *data set* for the study.

EXAMPLE 2.1

Table 2.1 shows a data set for a sample of six players selected from the 1988 national championship football team at the University of Notre Dame. The six players listed are referred to as the *elements* in the statistical study. The variables (characteristics of interest) for the elements are uniform number, position, height, weight, class, and hometown.

Most data that statisticians collect and analyze are numeric data. However, as the data set in Table 2.1 shows, data for some variables may be numeric and data for other variables may be nonnumeric. Specifically, the three variables position, class, and hometown have nonnumeric data. The nonnumeric symbols used are abbreviations for player's position, such as quarterback (QB) and linebacker (LB), and for the player's class standing, such as freshman (Fr), sophomore (So), junior (Jr), or senior (Sr). The other three variables—uniform number, height and weight—have numeric data.

■

The type of statistical analysis appropriate for the data on a particular variable depends upon the scale of measurement used for the variable. There are four scales of measurement: nominal, ordinal, interval and ratio. The scale of measurement determines the amount of information contained in the data and indicates the data summarization and statistical analyses that are most appropriate. We describe each of the four scales of measurement.

TABLE 2.1 A Data Set Selected From the 1988 University of Notre Dame Football Team Roster*

NAME	NO.	POSITION	HEIGHT	WEIGHT	CLASS	HOMETOWN
T. Rice	9	QB	6 ft 1 in.	198	Jr	Woodruff, SC
N. Bolcar	47	LB	6 ft 2 in.	232	Sr	Phillipsburg, NJ
A. Johnson	22	FB	6 ft 0 in.	225	Jr	South Bend, IN
M. Stonebreaker	42	LB	6 ft 1 in.	228	Jr	River Ridge, LA
T. Lyght	1	CB	6 ft 1 in.	181	So	Flint, MI
R. Ismail	25	SE	5 ft 10 in.	175	Fr	Wilkes-Barre, PA

*Source: They Wanted to Win—The 1988 Notre Dame Football National Championship Review, Host Communication Sports Publishing and the University of Notre Dame Athletic Department, 1989.

Nominal Scale

The scale of measurement for a variable is *nominal* when the data for the variable are simply labels used to identify an attribute of each element. For example, referring again to the data set in Table 2.1, we see that the first variable, uniform number, is measured on a nominal scale. This is so because the uniform numbers are simply labels used to identify the player. The second variable, position, is also measured on a nominal scale, with the abbreviation letters providing the position label for the player.

Other examples of variables where data have a nominal scale are as follows:

Sex (male, female)

Marital status (single, married, widowed, divorced)

Religious affiliation (many possibilities)

Part-identification code (A13622, 12B63)

Employment status (employed, unemployed)

House number (5654, 2712, 624)

Occupation (many possibilities)

As the preceding examples show, the key feature of the nominal scale is that the data are labels used to identify an attribute of the element.

Finally, it is important to note that arithmetic operations such as addition, subtraction, multiplication and division *do not* make sense for nominal data. Thus, even if the nominal data are numeric (e.g., uniform numbers) computations such as summing and averaging the data are inappropriate.

Ordinal Scale

The scale of measurement for a variable is *ordinal* when

1. The data have the properties of nominal data, and
2. The data can be used to rank or order the observations for the variable.

Referring to Table 2.1 again, we see that the variable *class* uses ordinal data. This is true because the data have the properties of nominal data in that Fr, So, Jr, and Sr are labels used to identify the player's class standing. In addition, the data permit ranking or ordering on the basis of class standing with the rank from lowest to highest class standing indicated by the ordering Fr, So, Jr, and Sr.

EXAMPLE 2.2

Let us use the following example as another illustration of the ordinal scale of measurement. Some restaurants place questionnaires on tables to solicit customer opinions concerning the restaurant's performance in terms of food, service, atmosphere, and so on. A questionnaire used by the Lobster Pot Restaurant in Redington Shores, Florida, is shown in Figure 2.1. Note that the customers completing the questionnaire are asked to provide ratings for six different variables: food, drinks, service, waiter, captain, and hostess. The response categories are excellent, good, and poor for each variable. The observations for each variable possess characteristics of nominal data (each response rating is a label for excellent, good, or poor quality). In addition, the observations can be ranked, or ordered, with respect to quality. For example, consider the variable on food quality. After collecting the data, we can rank the observations in order of food quality by beginning with the observations of excellent, followed by the observations of good,

	Excellent	**Good**	**Poor**	**Comments**
Food				
Drinks				
Service				
Waiter				
Captain				
Hostess				

Waiter's Name ⸻

Captain's Name ⸻

Other Comments ⸻

FIGURE 2.1   Customer Opinion Questionnaire Used by the Lobster Pot Restaurant, Redington Shores, Florida (Used with permission)

and, finally, by the observations of poor. With only three response categories, we can expect to see many observations with tied rankings. Nonetheless, the observations can be ranked in terms of food quality.

Like data obtained from a nominal scale, data obtained from an ordinal scale may be either nonnumeric or numeric. For the Lobster Pot questionnaire, the nonnumeric letters of E for excellent, G for good, and P for poor could be used to record the observations. Or, a numeric code with values of 1 for excellent, 2 for good, and 3 for poor could be used equally well. Finally, as with nominal data, it is important to remember that arithmetic operations *do not* make sense for ordinal data. Thus, even if the ordinal data are numeric, the computations of summing and averaging are inappropriate.

■

Interval Scale

The scale of measurement for a variable is *interval* when

1. The data have the properties of ordinal data, and
2. The interval between observations can be expressed in terms of a fixed unit of measure.

EXAMPLE 2.3

Temperature is a good example of a variable that uses an interval scale of measurement. The fixed unit of measure is a degree. An observation recorded at a particular point in time will be a numeric value specifying the amount, or quantity, of degrees. Such data possess the properties of ordinal data in that temperature observations of 35 degrees, 40 degrees, 85 degrees, and 90 degrees can be ranked, or ordered, from coldest to warmest. In addition, this interval scale has the property that the interval between observations can be expressed in terms of the fixed unit of measure (a degree). For example, the interval between 35 and 40 degrees is 5 degrees, the interval between 35 and 85 degrees is 50 degrees, and the interval between 85 and 90 degrees is 5 degrees. With nominal and ordinal data, such differences between observations are not meaningful.

The fixed unit of measure required by an interval scale means that the data *must always be numeric*. With interval data, the arithmetic operations of addition, subtraction, multiplication, and division are meaningful. As a result, data obtained using this scale lend themselves to more alternatives for statistical analysis than do data obtained from nominal or ordinal scales.

Ratio Scale

The scale of measurement for a variable is *ratio* when

1. The data have all the properties of interval data, and
2. The ratio of two observations is meaningful.

Variables such as distance, height, weight, and time use the ratio scale of measurement. A requirement of the ratio scale is that a zero value is inherently defined in the scale. Specifically, the zero value must indicate nothing exists for the variable at the zero point. Whenever we collect data on the cost of something, we use a ratio scale of measurement. Consider a variable indicating the cost of an automobile. The zero point is inherently defined in that a zero cost indicates the automobile is free (no cost). Then, comparing the $10,000 cost of one automobile with the $5,000 cost of a second automobile, the ratio property of the data shows that the first automobile is $10,000/5,000 = 2$ times, or twice, the cost of the second automobile. In Table 2.1, the variables height and weight have ratio-scale data.

Since ratio data have all of the properties of interval data, ratio data are *always numeric* and enable meaningful arithmetic operations such as addition, subtraction, multiplication, and division. As with interval data, ratio data lend themselves to more alternatives for statistical analysis than do data obtained from nominal or ordinal scales.

Table 2.2 provides a summary of the relationship between nonnumeric and numeric data and the four scales of measurement. We note that the nominal and ordinal scales can generate both nonnumeric and numeric data but that interval and ratio scales generate only numeric data.

The amount of information in the data varies with the scale of measurement. Nominal data contain the least amount of information, followed by ordinal, interval, and then ratio data. Since arithmetic operations are meaningful only for interval and ratio data, it is important to know the measurement scale used in order to employ the most appropriate statistical procedures. There are many statistical procedures that can be used with interval and ratio data that are not meaningful with nominal and ordinal data.

TABLE 2.2

The Relationship between Nonnumeric and Numeric Data and Scales of Measurement

SCALE OF MEASUREMENT	NONNUMERIC DATA	NUMERIC DATA
Nominal	Description indicates the category for the element	Numeric value indicates the category for the element
Ordinal	Description permits ranking or ordering of data	Numeric value permits ranking or ordering of data
Interval		Numeric values* are defined in fixed and equal units such that the interval between data values is meaningful
Ratio		Numeric values* have an inherently defined zero, and ratios of data values are meaningful

*Arithmetic operations are meaningful for these kinds of data.

Qualitative and Quantitative Data

Data can also be classified as being either qualitative or quantitative. *Qualitative data* provide labels or names for categories of like items. The categories for qualitative data may be identified by either nonnumeric descriptions or by numeric codes. Qualitative data are obtained from either a nominal or an ordinal scale of measurement. On the other hand, *quantitative data* indicate either "how much" or "how many" of something. Quantitative data are always numeric and are obtained from either an interval or a ratio scale of measurement.

In terms of the statistical methods used for summarizing data, qualitative data provided by nominal and ordinal scales employ similar methods, whereas quantitative data provided by interval and ratio scales employ similar methods. Tabular and graphical methods for summarizing qualitative data are presented in Section 2.2. Tabular and graphical methods for summarizing quantitative data are presented in Sections 2.3 and 2.4.

Notes and Comments

1. An element is the entity on which measurements are obtained. It could be a company, a person, an automobile, and so on. An observation is the set of measurements obtained for each element, so the number of observations and the number of elements in a data set will always be the same. The number of measurements obtained on each element is the number of variables. Thus, the total number of data items in a data set is the number of elements times the number variables. *Continued on next page.*

2. Data obtained from categorical responses are measured with either a nominal or ordinal scale. Such data are called qualitative data and are often nonnumeric. Assigning a numeric code to such data *does not* make it quantitative. Even when a numeric code is used, these data should not be subjected to arithmetic calculations.

3. For purposes of statistical analysis, the most important distinguishing characteristic of the types of data is that ordinary arithmetic operations are meaningful *only* with quantitative (interval and ratio-scaled) data. This is due to the fact that interval and ratio data share the property that the numeric values assigned are based on fixed and equal units of measurement.

EXERCISES

1. Based upon scores on a mathematics achievement test, students in a seventh grade class are assigned to one of three mathematics programs: remedial, regular, or advanced. Do the labels remedial, regular, and advanced represent nominal, ordinal, interval, or ratio data? Explain.

2. A poll conducted by the Gallup organization in July of 1988 asked a sample of voters to express an opinion on the United States space defense program referred to as the Star Wars program. Each individual responded to the Gallup poll question by selecting one of the following options: for the program, opposed to the program, no opinion. Was the measurement scale used in this poll nominal, ordinal, interval or ratio? Explain.

3. A California state agency classifies worker occupations as either professional, white-collar, or blue-collar. What type of measurement scale is being used? Explain.

4. A study of last-minute Christmas shoppers conducted by Best Products, Inc. (*USA Today,* December 22, 1988) asked individuals to indicate when they completed their holiday shopping. The response alternatives were

   Before Halloween
   Before Thanksgiving
   A few days before Christmas
   Christmas Eve

   Explain why this is an ordinal scale.

5. A study on the ages of individuals at the time of their first marriage (*The Wall Street Journal,* October 22, 1988) used data on the ages of men and women to indicate that men tend to be a little more than 2 years older than women at the time of their first marriage. Are the age data in this study obtained using a nominal, ordinal, interval, or ratio scale of measurement? Explain.

6. The 1989 television coverage of the Professional Golfers Association (PGA) Masters Golf Tournament used the following scoring scheme to represent a player's score on each hole: 0 for par, $+1$ for a score of 1 over par, $+2$ for a score of 2 over par, -1 for a score of 1 under par, -2 for a score of 2 under par, and so on. A particular golfer in the tournament obtained the following scores for the first nine holes: 0, $+1$, -1, 0, 0, $+2$, 0, $+1$, and 0.
   a. The total of the scores was used to obtain a measure of each golfer's overall performance in the tournament. What is the above golfer's total score for the first nine holes? What is your interpretation of this value?
   b. Does the scoring system use a nominal, ordinal, interval, or ratio scale? Explain.

2.2

Summarizing Qualitative Data

Frequency and Percent Frequency Distributions

We begin our discussion of methods for summarizing qualitative data by showing how to construct the tabular summaries known as a *frequency distribution* and a *percent frequency distribution*. These distributions are defined as follows:

> **Frequency Distribution**
>
> A frequency distribution is a tabular summary of a set of data showing the number of items in each of several nonoverlapping classes.

> **Percent Frequency Distribution**
>
> A percent frequency distribution is a tabular summary of a set of data showing the percent of items in each of several nonoverlapping classes.

The objective in developing these distributions is to provide insights about patterns in the data.

EXAMPLE 2.4

USA Today (December 13, 1988) reported on a study of the automobiles purchased by women in 1988. The qualitative data in the study consisted of a description of the top five automobiles purchased: Ford Escort, Honda Accord, Chevrolet Cavalier, Ford Taurus, and Hyundai Excel. The actual study summarized the purchases of several thousand women. Let us assume that the data shown in Table 2.3 represent a sample of 50 women who purchased one of these five cars.

TABLE 2.3

Data from a Sample of 50 New-car Purchases by Women in 1988

Honda Accord	Ford Escort	Ford Taurus
Ford Taurus	Chevrolet Cavalier	Honda Accord
Honda Accord	Ford Escort	Ford Taurus
Honda Accord	Hyundai Excel	Hyundai Excel
Ford Taurus	Hyundai Excel	Chevrolet Cavalier
Ford Taurus	Ford Escort	Ford Escort
Honda Accord	Chevrolet Cavalier	Chevrolet Cavalier
Ford Escort	Ford Escort	Chevrolet Cavalier
Honda Accord	Honda Accord	Hyundai Excel
Ford Taurus	Chevrolet Cavalier	Chevrolet Cavalier
Honda Accord	Hyundai Excel	Hyundai Excel
Honda Accord	Ford Escort	Ford Escort
Ford Escort	Honda Accord	Hyundai Excel
Chevrolet Cavalier	Chevrolet Cavalier	Ford Taurus
Hyundai Excel	Ford Escort	Ford Escort
Chevrolet Cavalier	Ford Escort	Honda Accord
Ford Taurus	Ford Taurus	

TABLE 2.4

**Frequency Distribution and Percent Frequency Distribution
of 50 New-car Purchases by Women in 1988**

CAR MODEL	FREQUENCY	PERCENT FREQUENCY
Chevrolet Cavalier	10	20
Ford Escort	12	24
Ford Taurus	9	18
Honda Accord	11	22
Hyundai Excel	8	16
Total	50	100

The frequency distribution for this data simply summarizes the count or frequency of each of the five purchase categories. A frequency distribution for the data in Table 2.3 is shown in Table 2.4. Note that the classes are nonoverlapping; each data item falls into one and only one of the five classes. Table 2.4 also shows the percent frequency distribution for the new car purchases. Note that the Ford Escort appears as the most frequently purchased automobile.

Compare the raw data in Table 2.3 with the summarized data in Table 2.4. This comparison shows that the tabular summaries provide more insight into the pattern of automobile purchasing behavior of women in 1988. As can be seen, the Ford Escort, with 24% of the purchases, and the Honda Accord, with 22% of the purchases, ranked number 1 and 2, respectively. Note also that Ford had two automobiles in the top five, with the Escort and Taurus accounting for 42% of the purchases in this sample.

■

Bar Graphs and Pie Charts

A *bar graph* is a graphical device for depicting qualitative data that have been summarized in a frequency distribution or a percent frequency distribution. On the horizontal axis of the graph we specify the labels that are used for each of the classes. Either a frequency scale or a percent frequency scale can be used for the vertical axis of the graph. Then, using a bar of fixed width drawn above each class label, we extend the height of the bar until we reach the frequency or percent frequency of the class as indicated by the vertical axis. The bars are separated to emphasize the fact that each class is a separate category. A bar graph of the frequency distribution for the 50 new-car purchases by women is shown in Figure 2.2. Note how the graphical presentation shows the Ford Escort and the Honda Accord to be the two most preferred models.

The pie chart is a commonly used graphical device for presenting percent frequency distributions for qualitative data. To draw a pie chart, first draw a circle; then, the percent frequency distribution is used to subdivide the circle into sectors, or parts, that correspond to the percent frequency for each class. For example, since there are 360 degrees in a circle and since the Ford Escort accounted for 24% of the new-car purchases, the sector of the pie chart labeled Ford Escort should consist of $.24 \times 360 = 86.4$ degrees. With similar calculations made for the other classes, the pie chart for the sample of new-car purchases by women in 1988 is shown in Figure 2.3.

The previous example used nominal data. In the next example, we will show that ordinal data can be summarized using the same tabular and graphical methods employed for nominal data.

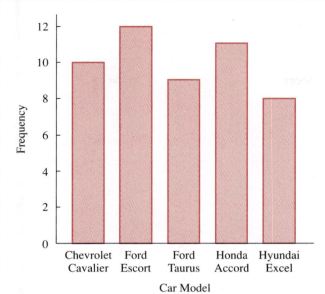

FIGURE 2.2   Bar Graph of 50 New-car Purchases by Women in 1988

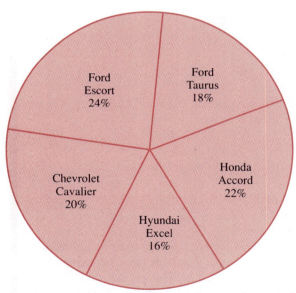

FIGURE 2.3   Pie Chart of 50 New-car Purchases by Women in 1988

EXAMPLE 2.5

Course evaluations at a major university are obtained by student responses to a questionnaire that is filled out on the last day of class. There are a variety of questions, which use a five-category response scale. One question is as follows:

Compared to other courses that you have taken, what is the overall quality of the course you are now completing?

‾‾‾‾   ‾‾‾‾   ‾‾‾‾   ‾‾‾‾‾   ‾‾‾‾‾
Poor   Fair   Good   Very Good   Excellent

TABLE 2.5

**Frequency Distribution and Percent Frequency
Distribution for 125 Course Evaluations**

RATING	FREQUENCY	PERCENT FREQUENCY
Poor	3	2.4
Fair	8	6.4
Good	24	19.2
Very Good	52	41.6
Excellent	38	30.4
Total	125	100.0

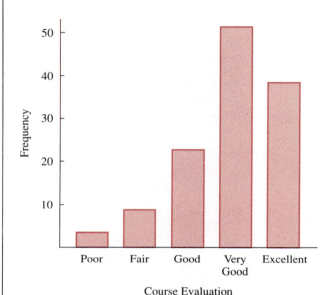

FIGURE 2.4   Bar Graph for 125 Course Evaluations

Data from the course evaluations measure the quality of the course on an ordinal scale using the five labels shown.

Tabular and graphical methods can be used to summarize the teaching evaluation data set. Table 2.5 and Figure 2.4 provide these summaries for the overall quality of the course question shown. Rating data were provided by 125 students. Since the scale is ordinal, we have ordered the labels for the five categories in both the table and the bar graph in Figure 2.4. A review of these summaries shows that the quality of the course is generally very good.

■

EXERCISES

Potential consumers of a new diet soft drink were asked to rate the taste of the product. Data were collected from a sample of 25 consumers. After tasting the new diet soft drink, the consumers were asked to rate the taste on the three-point scale of poor (P), good (G), or excellent (E). The following data were obtained:

$$
\begin{array}{ccccc}
G & P & G & E & G \\
G & E & P & G & G \\
G & G & E & P & E \\
E & G & P & G & G \\
P & G & G & E & E
\end{array}
$$

a. Provide a percent frequency distribution for the data.
b. Provide a bar graph for the data.
c. Using these data, what is your preliminary evaluation of the consumer attitude toward the taste of the new diet soft drink?

8. What are the favorite movies of the year? The biggest box-office successes for 1988 were listed in the *U.S. News and World Report,* December 26, 1988. Assume that sample data collected on movie preferences were summarized with the following letter codes:

A   Coming to America
B   Big
D   Crocodile Dundee II
R   Who Framed Roger Rabbit
V   Good Morning, Vietnam
O   Other motion picture preferred

The following sample data are available:

$$
\begin{array}{ccccccccccccccc}
V & R & O & R & B & A & D & V & V & R & R & D & A & B & V \\
R & R & A & D & V & A & B & R & R & V & R & A & V & B & V \\
D & R & A & V & B & A & O & R & R & B & R & A & D & R & R \\
B & A & R & O & A & A & V & A & D & A & D & R & B & R & B
\end{array}
$$

a. Prepare a frequency distribution for the data set.
b. Prepare a pie chart for the data set.
c. Rank-order the top five motion pictures for 1988. What motion picture appears to have been the most successful?

9. Voters participating in an exit poll for a recent election in Michigan were asked to state their political party affiliation. Coding the data using 1 for Democrat, 2 for Republican, and 3 for Independent, the data collected were as follows:

$$
\begin{array}{cccccccccccccccccccc}
1 & 2 & 2 & 1 & 3 & 1 & 2 & 2 & 2 & 1 & 2 & 3 & 2 & 3 & 2 & 1 & 1 & 2 & 1 & 2 \\
2 & 1 & 1 & 1 & 2 & 1 & 2 & 3 & 1 & 1 & 2 & 1 & 3 & 1 & 1 & 2 & 1 & 2 & 3 & 2
\end{array}
$$

a. Show a frequency distribution for the data.
b. Show a bar graph for the data.
c. Comment on what the data suggest about the strengths of the political parties in this voting area.

10. Freshmen entering the college of science at Eastern University were asked to indicate their preferred major. The following data was obtained:

Major	Chemistry	Physics	Biology	Geology
Number	112	33	65	21

Summarize the data by constructing

a. a percent frequency distribution
b. a bar graph
c. a pie chart

11. Employees at Electronics Associates are on a flextime system; under this system, the employees can begin their working day at 7:00, 7:30, 8:00, 8:30, or 9:00 A.M. The following data represent a sample of the starting times selected by the employees.

7:00	8:30	9:00	8:00	7:30	7:30	8:30	8:30	7:30	7:00
8:30	8:30	8:00	8:00	7:30	8:30	7:00	9:00	8:30	8:00

Summarize the data by constructing

a. a percent frequency distribution
b. a bar graph
c. a pie chart
d. Comment on what the data indicate about starting times.

12. A national restaurant chain provides a card on each table which contains questions regarding the customers' opinions about their meal, the service, and so on. One question asks the customer to rate the quality of the service as poor, below average, average, above average, or outstanding. The following data represent the results obtained for one restaurant in the chain.

Poor: 7
Below average: 14
Average: 33
Above average: 67
Outstanding: 19

Summarize the data by constructing

a. a percent frequency distribution
b. a bar graph
c. a pie chart
d. Comment on what the data indicate about the service at the restaurant.

2.3 Summarizing Quantitative Data: Tabular Methods

Stem-and-Leaf Display

A *stem-and-leaf display* is a simple method of summarizing quantitative data that provides information about the shape, or pattern, in the data set. Developing such a display also provides an easy way to arrange data in ascending or descending order.

EXAMPLE 2.6

A mathematics achievement test consisting of 100 questions was given to 50 sixth grade students at Maple Elementary School. The following data show the number of questions answered correctly by each student.

75	48	46	65	71
49	61	51	57	49
84	85	79	85	83
55	69	88	89	55
61	72	64	67	61
77	51	61	68	54
63	94	54	53	71
84	79	75	65	50
45	65	77	71	63
67	57	63	71	77

In order to create a stem-and-leaf display of these data, we will place the first digit of each data value to the left of a vertical line and the second digit to the right of the vertical line. The first digit forms the "stem" and the second digit is the "leaf." For the first item in the data set, 75, the stem-and-leaf numbers are arranged as follows:

Stem	Leaf
7	5

Scanning the data set, we find that the smallest value is 45, whereas the largest value is 94. Thus, the stems must range from 4 (for data in the 40s) to 9 (for data in the 90s). The complete set of stems for this data set is:

4	
5	
6	
7	
8	
9	

The stem-and-leaf display is completed by writing the second digit for each item as a leaf in the row containing the first digit. Thus, the first value of 75 is noted by writing a 5 as a leaf in the row headed by 7. The second item, with a value of 48, is noted by writing an 8 as a leaf in the row headed by 4. Each value in the data set can be added to the display by entering the second digit in the appropriate row. Adding the leaves for the entire data set provides the following:

4	8 6 9 9 5
5	1 7 5 5 1 4 4 3 0 7
6	5 1 9 1 4 7 1 1 8 3 5 5 3 7 3
7	5 1 9 2 7 1 9 5 7 1 1 7
8	4 5 5 3 8 9 4
9	4

Given the organization of the data, it is a simple matter to arrange the leaf values in each row in ascending order from smallest to largest. Doing so provides the following stem-and-leaf display:

```
4 | 5 6 8 9 9
5 | 0 1 1 3 4 4 5 5 7 7
6 | 1 1 1 1 3 3 3 4 5 5 5 7 7 8 9
7 | 1 1 1 1 2 5 5 7 7 7 9 9
8 | 3 4 4 5 5 5 8 9
9 | 4
```

The stem-and-leaf display is now complete. Using the first line of the display, we see that the ordered data set contains the data values 45, 46, 48, and two 49s. In addition, notice how the stem-and-leaf display provides information about the pattern, or distribution, of the test scores. For example, we see that the largest number of test scores are in the 60s and the second-largest number of test scores are in the 70's. Five students scored in the 40s, and one student scored above 90. Other observations and interpretations are up to the user.

■

The stem-and-leaf display of quantitative data provides visual insights concerning the pattern of the data as well as a means of arranging the data in ascending order. We note here that there is no one right way to develop and present a stem-and-leaf display. Personal preferences can lead to different displays for the same data set. For instance, if we believe that the six stems shown for the data set of Example 2.6 have condensed the data set too much, it is a simple matter to stretch the display by using two or more stems for each first digit. For example, to use two lines for each first digit, we place all data values ending in 0, 1, 2, 3, or 4 on one line and all values ending in 5, 6, 7, 8 and 9 on a second line.

EXAMPLE 2.6
(continued)

Using the preceding convention, the stems of the display for the 50 mathematics achievement test scores would be written as follows:

```
4 |
5 |
5 |
6 |
6 |
7 |
7 |
8 |
8 |
9 |
```

Using this stretched-stem format, the first stem row for 6 contains data values from 60 to 64 and the second stem row for 6 contains data values from 65 to 69. The complete stretched stem-and-leaf display appears as follows:

```
4 | 5 6 8 9 9
5 | 0 1 1 3 4 4
5 | 5 5 7 7
6 | 1 1 1 1 3 3 3 4
```

```
6 | 5 5 5 7 7 8 9
7 | 1 1 1 1 2
7 | 5 5 7 7 7 9 9
8 | 3 4 4
8 | 5 5 8 9
9 | 4
```

■

Since there is no one best way to set up a stem-and-leaf display, we are free to use any part of the number as the stem and the remaining part of the number as the leaf. For example, with data values in the thousands (such as 1644, 1765, 1852 and so on), stems could be expressed in terms of the first two digits: 16, 17, and 18. The next digits 4, 6, and 5, would be written as the leaves. Data of the form 22.75, 24.63, 25.30, and so on could be best summarized by using 22, 23, 24 and 25 as the stems with the next decimal digits 7, 6, and 3, appearing as the leaves.

Note that the above convention uses only single-digit leaves. Thus, when we have four digit numbers and stems involving two digits, the last digit is truncated. The numbers in the stem and leaf display will approximate the actual numbers to within ten units. For instance, the stem and leaf for the number 1644 would be 16 | 4. The stem and leaf for the number 22.75 would be 22 | 7. When digits have been truncated, the exact numerical values of the data do not appear in the display, but the display still provides a convenient summary for the data set.

Frequency Distribution

A frequency distribution provides another tabular method for summarizing quantitative data. We defined a frequency distribution in Section 2.2 to be a tabular summary of a set of data showing the number of items in each of several nonoverlapping classes. This definition holds for quantitative data as well as for qualitative data. However, with quantitative data, we have to be more careful in defining the classes. We do not have separate categories for the values of the data items. Three steps are necessary in order to define the classes for a frequency distribution with quantitative data: determine the number of classes, determine the width of each class, and determine the class limits for each class. These three steps are discussed next.

NUMBER OF CLASSES

Classes are formed by specifying the ranges of values that will be used to group items together in the frequency distribution. As a general guideline, we recommend using between 5 and 20 classes. Large data sets usually require more classes; small data sets can often be summarized quite nicely with as few as 5 or 6 classes. The goal is to use enough classes to show the variation in the data but not so many that there are only a few items in many of the classes.

WIDTH OF CLASSES

The second step in constructing a frequency distribution for quantitative data is to choose a width for each class. Choosing the number of classes and the class widths are not independent decisions. Since the smallest data value must be in the lowest class and the largest data value must be in the highest class, the fewer classes used, the wider they must be. The relationship between the number of classes and the class width can be written as follows:

$$\text{Approximate Class Width} = \frac{\text{Largest Data Value} - \text{Smallest Data Value}}{\text{Number of classes}}$$

Once the number of classes has been chosen, the approximate class width is determined. Experimenting with a few different numbers of classes and class widths is usually helpful in constructing a frequency distribution.

CLASS LIMITS

The last step in constructing a frequency distribution for quantitative data is to choose *class limits*. The class limits determine the range of data values that are grouped into each class. Care must be taken to be sure the limits are stated such that all data values fall in *one and only one class*.

For example, suppose we had integer data and were interested in defining class limits for class widths of 10. The class limits of 60–70, 70–80, 80–90, and so on would not be acceptable due to the fact that data values of 70 and 80 appear to belong to two classes. There is, however, flexibility in defining class limits. In this case, the class limits may be written as 60–69, 70–79, and 80–89 because with integer data, there are no items with values in the intervals from 69 to 70 and 79 to 80. Another alternative for writing class limits for quantitative data is 60–under 70, 70–under 80, and 80–under 90. An alternative such as this would be necessary if all values between 69 and 70, 79 and 80, and so on were possible. The final specification of the class limits is left to the user's discretion. However, the overriding consideration is that the limits must be defined so that all values clearly belong to one and only one class.

EXAMPLE 2.6 (continued)

Let us return to the data set containing the mathematics achievement test scores of 50 sixth grade students. The stem-and-leaf display completed earlier can help us define the number of classes, class widths and class limits. In this case, we have chosen to use 6 classes with class widths of 10. Since the data are integer values, the class limits of 40–49, 50–59, 60–69, 70–79, 80–89, and 90–99 are acceptable. The frequency distribution is formed by counting the number of items with values that fall into each class. Doing so provides the frequency distribution shown in Table 2.6.

■

TABLE 2.6

Frequency Distribution for 50 Sixth Grade Mathematics Achievement Test Scores

TEST SCORES	FREQUENCY
40–49	5
50–59	10
60–69	15
70–79	12
80–89	7
90–99	1
Total	50

TABLE 2.7

**Percent Frequency Distribution for
50 Sixth Grade Mathematics
Achievement Test Scores**

TEST SCORES	PERCENT FREQUENCY
40–49	10
50–59	20
60–69	30
70–79	24
80–89	14
90–99	2
Total	100

Percent Frequency Distribution

Data summarized in a frequency distribution can be converted to a percent frequency distribution by dividing each class frequency by the total number of data items and multiplying by 100 to convert the relative frequencies to percentages. The percent frequency distribution for the mathematics achievement test scores is shown in Table 2.7.

Referring to the tabular summaries shown in Tables 2.6 and 2.7, we can make some observations about the performance of the sixth graders on the achievement test. For example, note the following:

1. The most frequently occurring scores are in the interval 60–69, with 15 students, or 30%, getting this many questions correct.
2. Only 5 students, or 10%, answered fewer than 50 of the questions correctly.
3. Only 8 students, or 16% answered 80 or more questions correctly.

The value of frequency and percent frequency distributions is that they provide insights about the entire data set that cannot be easily obtained by viewing the individual items.

Cumulative Frequency and Cumulative Percent Frequency Distributions

Variations of the basic frequency and percent frequency distributions are the *cumulative frequency* and *cumulative percent frequency distributions*. The cumulative forms of these distributions provide additional information and insight about a data set. They contain the same number of classes as the frequency and percent frequency distributions; however, the cumulative forms show the total number of data items and the percentage of items with values *less than or equal to the upper limit* of the class.

**EXAMPLE 2.6
(continued)**

The frequency distribution and the cumulative frequency distribution for the mathematics achievement test score data presented in Example 2.6 are shown together in Table 2.8.

To see how the cumulative frequency distribution is constructed, consider the interval 60–69 in the frequency distribution. The upper limit for this class is 69. To determine the number of items with values less than or equal to 69, we simply sum the frequencies of the intervals 40–49, 50–59, and 60–69; doing so, we obtain 5 + 10 + 15 = 30.

TABLE 2.8

Frequency and Cumulative Frequency Distributions for the 50 Sixth Grade Mathematics Achievement Test Scores

FREQUENCY DISTRIBUTION		CUMULATIVE FREQUENCY DISTRIBUTION	
Test Scores	Frequency	Test Scores	Cumulative Frequency
40–49	5	Less than or equal to 49	5
50–59	10	Less than or equal to 59	15
60–69	15	Less than or equal to 69	30
70–79	12	Less than or equal to 79	42
80–89	7	Less than or equal to 89	49
90–99	1	Less than or equal to 99	50

TABLE 2.9

Percent Frequency and Cumulative Percent Frequency Distributions for the 50 Sixth Grade Mathematics Achievement Test Scores

PERCENT FREQUENCY DISTRIBUTION		CUMULATIVE PERCENT FREQUENCY DISTRIBUTION	
Test Scores	Percent Frequency	Test Scores	Cumulative Percent Frequency
40–49	10	Less than or equal to 49	10
50–59	20	Less than or equal to 59	30
60–69	30	Less than or equal to 69	60
70–79	24	Less than or equal to 79	84
80–89	14	Less than or equal to 89	98
90–99	2	Less than or equal to 99	100

Hence the value that we enter in the cumulative frequency distribution corresponding to a score of less than or equal to 69 is 30.

The cumulative percent frequency distribution is constructed in a similar fashion and is shown in Table 2.9

Several interpretations and insights are available from the cumulative forms of the distributions. For example, it is easy to see that 15 students, or 30%, scored less than or equal to 59 on the achievement test. Other observations concerning cumulative frequencies are possible.

■

Notes and Comments

There is no one correct way to present a stem-and-leaf display or the other tabular summaries presented in this section. The general procedures for constructing the displays and distributions have been described. However, user preferences and differences are permissible. Different individuals summarizing the same data set may present different, but acceptable, forms of stem-and-leaf displays and frequency distributions. The overriding concern is always to present the data in a summary form that is not misleading and can be easily interpreted and understood. *Continued on next page.*

2. The description of the classes for a cumulative frequency or cumulative percent frequency distribution will vary, depending upon how the class limits are defined for the frequency distribution. In Example 2.6, we used 40–49, 50–59, 60–69, and so on for the class limits. Using the data value 59 as an example, we see that 59 belongs to the second class. In the cumulative distribution, the description use for this class was "less than or equal to 59." If we had defined the class limits in the original frequency distribution as being 40–under 50, 50–under 60, 60–under 70, and so on, the description used for the classes of the cumulative distribution would have been "less than 50," "less than 60," "less than 70," and so on. In this case, the less than or equal to form would not have been used.

3. Given class limits for a frequency distribution, we may occasionally want to know the *class midpoints*. Each class midpoint is simply one-half of the distance between the class limits. For example, with a class limit denoted by 60–69, the class midpoint is 64.5. If the class limits are denoted as 60–under 70, the midpoint is one-half of the distance between 60 and 70, thus making the midpoint 65.

EXERCISES

13. The given data are the bowling scores Dave Axline obtained for the most recent 20 games:

$$
\begin{array}{ccccc}
180 & 190 & 201 & 176 & 196 \\
168 & 218 & 199 & 182 & 170 \\
182 & 176 & 199 & 198 & 171 \\
177 & 174 & 199 & 168 & 176 \\
\end{array}
$$

Summarize the data by constructing
a. a stem-and-leaf display
b. a frequency distribution
c. a percent frequency distribution
d. a cumulative frequency distribution
e. a cumulative percent frequency distribution

14. The conclusion from a 41-state poll conducted by the Joint Council on Economic Education (*Time,* January 9, 1989) was that students do not learn enough economics. The findings were based on test results from eleventh and twelfth grade students who took a 46-question multiple-choice test on basic economic concepts such as profit and the law of supply and demand. The following sample data represents a portion of the data on the number of questions answered correctly.

$$
\begin{array}{cccccccccc}
12 & 31 & 24 & 22 & 8 & 10 & 14 & 16 & 18 & 25 \\
16 & 15 & 21 & 30 & 22 & 24 & 19 & 13 & 16 & 15 \\
12 & 17 & 20 & 26 & 33 & 14 & 9 & 12 & 18 & 24 \\
18 & 19 & 22 & 16 & 17 & 23 & 28 & 18 & 14 & 19 \\
\end{array}
$$

Summarize the data using
a. a stem-and-leaf display
b. a frequency distribution

c. a percent frequency distribution

d. a cumulative frequency distribution

e. Based on these data, do you agree with the claim that students are not learning enough economics? Explain.

15. The given data show the number of automobiles arriving at a toll booth on the Kentucky Bluegrass Parkway during 20 intervals, each of 10 minutes duration.

26	26	38	24
32	22	15	33
19	27	21	28
16	20	34	24
27	30	31	33

Summarize the data by constructing

a. a frequency distribution

b. a percent frequency distribution

c. a cumulative frequency distribution

d. a cumulative percent frequency distribution

16. *Car and Driver* magazine (January, 1989) selected the Honda Civic as one of the ten best cars of 1989. Data provided about the Civic included fuel-economy information stated in miles per gallon. Assume that the following miles per gallon data were obtained from a sample of actual mileage tests with the Civic.

30.2	29.0	27.5	28.3	29.2	32.1	33.8	25.2	34.3	30.6
30.5	28.3	26.0	28.5	29.4	30.3	30.8	29.2	25.9	26.4
27.7	33.9	30.4	29.4	29.4	30.2	28.8	27.5	30.8	30.0

Summarize the data by constructing

a. a stem-and-leaf display

b. a frequency distribution

c. a cumulative frequency distribution

d. *Car and Driver* magazine reported the fuel economy of the Honda Accord as being in the range 22 to 27 miles per gallon. Does the Accord or the Civic appear to provide the lower cost transportation?

17. The Highway Loss Data Institute "Injury and Collision Loss Experience" (September 1988) rates car models based on the number of insurance claims filed after accidents. Index ratings near 100 are considered average. Lower ratings are better, and the car model is considered safer. Shown next are ratings for 20 midsize cars and 20 small cars.

Midsize cars:	81	91	93	127	68	81	60	51	58	75
	100	103	119	82	128	76	68	81	91	82
Small cars:	73	100	127	100	124	103	119	108	109	113
	108	118	103	120	102	122	96	133	80	140

Summarize the data for the midsize and small cars separately using

a. a stem-and-leaf display

b. a percent frequency distribution

c. Make a statement about what your summaries indicate about the safety of midsize cars compared to small cars.

18. National Airlines accepts flight reservations by phone. Shown below are the durations (in minutes) for a sample of 20 phone reservations. Construct frequency and percent frequency distributions for the data.

2.1	4.8	5.5	10.4
3.3	3.5	4.8	5.8
5.3	5.5	2.8	3.6
5.9	6.6	7.8	10.5
7.5	6.0	4.5	4.8

19. A psychologist asked 25 of her patients to take a written examination designed to measure a patient's depression level. The following data were obtained (higher scores indicate a higher level of depression).

75	63	33	69	26
62	77	54	61	96
87	57	61	56	79
78	67	78	68	75
28	89	61	51	41

Summarize the data by constructing
a. a frequency distribution
b. a percent frequency distribution
c. a cumulative percent frequency distribution

20. The personal computer has brought computer convenience and power into the home environment. But just how many hours a week are people actually using their personal computers at home? A study designed to determine the usage of personal computers at home (*U.S. News and World Report*, December 26, 1988) provided data in hours per week, as follows:

0.5	1.2	4.8	10.3	7.0	13.1	16.0	12.7	11.6	5.1
2.2	8.2	0.7	9.0	7.8	2.2	1.8	12.8	12.5	14.1
15.5	13.6	12.2	12.5	12.8	13.5	1.3	5.5	5.0	10.8
2.5	3.9	6.5	4.2	8.8	2.8	2.5	14.4	16.0	12.4
2.8	9.5	1.5	10.5	2.2	7.5	10.5	14.1	14.9	0.3

Summarize the data using a
a. stem-and-leaf display
b. frequency distribution (use a class width of 3 hours)
c. percent frequency distribution
d. Comment on what the data indicate about the usage of personal computers at home.

2.4 Summarizing Quantitative Data: Graphical Methods

Graphical summaries of quantitative data can be helpful in terms of providing a view showing the pattern of the entire data set. The most commonly used graphical presentations of quantitative data are histograms, frequency polygons, and ogives.

Histogram

Perhaps the most often used graphical presentation of quantitative data is a *histogram*. A histogram can be prepared for data that have previously been summarized in either a frequency distribution or a percent frequency distribution. A histogram is constructed by placing the variable of interest on the horizontal axis and the frequency or percent

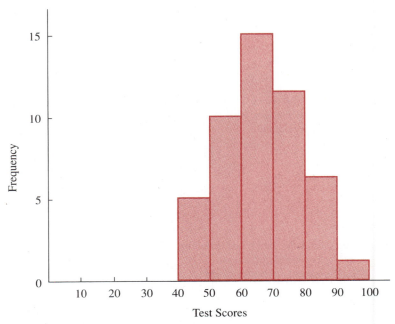

FIGURE 2.5 Histogram for 50 Sixth Grade Mathematics Achievement Test Scores

frequency values on the vertical axis. The frequency, or percent frequency, is shown by drawing a rectangle whose base is the class interval on the horizontal axis and whose height is the corresponding frequency or percent frequency.

**EXAMPLE 2.6
(continued)**

The frequency distribution of test scores for the 50 sixth grade students who completed the mathematical achievement test was shown in Table 2.6. A histogram of this data set is shown in Figure 2.5. Note that the highest frequency is shown by the rectangle appearing above the class interval 60–69. The height of the rectangle shows that the frequency of this class is 15. A histogram for the percent frequency distribution of this data set would look the same as the histogram in Figure 2.5 with the exception that the vertical axis would be labeled in terms of percentages rather than frequency values.

■

As Figure 2.5 shows, the rectangles of a histogram are drawn adjacent to one another, and there is no space between the rectangles of adjacent classes. This is the usual convention. Since the class limits of the frequency distribution were stated as 40–49, 50–59, 60–69, and so on, there appear to be intervals between classes corresponding to the one-unit intervals of 49 to 50, 59 to 60, 69 to 70, and so on. The spaces are eliminated in the histogram by drawing the vertical lines halfway between the class limits. For example, the vertical lines for the class 50–59 are drawn above the values 49.5 and 59.5 on the horizontal axis. Similarly, the vertical lines for the class 60–69 are shown above the values 59.5 and 69.5. This minor adjustment causes no interpretation problems, since it is clear in Figure 2.5 that the largest rectangle corresponds to scores in the 60s. In most histograms, this slight adjustment used to eliminate the spaces between classes goes unnoticed. If the class limits in the frequency distribution had been defined as 40–under

50, 50–under 60, and so on, the vertical lines would have been drawn above 40, 50, 60, and so on, and the one-half unit adjustment would be unnecessary.

Frequency Polygon

A *frequency polygon* provides an alternative to a histogram as a graphical method for presenting a frequency or a percent frequency distribution. Again, the possible values for the variable of interest are placed on the horizontal axis, and the frequency or percent frequency values are placed on the vertical axis. However, instead of using rectangles, as with the histogram, we find the class midpoints on the horizontal axis and then plot a point directly above each class midpoint at a height corresponding to the frequency or percent frequency of the class. Two extra classes are added, one at each end of the distribution. A frequency of zero for each of these classes means that the points plotted for the two added class midpoints will be on the horizontal axis. In this way, when the frequency polygon is drawn by connecting all the plotted points, both ends of the frequency polygon will touch the horizontal axis.

EXAMPLE 2.6 (continued)

A frequency polygon for the 50 mathematics achievement test scores is shown in Figure 2.6. The class midpoints are 44.5, 54.5, 64.5, 74.5, 84.5, and 94.5. The two added zero-frequency classes are 30–39 and 100–109, with corresponding midpoints of 34.5 and 104.5. The frequency polygon for the percent frequency distribution presented in Table 2.7 would look the same as the one shown in Figure 2.6 with the exception that the values on the vertical axis would be percent frequencies rather than frequencies.

■

Ogive

A graphical method used to present a cumulative frequency or a cumulative percent frequency distribution is called an *ogive*. The values for the variable of interest are again

FIGURE 2.6   Frequency Polygon for 50 Sixth Grade Mathematics Achievement Test Scores

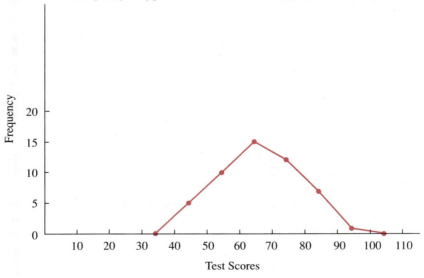

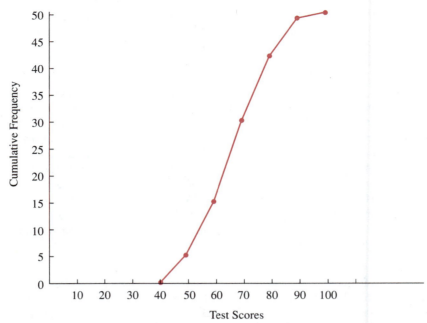

FIGURE 2.7 Ogive for 50 Sixth Grade Mathematics Achievement Test Scores

placed on the horizontal axis, with the cumulative frequencies or the cumulative percent frequencies on the vertical axis. A point is plotted for each class. The point is plotted above the upper limit of the class at a height corresponding to the cumulative frequency or cumulative percent frequency of the class. One additional point is then plotted above the lower limit of the first class at a height of zero. These points are then connected by straight-line segments to form the ogive.

Given any value, the ogive can be used to estimate the number of items or percentage of items with values less than or equal to the value being considered.

EXAMPLE 2.6 (continued)

The cumulative frequency distribution for the 50 mathematics achievement test scores is shown in Table 2.8. Figure 2.7 shows the ogive for this cumulative frequency distribution. The points are plotted above the upper class limit values of 49, 59, 69, and so on. Using the ogive, we could estimate that approximately 25 data values are less than or equal to 65.

■

EXERCISES

21. Provide a histogram, frequency polygon, and ogive for the 46-question multiple-choice test data from the Joint Council on Economic Education study in exercise 14. Use the ogive to estimate the percentage of students in the study who scored 50% or less on the examination.

22. Provide a histogram, frequency polygon, and ogive for the Honda Civic miles per gallon data in exercise 16.

23. Refer to the Highway Loss Data Institute safety index data from exercise 17.
   a. Construct a frequency polygon for the midsize and small cars on the same graph.
   b. Construct an ogive for the midsize and small cars on the same graph.
   c. Do these graphical summaries help indicate that midsize cars have a better safety index rating? Explain.

24. Develop a histogram and frequency polygon for the personal computer usage data in exercise 20. What are the values of the class midpoints in your data summary?

2.5   Summarizing Bivariate Data

In some statistical applications, data are collected on two attributes of the elements in a study. Such a data set contains *bivariate data*. By providing a data summary referred to as a *crosstabulation,* we are often able to gain some insight about the relationship between two variables. In Example 2.7 we consider the case of bivariate data when both variables are qualitative. In Example 2.8, we consider the case of bivariate data when one variable is qualitative and one variable is quantitative.

EXAMPLE 2.7

Alber's Brewery of Phoenix, Arizona, manufactures and distributes three types of beers: a low-calorie light beer, a regular beer, and a dark beer. The market research group has raised questions concerning differences in preferences for the three beers among male and female beer drinkers. A sample of 150 beer drinkers has been selected. After taste-testing each beer, the individuals in the sample are asked to state their preference or first choice. The two qualitative variables of interest in this study are gender of the beer drinker and beer preference. A partial listing of the data is shown next:

INDIVIDUAL	GENDER	BEER PREFERENCE
1	Female	Dark
2	Male	Light
3	Male	Regular
•	•	•
•	•	•
•	•	•
149	Female	Regular
150	Female	Light

Organizing the data for two qualitative variables into a table often provides valuable insights. Suppose we utilize the format of Table 2.10. Every individual in the sample can be classified as belonging to one of the six cells in the table. For example, an individual may be a male preferring regular beer (cell 2), a female preferring light beer (cell 4), a female preferring dark beer (cell 6), and so on. Since we have included all possible combinations of beer preferences and gender—or, in other words, listed all possible contingencies—Table 2.10 is called a *contingency table*.

TABLE 2.10

Format of a Contingency Table for Alber's Brewery

| | | VARIABLE 2: BEER PREFERENCE | | |
		Light	Regular	Dark
VARIABLE 1:	Male	Cell 1	Cell 2	Cell 3
GENDER	Female	Cell 4	Cell 5	Cell 6

EXAMPLE 2.7 (continued)

A summary of the data in the Alber's Brewery study in the form of a contingency table is shown next. Percentages are shown in parentheses below the cell frequencies.

| | | BEER PREFERENCE | | | |
		Light	Regular	Dark	**Total**
	Male	20 (13.3)	40 (26.7)	20 (13.3)	80 (53.3)
Gender	Female	30 (20.0)	30 (20.0)	10 (6.7)	70 (46.7)
	Total	50 (33.3)	70 (46.7)	30 (20.0)	150 (100.0)

Percentage ⟶ Number of individuals

In reviewing the contingency table or crosstabulation summary we see that of the 150 beer drinkers in the sample, 20, or 13.3%, were men who favored light beer; 30, or 20.0%, were women who favored light beer; and so on. The crosstabulation presentation facilitates inferences about the population of beer drinkers and their preferences. For example, the largest cell frequency of 40 (26.7%) suggests that the largest segment of beer drinkers consists of men who prefer regular beer. The smallest cell frequency of 10 (6.7%) suggests the smallest segment of beer drinkers consists of women who prefer dark beer.

Additional inferences are possible from the information in the margins of the contingency table. The entries in the bottom row show that 33.3% of beer drinkers prefer light beer, 46.7% prefer regular beer and 20% prefer dark beer. Information in the total column (the right margin) might be used to infer that 53.3 percent of beer drinkers are male and that 46.7 percent are female.

EXAMPLE 2.8

Let us now consider a bivariate data set in which one variable is quantitative and one variable is qualitative. In Table 1.1, we presented data from a sample of 50 accredited senior colleges and universities in the United States. The variable of interest was quantitative and provided the annual room and board cost at each of the sampled colleges and universities. Let us return to this data set and record the qualitative variable of school location. Each sampled college and university will be identified as being located in either the East, Midwest, South, or West. In this case we have bivariate data; each element in the sample (a school) provides data for two variables: annual room and board cost and location.

We will use a contingency table crosstabulation similar to the one shown in Table 2.10. It is necessary to develop class limits for the quantitative variable as well as specifying the classes for the qualitative variable, location. We will use class intervals of $1,500–1,999, $2,000–2,499, $2,500–2,999, and so on, for the annual room and board cost. The qualitative variable (location) has the natural classes of East, Midwest, South, and West. The crosstabulation process simply counts the number of data items for each combination of cost class and location. The crosstabulation summary including both frequencies and percentages is shown in Table 2.11. The percentages are shown in parentheses below each cell count or frequency.

■

Many interpretations are possible with this crosstabulation summary. Generally such a tabulation provides insights that are difficult to obtain by viewing the original bivariate data. Some interpretations based on the crosstabulation summary in Table 2.11 are as follows:

1. The East appears to be the most expensive region in terms of annual room and board cost. Eleven Eastern colleges and universities (22% of all schools in the study) show a room and board cost in the range from $3,000–3,999.
2. The only colleges and universities in the sample showing annual room and board costs under $2,000 per year are in the South. This suggests the South has the lowest room and board costs.
3. The West region is relatively inexpensive. Seven of the eight schools in the West region show an annual room and board cost of under $3,000.

TABLE 2.11

Crosstabulation of Annual College Room and Board Costs and College or University Location

ANNUAL COST	LOCATION				Total
	East	Midwest	South	West	
$1,500–1,999	0	0	3	0	3
	(0)	(0)	(6)	(0)	(6)
$2,000–2,499	1	4	0	3	8
	(2)	(8)	(0)	(6)	(16)
$2,500–2,999	1	5	4	4	14
	(2)	(10)	(8)	(8)	(28)
$3,000–3,499	7	1	6	0	14
	(14)	(2)	(12)	(0)	(28)
$3,500–3,999	4	2	1	0	7
	(8)	(4)	(2)	(0)	(14)
$4,000–4,499	1	1	1	0	3
	(2)	(2)	(2)	(0)	(6)
$4,500–4,999	0	0	0	1	1
	(0)	(0)	(0)	(2)	(2)
Total	14	13	15	8	50
	(28)	(26)	(30)	(16)	(100)

(Percentages are in parentheses)

4. The most expensive school in the sample is in the West. Reviewing the original data in Table 1.1, we see that this school is Pepperdine, with an annual room and board cost of $4,535, reflecting the high cost of living in Malibu, California.

Other interpretations are possible from the information in the margins. We see that 28% of the schools sampled were in the East, 26% were in the Midwest, 30% were in the South, and 16% were in the West. The information in the bottom row is essentially a frequency and percent frequency distribution for school location. The information in the right-most column is a frequency distribution and percent frequency distribution for annual room and board cost.

EXERCISES

25. A large amusement park surveyed park visitors at the end of the day in order to investigate what effect (if any) the distance traveled had on how satisfied visitors are with regard to the food service facilities at the park. Responses were coded as S for satisfied and NS for not satisfied. The classes for distance traveled were summarized as follows: less than 25 miles, 25 to 100 miles, and over 100 miles. The data collected are as follows:

OPINION	DISTANCE (MILES)
S	50
S	15
NS	10
NS	120
S	115
NS	12
S	20
S	18
S	140
NS	40
NS	15
S	5
NS	175
S	120
S	35
S	10
NS	70
S	14
NS	120
S	20

Summarize the distance traveled using the classes 0-24, 25-99, and 100 or more. Construct a contingency table for the data and develop whatever conclusions appear to be appropriate.

26. Eastern Pharmaceutical Corporation is testing a new drug intended to help relieve the symptoms associated with hay fever. One hundred patients were given different levels of the drug (A, B, or C) and then observed for any possible side effects. The following results were obtained: 60 patients experienced no side effects, and of this group 25 had been given level A of the drug and 30 had been given level B; 20 of the patients that were given level B and

15 of the patients that were given level C experienced some side effects. Construct a contingency table for these data and develop whatever conclusions appear to be appropriate.

SUMMARY

Data are the measurements that are collected and processed in a statistical study. They are the raw material of statistics and are the basis for learning about a phenomenon being studied. Data may be nonnumerical or numerical.

Four scales of measurement are used for data. They are the nominal, ordinal, interval and ratio scales. The amount of information in the data depends upon the measurement scale used. Nominal data contain the least amount of information, followed by ordinal, interval, and ratio data. Nominal and ordinal scales provide qualitative data. Interval and ratio scales provide quantitative data. Ordinary mathematical operations of addition, subtraction, multiplication and division are meaningful only if the interval or ratio measurement scale has been used.

Data are often difficult to interpret directly in raw form. Tabular and graphical procedures of descriptive statistics are used to summarize data in order to enhance insight and interpretation. Stem-and-leaf displays, frequency distributions, percent frequency distributions, cumulative frequency distributions, and cumulative percent frequency distributions are common tabular summaries of data. Graphical methods of bar graphs and pie charts may be used to summarize qualitative data, whereas graphical methods of histograms, frequency polygons and ogives may be used to summarize quantitative data. Figure 2.8 shows the tabular and graphical procedures that may be used to summarize data.

FIGURE 2.8 Tabular and Graphical Procedures for Summarizing Data

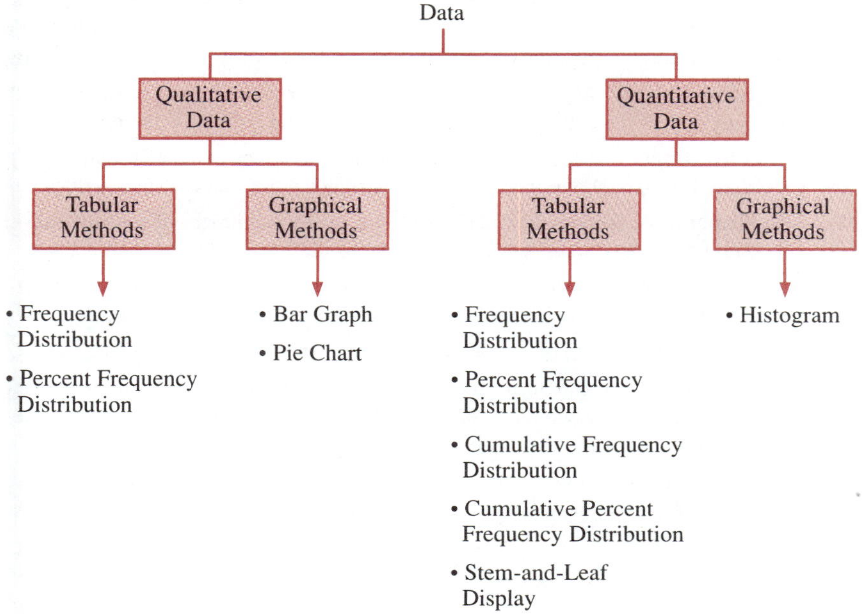

The chapter concluded with a discussion of bivariate data and showed how contingency table or crosstabulation summaries can provide insights and interpretations for bivariate data.

GLOSSARY

Data Facts and figures collected in a statistical study. Data are the raw material for statistical analysis and summarization.

Nominal scale A scale of measurement that uses a label or category to define an attribute of an element. Nominal data may be recorded with a nonnumeric description or with a numeric code.

Ordinal scale A scale of measurement that has the properties of a nominal scale and can be used to rank or order the observations. Ordinal data may be recorded with a nonnumeric description or with a numeric code.

Interval scale A scale of measurement that has the properties of an ordinal scale and the interval between observations is expressed in terms of a fixed unit of measure. Interval data are always numeric.

Ratio scale A scale of measurement that has the properties of an interval scale and the ratio of observations is meaningful. Ratio data are always numeric.

Qualitative data Data obtained with a nominal or ordinal scale of measurement. Qualitative data may be recorded with a nonnumeric description or with a numeric code.

Quantitative data Data obtained with an interval or ratio scale of measurement. Quantitative data are always numeric and indicate how much or how many for the variable of interest.

Frequency distribution A tabular summary of a set of data showing the number of items in each of several nonoverlapping classes.

Percent frequency distribution A tabular summary of a set of data showing the percent of items in each of several nonoverlapping classes.

Bar graph A graphical device for depicting the information presented in a frequency distribution or percent frequency distribution of qualitative data.

Pie chart A graphical device for presenting a summary of qualitative data based upon subdividing a circle into sectors that correspond to the percent frequency of each class.

Stem-and-leaf display A method for summarizing interval or ratio data that simultaneously rank-orders the data and provides insight into the pattern of the data.

Class limits The upper and lower numerical values associated with a frequency distribution class for quantitative data. Class limits indicate the range of data values belonging to a class.

Cumulative frequency distribution A tabular summary of a set of data showing the total number of data items less than or equal to the upper limit of each class.

Cumulative percent frequency distribution A tabular summary of a set of data showing the percentage of items less than or equal to the upper limit of each class.

Class midpoint The average of the lower and upper class limits.

Histogram A graphical presentation of a frequency distribution or percent frequency distribution for quantitative data. A histogram is constructed by placing the class intervals on the horizontal axis and the frequencies or percent frequencies on the vertical axis.

Frequency polygon A graphical presentation of a frequency distribution or percent frequency distribution for quantitative data. Points used for the frequency polygon are plotted above the class midpoints.

Ogive A graphical presentation of a cumulative frequency or a cumulative percent frequency distribution.

Bivariate data   Two data items collected for each element.

Crosstabulation   A table that summarizes the frequency and percent frequency for the various categories of bivariate data.

Contingency table   A table that provides cells or categories for each possible combination of bivariate data.

REVIEW QUIZ

TRUE/FALSE

1. The speed of an airplane in miles per hour is ordinal data.

2. A football player's uniform number is nominal data.

3. Room numbers in a building are examples of interval data.

4. The place a person finishes in a golf tournament (first, second, third, etc.) is an example of ordinal data.

5. Interval and ratio data must be numeric data.

6. Nominal and ordinal data must be numeric data.

7. A frequency distribution is a graphical summary of a set of data.

8. A percent frequency distribution has twice as many classes as a frequency distribution.

9. In a bar graph the width of the bar is proportional to the number of items in the class.

10. The class midpoint is halfway between the class limits.

11. A histogram is a graphical presentation of a frequency distribution for quantitative data.

12. A frequency polygon is a graphical presentation of a cumulative frequency distribution.

13. A stem-and-leaf display provides insight into the shape of the distribution of a set of data.

MULTIPLE CHOICE

14. Which of the following measures involve nominal data?
    a. the test score on an exam
    b. the number on a basketball player's jersey
    c. the speed of an automobile
    d. the class rank of a college student

15. Which of the following measures involve ordinal data?
    a. the score in a baseball game
    b. the place of a baseball team in the league standings
    c. the height of a flagpole
    d. the weight of a fish

16. A group of union members indicated what they felt was the most important issue in the upcoming labor management negotiations; the responses are summarized in the accompanying frequency distribution.

Issue	Frequency
Wages	22
Medical benefits	10
Retirement	15
Working conditions	13
Total	60

The percent frequency of the wages issue is closest to
a. 22
b. 15
c. 40
d. 60

17. Refer to the frequency distribution in question 16. Fringe benefits include both medical benefits and retirement. What percentage of the union membership feels that fringe benefits are the most important issue?
a. 16.7%
b. 25%
c. 63%
d. 41.7%

18. Consider the following frequency distribution.

Completion Time (min)	Frequency
8–10	5
11–13	8
14–16	15
17–19	12
20–22	7
23–25	3
Total	50

The width of the class intervals is
a. 3
b. 6
c. 2
d. 2.5

19. Refer again to the frequency distribution in question 18. The midpoint for the second class is
a. 11
b. 11.5
c. 12
d. 12.5

20. Consider the following stem-and-leaf display.

$$
\begin{array}{c|cccccc}
6 & 2 & 2 & 4 \\
7 & 3 & 6 & 6 & 7 & 8 & 9 \\
8 & 1 & 5 & 7 \\
9 & 0 & 1 \\
\end{array}
$$

The data set consists of how many items?
a. 14
b. 18
c. 6
d. 20

27. Determine whether the following represent nominal, ordinal, interval, or ratio data. Explain.
a. religious preference
b. age (in years)
c. student grade point average
d. credit card numbers
e. SAT math scores
f. marital status (single, married, divorced)

28. Frequent airline travelers were asked to indicate the airline they believed offered the best overall service. The four choices were American Air(A), East Coast Air(E), Suncoast(S), and Great Western(W). The following data were obtained.

```
E  A  E  S  W  W  E  S  W  E
W  E  E  A  S  S  W  E  A  W
W  S  E  E  A  E  E  S  W  A
S  E  A  W  A  A  W  E  S  W
```

Summarize the data by constructing
a. a frequency distribution
b. a percent frequency distribution
c. a bar graph
d. a pie chart

29. The administrator of a large city hospital has been collecting data regarding the number of patients treated in the emergency room on weekends. The following data are the numbers treated for each of the previous 15 weeks.

```
154  177  164  145  110
214  131  122  180  191
172  148  157  174  160
```

Summarize the data by constructing
a. a frequency distribution
b. a percent frequency distribution
c. a cumulative percent frequency distribution
d. a histogram

30. A nursery school offers programs for 4-year-olds ranging from a 1-day-a-week program to a 5-day-a-week program. To help in planning, the school's director surveyed parents regarding the type of program they preferred. The following data, which represents the number of days, were obtained

```
3  3  1  2  2  4  4  2  3  3
3  5  3  3  2  2  1  5  3  3
2  4  5  3  3  3  4  2  2  4
```

Summarize the data by constructing
a. a frequency distribution
b. a percent frequency distribution
c. a histogram
d. a frequency polygon

31. Dinner check amounts for La Maison's French Restaurant are shown below:

```
42.65  36.12  52.90  44.26  52.00
34.10  39.86  29.40  48.75  82.00
38.40  44.50  79.80  74.45  71.81
46.62  56.12  63.00  63.06  59.42
```

a. Construct frequency and percent frequency distributions for the data.
b. Construct a cumulative percent frequency distribution for the data.

32. The following data represent quarterly sales volumes for 40 selected corporations.

```
17,864,000  15,065,000  42,200,000  13,523,000
49,747,000  20,510,000   5,520,000   7,985,000
 3,624,000  11,556,000   1,855,000   9,023,000
 3,804,000   5,933,000  23,900,000   6,145,000
 9,232,000   2,979,000   1,059,000  42,789,000
 5,143,000  33,380,000  20,779,000   6,145,000
 2,141,000  17,768,000  18,017,000  42,800,000
 5,090,000  41,626,000  12,003,000   6,840,000
 3,669,000  37,738,000  40,765,000  21,946,000
13,614,000  39,914,000   7,846,000  25,837,000
```

a. Construct a frequency distribution to summarize these data. Use a class width of $5,000,000.
b. Develop a percent frequency distribution for the data.
c. Construct a cumulative frequency distribution for the data.
d. Construct a cumulative percent frequency distribution for the data.

33. Use the data in exercise 32 for the following.
a. Construct a histogram as a graphical representation of the data.
b. Construct a frequency polygon for the data.
c. Construct an ogive for the data.

34. The given data show home mortgage loan amounts (in dollars) handled by a particular loan officer in a savings and loan company. Use a frequency distribution, percent frequency distribution, and histogram to help summarize these data.

20,000	38,500	33,000	27,500	34,000
12,500	25,999	43,200	37,500	36,200
25,200	30,900	23,800	28,400	13,000
31,000	33,500	25,400	33,500	29,200
39,000	38,100	30,500	45,500	30,500
52,000	40,500	51,600	42,500	44,800

35. Given below are the closing prices at week's end for 40 common stocks.

$7\frac{1}{2}$	$16\frac{1}{4}$	$19\frac{3}{4}$	$7\frac{3}{8}$	$10\frac{1}{8}$
$5\frac{3}{4}$	$7\frac{7}{8}$	$6\frac{7}{8}$	$24\frac{5}{8}$	$17\frac{1}{8}$
$12\frac{3}{4}$	11	5	$24\frac{3}{4}$	$34\frac{3}{4}$
$14\frac{3}{4}$	$10\frac{3}{8}$	$35\frac{1}{8}$	$20\frac{1}{4}$	$12\frac{5}{8}$
42	$10\frac{3}{8}$	57	$19\frac{3}{4}$	28
$63\frac{7}{8}$	$10\frac{3}{4}$	$7\frac{7}{8}$	$17\frac{5}{8}$	48
$17\frac{1}{4}$	$9\frac{3}{4}$	44	$11\frac{1}{8}$	20
$41\frac{3}{4}$	$16\frac{5}{8}$	$21\frac{1}{4}$	$8\frac{5}{8}$	$16\frac{3}{8}$

a. Construct frequency and percent frequency distributions for these data.
b. Construct cumulative frequency and cumulative percent frequency distributions for these data.

36. Use the data in exercise 35 for the following:
a. Construct a histogram for the data. Plot percent frequency on the vertical axis.
b. Construct a frequency polygon for the data.
c. Construct an ogive for the data.

37. The grade point averages for 30 students majoring in economics are given.

2.21	3.01	2.68	2.68	2.74
2.60	1.76	2.77	2.46	2.49
2.89	2.19	3.11	2.93	2.38
2.76	2.93	2.55	2.10	2.41
3.53	3.22	2.34	3.30	2.59
2.18	2.87	2.71	2.80	2.63

a. Construct a percent frequency distribution for the data.
b. Construct a cumulative percent frequency distribution for the data.

38. Use the data in exercise 37 for the following:
a. Construct a histogram for the data. Plot percent frequency on the vertical axis.
b. Construct an ogive for the data.

39. Points scored by the winning team in 25 A.C.C. (Atlantic Coast Conference) basketball games are shown.

86	79	74	72	91
82	64	75	72	74
63	80	78	95	82
86	77	73	69	72
81	85	92	62	90

a. Construct frequency and percent frequency distributions for the data.
b. Construct a cumulative percent frequency distribution for the data.
c. Construct a histogram.

40. The final examination scores in a section of calculus resulted in the following data.

$$
\begin{array}{ccccc}
56 & 77 & 84 & 82 & 42 \\
61 & 44 & 95 & 98 & 84 \\
93 & 62 & 96 & 78 & 88 \\
58 & 62 & 79 & 85 & 89 \\
89 & 97 & 53 & 76 & 75 \\
\end{array}
$$

Show a stem-and-leaf display for these data.

41. A 150-question social-awareness test was given to a group of college freshmen. The following data show the number of questions answered correctly by each of the students.

$$
\begin{array}{cccccccccc}
121 & 114 & 94 & 136 & 144 & 126 & 98 & 103 & 118 & 127 \\
135 & 97 & 119 & 117 & 122 & 138 & 142 & 141 & 102 & 105 \\
\end{array}
$$

Show a stem-and-leaf display for these data.

42. The following data show total yardage accumulated over the football season for 20 receivers.

$$
\begin{array}{cccccccccc}
744 & 652 & 576 & 1112 & 971 & 451 & 1023 & 852 & 809 & 596 \\
941 & 975 & 400 & 711 & 1174 & 1278 & 820 & 511 & 907 & 1251 \\
\end{array}
$$

Show a stem-and-leaf display for these data.

COMPUTER EXERCISE

Consolidated Foods, Inc. has opened several new grocery stores at a variety of locations over the past 2 years. One of the special services at these new stores is that customers may pay for their purchases using Visa or Mastercard credit cards as well as using either cash or an approved check. In order to better understand how customers are using the new payment feature, a sample of 100 customers was selected over a 1-week period. Data collected for each customer included how much was spent during the shopping trip and the method of payment. The data collected are shown in the accompanying table.

QUESTIONS

1. Develop tabular and graphical summaries of the method of payment data that would be helpful to the management of Consolidated Foods, Inc.

2. Develop tabular and graphical summaries of the amount spent data that would be helpful to the management of Consolidated Foods, Inc.

3. Combine the tabular summaries in questions 1 and 2 to present a contingency table of the bivariate data set.

4. Provide a verbal statement and discussion of what the tabular and graphical summaries indicate about the method of payment, the amount spent, and the relationship between the method of payment and the amount spent.

AMOUNT SPENT ($)	METHOD OF PAYMENT	AMOUNT SPENT ($)	METHOD OF PAYMENT
84.12	Check	86.34	Check
34.66	Credit card	20.23	Credit card
37.27	Credit card	108.70	Check
38.82	Credit card	45.36	Credit card
46.50	Credit card	83.31	Check
99.67	Check	64.45	Credit card
70.18	Check	54.33	Credit card
99.21	Check	16.78	Cash
138.42	Check	115.96	Check
93.68	Check	95.83	Check
120.89	Check	19.76	Cash
10.14	Cash	35.37	Cash
74.51	Check	111.98	Check
17.91	Check	103.95	Check
49.59	Check	90.40	Credit card
4.74	Cash	6.68	Cash
48.14	Cash	32.09	Credit card
65.67	Credit card	79.70	Credit card
89.66	Check	96.08	Credit card
96.40	Check	20.60	Cash
54.16	Credit card	78.81	Check
79.55	Check	123.62	Check
67.95	Check	125.01	Check
30.69	Cash	41.58	Credit card
151.89	Check	36.73	Credit card
130.41	Check	52.07	Credit card
98.80	Check	19.78	Cash
23.59	Cash	66.44	Check
104.67	Check	5.08	Cash
90.04	Check	50.15	Credit card
77.62	Check	114.42	Check
36.01	Cash	97.26	Credit card
88.17	Check	22.75	Cash
66.76	Credit card	53.63	Credit card
23.50	Cash	132.31	Check
127.34	Check	105.54	Check
26.02	Cash	66.09	Check
79.77	Check	62.24	Check
29.35	Check	97.93	Check
71.31	Credit card	10.57	Cash
43.57	Credit card	51.21	Credit card
76.18	Credit card	90.17	Check
59.38	Credit card	24.08	Credit card
72.99	Credit card	42.72	Cash
19.24	Cash	97.72	Check
80.20	Check	112.67	Check
55.79	Cash	14.30	Cash
134.27	Check	28.76	Credit card
64.68	Credit card	81.85	Check
75.54	Check	56.84	Credit card

3

Descriptive Statistics II: Measures of Location and Dispersion

The American Dream: Is it Still Affordable?

Home ownership is said to be part of the American dream. It is the symbol for a way of life sought by most Americans. However, with the continued rise in the cost of housing, the dream of home ownership is beginning to fade for a significant portion of the population. In 1980, 64.4% of all households were homeowners; since then the percentage has declined steadily.

Statistics that summarize the trend in the increasing cost of homeownership point to the fact that in 1970 the median price of a new house in the United States was $23,400. In 1988, this median price of a new house had risen to $123,500. The higher housing costs and higher interest rates are forcing individuals in the lower- and middle-income brackets out of the housing market. If the trend continues, home ownership will be out of reach of all but the wealthiest of American families.

The upward trend in housing costs affects not only families, builders, construction workers, building-materials suppliers, and realtors but, as some suggest, our democracy itself. Anything can happen when a human dream goes unfulfilled. And the evidence is that the American dream of home ownership may be unfulfilled in the future.

Median resale prices in some major metropolitan areas are as follows:

Stock, Boston

Orange County California	$204,000
San Francisco	$196,300
New York	$191,900
Boston	$182,900
Hartford	$169,000
Washington D.C.	$131,600
Providence	$130,400
Chicago	$ 99,300
Philadelphia	$ 97,600
Seattle	$ 93,600
Dallas/Fort Worth	$ 85,500
Minneapolis/St. Paul	$ 84,300

Phoenix	$ 79,100
Birmingham	$ 76,500
Detroit	$ 72,300
Indianapolis	$ 66,700
Buffalo	$ 64,800
Houston	$ 63,500
Des Moines	$ 59,900

Based on "Affording the Unaffordable," *U.S. News & World Report,* September 19, 1988.

In this chapter we continue the discussion of descriptive statistics by introducing several numerical measures of location and dispersion: for example, the mean, the median, the variance, and the standard deviation. Numerical measures that are computed for a population are called *population parameters;* when they are computed for a sample, they are called *sample stastistics.*

As we showed in the "Statistics in the News" article on housing costs, numerical measures are an important part of many statistical presentations; for example, in that article, median resale prices are reported as a measure of housing costs in various cities.

3.1   Measures of Location

Mean

Perhaps the most important numerical measure is the *mean,* or average value, of the data. The mean provides a good measure of central location for a data set. It is obtained by adding all the data values and dividing by the number of items. If the data set is a sample the mean is denoted by $\bar{x}$ (pronounced x bar); if the data set is a population the mean is denoted by the Greek letter μ (pronounced $m\bar{u}$).

EXAMPLE 3.1

A sample of five class sections in the College of Arts and Science provided the following data on the number of students in each class section.

$$46, \ 54, \ 42, \ 46, \ 32$$

The sample mean for this data is computed as follows:

$$\bar{x} = \frac{46 + 54 + 42 + 46 + 32}{5}$$

$$= \frac{220}{5} = 44$$

Thus for the five sections sampled, the mean is 44 students per section.

■

In specifying general statistical formulas, it is customary to denote the value of the first data item by x_1, the value of the second data item by x_2, and so on. Using this notation, the general formula for the sample mean is

---

Sample Mean

$$\bar{x} = \frac{x_1 + x_2 + \cdots + x_n}{n} = \frac{\sum_{i=1}^{n} x_i}{n}$$

(3.1)

where n = number of items in the sample.

In this formula, the numerator denotes the sum of the data values starting with $i = 1$ and ending with $i = n$. That is,

$$\sum_{i=1}^{n} x_i = x_1 + x_2 + \cdots + x_n$$

The uppercase Greek letter Σ (sigma) is used as a summation sign.

**EXAMPLE 3.1
(continued)**

Using the notation x_1, x_2, x_3, x_4, x_5 to represent the number of students in each of the 5 sections sampled, we have:

$$x_1 = 46$$
$$x_2 = 54$$
$$x_3 = 42$$
$$x_4 = 46$$
$$x_5 = 32$$

Thus to compute the sample mean, we can write

$$\bar{x} = \frac{\sum_{i=1}^{n} x_i}{n} = \frac{\sum_{i=1}^{5} x_i}{5} = \frac{x_1 + x_2 + x_3 + x_4 + x_5}{5}$$

$$= \frac{46 + 54 + 42 + 46 + 32}{5} = 44$$

When we want to sum all the values in a data set, we use the following abbreviated summation notation:

$$\sum x_i = \sum_{i=1}^{n} x_i$$

The starting and ending points for the summation, shown below and above Σ, respectively, are dropped in the abbreviated notation. It is understood that all the data values are to be included in the sum. A summary of the summation notation and operations used in this text is contained in Appendix C.

EXAMPLE 3.2

The Nielsen organization provides data on television viewing in the United States. A study (*USA Today,* December 13, 1988) reported that the mean number of hours of television viewing per week was increasing. Suppose that the following data provide the hours of television viewing per week for a sample of 16 college students.

14, 9, 12, 4, 20, 26, 17, 15, 18, 15, 10, 6, 16, 15, 8, 5

Since there are 16 sample observations, $n = 16$. The sample mean is given by

$$\bar{x} = \frac{\Sigma x_i}{n} = \frac{14 + 9 + \cdots + 5}{16} = \frac{210}{16} = 13.125$$

Hence, the average time spent viewing television for the 16 students is 13.125 hours per week.

■

Equation (3.1) shows how the mean is computed for a sample of n items. The formula for computing the mean of a population is the same, but we use different notation to indicate that we are dealing with the entire population. The number of items in the population is denoted by N, and, as we mentioned previously, the symbol for the population mean is μ.

Population Mean

$$\mu = \frac{\Sigma x_i}{N} \qquad\qquad (3.2)$$

A data set occasionally contains data items with unusually small or unusually large values (or both). Such items can have a significant influence on the value of the mean and cause it to provide a poor description of the central location of the data set. In such cases, removing, or "trimming," the unusually small or unusually large values from the data set can provide a better measure of central location. Trimming the data set by removing the smaller and larger values provides what is called the *trimmed mean* as a measure of central location. For example, a 5% trimmed mean removes the smallest 5% of the values and the largest 5% of the values from the data set. The 5% trimmed mean is then computed as the mean of the middle 90% of the data. In general, an α% trimmed mean is obtained by trimming α% of the values from each end of the data set and computing the mean of the remaining items.

Another measure of central location that is not influenced by extreme values is the median. Let us now see how the median is computed.

Median

The *median* is another numerical measure of central location for a set of data. The median for a set of data is that value falling in the middle when the data items are arranged in ascending order (rank ordered from smallest to largest). If there are an odd number of data items, the median is the middle item. If the number of data items is even, there is no single middle value. We follow the convention of defining the median to be the average of the middle two values in this case. For convenience this definition is restated below.

Median

If there is an odd number of items in the data set, the median is the value of the
    middle item when all items are arranged in ascending order.
If there is an even number of items in the data set, the median is the average value
    of the two middle items when all items are arranged in ascending order.

EXAMPLE 3.1 (continued)

Let us apply the above definition to compute the median for the sample of five class sections from the College of Arts and Science. Arranging the five data values in ascending order provides the following rank-ordered list:

$$32 \quad 42 \quad 46 \quad 46 \quad 54$$

Since $n = 5$ is odd, the median is the middle item in the above rank-ordered list. That is, the median is the value of the third item, 46. Even though there are two values of 46 for this data set, each value is treated as a separate item when we place the data in rank order and hence when we determine the median.

■

EXAMPLE 3.2 (continued)

A rank ordering of the television-viewing times for the 16 students in Example 3.2 produced the following list:

$$4 \quad 5 \quad 6 \quad 8 \quad 9 \quad 10 \quad 12 \quad 14 \quad 15 \quad 15 \quad 15 \quad 16 \quad 17 \quad 18 \quad 20 \quad 26$$

$$\uparrow$$
$$\text{Median} = 14.5$$

Since $n = 16$ is even, the median is the average value of the two middle items, that is, the eighth and ninth items. Hence the median is $(14 + 15)/2 = 14.5$.

■

Although the mean is the most commonly used measure of central location, there are a number of situations in which the median is a better measure. The mean is influenced by extreme values in a data set, but the median is not. The following example demonstrates this situation.

EXAMPLE 3.3

A sample of five families in Herrold, Iowa, showed the following annual family incomes:

$$\$17,500 \quad \$23,000 \quad \$24,000 \quad \$26,000 \quad \$320,000$$

The median annual income is \$24,000 (the third item). The mean is given by

$$\bar{x} = \frac{\$17,500 + \$23,000 + \$24,000 + \$26,000 + \$320,000}{5} = \$82,100$$

In this case, the median is a better indication of typical incomes than the mean. The mean has been substantially influenced by the single family with a very large income. Indeed, the mean annual income for the sample is over three times the annual income for 4 of the 5 families.

■

Mode

A third numerical measure, the *mode,* provides the location of the most frequently occurring value in the data set. The mode is defined as follows:

EXAMPLE 3.1 (continued)

Referring to the sample of five class section sizes with data values 46, 54, 42, 46, and 32, we see that the only value that occurs more than once is 46. Since this value, occurring with a frequency of 2, has the greatest frequency in the data set, it is the mode.

■

EXAMPLE 3.2 (continued)

Referring to the 16 weekly television-viewing times for college students, we see that 15 hours of viewing time occurs with the greatest frequency. Thus 15 (occurring with a frequency of 3) is the mode.

■

Although there is a single value that occurs with the greatest frequency in both of the above examples, situations can arise for which the greatest frequency occurs at two or more different values. The data set introduced in Example 2.10 of Chapter 2 gives the scores of 50 sixth-grade students on a mathematics achievement test. The score of 61 occurred 4 times and the score of 71 occurred 4 times. Thus both 61 and 71 are modes for this data set; we say that the data set is *bimodal*. If a data set has more than two modes it is said to be *multimodal;* in such cases the mode is almost never reported, since listing three or more modes would not do a very good job of describing a central location for the data. In the extreme case, every data value can be different, and an argument can be made that every observation is a mode. In such cases the mode is not an appropriate measure of central location, and thus no mode would be reported.

The type of data for which the mode is often a good measure of location is nominal data. For example, the nominal data set introduced in Example 2.8 of Chapter 2 resulted in the following frequency distribution for automobile purchases by women during 1988.

CAR MODEL	FREQUENCY
Chevrolet Cavalier	9
Ford Escort	14
Ford Taurus	8
Honda Accord	11
Hyundai Excel	8

The mode, or most frequently preferred automobile, is the Ford Escort. For this type of data it obviously makes no sense to speak of the mean or median automobile purchase. But the mode does provide a good indicator of what we are interested in, the automobile preferred by the greatest number of women.

Percentiles

A *percentile* is a numerical measure that also locates values of interest in a data set. A percentile provides information regarding how the data items are spread over the interval from the lowest value to the highest value. It is a measure of relative location. In large data sets that do not have numerous repeated values, the *p*th percentile is a value that divides the data set into two parts. Approximately *p* percent of the items take on values less than the *p*th percentile; approximately $(100 - p)$ percent of the items take on values greater than the *p*th percentile. The definition of the *p*th percentile is as follows.

Percentile

The *p*th percentile of a data set is a value such that *at least p* percent of the items take on this value or less and *at least* $(100 - p)$ percent of the items take on this value or more.

Admission test scores for colleges and universities are frequently reported in terms of percentiles. For instance, suppose an applicant has a raw score of 542 on the verbal portion of an admissions test. It may not be readily apparent how this student performed relative to other students taking the same test. However, if the raw score of 542 corresponds to the 70th percentile, then we know that approximately 70% of the students had scores less than this individual and approximately 30% scored better. The following three-step procedure can be used to calculate the *p*th percentile of a data set.

CALCULATING THE *p*TH PERCENTILE

STEP 1 Arrange the data values in ascending order.

STEP 2 Compute an index *i* as follows:

$$i = \left(\frac{p}{100}\right) n$$

where *p* is the percentile of interest and *n* is the number of data values.

STEP 3 (a) If *i is not an integer,* the next integer *greater than i* denotes the position of the pth percentile.

(b) If *i is an integer,* the *p*th percentile is the average of the data values in positions *i* and $i + 1$.

EXAMPLE 3.2 (continued)

As an illustration, let us determine the 90th percentile for the television-viewing times of Example 3.2

STEP 1 Arrange the 16 data values in ascending order:

4 5 6 8 9 10 12 14 15 15 15 16 17 18 20 26

STEP 2 Compute *i* as follows:

$$i = \left(\frac{90}{100}\right) 16 = 14.4$$

STEP 3 Since *i* is not an integer, the position of the 90th percentile is the next integer value greater than 14.4, the fifteenth position. Thus the 90th percentile, in the fifteenth position for the data shown in step 1, is 20.

As another illustration, let us compute the 50th percentile for the data set. Applying step 2, we obtain

$$i = \left(\frac{50}{100}\right) 16 = 8$$

Since i is an integer, step 3(b) indicates that the 50th percentile is the average of the eighth and ninth data values; thus the 50th percentile is $(14 + 15)/2 = 14.5$.

◼

The 50th percentile has divided the data set into two parts; 50% of the data values are less and 50% of the data values are greater. This is exactly what is accomplished by finding the median of a data set. In other words, *the median and the 50th percentile are the same*. Refer again to the section on the median to see that the median for the television-viewing data set was computed to be 14.5.

Quartiles

It is often desired to divide a data set into four parts with each part containing one-fourth, or 25%, of the data values. Figure 3.1 shows a data set divided into four parts. The division points are referred to as the *quartiles* and are defined as follows:

Q_1 = First Quartile, or 25th Percentile
Q_2 = Second Quartile, or 50th Percentile (also the Median)
Q_3 = Third Quartile, or 75th Percentile.

EXAMPLE 3.2 (continued)

The television-viewing data are again arranged in ascending order. Q_2, the median, has already been identified as 14.5.

4   5   6   8   9   10   12   14   15   15   15   16   17   18   20   26

The computations of Q_1 and Q_3 require the use of the rule for finding the 25th and 75th percentiles of a set of data. These calculations are as follows:

For Q_1,

$$i = \left(\frac{25}{100}\right) 16 = 4$$

FIGURE 3.1 Quartiles of a Data Set

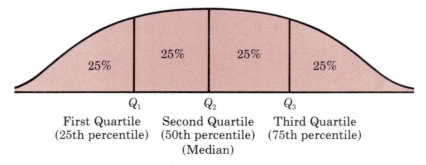

Q_1 Q_2 Q_3

First Quartile   Second Quartile   Third Quartile
(25th percentile)   (50th percentile)   (75th percentile)
                (Median)

Since i is an integer, step 3(b) indicates that the first quartile, or 25th percentile, is the average of the fourth and fifth data values; thus, $Q_1 = (8 + 9)/2 = 8.5$.

For Q_3,

$$i = \left(\frac{75}{100}\right) 16 = 12$$

Again, since i is an integer, step 3(b) indicates that the third quartile, or 75th percentile, is the average of the twelfth and thirteenth data values; thus, $Q_3 = (16 + 17)/2 = 16.5$.

Thus, as shown below, the first quartile, the median, and the third quartile have divided the 16 data values into four parts, with each part consisting of 25% of the data values.

$$
\begin{array}{ccccccccc}
4 & 5 & 6 & 8 \,\big|\, 9 & 10 & 12 & 14 \,\big|\, & 15 & 15 & 15 & 16 \,\big|\, & 17 & 18 & 20 & 26 \\
& & Q_1 = 8.5 & & \text{Median} = 14.5 & & & Q_3 = 16.5
\end{array}
$$

We have defined the quartiles as the 25th, 50th, and 75th percentiles. Thus, we computed their values in the same fashion as for that of other percentiles. However, there is some variation in the conventions followed in computing quartiles. The actual value computed may vary slightly depending on the approach used. Nevertheless, the objective of all procedures for computing quartiles is to divide a data set into roughly four equal parts.

One approach to dividing a data set into four equal parts has been recently developed by proponents of exploratory data analysis. A *lower hinge* (lower 25%) and an *upper hinge* (upper 25%) are computed. To find these hinges, the median position is found for the data set. Then the data set is divided into two equal parts: data in positions less than or equal to the median position and data in positions greater than or equal to the median position. The median for the data *less than or equal to the median position* is the lower hinge. The median for the data *greater than or equal to the median position* is the upper hinge.

EXAMPLE 3.2 (continued)

Referring again to the television-viewing data with 16 items, we find that the median position is halfway between the 8th and 9th items. The data values in positions less than or equal to the median position are

$$4 \quad 5 \quad 6 \quad 8 \quad 9 \quad 10 \quad 12 \quad 14$$

The median of these data values is 8.5. Thus, the lower hinge is 8.5. The data values in positions greater than or equal to the median position are

$$15 \quad 15 \quad 15 \quad 16 \quad 17 \quad 18 \quad 20 \quad 26$$

The median of these data values is 16.5. Thus, the upper hinge is 16.5.

1. A hospital emergency room recorded the number of patients treated on Sunday for each of the past 16 weeks.

    $$11, 14, 18, 14, 21, 17, 13, 21, 25, 19, 17, 13, 28, 13, 17, 18$$

    Compute the mean, median, and mode for these data.

2. Japanese automobile manufacturers established export quotas for automobiles to be shipped to the United States. Although the quotas pleased the Detroit automobile manufacturers, the quotas meant Japanese cars would be in short supply and more expensive for the U.S. consumer. *The Wall Street Journal* (January 11, 1989) listed the Japanese export quotas for each of the 12 months of 1988. The data shown below are in terms of thousands of automobiles.

    $$145, 135, 100, 220, 170, 145, 190, 155, 210, 200, 205, 180$$

    a. What are the mean, median, and mode for the automobile quota data?
    b. Compute and interpret the first and third quartiles for this data set.

3. Monthly sales data for car telephone units for the RC Radio Corporation are:

    $$80, 115, 82, 102, 94, 90, 88, 91, 89, 95, 105, 108$$

    Compute the mean, median, and mode for monthly sales volumes.

4. A sample of 15 college seniors showed the following credit hours taken during the final term of the senior year:

    $$18, 15, 21, 18, 16, 18, 21, 19, 15, 14, 18, 17, 20, 15, 16$$

    What are the mean, median, and mode for credit hours? Compute and interpret the 70th percentile for these data.

5. The data given below show the number of automobiles arriving at a toll booth during 20 intervals, each of 10 minutes duration. Compute the mean, median, mode, first quartile, and third quartile for the data.

26	26	38	24
32	22	15	33
19	27	21	28
16	20	34	24
27	30	31	33

6. In automobile mileage and gasoline consumption testing, 13 automobiles were road tested for 300 miles in both city and country driving conditions. The following data were recorded for miles-per-gallon performance:

    City: 16.2, 16.7, 15.9, 14.4, 13.2, 15.3, 16.8, 16.0, 16.1, 15.3, 15.2, 15.3, 16.2
    Country: 19.4, 20.6, 18.3, 18.6, 19.2, 17.4, 17.2, 18.6, 19.0, 21.1, 19.4, 18.5, 18.7

    Use the mean, median, and mode to make a statement about the difference in performance for city and country driving.

7. *American Demographics* (December 1988) reported that 25 million Americans get up each morning and go to work in their offices at home. The growing availability and use of personal computers is suggested as one of the reasons more people can operate at-home businesses. The article presented data on the ages of individuals who work at home. Assume the following is a sample of age data for these individuals.

$$22, \quad 58, \quad 24, \quad 50, \quad 29, \quad 52, \quad 57, \quad 31, \quad 30, \quad 41,$$
$$44, \quad 40, \quad 46, \quad 29, \quad 31, \quad 37, \quad 32, \quad 44, \quad 49, \quad 29,$$

a. Compute the mean and mode for the data.
b. The median age of the population of all adults is 40.5 years. Use the median age of the above data to comment on whether the at-home workers tend to be younger or older than the median of all adults.
c. Compute the first and third quartiles for the data set.
d. Compute and interpret the 32nd percentile for the data set.

8. *The Los Angeles Times* (January 6, 1989) reported the air-quality index for various areas of Southern California. Index ratings of 0–50 are considered good; 51–100, moderate; 101–200, unhealthy; 201–275, very unhealthy; and over 275, hazardous. Recent air-quality indexes for Pomona were 28, 42, 58, 48, 45, 55, 60, 49, and 50.
a. Compute the mean, median and mode for the data. Could the Pomona air-quality index be considered good?
b. Compute the 25th percentile and 75th percentile for the air-quality index data.

3.2 Measures of Dispersion

It is often desirable to consider the dispersion or variability in the values of a data set. For example, assume that an individual has the choice of using public transportation or driving an automobile to work. A major consideration would likely be the travel time associated with the two methods of transportation. Suppose that over a period of several months the individual used public transportation and driving an automobile the same number of times and found that the mean time to get to work was 32 minutes for both alternatives. At first glance, then, it would appear that both alternatives offer comparable service. However, before reaching a final conclusion, consider the histograms shown in Figure 3.2. Although the mean travel time is 32 minutes for both methods of transportation, do both alternatives possess the same degree of reliability in terms of getting to work on time? Note the dispersion, or variability, in the data. Which method of transportation would you prefer?

For many people the greater variability exhibited in the times for the public transportation system would be a major concern. That is, to protect against arriving late, one would have to allow for the maximum possible travel times of approximately 40 minutes for public transportation. With a car, one would only need to allow for maximum times up to approximately 34 minutes. Of even more concern, however, are the wide extremes and variation that must be expected when using public transportation. This illustration shows that although the average, or mean, travel time is an important consideration, the dispersion, or variability, in the travel times is at least as important and might in this case be an overriding consideration. We turn now to a discussion of some commonly used numerical measures of the dispersion, or variability, in a set of data.

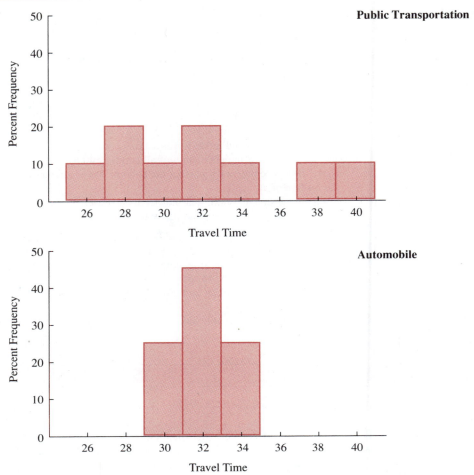

FIGURE 3.2 Histograms of Travel Times using Public Transportation and an Automobile

Range

The range is perhaps the simplest measure of variability to compute.

> **Range**
>
> The *range* for a set of data is the difference between the largest and smallest values.

EXAMPLE 3.1
(continued)

The data on class section size from Example 3.1 are 46, 54, 42, 46 and 32. The largest section size is 54 and the smallest section size is 32. Thus the range is 54 − 32 = 22.

■

Although the range is the easiest of the numerical measures of dispersion to compute, it is not widely used because the range is based on only two of the items in the data set and thus is influenced too much by extreme data values.

**EXAMPLE 3.3
(continued)**

The annual incomes for the sample of five families in Herrold, Iowa, are

$17,500 $23,000 $24,000 $26,000 $320,000

The range for this data set is $320,000 − 17,500 = $302,500, showing a large dispersion in the data set. In this case the extreme value of $320,000 leads to the large value for the range.

■

Interquartile Range

A form of the range that avoids the dependence on extreme values in the data set is the *interquartile range* (IQR).* This descriptive measure of dispersion is simply the difference between the third quartile (Q_3) and the first quartile (Q_1). In effect, it is showing the range for the middle 50% percent of the data and, as such, is not affected by the extreme values in the data set.

**EXAMPLE 3.2
(continued)**

Previously we considered the data showing television-viewing time for a sample of 16 college students. The largest viewing time was 26 hours per week, whereas the smallest viewing time was 4 hours per week. In the discussion of quartiles we found the first quartile for this data set was $Q_1 = 8.5$ and the third quartile was $Q_3 = 16.5$. As measures of dispersion in the data set, we could provide the following:

$$\text{Range} = 26 - 4 = 22 \text{ hours}$$

$$\text{IQR} = Q_3 - Q_1$$

$$= 16.5 - 8.5 = 8 \text{ hours}$$

Thus, we know that the entire data set ranges over 22 hours, whereas the middle 50% of the data values range over 8 hours.

■

Variance

A key step in computing the variance of a data set involves the computation of the difference between each data value and the mean for the data set. The difference between each observation x_i and the mean ($\bar{x}$ for a sample, μ for a population) is called a *deviation about the mean*.

**EXAMPLE 3.1
(continued)**

The sample mean for the data set of Example 3.1 is a section size of 44 students. A summary of the data, including the computation of the deviations from the sample mean, is shown below.

*Another term used for the interquartile range is the *Q-spread* of the data set.

NUMBER OF STUDENTS IN SECTION (x_i)	MEAN SECTION SIZE ($\bar{x}$)	DEVIATION FROM SAMPLE MEAN ($x_i - \bar{x}$)
46	44	2
54	44	10
42	44	-2
46	44	2
32	44	-12
	Total	0

Note: Sum of deviations about the mean is 0.

■

We might first think of summarizing the dispersion in a data set by computing the average deviation about the mean. However, a little reflection based upon a study of the above table would lead us to discard that idea—the sum of the deviations about the mean for the five sections sampled in equal to zero. This is true for any data set; that is

$$\Sigma\,(x_i - \bar{x}) = 0$$

The positive and negative deviations cancel each other, causing the average deviation for any data set to equal zero. Thus the average deviation cannot measure the variability in a data set since it will always be equal to $0/n = 0$.

One approach to preventing the positive and negative deviations from canceling out is to take the absolute value of each deviation. The average absolute deviation can then be computed as a measure of variability. While this measure is sometimes used, the most common approach to preventing the cancellation of deviations is to square them.

EXAMPLE 3.1 (continued)

The squared deviations and their sum are shown below.

NUMBER OF STUDENTS IN SECTION (x_i)	DEVIATION FROM SAMPLE MEAN ($x_i - \bar{x}$)	SQUARED DEVIATION ($x_i - \bar{x}$)^2
46	2	4
54	10	100
42	-2	4
46	2	4
32	-12	144
	Totals 0	256
	$\Sigma(x_i - \bar{x})$	$\Sigma(x_i - \bar{x})^2$

■

A measure of variability based on the squared deviations is the *average squared deviation*. If the data set involved is a population, the average of the squared deviations is called the *population variance*. The population variance is denoted by the Greek symbol σ^2 (pronounced sigma squared). Given a population of N items and using μ to represent the population mean, the definition of the population variance is given by (3.3).

In most statistical applications, the data set being analyzed is a sample. When we compute the variance for a sample, we are often interested in using the sample variance as an estimate of the population variance σ^2. At this point it might seem that the average of the squared deviations of the sample values from $\bar{x}$ would provide a good estimate of the population variance. However, statisticians have found that the average squared deviation for the sample has the undesirable feature of providing a biased estimate of the population variance σ^2; specifically, it tends to underestimate the population variance.

Although it is beyond the scope of this text, it can be shown that if the sum of the squared deviations in a sample is divided by $n - 1$, and not n, then the resulting sample statistic provides an unbiased estimate of the population variance. For this reason, the *sample variance* is not defined as the average squared deviation in the sample. Rather, it is denoted by s^2 and is defined as follows.

EXAMPLE 3.1 (continued)

Let us now compute the sample variance for the data on the number of students in the College of Arts and Science classes. Recall that for this data set, $\Sigma(x_i - \bar{x})^2 = 256$. Hence with $n - 1 = 4$, we obtain

$$s^2 = \frac{\Sigma(x_i - \bar{x})^2}{n - 1} = \frac{256}{4} = 64$$

■

While admittedly it is difficult to obtain an intuitive feel for the meaning of the variance, we can note that larger variances could be obtained only from data sets with larger deviations about the mean and, therefore, more dispersion.

EXAMPLE 3.2 (continued)

Recall that the sample mean for average weekly television-viewing time of a sample of college students is $\bar{x} = 13.125$. Let us use (3.4) to compute the sample variance for this data set. The data set and computations are as follows:

x_i	$(x_i - \bar{x})$	$(x_i - \bar{x})^2$
14	.875	.765625
9	−4.125	17.015625
12	−1.125	1.265625
4	−9.125	83.265625

x_i	$(x_i - \bar{x})$	$(x_i - \bar{x})^2$
20	6.875	47.265625
26	12.875	165.765625
17	3.875	15.015625
15	1.875	3.515625
18	4.875	23.765625
15	1.875	3.515625
10	−3.125	9.765625
6	−7.125	50.765625
16	2.875	8.265625
15	1.875	3.515625
8	−5.125	26.265625
5	−8.125	66.015625
Totals 210	0.000	525.750000
Σx_i	$\Sigma(x_i - \bar{x})$	$\Sigma(x_i - \bar{x})^2$

Using (3.4) we obtain a sample variance of

$$s^2 = \frac{\Sigma(x_i - \bar{x})^2}{n - 1} = \frac{525.75}{15} = 35.05$$

Standard Deviation

The *standard deviation* of a data set is defined to be the positive square root of the variance. Following the notation we adopted for a sample variance and a population variance, we use s to denote the sample standard deviation and σ to denote the population standard deviation. The standard deviation is derived from the variance in the following manner:

Standard Deviation

$$\text{Sample Standard Deviation} = s = \sqrt{s^2} \qquad (3.5)$$

$$\text{Population Standard Deviation} = \sigma = \sqrt{\sigma^2} \qquad (3.6)$$

EXAMPLE 3.1 (continued)

Recall that the sample variance for the sample of section sizes in the College of Arts and Science is $s^2 = 64$. Thus the sample standard deviation is $s = \sqrt{64} = 8$.

Obviously, the standard deviation is also a measure of dispersion, since the square root of a larger variance will provide a larger standard deviation. However, the standard deviation is more often used as a measure of dispersion because it is in the same units as the data. For instance, in the example just considered, the variance of section sizes is 64 "students squared." The standard deviation is 8 students. The fact that variance is reported in units of the original data squared makes it difficult to obtain an intuitive feel for it as a measure of variability.

EXAMPLE 3.2
(continued)

Referring to the data on television viewing time in Example 3.2, we see that with $s^2 = 35.05$, the sample standard deviation measure of dispersion is $s = \sqrt{35.05} = 5.92$ hours.

■

Coefficient of Variation

Often we are interested in a relative measure of the variability in a data set. For example, a standard deviation of 1 inch would be considered very large for a batch of motor-mount bolts used in automobiles. However, a standard deviation of 1 inch would be considered small for the length of a telephone pole. When the means for data sets differ greatly, we do not get an accurate picture of the relative variability in the two data sets by comparing the standard deviations. A measure of variability that can be used for such comparisons is the *coefficient of variation*. The formula for computing the coefficient of variation is given by (3.7).

Coefficient of Variation

$$CV = \frac{\text{Standard Deviation}}{\text{Mean}} \times 100 \qquad (3.7)$$

As shown above, the coefficient of variation, CV, is the ratio of the standard deviation to the mean, expressed as a percentage.

EXAMPLE 3.1
(continued)

Recall that for the sample of section sizes in the College of Arts and Science, the sample mean and sample standard deviation are 44 and 8, respectively. Thus for this data set the coefficient of variation is $(8/44) \times 100 = 18.18$. In other words, we could say that the standard deviation of the sample is 18.18% of the value of the sample mean.

■

Notes and Comments

1. Rounding the values of the sample mean $\bar{x}$ and the values of the squared deviations $(x_i - \bar{x})^2$ may introduce rounding errors in the values of the variance and standard deviation. Generally, you can minimize the effect of rounding errors by carrying six significant digits during the calculation process.

2. An alternate formula is available for the computation of the variance of a data set. This alternate formula for the sample variance is written as follows:

$$s^2 = \frac{\sum x_i^2 - (\sum x_i)^2/n}{n - 1} \qquad (3.8)$$

This formula will ease the computational effort slightly and help reduce rounding errors.

---

9. The number of freshmen majoring in computer science at Maine Institute of Technology for the past 5 years is:

$$390, 380, 400, 410, 420$$

Compute the sample mean, variance, and standard deviation for this data set.

10. *The Washington Post* (January 7, 1989) reported on overcrowding in the Virginia prison system. Use the data below as the population of capacities of the five Virginia state prisons:

$$233, 164, 587, 52, 175$$

Compute the variance and standard deviation for this population.

11. Exercise 8 summarized data from the *Los Angeles Times,* which reported the air-quality index for various areas of Southern California. A sample of air quality index values for Pomona provided the following data: 28, 42, 58, 48, 45, 55, 60, 49 and 50.
    a. Compute the sample variance and sample standard deviation for these data.
    b. A sample of air-quality index readings for Anaheim provided a sample mean of 48.5, a sample variance of 136, and sample standard deviation of 11.66. What comparisons can you make between the air quality in Pomona and Anaheim based on these descriptive statistics?

12. Given are the annual household incomes for 10 families in Grimes, Nebraska.

20,648	27,416
16,517	23,555
24,821	19,226
162,936	21,800
28,527	22,222

    a. Compute the range as a measure of variability.
    b. Compute the interquartile range as a measure of variability.
    c. Compute the standard deviation as a measure of variability.

13. A production department uses a sampling procedure to test the quality of newly produced items. The department employs the following decision rule at an inspection station: If a sample of 14 items has a variance of more than .01 for a certain characteristic, the production line must be shut down for repairs. Suppose the following data have just been collected:

3.43	3.45	3.43
3.48	3.52	3.50
3.39	3.48	3.41
3.38	3.49	3.45
3.51	3.50	

Should the production line be shut down? Why or why not?

14. The following times were recorded by the quarter-mile and mile runners of a university track team (times are in minutes):

Quarter-Mile Times:	.92	.98	1.04	.90	.99
Mile Times:	4.52	4.35	4.60	4.70	4.50

After viewing this sample of running times, the coach commented that the quarter-milers turned in the more consistent times. Use the standard deviation and the coefficient of variation to summarize the variability in the data sets. Does the use of the coefficient of variation measure indicate that the coach's statement should be qualified? Why?

3.3   Some Uses of the Mean and Standard Deviation

We have developed several measures of location and variability for a set of data. The mean is the most widely used measure of location, whereas the standard deviation and variance are the most widely used measures of variability. Using only the mean and the standard deviation, we can learn much about a data set.

Z-Scores

Using the mean and standard deviation together enables us to make statements about the relative location of any item in a data set. Suppose we have a sample of n items, with the values denoted by $x_1, x_2, \ldots, x_n$, and that we have already computed the sample mean, $\bar{x}$, and the sample standard deviation, s. Associated with each item is a value, called a z-score. Equation (3.9) shows how the z-score is computed for value x_i.

Z-Score

$$z_i = \frac{x_i - \bar{x}}{s} \qquad\qquad (3.9)$$

where z_i is the z-score for item i
           $\bar{x}$ is the sample mean
           s is the sample standard deviation

The z-score for an item is often referred to as its standardized value. For instance, the standardized value, z_1, can be interpreted as the *number of standard deviations x_1 is from the mean $\bar{x}$*. For example, $z_1 = 1.2$ would indicate x_1 is 1.2 standard deviations above, or larger than, the sample mean. Similarly, $z_2 = -0.5$ would indicate data value x_2 is .5, or ½, standard deviation below, or less than, the sample mean. As can be seen from equation (3.9), z-scores greater than zero occur for items with values larger than the sample mean, and z-scores less than zero occur for items with values smaller than the mean. A z-score of zero indicates that the value of the item is equal to the mean.

From this discussion, it should be clear that the z-score can be interpreted as a measure of the relative location of an item in a data set. Indeed, items in two different data sets with the same z-score can be said to have the same relative location.

Chebyshev's Theorem

Chebyshev's theorem permits us to make statements about the percentage of items that must lie within a certain number of standard deviations from the mean for any data set.

Some of the statements that can be made as a result of Chebyshev's Theorem are the following:

Chebyshev's Theorem

For any set of data, *at least*

- 75% of the items must lie within two standard deviations of the mean
- 89% of the items must lie within three standard deviations of the mean
- 94% of the items must lie within four standard deviations of the mean

---

EXAMPLE 3.4

Midterm examination scores for 100 students in a college statistics course had a mean of 70 and a standard deviation of 5. How many students scored between 60 and 80? How many scored between 50 and 90?

For the scores between 60 and 80, we note that the value of 60 is two standard deviations below the mean and the value of 80 is two standard deviations above the mean. Using Chebyshev's theorem we see that at least 75% of the items must have values within two standard deviations of the mean. Thus, at least 75 of the students must have scored between 60 and 80.

For the range 50 to 90, we see that 50 is four standard deviations below the mean and 90 is four standard deviations above the mean. Thus, at least 94% of the students must have scored between 50 and 90.

■

In the next subsection we will see that when the distribution of data is known to be mound-shaped, larger percentages of the data can be said to lie within two and three standard deviations of the mean. Indeed, we will see that then it is even possible to estimate that a significant portion of the data lies within one standard deviation of the mean.

The Empirical Rule

One of the advantages of Chebyshev's theorem is that it applies to any data set, regardless of the shape of the distribution of the data. In practical applications, however, it has been found that many data sets have a mound-shaped, or bell-shaped, distribution like the one shown in Figure 3.3. When it is believed that the data set approximates this pattern, the *empirical rule* can be used to estimate the percentage of items that fall within a specified number of standard deviations of the mean.[1]

EXAMPLE 3.5

Liquid detergent cartons are filled automatically on a production line. Filling weights frequently have a bell-shaped distribution. If the mean filling weight is 16 ounces and the standard deviation is 0.25 ounces, we can use the empirical rule to conclude the following:

---

[1] The empirical rule is based on the normal probability distribution, which is presented in detail in Chapter 6.

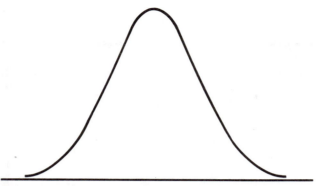

FIGURE 3.3 A Mound-Shaped, or Bell-Shaped, Distribution

Empirical Rule

For a data set having a bell-shaped distribution similar to the one shown in Figure 3.3.

■ Approximately 68% of the data items will fall within one standard deviation of the mean

■ Approximately 95% of the data items will fall within two standard deviations of the mean

■ Almost all the data will fall within three standard deviations of the mean

---

Approximately 68% of the filled items will have weights between 15.75 and 16.25 ounces (that is, within one standard deviation of the mean).

Approximately 95% of the filled items will have weights between 15.50 and 16.50 ounces (that is, within two standard deviations of the mean).

Almost all filled items will have weights between 15.25 and 16.75 ounces (that is, within three standard deviations of the mean).

■

Detecting Outliers

Sometimes a set of data will have one or more items with unusually large or unusually small values. Extreme values such as these are called *outliers*. Experienced statisticians take steps to identify outliers and then review each one carefully. An outlier may be an item for which the value has been incorrectly recorded. If so, the value can be corrected before proceeding with further analysis. An outlier may also be an item that was incorrectly included in the data set; if so it can be removed. Finally, an outlier may just be an unusual item that has been correctly recorded and does belong in the data set. In such cases the item should remain in the data set, but care should be exercised to ensure that the outlier does not cause misleading inferences to be made.

Standardized data values (z-scores) can be used to detect outliers. The farther a standardized data value is from zero, the more likely the item is an outlier. Recall that the empirical rule allows us to conclude that for many data sets encountered in practice, almost all the items lie within three standard deviations of the mean. Thus, when using z-scores to identify outliers, we recommend treating any item with a z-score less than -3 or greater than 3 as an outlier. Such items can then be reviewed for accuracy and for whether or not they belong in the data set.

EXAMPLE 3.1 (continued)

Let us return to the sample of five class sections in the College of Arts and Sciences; the class sizes were 46, 54, 42, 46 and 32. The sample mean is $\bar{x} = 44$ and the sample standard deviation is $s = 8$. The calculations shown below provide the corresponding standardized values, or z-scores.

x_i	$x_i - \bar{x}$	$z_i = (x_i - \bar{x})/s$
46	2	$2/8 = +0.25$
54	10	$10/8 = +1.25$
42	-2	$-2/8 = -0.25$
46	2	$2/8 = +0.25$
32	-12	$-12/8 = -1.50$

The class size of 32 provides the most extreme value, occurring at $z = -1.50$, or 1.50 standard deviations below the mean. However, this standardized value is within the -3 to $+3$ guideline and thus the class size of 32 is not identified as an outlier.

■

Notes and Comments

1. Before analyzing a data set, statisticians usually make a variety of checks to ensure the validity of data. In a large study it is not uncommon for errors to be made recording data values or inputting the values at a computer terminal. Identifying outliers is one tool used to catch such errors and ensure the validity of data.

2. Chebyshev's theorem is applicable for any data set; it makes a statement about the minimum number of items that lie within a certain number of standard deviations of the mean. If the data set is known to be approximately mound-shaped, more can be said. For instance, the empirical rule allows us to say that *approximately* 95% of the items lie within two standard deviations of the mean; Chebyshev's theorem allows us to conclude only that at least 75% of the items lie in the same interval.

EXERCISES

15. A 1989 salary survey (*Working Woman*, January 1989) listed the average salary of elementary and secondary school teachers as $28,085. Assume that the standard deviation of salaries is $4500.
   a. Janice Herbranson was identified as a teacher in a one-room schoolhouse in McLeod, North Dakota. Her salary was reported to be $8100 per year. What is the standardized

value associated with $8100? Comment on whether or not this salary figure might be an outlier.

b. Compute the z-score for each of the following salaries: $33,500, $25,200, $28,985 and $39,000. Should any of these be reviewed as possible outliers?

16. Use the salary data in exercise 15 and Chebyshev's theorem to find the percentage of elementary school teachers that must have salaries in the following ranges:

    a. $19,085 to $37,085

    b. $14,585 to $41,585

    c. Repeat parts a and b if it can be assumed that the distribution of teacher salaries is approximately bell-shaped.

17. Use the data on household incomes for 10 families in Grimes, Nebraska (Exercise 12). What is the z-score for the income of $162,936? Interpret this standardized value and comment on whether or not this income value should be considered an outlier.

18. A sample of 10 NCAA men's college basketball scores (January 5, 1989) provided the following winning teams and the number of points scored:

Rutgers	87
Niagara	79
Mississippi	80
Western Kentucky	64
Purdue	75
Tulsa	70
Texas-El Paso	82
Stanford	83
Iowa	93
Montana	62

    a. Compute the sample mean and sample standard deviation for this data.

    b. In another game, Penn State beat Massachusetts by a score of 117 to 79. Should the Penn State score be considered an outlier? Explain.

    c. Assume that the distribution of the points scored by winning teams has a mound-shaped distribution. Estimate the percentage of all NCAA basketball games in which the winning team will score 87 or more points. Estimate the percentage of all NCAA basketball games in which the winning team will score 58 or less points.

3.4   Measures of Location and Dispersion for Grouped Data

In most cases measures of location and dispersion for a data set are computed using the individual data values. However, sometimes we are presented with data in grouped, or frequency distribution, form. This section describes how approximation of the mean, variance, and standard deviation can be obtained from a frequency distribution.

Mean

Recall that in order to compute the sample mean using the individual data items, we simply sum all the values and divide by *n*, the sample size. If the data are available only in frequency distribution form, we can approximate the sum of the values by first finding an approximation for the sum of the values for each class and then adding these for all classes.

To do this, we treat the midpoint of each class as if it were the mean of the items in the class. Let M_i denote the midpoint for class i and f_i denote the frequency of the class. Then an approximation to the sum of the items in class i is given by $f_i M_i$. Summing these approximations over all classes, we obtain $\Sigma f_i M_i$, which approximates the sum of all the data values.

Once this approximation of the sum of all the data values is obtained, an approximation of the mean is computed by dividing this sum by the total number of data items. The following formula is thus used to compute the sample mean from grouped data.

Sample Mean for Grouped Data

$$\bar{x} = \frac{\Sigma f_i M_i}{n} \tag{3.10}$$

As we indicated at the beginning of this section, we do not expect the calculations based on the individual data values and grouped data to provide exactly the same numerical result. Thus $\bar{x}$ calculated using (3.10) is an approximation of $\bar{x}$ calculated when all data values are known.

EXAMPLE 3.6

In Example 2.10 of Chapter 2 we analyzed a data set involving the scores obtained by each of 50 sixth graders on a mathematics achievement test. We developed a frequency distribution using six class intervals shown in Table 3.1. Note that we have added two columns to the frequency distribution. One column is for the class midpoints and the other is for the approximation of the sum of data values in each class, $f_i M_i$. The sum of the items in the last column provides the approximation to the sum of the 50 data values. Using this sum we compute the mean for the grouped data as follows:

$$\bar{x} = \frac{\Sigma f_i M_i}{50} = \frac{3315}{50} = 66.3$$

TABLE 3.1

Computation of the Sample Mean for Mathematics Achievement Test Score Using Grouped Data

FREQUENCY DISTRIBUTION			
Class Interval	Frequency (f_i)	Class Midpoint (M_i)	$f_i M_i$
40–49	5	44.5	222.5
50–59	10	54.5	545.0
60–69	15	64.5	967.5
70–79	12	74.5	894.0
80–89	7	84.5	591.5
90–99	1	94.5	94.5
Total	50	$\Sigma f_i M_i$	3315.0

Referring to the original data as presented in Example 2.10 and computing the sample mean using the ungrouped data, we find $\bar{x}$ = 66.5. Thus, the approximation error using the grouped data computation is only 66.5 − 66.3 = .2. In this case, the grouped data approximation of the sample mean is very good.

■

Variance

The approach to computing the sample variance for a set of grouped data is to use a slightly altered form of the formula for the sample variance as provided in (3.4). Since we no longer have the individual data values, we treat the class midpoint as being a representative value for the data items in each class. Then, we weight the squared deviation from each midpoint by the frequency of its corresponding class. Using this approach (3.4) is modified as follows.

Sample Variance for Grouped Data

$$s^2 = \frac{\Sigma f_i (M_i - \bar{x})^2}{n - 1}$$ (3.11)

EXAMPLE 3.4 (continued)

The calculation of the sample variance for the grouped data of Example 3.6 is shown in Table 3.2. Using the grouped data sample mean of $\bar{x}$ = 66.3, we see that the sample variance is 157.92.

■

TABLE 3.2   **Computation of the Sample Variance for Mathematics Achievement Test Scores Using Grouped Data**

FREQUENCY DISTRIBUTION

Class Interval	Frequency (f_i)	Class Midpoint (M_i)	$(M_i - \bar{x})$	$(M_i - \bar{x})^2$	$f_i(M_i - \bar{x})^2$
40−49	5	44.5	−21.8	475.24	2376.20
50−59	10	54.5	−11.8	139.24	1392.40
60−69	15	64.5	− 1.8	3.24	48.60
70−79	12	74.5	8.2	67.24	806.88
80−89	7	84.5	18.2	331.24	2318.68
90−99	1	94.5	28.2	795.24	795.24
Total	50			$\Sigma f_i(M_i - \bar{x})^2$ =	7738.00

$$s^2 = \frac{\Sigma f_i (M_i - \bar{x})^2}{n - 1} = \frac{7738.00}{49} = 157.92$$

Standard Deviation

The standard deviation computed from grouped data is simply the square root of the variance computed from grouped data. For the data of Example 3.6, the sample standard deviation computed from grouped data is $s = \sqrt{157.92} = 12.57$.

19. The following frequency distribution for the first examination in sociology was posted on the department bulletin board.

EXAMINATION GRADE	FREQUENCY
40–49	3
50–59	5
60–69	11
70–79	22
80–89	15
90–99	6
Total	62

Treating these data as a sample, compute the mean, variance, and standard deviation.

20. *The Journal of Personal Selling and Sales Management* (August 1988) reported a study investigating the use of a persuasion technique in selling. A sample of 56 participants was used in the study. Assume that the frequency distribution below shows the number of sales presentations made per week by the 56 participants.

NUMBER OF PRESENTATIONS	FREQUENCY
10–12	5
13–15	9
16–18	22
19–21	12
22–24	8
Total	56

a. What is the mean number of sales presentations made per week for participants in this study?

b. What is the variance? Standard deviation?

21. A service station has recorded the following frequency distribution for the number of gallons of gasoline (rounded to the nearest gallon) sold per car on a given day.

GASOLINE (GALLONS)	FREQUENCY
0–4	74
5–9	192
10–14	280
15–19	105
20–24	23
25–29	6
Total	680

Compute the standard measures of central location and dispersion for these grouped data. If the service station expects to service about 120 cars on a given day, what is an estimate of the total number of gallons that will be sold?

22. *Psychology Today* (November 1988) reported on the characteristics of individuals caught in the act of shoplifting in a supermarket. Data from 9832 cases showed 56% of the shoplifters were male and 44% were female. The value of most items stolen was in the range $1 to $5. The following frequency distribution shows the age of the shoplifters.

AGE OF SHOPLIFTER	FREQUENCY
6–11	364
12–17	1249
18–29	3392
30–59	3962
60–80	865
Total	9832

a. What is the sample mean age of supermarket shoplifters?
b. What are the variance and standard deviation of the ages?

Exploratory Data Analysis

In Chapter 2 we introduced exploratory data analysis with the use of stem-and-leaf displays. The focus of exploratory data analysis is on using simple arithmetic and easy-to-draw pictures to summarize data. In this section we continue our introduction of exploratory data analysis by considering 5-number summaries and box plots.

5-Number Summaries

In a 5-number summary, the following 5 numbers are used to summarize a data set:

1. Smallest value in the data set
2. First quartile (Q_1)
3. Median (2nd quartile)
4. Third quartile (Q_3)
5. Largest value in the data set

EXAMPLE 3.2 (continued)

Referring to the data and computations we have made with the sample of 16 television-viewing times, the smallest value in the data set is 4 and the largest value is 26. Previous analysis provided $Q_1 = 8.5$, median $= 14.5$, and $Q_3 = 16.5$. Thus the 5-number summary of the television-viewing time data is 4, 8.5, 14.5, 16.5, and 26. One-fourth, or 25%, of the data values fall between adjacent numbers in the 5-number summary.

Box Plots

When there are no outliers in the data set, a *box plot* is a graphical presentation of a 5-number summary. Figure 3.4 shows the box plot for television-viewing times from Example 3.2. The dashed lines are referred to as *whiskers* and extend from Q_1 to the smallest value in the data set and from Q_3 to the largest value in the data set. Each whisker contains 25% of the data. The *box* extends from Q_1 to Q_3 and includes the middle 50% of the data. The vertical line drawn within the box identifies the median of the data set. Basically, the box plot provides an easy-to-interpret graphical presentation of the information contained in the 5-number summary. When outliers are present, the construction of the box plot is modified somewhat. To see how, we must first discuss the notion of fences.

Fences and the Detection of Outliers

Previously we defined outliers as unusually small or large values in a data set that should be reviewed to be sure that they are correctly recorded or that they belong in the data set. We showed how z-scores could be used to identify outliers. In exploratory data analysis, fences are often used to identify outliers.

In Figure 3.5 we have added four vertical lines to the box plot for the television-viewing time data. Two of the lines are identified as *inner fences,* and two are identified as *outer fences*. The location of the fences depends on the location of the first and third quartiles, Q_1 and Q_3, as well as the size of the interquartile range (IQR) for the data set. For the television-viewing-time data, IQR $= Q_3 - Q_1 = 16.5 - 8.5 = 8$. The inner fences are located at a distance of 1.5 IQR below Q_1 and at a distance of 1.5 IQR above Q_3. The outer fences are located 3 IQR below Q_1 and 3 IQR above Q_3.

Fences are used to determine how far data values must be from the first and third quartiles before they should be considered outliers. Data values falling between the inner and outer fences are considered mild outliers. Data values falling outside the outer fences are considered extreme outliers. Both mild and extreme outliers should be reviewed for to ensure validity. As Figure 3.5 shows, the television-viewing-time data all fall within the inner fences. As such, the fences provide no indication of outliers being present in the data set.

As stated above, when outliers are present, a box plot is somewhat differently constructed. Figure 3.6 shows what a box plot would look like for a data set with outliers. When outliers exist, the whiskers are extended to the smallest and largest data values

FIGURE 3.4 Box Plot of the Television-Viewing Time Data

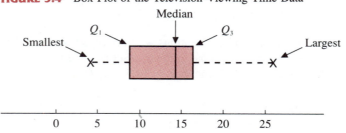

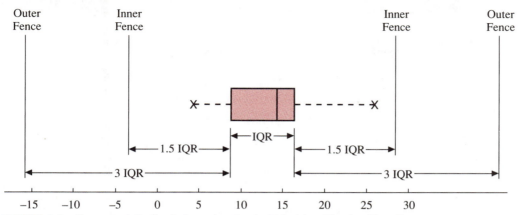

FIGURE 3.5 Fences and Outlier Information for the Television-Viewing Time Data

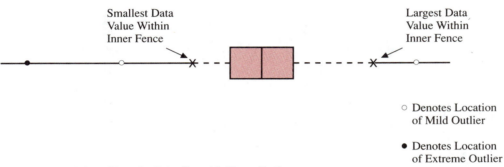

Smallest Data
Value Within
Inner Fence

Largest Data
Value Within
Inner Fence

○ Denotes Location
of Mild Outlier

● Denotes Location
of Extreme Outlier

FIGURE 3.6 A Box Plot of a Data Set with Three Outliers

within the inner fences. Mild outliers are then denoted by circles, and extreme outliers are denoted with dots. The data set in Figure 3.6 can be seen to have two mild outliers and one extreme outlier.

Notes and Comments

1. Using the fences to identify outliers, one will not always identify the same items as when using z-scores and identifying those items with z-scores less than -3 and greater than 3 as outliers. However, the objective of both approaches is simply to identify items that should be reviewed to ensure the validity of a data set. Thus, items identified using either procedure should be reviewed.

2. An advantage of the exploratory data analysis procedures is that they are easy to use; few numerical calculations are necessary. One simply needs to put the items into ascending order in order to construct the 5-number summary, the fences, and the box plot. It is not necessary to compute the mean and standard deviation.

23. Data on Japanese automobile manufacturer quotas were presented in Exercise 2. Provide a 5-number summary and box-and-whisker plot for this data.

24. Exercise 12 provides the annual household incomes for 10 families in Grimes, Nebraska. The data is repeated below:

20,648	27,416
16,517	23,555
24,821	19,226
162,936	21,800
28,527	22,222

a. Provide a 5-number summary for the data.
b. Compute the location of the fences. Does the income of $162,936 appear as a mild or extreme outlier?
c. Show the box plot for this data.

25. *Forbes* (January 9, 1979) published its 41st annual report on American industry. Data on the percent growth in sales for the past 12 months were presented for 26 companies in the paper industry. These data are as follows:

16.1	49.9	23.7	15.6	1.9
10.8	20.4	12.2	22.4	4.9
13.4	6.8	15.8	12.1	19.3
10.0	46.1	27.0	7.0	12.5
6.1	15.6	6.3	16.7	10.1
55.9				

a. Provide a 5-number summary of these data.
b. Compute the inner and outer fences.
c. Do there appear to be outliers? How would this information be helpful to a financial analyst?
d. Show a box plot.

26. The *Forbes* annual report mentioned in Exercise 25 also provided annual sales data in millions of dollars for 17 companies in the chemical industry. These data are as follows:

484	2,731	598	2,472
3,261	1,220	4,514	32,249
15,980	8,030	3,122	8,258
1,061	2,188	2,366	1,049
636			

a. Provide a 5-number summary of these data.
b. Compute the inner and outer fences.
c. Do there appear to be outliers? What does outlier information tell you in this case? *Note:* Companies associated with some of the data are as follows: Valspar (484), Dow Chemical (15,980) and DuPont (32,249).
d. Show a box plot.

3.6 The Role of the Computer in Descriptive Statistics

In this section we describe the role of computers and computer software packages in statistical analysis by showing how a statistical package can be used to generate descriptive statistics for a data set. In future chapters we provide further illustrations showing how statistical computing systems can support the analysis and interpretation of data.

In the 1960s there were relatively few computer software packages for analyzing data. Since that time, however, the situation has changed dramatically. Today the user has a choice of packages such as SAS (Statistical Analysis System), SPSS (Statistical Package for the Social Sciences), BMDP (UCLA Biomedical statistical package), and Minitab, to name just a few. In addition, the microcomputer statistical software package known as *The Data Analyst* has been specifically developed to accompany this text. Throughout the text we will be using Minitab to illustrate the application of statistical software packages. Minitab is a general-purpose statistical computing system that can be used on a variety of mainframe or personal computers. *The Data Analyst* was developed by the authors of this text specifically for personal computer applications. Both systems have been designed for users who have had little or no previous computer experience. The systems are easy to use and offer a great deal of power and computational aid for data summarization and statistical analysis.

In the illustrations of Minitab, we describe its use in what is referred to as *interactive mode*. In this mode the user enters data and commands from a computer terminal or personal computer keyboard; Minitab carries out each user command as soon as it is given. In the figures containing the steps of a Minitab session, we show the computer responses from the Minitab system in black and the input by the user in color.

Minitab consists of a worksheet of rows and columns in which data are stored. The columns are denoted with labels c1, c2, and so on, unless the user elects to name the columns with specific labels. The rows of the worksheet correspond to the individual items or elements of the data set. That is, a separate row is used for each item. The Minitab system consists of about 150 commands, which can be used to analyze the data stored in the worksheet. To provide an illustration of how Minitab works, we use the data set originally presented in Example 2.6. Recall that this data showed the number of questions answered correctly by each of 50 students who took a 100-question mathematical achievement test. For convenience, this data set is shown in Table 3.3.

Referring to Figure 3.7, we assume that the user has loaded the Minitab system into the computer. The MTB > symbol appears on the terminal screen, which indicates that

TABLE 3.3

Fifty Mathematics Achievement Test Scores

75	48	46	65	71
49	61	51	57	49
84	85	79	85	83
55	69	88	89	55
61	72	64	67	61
77	51	61	68	54
63	94	54	53	71
84	79	75	65	50
45	65	77	71	63
67	57	63	71	77

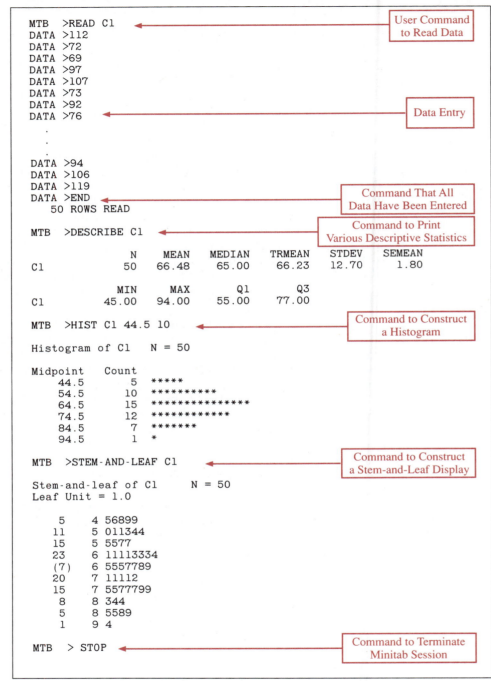

```
MTB  >READ C1                          ◄────────  User Command
DATA >112                                          to Read Data
DATA >72
DATA >69
DATA >97
DATA >107
DATA >73
DATA >92                               ◄────────  Data Entry
DATA >76
        .
        .
        .
DATA >94
DATA >106
DATA >119
DATA >END                              ◄────────  Command That All
   50 ROWS READ                                    Data Have Been Entered

MTB  >DESCRIBE C1                       ◄────────  Command to Print
                                                   Various Descriptive Statistics
              N     MEAN    MEDIAN   TRMEAN   STDEV   SEMEAN
C1           50    66.48    65.00    66.23   12.70    1.80

             MIN     MAX      Q1      Q3
C1         45.00   94.00   55.00   77.00

MTB  >HIST C1 44.5 10                   ◄────────  Command to Construct
                                                   a Histogram
Histogram of C1   N = 50

Midpoint   Count
   44.5        5   *****
   54.5       10   **********
   64.5       15   ***************
   74.5       12   ************
   84.5        7   *******
   94.5        1   *

MTB  >STEM-AND-LEAF C1                  ◄────────  Command to Construct
                                                   a Stem-and-Leaf Display
Stem-and-leaf of C1      N = 50
Leaf Unit = 1.0

    5      4 56899
   11      5 011344
   15      5 5577
   23      6 11113334
   (7)     6 5557789
   20      7 11112
   15      7 5577799
    8      8 344
    5      8 5589
    1      9 4

MTB  > STOP                             ◄────────  Command to Terminate
                                                   Minitab Session
```

FIGURE 3.7 Computer Analysis of Mathematics Achievement Test Scores Using Minitab

the Minitab system is waiting for a command from the user. The user then inputs READ C1, indicating that the system is to take the data from the lines that follow and store the data in column 1 of the worksheet. Note then that the next line shows that Minitab's response is DATA>; thus the user enters the first data value. This process continues with one data value being entered per line. When all the data values have been input, the data entry process is concluded when the user responds to the request for more data with the command end. The data set now resides in the Minitab worksheet in column 1.

The minitab command DESCRIBE C1 is the user request to obtain descriptive statistics for the mathematics achievement test scores stored in column 1. The descriptive statistics provided show that there are 50 data items with a mean score of 66.48 and a median score of 65. The standard deviation is shown to be 12.7. In addition, the maximum data value, the minimum data value, and the quartiles Q_1 and $Q3$ can be combined with the median to provide the 5-number summary of 45, 55, 65, 77, and 94.

Next the user command HIST C1 44.5 10 requests a histogram for the data in column 1 using a starting class midpoint of 44.5 and a class width of 10. The Minitab system then provides the information for the histogram, as shown. The analysis of the mathematical achievement test scores is completed by using the command STEM-AND-LEAF C1 to obtain the Minitab stem-and-leaf display of the data.

The Data Analyst software package is what is called a menu-driven system. At each step the system provides the user with a list of options such as create a data set, edit a data set, obtain descriptive statistics, obtain a frequency distribution, and so on. The software will run on any IBM-compatible personal computer. After loading DOS and inserting *The Data Analyst* diskette, the command DA begins operation. By choosing menu options, the appropriate statistical methodology can be selected and data can easily be entered into the computer.

To obtain the computer output shown in Figure 3.8 we proceeded as follows: (1) Selected Descriptive Statistics, (2) selected Create a Data Set and input the test scores,

FIGURE 3.8 Computer Output concerning Mathematics Achievement Test Scores Obtained Using *The Data Analyst*

```
DESCRIPTIVE MEASURES OF LOCATION AND DISPERSION
-------------------------------------------------

DATA SET NAME:   MATHACH

VARIABLE:   Scores                    OBSERVATIONS:   50
```

Mean	Variance	Standard Deviation	Coefficient of Variation
66.48	161.28	12.70	19.10

Minimum Value	25th Percentile	Median	75th Percentile	Maximum Value
45.00	55.00	65.00	77.00	94.00

```
FREQUENCY DISTRIBUTION
----------------------

DATA SET NAME:  MATHACH

VARIABLE:  Scores                     OBSERVATIONS:  50
```

LOWER LIMIT	UPPER LIMIT	FREQUENCY
40	49	5
50	59	10
60	69	15
70	79	12
80	89	7
90	99	1

(3) selected Measures of Location and Dispersion, and (4) selected Frequency Distribution. Note in Figure 3.8, that the data set has been named "MATHACH" and that the variable name we have chosen is "Scores." Section 1.5 provides more detail on the use of *The Data Analyst*.

SUMMARY

In this chapter we introduced several statistical measures that can be used to describe the location and dispersion of a data set. Unlike the tabular and graphical procedures for summarizing data presented in Chapter 2, the measures introduced in this chapter summarize the data in terms of numerical values. When the numerical values obtained are for a sample, they are called sample statistics. When they are for a population, they are called population parameters.

As measures of location, we defined the mean, median, and mode. Then the concepts of percentiles and quartiles were used to describe the location of other values in the data set. Next, we presented the range, interquartile range, variance, standard deviation, and coefficient of variation as statistical measures of variability or dispersion in a data set. We described how the mean, variance, and standard deviation could be computed for grouped data. However, we recommend using the measures based on the individual data values unless the grouped format is the only manner in which the data are available.

A discussion of two exploratory data analysis techniques that can be used to summarize data more effectively was included in Section 3.5. Specifically, we showed how to develop a 5-number summary and a box plot in order to provide simultaneous information about the location, dispersion, and shape of the underlying distribution. The chapter concluded with a discussion of the role of the computer in descriptive statistics. Software packages of Minitab and *The Data Analyst* were used to illustrate how statistical computing systems can support the analysis and interpretation of data.

GLOSSARY

Population parameter A numerical value used as a summary measure for a population of data (e.g., the population mean, μ, the population variance, σ^2, and the population standard deviation, σ).

Sample statistic A numerical value used as a summary measure for a sample (e.g., the sample mean, $\bar{x}$, the sample variance, s^2, and the sample standard deviation, s).

Mean A measure of the central location of a data set. It is computed by summing all the values in the data set and dividing by the number of items.

Median A measure of central location of a data set. It is the value which splits the data set into two equal groups—one with values greater than or equal to the median, and one with values less than or equal to the median.

Mode A measure of location for a data set, defined as the most frequently occurring data value.

Percentile A value such that at least $p\%$ of the items in the data set are less than or equal to its value and at least $(100 - p)\%$ of the items are greater than or equal to it.

Quartiles The division points of a data set such that ¼, or 25%, of the data are in each section. The first quartile is the 25th percentile; the second quartile is the 50th percentile, or median; the third quartile is the 75th percentile.

Range   A measure of dispersion for a data set, defined to be the difference between the highest and lowest values.

Interquartile range   A measure of dispersion for a data set, defined to be the difference between the third and first quartiles.

Variance   A measure of dispersion for a data set, found by summing the squared deviations of the data values about the mean and then dividing the total by N if the data set is a population or by $n - 1$ if the data set is from a sample.

Standard deviation   A measure of dispersion for a data set, found by taking the square root of the variance.

Coefficient of variation   A measure of relative dispersion for a data set, found by dividing the standard deviation by the mean and multiplying by 100.

z-Score   A standardized value computed by subtracting the mean from the data value x_i and then dividing the result by the standard deviation. The z-score can be interpreted as the number of standard deviations the data value x_i is from its mean.

Outlier   An unusually large or small data value that should be reviewed for data accuracy and for whether or not it belongs in the data set.

Chebyshev's theorem   A theorem that allows the use of knowledge of the standard deviation and mean to draw conclusions about the fraction of data items within k standard deviations of the mean for any data set.

Empirical rule   A rule that allows the use of knowledge of the standard deviation and mean to draw conclusions about the fraction of data items within one, two, and three standard deviations of the mean for a data set that is mound-shaped.

Grouped data   Data available in class intervals as summarized by a frequency distribution. Individual values of the original data are not recorded.

5-number summary   An exploratory data analysis technique that uses the following 5 numbers to summarize the data set: smallest value, first quartile, median, third quartile, largest value.

Box plot   A graphical presentation for the data within the inner fences.

Fences   Values located 1.5 and 3 times the interquartile range below the first quartile and 1.5 and 3 times the interquartile range above the third quartile. Data values outside the inner fences are considered mild outliers; data values outside the outer fences are considered extreme outliers.

KEY FORMULAS

SAMPLE MEAN

$$\bar{x} = \frac{\Sigma x_i}{n} \tag{3.1}$$

POPULATION MEAN

$$\mu = \frac{\Sigma x_i}{N} \tag{3.2}$$

POPULATION VARIANCE

$$\sigma^2 = \frac{\Sigma(x_i - \mu)^2}{N} \tag{3.3}$$

SAMPLE VARIANCE

$$s^2 = \frac{\Sigma(x_i - \bar{x})^2}{n - 1} \qquad (3.4)$$

STANDARD DEVIATION

$$\text{Sample Standard Deviation} = s = \sqrt{s^2} \qquad (3.5)$$

$$\text{Population Standard Deviation} = \sigma = \sqrt{\sigma^2} \qquad (3.6)$$

COEFFICIENT OF VARIATION

$$\left(\frac{\text{Standard Deviation}}{\text{Mean}}\right) \times 100 \qquad (3.7)$$

z-SCORE

$$z_i = \frac{(x_i - \bar{x})}{s} \qquad (3.9)$$

SAMPLE MEAN FOR GROUPED DATA

$$\bar{x} = \frac{\Sigma f_i M_i}{n} \qquad (3.10)$$

SAMPLE VARIANCE FOR GROUPED DATA

$$s^2 = \frac{\Sigma f_i (M_i - \bar{x})^2}{n - 1} \qquad (3.11)$$

REVIEW QUIZ

TRUE/FALSE

1. The mean for a set of data is found by adding all the data values and dividing by the number of items minus 1.

2. The mean and median can never be equal.

3. The median and the 50th percentile are the same.

4. The mode of a set of data is the value that occurs with greatest frequency.

5. The standard deviation is the average of the differences between the data values and the mean.

6. The variance is the square root of the standard deviation.

7. The interquartile range is the difference between the fourth and second quartiles.

8. The sample variance is the average of the squared deviations in a sample.

9. The coefficient of variation is a measure of the relative variability in a data set.

10. Fences are used to detect outliers in a data set.

MULTIPLE CHOICE

11. The mean for the data 6, 9, 10, 12, 13 is closest to
  a. 9.5
  b. 9.9
  c. 10.5
  d. 11.0

12. The median for the data set in question 11 is
  a. 9.5
  b. 10
  c. 11
  d. 12

13. The standard deviation for the data set in question 11 is closest to
  a. 3
  b. 5
  c. 8
  d. 10

14. The range for the data set in question 11 is closest to
  a. 3
  b. 4
  c. 5
  d. 6

15. The coefficient of variation for the data set in question 11 is closest to
  a. 27
  b. 2
  c. 5
  d. 20

16. The interquartile range for the data set in question 11 is closest to
  a. 4
  b. 3
  c. 1
  d. 7

SUPPLEMENTARY EXERCISES

27. A sample of six recent home mortgage loans showed the following interest rates.

12.5, 13.2, 11.2, 13.0, 12.0, 12.5

Compute the following descriptive statistics for the data set
  a. mean
  b. median
  c. mode
  d. 25th percentile
  e. range
  f. interquartile range
  g. variance
  h. standard deviation
  i. coefficient of variation

28. *Newsweek* (January 9, 1989) reported statistics on the number of visits a couple makes to a therapist while undergoing marriage counseling. With data based on this article, assume that the following shows the number of visits from a sample of 9 couples:

$$12, 8, 3, 13, 18, 20, 10, 9, 18$$

Compute the following descriptive statistics for this data.

a. mean
b. median
c. mode
d. 40th percentile
e. range
f. variance
g. standard deviation

29. The following data show home mortgage loan amounts handled by a particular loan officer in a savings and loan association:

20,000	38,500	33,000	27,500	34,000
12,500	25,900	43,200	37,500	36,200
25,200	30,900	23,800	28,400	13,000
31,000	33,500	25,400	33,500	20,200
39,000	38,100	30,500	45,500	30,500
52,000	40,500	51,600	42,500	44,800

Find the mean, median, and mode for these data.

30. Calculate the variance, standard deviation, and range for the mortgage amounts shown in exercise 29. Conversion of the data to 1000s (e.g., 38,500 is listed as 38.5) may ease the burden of having to work with large numbers.

31. *Time* (January 9, 1989) published an article on the academic ability of college athletes. The article noted that some of the most successful athletic programs (citing the University of Notre Dame and Duke University) have athletes with very good college board scores. Assume that the following sample data are typical of college board scores for Notre Dame football players:

$$1100, 970, 1000, 1250, 880, 790, 1300, 1050, 900, 950, 1120$$

a. Compute the mean, median and mode.
b. Compute the range and interquartile range.
c. Compute the variance and standard deviation.
d. Using *z*-scores, state whether or not there are any outliers in this data set.

32. Soft-drink purchases at the Wright Field concession stands show the following 1-day totals:

DRINK	UNITS PURCHASED
Cola	4553
Diet cola	2125
Uncola	1850
Orange soda	1288
Root beer	1572

What is the mode for the above sample data?

33. Light bulbs manufactured by a well known electrical equipment firm are known to have a mean life of 800 hours, with a standard deviation of 100 hours. Use Chebyshev's theorem to answer the following questions.

a. What percentage of the light bulbs will have a life of 600 to 1000 hours?

b. What percentage of the light bulbs will have a life of 500 to 1100 hours?

34. Daily volume for the stock market over a 6-month period showed a mean of 150 million shares with a standard deviation of 25 million shares. Use the empirical rule to answer the following questions.

a. What percentage of the days showed volumes between 100 and 200 million shares?

b. What percentage of the days showed volumes between 75 and 225 million shares?

c. Provide an interval for daily volume that will include approximately two-thirds of the days.

35. A frequency distribution for the duration of 20 long-distance telephone calls (rounded to the nearest minute) is shown below:

CALL DURATION	FREQUENCY
4–7	4
8–11	5
12–15	7
16–19	2
20–23	1
24–27	1
	Total 20

Compute the mean, variance, and standard deviation for the above data.

36. Dinner check amounts at La Maison French Restaurant (rounded to the nearest dollar) have the following frequency distribution:

DINNER CHECK (DOLLARS)	FREQUENCY
25–34	2
35–44	6
45–54	4
55–64	4
65–74	2
75–84	2
	Total 20

Compute the mean, variance, and standard deviation for the given data.

37. Automobiles traveling on the New York State Thruway are checked for speed by a state police radar system. A frequency distribution of speeds is shown.

SPEED (MILES PER HOUR)	FREQUENCY
40–44	10
45–49	40
50–54	150
55–59	175
60–64	75
65–69	15

SPEED (MILES PER HOUR)	FREQUENCY
70–74	7
75–79	3
	Total 475

a. What is the mean speed of the automobiles traveling on the New York State Thruway?
b. Compute the variance and the standard deviation.

38. The following data are the final examination scores in a section of calculus.

$$
\begin{array}{ccccc}
56 & 77 & 84 & 82 & 42 \\
61 & 44 & 95 & 98 & 84 \\
93 & 62 & 96 & 78 & 88 \\
58 & 62 & 79 & 85 & 89 \\
89 & 97 & 53 & 76 & 75
\end{array}
$$

a. Provide a 5-number summary.
b. Locate the inner and outer fences.
c. Identify any outliers.
d. Provide a box plot.

39. The following data show the results of a 150-question social-awareness test that was given to a group of first-year college students. The data are number of questions answered correctly by each of the students.

$$
\begin{array}{cccccccccc}
121 & 114 & 94 & 136 & 144 & 126 & 98 & 103 & 118 & 127 \\
135 & 97 & 119 & 117 & 122 & 138 & 142 & 141 & 102 & 105
\end{array}
$$

a. Provide a 5-number summary.
b. Locate the inner and outer fences.
c. Identify any outliers.
d. Provide a box plot.

40. The following data show the total yardage accumulated over the football season for 20 receivers.

$$
\begin{array}{cccccccccc}
744 & 652 & 576 & 1112 & 920 & 451 & 1023 & 852 & 809 & 640 \\
900 & 975 & 400 & 711 & 1174 & 1278 & 820 & 211 & 897 & 1451
\end{array}
$$

a. Provide a 5-number summary.
b. Locate the inner and outer fences.
c. Identify any outliers.
d. Provide a box plot.

COMPUTER EXERCISE

A national association of nurses has sponsored a study to determine the job satisfaction of nurses employed in hospitals. As part of the study, 50 nurses were asked to indicate their degree of satisfaction in their work, in their pay, and in their opportunities for promotion. Each of the three aspects of satisfaction were measured on a scale of 0 to 100, with larger values indicating higher degrees of satisfaction. Data were also collected on the type of hospital where each nurse was employed. The hospital types considered in the study were investor-owned hospitals, Veterans Administration (VA) hospitals, and university hospitals. The following data were collected:

SATISFACTION SCORES			TYPE OF HOSPITAL	SATISFACTION SCORES			TYPE OF HOSPITAL
Work	Pay	Promotion		Work	Pay	Promotion	
71	49	58	VA	72	76	37	VA
84	53	63	University	71	25	74	Investor-owned
84	74	37	VA	69	47	16	University
87	66	49	University	90	56	23	University
72	59	79	University	84	28	62	Investor-owned
72	37	86	VA	86	37	59	VA
72	57	40	Investor-owned	70	38	54	Investor-owned
63	48	78	VA	86	72	72	VA
84	60	29	VA	87	51	57	Investor-owned
90	62	66	Investor-owned	77	90	51	University
73	56	55	VA	71	36	55	University
94	60	52	VA	75	53	92	University
84	42	66	Investor-owned	74	59	82	Investor-owned
85	56	64	Investor-owned	76	51	54	University
88	55	52	University	95	66	52	VA
74	70	51	University	89	66	62	Investor-owned
71	45	68	Investor-owned	85	57	67	Investor-owned
88	49	42	Investor-owned	65	42	68	VA
90	27	67	VA	82	37	54	VA
85	89	46	University	82	60	56	VA
79	59	41	University	89	80	64	University
72	60	45	Investor-owned	74	47	63	Investor-owned
88	36	47	Investor-owned	82	49	91	Investor-owned
77	60	75	Investor-owned	90	76	70	VA
64	43	61	Investor-owned	78	52	72	VA

QUESTIONS

1. Develop descriptive measures of location for each of the three job-satisfaction variables. What aspect of the job is the most satisfying for the nurses? What appears to be the most critical issue, or issue of lowest satisfaction for nurses? Explain.
2. Develop descriptive measures of dispersion for each of the three measures of job satisfaction. Which measure of satisfaction appears to have the greatest difference of opinion among the nurses?
3. Develop summary information for the type of hospital variable.
4. Show how exploratory data analysis can help summarize these data.

4 Introduction to Probability

What You Will Learn in This Chapter:

- Three methods commonly used for assigning probabilities to experimental outcomes
- How to compute the probability of an event from its experimental outcomes
- How to use the rules of probability to compute the probability of an event from known probabilities of related events
- The purpose and use of Bayes' theorem to revise probabilities
- Important terms such as experiment, sample space, event, mutually exclusive event, independent event, and conditional probability

Contents

Probability: What It Tells About the Typical American

Americans are the most widely researched people on earth. Every day, between 20,000 and 30,000 of us participate in some sort of survey. Based on these surveys, estimates are made of the probability that a "typical American" will engage in some event or activity, have certain likes, dislikes, habits, and beliefs, and so on. For example, the probability that a "typical American" believes he or she exercises enough is .43, and the probability that a "typical American" watches TV almost every evening is .63. Some other common events and their probability estimates are shown below.

Are any of these people "typical Americans"?

EVENT	PROBABILITY
Goes to McDonald's every day	.05
Water skis	.07
Has high blood pressure	.12
Jogs	.13
Is underweight	.18
Never exercises at all	.25
Has never smoked	.46
Daydreams about being rich	.52

EVENT	PROBABILITY
Has experienced ESP	.67
Lives in a metropolitan area	.77
Considers themselves happy	.90

From *100% American* by Daniel Evan Weiss. New York: Poseidon Press, 1988.

Life is filled with uncertainty; some examples of events involving uncertainty are referred to in "Statistics in the News." Some others are the following:

1. What is the chance that your car will be towed away if you park illegally in the visitors' parking lot?
2. What is the likelihood of receiving an A on the first exam if you wait until the night before to begin studying?
3. How likely is completing your college program on time if you take a part-time job that requires you to work 20 hours per week?
4. What are the chances you will get a 7 on the first roll of the dice at a casino in Las Vegas?

Probabilities are most useful in effectively dealing with such uncertainties. In everyday terminology, *probability* can be thought of as a numerical measure of the chance, or likelihood, that a particular event will occur. The probabilities for several events involving typical Americans are presented at the end of "Statistics in the News."

Probability values are always assigned on a scale of 0 to 1. A probability near 0 indicates that an event is very unlikely to occur; a probability near 1 indicates that an event is almost certain to occur. Other probabilities between 0 and 1 represent varying degrees of likelihood that an event will occur. Figure 4.1 depicts this view of probability as a numerical measure of the likelihood an event will occur.

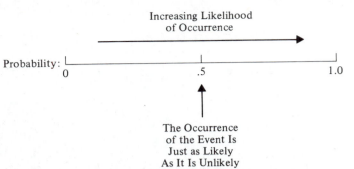

FIGURE 4.1 Probability as a Numerical Measure of the Likelihood of an Event's Occurrence

EXAMPLE 4.1

If we consider the event "rain tomorrow," we understand that when the television weather report indicates a near-zero probability of rain, there is almost no chance of rain. However, if a .90 probability of rain is reported, we know that it is very likely or almost certain that rain will occur. A .50 probability indicates that rain is just as likely to occur as not.

■

4.1   Experiments, the Sample Space, and Probability

Using the terminology of probability, we define an *experiment* to be any process that generates well-defined outcomes. On any single repetition of the experiment, one and only one of the possible experimental outcomes will occur. Several examples of experiments and their associated outcomes are as follows:

EXPERIMENT	EXPERIMENTAL OUTCOMES
Toss a coin	Head, tail
Apply for a job	Hired, not hired
Roll a die	1, 2, 3, 4, 5, 6
Play a football game	Win, lose, tie

When we have specified all possible experimental outcomes, we have identified what statisticians call the *sample space* for the experiment.

Sample Space

The sample space for an experiment is the set of all experimental outcomes.

EXAMPLE 4.2

Consider the experiment of tossing a coin. The experimental outcomes are determined by the upward face of the coin—a head or a tail. If we let *S* denote the sample space, we can use the following notation to describe the sample space:

$$S = \{\text{head, tail}\}$$

■

EXAMPLE 4.3

Consider the experiment of rolling a die, with the experimental outcomes defined as the number of dots appearing on the upward face of the die. In this experiment, the numerical values 1, 2, 3, 4, 5, and 6 represent the possible experimental ocomes. We can denote the sample space for this experiment as follows:

$$S = \{1, 2, 3, 4, 5, 6\}$$

■

For some experiments, it is not easy to determine the number of experimental outcomes. This is especially true when the experiment involves multiple steps.

EXAMPLE 4.4

Consider the experiment of tossing two coins. Let the experimental outcomes be defined in terms of the pattern of heads and tails appearing on the upward faces of the two coins. How many experimental outcomes are possible for this experiment? This can be thought of as a two-step experiment in which step 1 is the tossing of the first coin and step 2 is the tossing of the second coin. If we use H to denote a head and T to denote a tail, (H, H) indicates the experimental outcome with a head on the first coin and a head on the second coin. Continuing this notation, we can describe the sample space (S) for the two-coin-tossing experiment as follows:

$$S = \{(H, H), (H, T), (T, H), (T, T)\}$$

Thus we see that there are four outcomes for this experiment.

■

Example 4.4 describes an experiment consisting of two steps, in which there are two possible outcomes (head or tail) on the first step and two possible outcomes (head or tail) on the second step. Let us introduce a rule that is helpful in determining the number of outcomes for an experiment consisting of multiple steps.

A Counting Rule for Multiple-Step Experiments

If an experiment can be described as a sequence of k steps in which there are n_1 possible outcomes on the first step, n_2 possible outcomes on the second step, and so on, then the total number of experimental outcomes is given by $(n_1) (n_2) \ldots (n_k)$.

EXAMPLE 4.4
(continued)

Looking at the experiment of tossing two coins as a sequence of first tossing one coin $(n_1 = 2)$ and then tossing the other coin $(n_2 = 2)$, we can see from the counting rule that there must be $(2) (2) = 4$ distinct experimental outcomes.

■

A graphical device that is helpful in visualizing an experiment and enumerating outcomes in a multiple-step experiment is a *tree diagram*. Figure 4.2 shows a tree diagram for the two-coin-tossing experiment described in Example 4.4. The sequence of steps is depicted by moving from left to right through the tree. Step 1 corresponds to tossing the first coin, and there are two branches corresponding to the two possible outcomes. Step 2 corresponds to tossing the second coin, and for each possible outcome at step 1, there

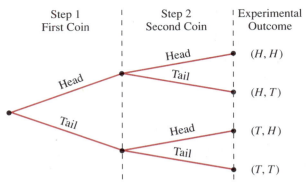

FIGURE 4.2　Tree Diagram for the Experiment of Tossing Two Coins

are two branches corresponding to the two possible outcomes at step 2. Finally, each of the points on the right-hand end of the tree corresponds to an experimental outcome. Each path through the tree from the leftmost node to one of the nodes at the right hand side of the tree corresponds to a unique sequence of outcomes for each step.

Now let us see how probabilities can be assigned to experimental outcomes. The three approaches most frequently used are the classical, relative frequency, and subjective methods. Regardless of the method used, the probabilities assigned must satisfy two basic requirements.

Basic Requirements for Assigning Probabilities

1. The probability assigned to each experimental outcome must be between 0 and 1, inclusively. If we let E_i denote the ith experimental outcome and $P(E_i)$ be its probability, then this requirement can be written as

$$0 \leq P(E_i) \leq 1 \text{ for all } i \tag{4.1}$$

2. The sum of the probabilities for all the experimental outcomes must equal 1. If there are n experimental outcomes, this requirement can be written as

$$P(E_1) + P(E_2) + \cdots + P(E_n) = 1 \tag{4.2}$$

The *classical method* of assigning probabilities is appropriate when all the experimental outcomes are equally likely. If there are n experimental outcomes, a probability of $1/n$ is assigned to each experimental outcome. Note that when using this approach, the two basic requirements for assigning probabilities are automatically satisfied.

EXAMPLE 4.2 (continued)

For the experiment of tossing a single coin, the two experimental outcomes—head and tail—are equally likely. Therefore, since one of the two equally likely outcomes is a head, the probability of observing a head is ½, or .50. Similarly, the probability of observing a tail is also ½, or .50.

■

**EXAMPLE 4.3
(continued)**

For the experiment of rolling a die, it would also seem reasonable to conclude that the six possible outcomes are equally likely, and hence each outcome is assigned a probability of $\frac{1}{6}$. If $P(1)$ denotes the probability that one dot appears on the upward face of the die, then $P(1) = \frac{1}{6}$. Similarly, $P(2) = \frac{1}{6}$, $P(3) = \frac{1}{6}$, $P(4) = \frac{1}{6}$, $P(5) = \frac{1}{6}$, and $P(6) = \frac{1}{6}$.

■

**EXAMPLE 4.4
(continued)**

Recall that there are four experimental outcomes for the experiment involving the tossing of two coins. Since it seems reasonable to assume that the four possible experimental outcomes are equally likely, we assign a probability of $\frac{1}{4}$ to each; thus the probability of experimental outcome *(H, H)* is $\frac{1}{4}$, the probability of experimental outcome *(H, T)* is $\frac{1}{4}$, the probability of experimental outcome *(T, H)* is $\frac{1}{4}$, and the probability of experimental outcome *(T, T)* is $\frac{1}{4}$.

■

The *relative frequency method* of assigning probabilities is appropriate when data are available to estimate the proportion of the time the experimental outcome will occur when the experiment is repeated a "large number" of times.

EXAMPLE 4.5

As part of a study of the X-ray department for a local hospital, the number of patients waiting for service at 9:00 A.M. was recorded for 20 successive days. The following results were obtained.

NUMBER WAITING	NUMBER OF DAYS OUTCOME OCCURRED
0	2
1	5
2	6
3	4
4	3
Total	20

These data show that on 2 of the 20 days, 0 patients were waiting for service, on 5 of the days, 1 patient was waiting for service, and so on. Using the relative frequency method, we would assign a probability of $\frac{2}{20} = .10$ to the experimental outcome of 0 patients waiting for service, $\frac{5}{20} = .25$ to the experimental outcome of 1 patient waiting, $\frac{6}{20} = .30$ to 2 patients waiting, $\frac{4}{20} = .20$ to 3 patients waiting, and $\frac{3}{20} = .15$ to 4 patients waiting.

■

EXAMPLE 4.6

In the test market evaluation of a new product, 400 potential customers were contacted; 100 actually purchased the product, but 300 did not. In effect, then, we have repeated an experiment of contacting a customer 400 times and have found that the product was purchased 100 times. Thus using the relative frequency approach, we assign a probability of $100/400 = .25$ to the experimental outcome of a potential customer purchasing the product. Similarly, $300/400 = .75$ is assigned to the experimental outcome of a potential customer not purchasing the product.

■

As these two examples illustrate, the two basic requirements of probability are also satisfied automatically when the relative frequency approach is used.

The *subjective method* of assigning probabilities is most appropriate when it is unrealistic to assume that the experimental outcomes are equally likely and when little relevant data are available. When the *subjective method* is used to assign probabilities to the experimental outcomes, we may use any information available, such as our experience, intuition, etc. After considering all available information, a probability value that expresses our degree of belief that the experimental outcome will occur is specified. Since subjective probability expresses a person's degree of belief, it is personal. Using the subjective method, different people can be expected to assign different probabilities to the same experimental outcome.

When using the subjective probability assignment method, extra care must be taken to ensure that requirements (4.1) and (4.2) are satisfied. Regardless of a person's degree of belief, the probability value assigned to each experimental outcome must be between 0 and 1, inclusive, and the sum of all the experimental outcome probabilities must equal 1.

EXAMPLE 4.7

Consider the next football game that the Pittsburgh Steelers will play. What is the probability that the Steelers will win? The experimental outcomes of win, lose, and tie are obviously not equally likely. Also, since the teams involved will not have played several times previously in the same year, there are no relative frequency data available relevant to the game. Thus if we want an estimate of the probability of the Steelers' winning, we must use the subjective method and state a value that expresses our degree of belief that they will win.

EXAMPLE 4.8

Tom and Judy Elsbernd have just made an offer to purchase a house. Two outcomes are possible:

$$E_1 = \text{their offer is accepted}$$
$$E_2 = \text{their offer is rejected}$$

Judy believes that the probability their offer will be accepted is .8; thus, Judy would set $P(E_1) = .8$ and $P(E_2) = .2$. Tom, however, believes that the probability that their offer will be accepted is .6; hence, Tom would set $P(E_1) = .6$ and $P(E_2) = .4$. Note that Tom's probability estimate for E_1 reflects the fact that he is a bit more pessimistic than Judy is about their offer being accepted.

Both Judy and Tom have assigned probabilities that satisfy the two requirements for assigning probabilities. The fact that their probability estimates are different emphasizes the personal nature of the subjective method.

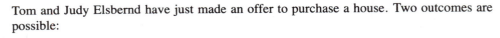

EXERCISES

Consider the experiment of drawing 1 card from a deck of 52 playing cards and observing whether the card is red or black.

a. What are the experimental outcomes?

b. Assign probabilities to the experimental outcomes.

2. Consider the experiment of drawing 1 card from a deck of 52 playing cards and observing whether or not the card is a spade.

a. What is the probability of drawing a spade?

b. What is the probability of not drawing a spade?

3. Consider the experiment of tossing a coin three times.

a. Develop a tree diagram for the experiment.

b. List the experimental outcomes.

c. What is the probability for each outcome?

4. Consider the experiment of administering a true-false exam consisting of ten questions. Each different sequence of answers is an experimental outcome.

a. How many experimental outcomes are there?

b. If a student guesses on every question, what is the probability of any particular experimental outcome?

5. Consider the experiment of rolling a pair of dice. Each die has six possible results (the number of dots on its face).

a. How many outcomes are possible for this experiment?

b. Show a tree diagram for the experiment.

c. How many experimental outcomes provide a sum of 7 for the dots on the dice?

6. Many states design their automobile license plates such that space is available for up to six letters or numbers.

a. If a state decides to use only numerical values for the license plates, how many different license plate numbers are possible? Assume that 000000 is an acceptable license plate number, although it will be used only for display purposes at the license bureau. (*Hint:* Use the counting rule.)

b. If the state decides to use two letters followed by four numbers, how many different license plate numbers are possible? Assume that the letters I and O will not be used because of their similarity to numbers 1 and 0.

c. Would larger states, such as New York or California tend to use more or fewer letters in license plates? Explain.

7. The final exam in a course in contemporary science resulted in the following grades.

Grade	A	B	C	D	F
Number	7	12	16	5	3

a. What is the probability that a randomly selected student received an A?

b. What is the probability that a randomly selected student received a C?

8. Faced with the question of determining the probability of obtaining either 0 heads, 1 head, or 2 heads when flipping a coin twice, an individual argued that it seems reasonable to treat the outcomes as equally likely and that the probability of each event is ⅓. Do you agree? Explain.

9. An investor forecasts that the probabilities that a certain stock will either go down, remain the same, or go up are .20, .60, and .30, respectively. Does this seem reasonable? Explain.

10. Planes flying from New York City to Chicago are listed as either arriving early, on time, or late. Discuss how you would develop estimates of the probabilities for each of these events.

11. A company that manufactures toothpaste has five different package designs they want to study. Assuming that one design is just as likely to be preferred by a consumer as any other design, what probability would you assign to a randomly selected consumer preferring each of the package designs? In an actual experiment, 100 consumers were asked to pick the design they preferred. The following data were obtained.

Design	1	2	3	4	5
Total	5	15	30	40	10

Do the data appear to confirm the belief that one design is just as likely to be selected as another? Explain.

12. A small-appliance store in Madeira has collected data on refrigerator sales for the last 50 weeks.

NUMBER OF REFRIGERATORS SOLD	NUMBER OF WEEKS
0	6
1	12
2	15
3	10
4	5
5	2
	50

Suppose that we are interested in the experiment of observing the number of refrigerators sold in 1 week of store operations.
   a. How many experimental outcomes are there?
   b. Which approach would you recommend for assigning probabilities to the experimental outcomes?
   c. Assign probabilities to the experimental outcomes and verify that your assignments satisfy the two basic requirements.

13. Strom Construction has made a bid on two contracts. The owner has identified the possible outcomes and subjectively assigned probabilities as follows:

EXPERIMENTAL OUTCOME	OBTAIN CONTRACT 1	OBTAIN CONTRACT 2	PROBABILITY
1	Yes	Yes	.15
2	Yes	No	.15
3	No	Yes	.30
4	No	No	.25

   a. Are these valid probability assignments? Why or why not?
   b. What would have to be done to make the probability assignments valid?

Events and Their Probabilities

In the introduction to this chapter we used the term *event* much as it would be used in everyday conversation. However, in order to show how the probability of an event is determined, we must now introduce a formal definition of this concept.

One possible subset, event, consists of all the experimental outcomes. However, since the sample space *is* the set of all experimental outcomes, this means that the sample space is, itself, an event. Note also that a subset of the sample space can consist of just one experimental outcome; in such cases we refer to the resulting event as a *simple event*.

EXAMPLE 4.3 (continued)

Recall that for the experiment of rolling a die, the sample space was $S = \{1, 2, 3, 4, 5, 6\}$. There are six simple events, each of which corresponds to one of the experimental outcomes.

$$E_1 = \{1\} \quad E_4 = \{4\}$$
$$E_2 = \{2\} \quad E_5 = \{5\}$$
$$E_3 = \{3\} \quad E_6 = \{6\}$$

If we define A to be the event that an even number of dots appear on the upward face of the die, then we can describe event A as follows:

$$A = \{2, 4, 6\}$$

Similarly, if B is the event that an odd number of dots appear on the upward face of the die, then

$$B = \{1, 3, 5\}$$

We see that events A and B are simply different subsets of the sample space.

■

EXAMPLE 4.9

A cab company has analyzed its operating records for the past 20 days. On 8 of these days, no vehicle breakdowns were observed; on 6 of the days one cab had a breakdown; on 3 days there were 2 breakdowns; on 2 days there were 3 breakdowns; and on 1 of the days 4 cabs had breakdowns. Let

$$S = \{0, 1, 2, 3, 4\}$$

denote the sample space for the experiment of observing the number of cab breakdowns on a day. The numerical values, 0, 1, 2, 3, and 4 denote the number of breakdowns (the experimental outcomes). If A is defined as the event that 2 or more vehicle breakdowns are observed on a typical day, then

$$A = \{2, 3, 4\}$$

If B is defined as the event that less than 2 breakdowns are observed, then

$$B = \{0, 1\}$$

Similarly, if C is defined as the event that no breakdowns are observed, then

$$C = \{0\}$$

We see that the event C consists of just one experimental outcome; thus, C is a simple event.

■

Given the probabilities of the experimental outcomes, we can use the following definition to compute the probability of any event.

Using this definition, we calculate the probability of a particular event by adding the probabilities of the experimental outcomes that make up the event.

EXAMPLE 4.3 (continued)

Recall that for the experiment of rolling a die, we used the classical approach to assign probabilities of $P(1) = \frac{1}{6}, P(2) = \frac{1}{6}, \ldots, P(6) = \frac{1}{6}$. Thus to compute the probability of event $A = \{2, 4, 6\}$, we sum the probabilities of the experimental outcomes 2, 4, and 6:

$$P(A) = P(2) + P(4) + P(6)$$
$$= \frac{1}{6} + \frac{1}{6} + \frac{1}{6} = \frac{3}{6} = \frac{1}{2}$$

Similarly, for $B = \{1, 3, 5\}$,

$$P(B) = P(1) + P(3) + P(5)$$
$$= \frac{1}{6} + \frac{1}{6} + \frac{1}{6} = \frac{3}{6} = \frac{1}{2}$$

■

EXAMPLE 4.9 (continued)

Using the relative frequency method, we can use the data provided to estimate the probability of a specific number of breakdowns on a day selected at random. The probabilities assigned to the experimental outcomes are shown below.

NUMBER OF BREAKDOWNS	NUMBER OF OCCURRENCES	PROBABILITY
0	8	8/20 = .40
1	6	6/20 = .30
2	3	3/20 = .15
3	2	2/20 = .10
4	1	1/20 = .05
	Totals 20	1.00

Thus, the probability of event $A = \{2, 3, 4\}$ is given by

$$P(A) = P(2) + P(3) + P(4)$$
$$= .15 + .10 + .05 = .30$$

Similarly, the probability of event $B = \{0, 1\}$ is

$$P(B) = P(0) + P(1)$$
$$= .40 + .30 = .70$$

Finally, we see that for event $C = \{0\}$, $P(C) = P(0) = .40$.

■

Any time we can identify all the experimental outcomes and assign the corresponding probabilities, we can use the definition of this section to compute the probability of an event of interest. However, in many experiments the number of experimental outcomes is large, and their identification—as well as determining their associated probabilities— becomes extremely cumbersome, if not impossible. In the remaining sections of this chapter we present rules that can often be used to compute the probability of an event without knowledge of the probability of each experimental outcome. These probability relationships require a knowledge of the probabilities for related events. Probabilities of events are then computed directly from the related event probabilities using one or more of the probability rules.

Notes and Comments

1. As noted, the sample space is, itself, an event. Since it contains all the experimental outcomes, it has a probability of 1; that is $P(S) = 1$.

2. When the classical method is used to assign probabilities, the assumption is that the experimental outcomes are equally likely. In such cases the probability of any event of interest can be computed by dividing the number of experimental outcomes in the event by the total number of experimental outcomes. For instance, recall Example 4.3, when $A = \{2, 4, 6\}$. Since we assumed that the experimental outcomes are equally likely and since the total number of experimental outcomes is 6, $P(A) = 3/6 = .5$.

14. Consider the experiment of tossing a coin twice.
   a. What is the probability of the event "no heads"?
   b. What is the probability of the event "one head"?

15. Consider the experiment of drawing one card from a deck of 52 playing cards and observing what it is.
   a. What is the probability of the event "draw on ace"?
   b. What is the probability of the event "draw a face card"?

16. Consider the experiment of rolling a pair of dice. Suppose that we are interested in the sum of the face values showing on the dice.
   a. How many experimental outcomes are possible? (*Hint:* Use the counting rule.)
   b. List the experimental outcomes.
   c. What is the probability of obtaining a value of 7?
   d. What is the probability of obtaining a value of 9 or greater?
   e. Since there are six possible even values (2, 4, 6, 8, 10, and 12) and only five possible odd values (3, 5, 7, 9, and 11), the dice should show even values more often than odd values. Do you agree with this statement? Explain.

17. Suppose that a manager of a large apartment complex provides the following subjective probability estimate about the number of vacancies that will exist next month.

VACANCIES	PROBABILITY
0	.05
1	.15
2	.35
3	.25
4	.10
5	.10

List the experimental outcomes in each of the following events and provide the probability of each event.
   a. no vacancies
   b. at least four vacancies
   c. two or fewer vacancies

18. A sample of 100 customers of Montana Gas and Electric resulted in the following frequency distribution of monthly charges.

AMOUNT $	NUMBER
0–49	13
50–99	22
100–149	34
150–199	26
200–249	5

   a. Let A be the event that monthly charges are $150 or more. Find $P(A)$.
   b. Let B be the event that monthly charges are less than $150. Find (B).

19. A survey of 50 students at Tarpon Springs College regarding the number of extracurricular activities resulted in the following data.

Number of activities	0	1	2	3	4	5
Frequency	8	20	12	6	3	1

a. Let A be the event that a student participates in at least 1 activity. Find $P(A)$.
b. Let B be the event that a student participates in 3 or more activities. Find $P(B)$.
c. What is the probability a student participates in exactly 2 activities?

20. A marketing manager is attempting to assign probability values to the possible profits and losses resulting from a new product. Relying on subjective probabilities, the manager's probability estimates are as follows.

$$P(\text{profit over } \$10,000) = .25$$

$$P(\text{profit from } \$0 \text{ to } \$10,000) = .50$$

$$P(\text{loss}) = .15$$

What advice would you offer before the manager uses these estimates to perform further probability calculations?

21. A telephone survey was used to determine viewer response to a new television show. The following data were obtained.

RATING	FREQUENCY
Poor	4
Below average	8
Average	11
Above average	14
Excellent	13

a. What is the probability that a randomly selected viewer rates the new show as average or better?
b. What is the probability that a randomly selected viewer rates the new show below average or worse?

22. A bank has observed that credit card account balances have been growing over the past year. A sample of 200 customer accounts resulted in the following data.

AMOUNT OWNED $	FREQUENCY
0-99	62
100-199	46
200-299	24
300-399	30
400-499	26
500-599	12

a. Let A be the event that a customer's balance is less than $200. Find $P(A)$.
b. Let B be the event that a customer's balance is $300 or more. Find $P(B)$.

Rules for Computing Event Probabilities

Complement of an Event

Given an event *A*, the *complement* of *A* is defined to be the event consisting of all sample points that are *not* in *A*. The complement of *A* is denoted by A^c. Figure 4.3 provides a diagram known as a *Venn diagram*, which illustrates the concept of an event and its complement. The rectangular area represents the sample space for the experiment and, as such, contains all possible experimental outcomes. The circle represents event *A* and contains only the experimental outcomes in *A*. The shaded region of the diagram contains all experimental outcomes not in event *A*, which is the definition of event A^c, the complement of *A*.

In any probability application, event *A* and its complement A^c must satisfy

$$P(A) + P(A^c) = 1 \qquad (4.3)$$

Solving for *P(A)*, we obtain the following result:

> ### Computing Probability Using the Complement
>
> $$P(A) = 1 - P(A^c) \qquad (4.4)$$

Equation (4.4) shows that the probability of an event *A* can easily be computed if the probability of its complement, $P(A^c)$, is known.

EXAMPLE 4.10

Based upon an analysis of student records, the placement director at a university states that 98% of the students who interview with a particular firm are not given a job offer. Letting *A* denote the event of a job offer and A^c denote the event of no job offer, the placement director is stating that $P(A^c) = .98$. Using Equation (4.4), we see that

$$P(A) = 1 - P(A^c) = 1 - .98 = .02$$

FIGURE 4.3   Complement of Event *A*

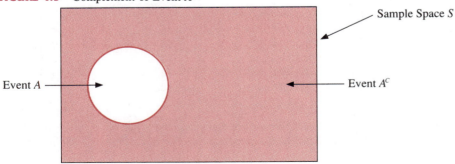

**EXAMPLE 4.10
(continued)**

This shows that there is a .02 probability that a student who interviews with the firm will receive a job offer.

■

EXAMPLE 4.11

A purchasing agent states that there is a .90 probability that a supplier will send a shipment that is free of defective parts. Using the complement, we can conclude that there is a $1 - .90 = .10$ probability that the shipment will contain at least one defective part.

■

Union and Intersection of Events

Given two or more events, we often are interested in the events obtained by combining two or more of the original events. To begin with let us consider the *union* of two events.

> **Union of Two Events**
>
> Given two events A and B, the *union* of A and B is the event containing all experimental outcomes belonging to A *or* B *or both*. The union is denoted by $A \cup B$.

The Venn diagram shown in Figure 4.4 depicts the union of events A and B. Note that the shaded region contains all experimental outcome in event A as well as all the experimental outcomes in event B. The fact that the circles overlap indicates that there are some experimental outcomes contained in both A and B.

The intersection of two events is defined as follows:

> **Intersection of Two Events**
>
> Given two events A and B, the *intersection of A and B* is the event containing the experimental outcomes belonging to *both A and B*. The intersection is denoted by $A \cap B$.

FIGURE 4.4   Union of Events A and B

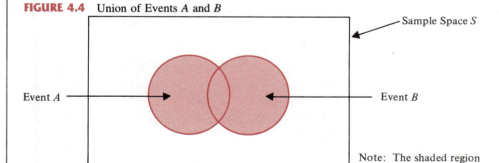

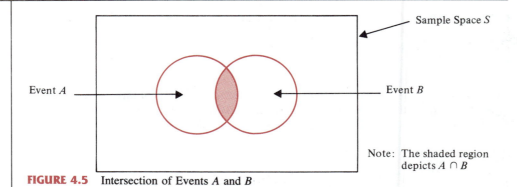

Sample Space *S*

Event *A*

Event *B*

Note: The shaded region depicts *A* ∩ *B*

FIGURE 4.5   Intersection of Events *A* and *B*

A Venn diagram depicting the intersection of the events *A* and *B* is shown in Figure 4.5. The area where the two circles overlap is the intersection; it is an event containing the experimental outcomes that are in both *A* and *B*.

Addition Rule

The *addition rule* is used to compute the probability of the union of two events.

Addition Rule

$$P(A \cup B) = P(A) + P(B) - P(A \cap B) \qquad (4.5)$$

To obtain an intuitive understanding of the addition rule, refer to Figure 4.4; note that the first two terms in the addition rule, *P(A)* + *P(B)*, account for all the experimental outcomes in *A* and all the experimental outcomes in *B*. However, since the experimental outcomes in the intersection, *A* ∩ *B*, are in both *A* and *B*, when we compute *P(A)* + *P(B)* we are in effect counting each of the experimental outcomes in *A* ∩ *B* twice. The addition rule corrects for this by subtracting *P(A* ∩ *B)*.

EXAMPLE 4.12

As an illustration of the addition rule, consider the following grades obtained in an introductory psychology course. Of 200 students taking the course, 160 passed the midterm exam and 140 passed the final exam; 124 students passed both exams. Letting

$$M = \text{event of passing the midterm exam}$$

$$F = \text{event of passing the final exam}$$

$$M \cap F = \text{event of passing both exams}$$

the given relative frequency information leads to the following event probabilities:

$$P(M) = \frac{160}{200} = .80$$

$$P(F) = \frac{140}{200} = .70$$

$$P(M \cap F) = \frac{124}{200} = .62$$

After reviewing the grades, the professor of the course decides to give a passing grade to any student who passed at least one of the two exams. That is, a passing grade will be given to any student who passes the midterm, to any student who passes the final, and to any student who passes both exams. What is the probability of receiving a passing grade in this course?

While your first reaction may be to try to count how many of the 200 students passed at least one exam, that information is not available; even if it were, the counting process would be tedious. However, note that the question concerns the union of the events M and F. That is, we want to know the probability a student passes the midterm (M), passes the final (F), or both. Thus we want to know $P(M \cup F)$. Here is where the addition rule can be helpful.

$$P(M \cup F) = P(M) + P(F) - P(M \cap F)$$

Knowing the three probabilities on the right-hand side of the above expression, we can write

$$P(M \cup F) = .80 + .70 - .62 = .88$$

Thus, there is a .88 probability of passing the course because there is a .88 probability of passing at least one of the exams.

EXAMPLE 4.13

A study involving the television-viewing habits of married couples found that 30% of the husbands and 20% of the wives were regular viewers of a particular Friday evening program. Both husband and wife were regular viewers in 12% of the households. What is the probability that at least one of the two individuals is a regular viewer of the program?

Letting

$$H = \text{husband is a regular viewer}$$

$$W = \text{wife is a regular viewer}$$

we have $P(H) = .30$, $P(W) = .20$, and $P(H \cap W) = .12$. Using the addition rule, we have

$$P(H \cup W) = P(H) + P(W) - P(H \cap W) = .30 + .20 - .12 = .38$$

This shows there is a .38 probability that at least one of the spouses is a regular viewer of the program.

Mutually Exclusive Events

Let us now see how the addition rule is applied to *mutually exclusive events*. First, we define mutually exclusive events.

Mutually Exclusive Events

Two or more events are said to be *mutually exclusive* if the events do not have any experimental outcomes in common.

Events A and B are said to be mutually exclusive if when one event occurs the other cannot occur. Thus if A and B are mutually exclusive, then their intersection contains no experimental outcome. A Venn diagram for the mutually exclusive events A and B is shown in Figure 4.6. Since $P(A \cap B) = 0$, the addition rule for mutually exclusive events can be shortened to:

Addition Rule for Mutually Exclusive Events

$$P(A \cup B) = P(A) + P(B) \qquad (4.6)$$

EXAMPLE 4.14

If A denotes the event that you get a grade of A for a term paper and B denotes the event that you get a grade of B, then A and B are mutually exclusive events. Thus if $P(A) = .20$ and $P(B) = .50$, then

$$P(A \cup B) = P(A) + P(B)$$
$$= .20 + .50 = .70$$

■

FIGURE 4.6 Mutually Exclusive Events

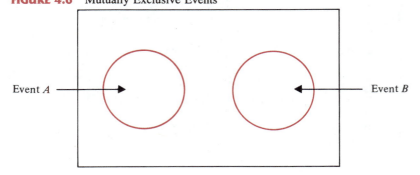

EXERCISES

23. Suppose $P(A) = .13$. Find $P(A^c)$.

24. Suppose $P(A) = .70$, $P(B) = .20$, and $P(A \cap B) = .15$.
   a. Find $P(A \cup B)$.
   b. Find $P(B^c)$.
   c. Find $P(A \cap B^c)$.

25. Suppose that we have a sample space $S = \{E_1, E_2, E_3, E_4, E_5, E_6, E_7\}$, where $E_1, E_2, \ldots E_7$ denote the experimental outcomes (simple events) and the following probability assignments are given for the simple events.

$$P(E_1) = .05$$
$$P(E_2) = .20$$
$$P(E_3) = .20$$
$$P(E_4) = .25$$
$$P(E_5) = .15$$
$$P(E_6) = .10$$
$$P(E_7) = \underline{.05}$$
$$\text{Total} \quad 1.00$$

Let

$$A = \{E_1, E_4, E_6\}$$
$$B = \{E_2, E_4, E_7\}$$
$$C = \{E_2, E_3, E_5, E_7\}$$

   a. Find $P(A)$, $P(B)$, and $P(C)$.
   b. Find $A \cup B$ and $P(A \cup B)$.
   c. Find $A \cap B$ and $P(A \cap B)$.
   d. Are events A and C mutually exclusive?
   e. Find B^c and $P(B^c)$.

26. Let

M = person interviewed is male

F = person interviewed is female

D = person believes the man should always pay on first date

W = person believes one who requests the date should pay

E = person believes the one who earns more money should pay on first date

As reported in *Money* magazine (November, 1988), $P(M) = .61$, $P(F) = .39$, $P(D) = .41$, $P(W) = .41$, $P(E) = .02$, and $P(D \cap M) = .29$.

a. Are M and D mutually exclusive events?

b. Find $P(D \cup M)$.

c. Find $P(D \cup W)$.

d. Find $P(D \cap F)$.

e. Find $P(D \cup F)$.

27. Consider an experiment where eight possible outcomes exist. We will denote the experimental outcomes as $E_1, E_2, \ldots, E_8$. Suppose the following events are defined.

$$A = (E_1, E_2, E_3, E_5)$$

$$B = (E_2, E_4, E_5, E_8)$$

Note that experimental outcomes E_6 and E_7 are in neither event A nor B. List the experimental outcomes making up the following events:

a. $A \cup B$

b. $A \cap B$

c. A^c

d. Are A and B mutually exclusive events? Explain.

28. In a study conducted to evaluate the effect of an allergy relief medicine, 250 patients with symptoms that included itchy eyes and a skin rash were given the new drug. The results of the study are as follows: 90 of the patients treated experienced eye relief, 135 had their skin rash clear up, and 45 experienced both relief from itchy eyes and the skin rash. What is the probability that a patient that takes the drug will experience relief for at least one of the two symptoms?

29. In a study of 100 students that had been awarded university scholarships, it was found that 40 had part-time jobs, 25 had made the dean's list the previous semester, and 15 had both a part-time job and had made the dean's list. What was the probability that a student had a part-time job or was on the dean's list?

30. During winter in Cincinnati, Mr. Krebs experiences difficulty in starting his two cars. The probability that the first car starts is .80, and the probability that the second car starts is .40. There is a probability of .30 that both cars start.

a. Define the events involved and use probability notation to show the probability information given above.

b. What is the probability that at least one car starts?

c. What is the probability that Mr. Krebs cannot start either of the two cars?

31. From "Statistics in the News" at the beginning of the chapter, we can define the following events and probabilities.

$$H = \text{person has high blood pressure}$$

$$E = \text{person never exercises at all}$$

$$S = \text{person has never smoked}$$

$$P(H) = .12,\ P(E) = .25,\ P(S) = .46$$

a. Are any of these events mutually exclusive?
b. What is the probability a randomly selected person has smoked?
c. Is the probability that a person never exercises at all or has high blood pressure greater than .40? Explain.

4.4   Conditional Probability, Independence, and the Multiplication Rule

Conditional Probability

Often, the probability of an event is influenced by whether or not a related event has occurred. For instance, most people believe that the probability of an accident increases if the driver has been drinking. In such cases, the concept of conditional probability can be used to compute the probability of one event given that another related event is known to have occurred.

EXAMPLE 4.15

A major metropolitan police force in the Eastern United States consists of 1200 officers— 960 men and 240 women. Over the past 2 years, 324 officers on the police force have been awarded promotions. The breakdown of promotions for male and female officers is shown in Table 4.1.

After reviewing the data in Table 4.1, a committee of female officers charged discrimination on the basis that 288 male officers had received promotions, whereas only 36 female officers had received promotions. The police administration countered with the argument that the relatively low number of promotions for female officers was not due to discrimination but due to the fact that there are fewer female officers on the police force.

After reflecting on this situation, we see that the real issue involves not the number of promotions but the probability of promotion given that the officer is male and the probability of promotion given that the officer is female.

■

TABLE 4.1

Promotional Status of Police Officers Over the Past 2 Years

	MEN	WOMEN	Totals
Promoted	288	36	324
Not Promoted	672	204	876
Totals	960	240	1200

In dealing with conditional probabilities, our interest is in computing probabilities that can be used to answer questions such as those raised in Example 4.15. Suppose that we have an event A with probability denoted by $P(A)$. If we should learn that a related event, B, has occurred, we would want to take advantage of this additional information in computing the probability for event A.

The probability of event A *given* that another event B is known to have occurred is written $P(A|B)$. The vertical line, $|$, between A and B is used to denote the fact that we are considering the probability of event A *given* the condition that event B has occurred. The notation $P(A|B)$ is read "the probability of A given B." Shown below are the mathematical definitions for the conditional probability of A given B and the conditional probability of B given A. These formulas are used to compute conditional probabilities.

Conditional Probability

$$P(A|B) = \frac{P(A \cap B)}{P(B)} \qquad (4.7)$$

or

$$P(B|A) = \frac{P(A \cap B)}{P(A)} \qquad (4.8)$$

EXAMPLE 4.15 (continued)

Returning to the police department discrimination case, we let

M = event that a randomly selected officer is a man

W = event that a randomly selected officer is a woman

A = event that a randomly selected officer is promoted

Dividing the data values in Table 4.1 by the total of 1200 officers permits us to summarize the available information in the following probability values:

$P(M \cap A) = \dfrac{288}{1200} = .24 =$ probability that an officer is a man *and* is promoted

$P(M \cap A^c) = \dfrac{672}{1200} = .56 =$ probability that an officer is a man *and* is not promoted

$P(W \cap A) = \dfrac{36}{1200} = .03 =$ probability that an officer is a woman *and* is promoted

$P(W \cap A^c) = \dfrac{204}{1200} = .17 =$ probability that an officer is a woman *and* is not promoted

Since each of these values gives the probability of the intersection of two events, the probabilities are given the name of *joint probabilities*. Table 4.2, which provides a summary of the probability information for the police officer promotion situation, is referred to as a *joint probability table*.

The values in the margins of the joint probability table provide the probabilities of each event separately. That is, $P(M) = .80$, $P(W) = .20$, $P(A) = .27$, and $P(A^c) = .73$. Thus we see that 80% of the force is male, 20% of the force is female, 27% of all officers received promotions, and 73% were not promoted. These probabilities are referred to as *marginal probabilities* because of their location in the margins of the joint probability table.

The conditional probabilities of relevance in the discrimination charge are $P(A|M)$, the probability of an officer being promoted given the officer is a man, and $P(A|W)$, the probability that an officer is promoted given the officer is a woman. Using the joint and marginal probabilities, we can apply the definition of conditional probability to find the probability of promotion given a male officer and the probability of promotion given a female officer.

$$P(A|M) = \frac{P(A \cap M)}{P(M)} = \frac{.24}{.80} = .30$$

$$P(A|W) = \frac{P(A \cap W)}{P(W)} = \frac{.03}{.20} = .15$$

We see that the probability of promotion is twice as great for male officers. These conditional probabilities do not necessarily prove that the female officers have been discriminated against, but they do support the female officers' argument.

■

EXAMPLE 4.16

A research study concerning the relationship between smoking and heart disease in men over 50 years old led to the finding that 10% of the men smoked and had experienced heart disease. Furthermore, it was known that 30% of the men in the study were smokers. Let

$$H = \text{has experienced heart disease}$$

$$S = \text{smoker}$$

Using Equation (4.7) we can compute the conditional probability of a man over 50 experiencing heart disease given that he smokes.

$$P(H|S) = \frac{P(H \cap S)}{P(S)} = \frac{.10}{.30} = .33$$

■

TABLE 4.2

Joint Probability Table for Promotion of Police Officers

	MEN (M)	WOMEN (W)	Totals
Promoted (A)	.24	.03	.27
Not Promoted (A^c)	.56	.17	.73
Totals	.80	.20	1.00

Independent Events

In Example 4.15 involving promotional practices for male and female police officers, we saw that $P(A) = .27$, $P(A|M) = .30$, and $P(A|W) = .15$. As you will recall, the events were defined as follows: A = promotion, M = male officer, and W = female officer. These data show that the probability of a promotion (event A) is affected, or influenced by, whether the officer is male or female. In particular, since $P(A|M) \neq P(A)$, we say events A and M are *dependent* events. That is, the probability of event A (promotion) is influenced by knowing whether or not M (the officer is male) occurs. Similarly, with $P(A|W) \neq P(A)$, we say events A and W are *dependent* events. On the other hand, if the probability of event A is not affected by the occurrence of event M, that is, $P(A|M) = P(A)$, we say events A and M are *independent* events. This leads us to the following definition of the independence of two events.

Independent Events

Two events A and B are independent if

$$P(A|B) = P(A) \qquad (4.9)$$

or

$$P(B|A) = P(B) \qquad (4.10)$$

Otherwise, the events are dependent.

EXAMPLE 4.17

Suppose that $P(A) = .30$, $P(B) = .25$, and $P(A|B) = .20$. Are events A and B independent? Since $P(A|B) \neq P(A)$, the events are dependent.

■

Multiplication Rule

Recall that the addition rule is used to compute the probability of the union of two events. We now show how the *multiplication rule* can be used to find the probability of the intersection of two events (i.e., the joint probability of the two events). The multiplication rule follows from the definition of conditional probability. Using (4.7) and (4.8) and solving for $P(A \cap B)$, we obtain the multiplication rule:

Multiplication Rule

$$P(A \cap B) = P(B)P(A|B) \qquad (4.11)$$

or

$$P(A \cap B) = P(A)P(B|A) \qquad (4.12)$$

EXAMPLE 4.18

A newspaper circulation department knows that 84% of its customers subscribe to the daily edition of the paper. Letting D denote the event that a customer subscribes to the daily edition, we set $P(D) = .84$. In addition, it is known that the probability that a customer already holding a daily subscription also subscribes to the Sunday edition (event S) is .75; that is, $P(S|D) = .75$. What is the probability that a customer subscribes to both the daily and Sunday editions of the newspaper? Using the multiplication rule, we compute the desired probability, $P(D \cap S)$, as follows:

$$P(D \cap S) = P(D)P(S|D) = (.84)(.75) = .63$$

This tells us that 63% of the newspaper's customers take both the daily and Sunday editions.

■

Before concluding this section, let us consider the special case of the multiplication rule for independent events. Recall that events A and B are independent whenever $P(A|B) = P(A)$ or $P(B|A) = P(B)$. Applying equations (4.11) and (4.12) for the special case of independent events, we obtain (4.13).

Multiplication Rule for Independent Events

$$P(A \cap B) = P(A)P(B) \qquad (4.13)$$

Thus to compute the probability of the intersection of two independent events, we simply multiply the probabilities of the two events. The multiplication rule for independent events provides another means of determining if two events are independent. For instance, if $P(A \cap B) = P(A)P(B)$, then A and B are independent; if $P(A \cap B) \neq P(A)P(B)$, then A and B are dependent.

EXAMPLE 4.19

A service station manager knows from past experience that 80% of the customers use a credit card when purchasing gasoline. What is the probability that the next two customers purchasing gasoline will both use credit cards? If we let

A = event that the first customer uses a credit card

B = event that the second customer uses a credit card

then the event of interest is $A \cap B$. It seems reasonable to assume that A and B are independent events. Thus, $P(A \cap B) = P(A)P(B) = (.80)(.80) = .64$.

■

Notes and Comments

Do not confuse the concepts of mutually exclusive events and independent events. For example, consider two mutually exclusive events, A and B, both of which have a nonzero probability. Are these two events independent? Since A and B are mutually exclusive events, we know that if one event occurs, the other event cannot occur. This means that the probability of A depends upon the occurrence or nonoccurrence of B. Thus, A must be dependent upon B. In general, two mutually exclusive events are dependent.

EXERCISES

32. Let $P(A) = .60$, $P(B) = .45$, and $P(A \cap B) = .30$.
  a. Find $P(A \cup B)$.
  b. Find $P(A \mid B)$.
  c. Find $P(B \mid A)$.
  d. Are events A and B independent?
  e. Are events A and B mutually exclusive?

33. Let $P(A) = .70$, $P(B) = .20$, $P(C) = .40$, $P(A \cap B) = .10$, and $P(B \cap C) = .08$.
  a. Find $P(A \mid B)$.
  b. Are A and B mutually exclusive events?
  c. Are A and B independent events?
  d. Find $P(B \mid C)$.
  e. Are B and C independent events?

34. Assume that we have two events, A and B, which are mutually exclusive and that it is known that $P(A) = .30$ and $P(B) = .40$.
  a. What is $P(A \cap B)$?
  b. What is $P(A \mid B)$?
  c. A student in statistics argues that the concepts of mutually exclusive events and independent events are really the same and that if events are mutually exclusive they must be independent. Do you agree with this statement? Use the probability information in this problem to justify your answer.
  d. What general conclusion would you make about mutually exclusive and independent events given the results of this problem?

35. A Daytona Beach nightclub has collected the following data on the age and marital status of 140 customers.

		MARITAL STATUS	
		Single	Married
Age	Under 30	77	14
	30 or Over	28	21

  a. Develop a joint probability table using the given data.
  b. Use the marginal probabilities to comment on the ages of customers attending the club.
  c. Use the marginal probabilities to comment on the marital status of customers attending the club.
  d. What is the probability of finding a customer who is single and under the age of 30?

e. If a customer is under 30, what is the probability that he or she is single?

f. Is marital status independent of age? Explain, using probabilities.

36. A survey of automobile ownership was conducted for 200 families in Houston. The results of the study showing ownership of automobiles of United States and foreign manufacture are summarized in the following table.

| | | DO YOU OWN A U.S. CAR? | | |
		Yes	No	Totals
Do you own	Yes	30	10	40
a foreign car?	No	150	10	160
	Totals	180	20	200

a. Show the joint probability table for the given data.

b. Use the marginal probabilities to compare U.S. and foreign car ownership.

c. What is the probability that a family will own both a U.S. car and a foreign car?

d. What is the probability that a family owns a car, U.S. or foreign?

e. If a family owns a U.S. car, what is the probability that it also owns a foreign car?

f. If a family owns a foreign car, what is the probability that it also owns a U.S. car?

g. Are U.S. and foreign car ownership independent events? Explain.

37. Shown are data from a sample of 80 families in a midwestern city. The data shows the record of college attendance by fathers and their oldest sons.

| | | SON | |
		Attended College	Did Not Attend College
Father	Attended College	18	7
	Did Not Attend College	22	33

a. Show the joint probability table.

b. Use the marginal probabilities to comment on the comparison between fathers and sons in terms of attending college.

c. What is the probability that a son attends college given that his father attended college?

d. What is the probability that a son attends college given that his father did not attend college?

e. Is attending college by the son independent of whether or not his father attended college? Explain, using probability values.

38. The Texas Oil Company provides a limited partnership arrangement whereby small investors can pool resources in order to invest in large scale oil exploration programs. In the exploratory drilling phase, locations for new wells are selected based on the geologic structure of the proposed drilling sites. Experience shows that there is a .40 probability of a type A structure present at the site given a productive well. It is also known that 50% of all wells are drilled in locations with type A structure. Finally, 30% of all wells drilled are productive.

a. What is the probability of a well being drilled in a type A structure *and* being productive?

b. If the drilling process begins in a location with a type A structure, what is the probability of having a productive well at the location?

c. Is finding a productive well independent of the type A geologic structure? Explain.

39. In a study involving a manufacturing process, 10% of all parts tested were defective, and 30% of all parts were produced on machine A. Given that a part was produced on machine A, there is a .15 probability that it is defective.

a. What is the probability that a part tested is both defective and produced by machine A?

b. If a part is found to be defective, what is the probability that it came from machine A?

c. Is finding a defective part independent of its being produced on machine A? Explain.

d. What is the probability of the part being either defective or produced by machine A?

e. Are the events "a defective part" and "produced by machine A" mutually exclusive events? Explain.

40. A hospital has placed two rush orders for a particular drug from two different suppliers, A and B. If neither order arrives in 4 days, a research project must be stopped until at least one of the orders arrives. The probability that supplier A can deliver the material in 4 days is .55. The probability that supplier B can deliver the material in 4 days is .35.

a. What is the probability that both suppliers deliver the material in 4 days? Since two separate suppliers are involved, we are willing to assume independence.

b. What is the probability that at least one supplier delivers the material in 4 days?

c. What is the probability the research project is shut down in 4 days because of a shortage in raw material (i.e., both orders are late)?

4.5 Bayes' Theorem

The discussion of conditional probability suggests that it is possible to revise or update probabilities given new information. Often, we begin a probability analysis with *prior probability* estimates for specific events of interest. Then, from sources such as a sample, a special report, a product test, and so on, we obtain additional information affecting the probabilty of the events. Given this new information, we want to revise or update the prior probability values. The updated, or revised, probabilities for the events are referred to as *posterior* probabilities. *Bayes' theorem* provides a means for computing posterior probabilities. The steps of the Bayesian probability revision process are shown in Figure 4.7.

FIGURE 4.7 Probability Revision Using Bayes' Theorem

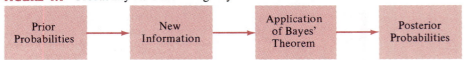

EXAMPLE 4.20

A manufacturing firm receives 65% of its parts from one supplier and 35% from a second supplier. The quality of the purchased parts varies with the supplier. Table 4.3 shows the percentages of good and bad parts received from the two suppliers. Let A_1 denote the

TABLE 4.3

Percentages of Good and Bad Parts Received from Suppliers

	PERCENTAGE GOOD PARTS	PERCENTAGE DEFECTIVE PARTS
Supplier 1	98%	2%
Supplier 2	95%	5%

event a part comes from supplier 1 and A_2 the event that a part comes from supplier 2. If we let G denote the event that a part is good and D denote the event that a part is defective, the information in Table 4.3 leads to the following conditional probability values:

$$P(G|A_1) = .98 \qquad P(D|A_1) = .02$$

$$P(G|A_2) = .95 \qquad P(D|A_2) = .05$$

Furthermore, given the percentage of parts received from each supplier, if a part is selected at random, the probability it came from supplier 1 is $P(A_1) = .65$ and the probability it came from supplier 2 is $P(A_2) = .35$. We note here that events A_1 and A_2 are mutually exclusive and collectively exhaustive events. That is, (1) if one event occurs the other event cannot occur, and (2) the two events represent the only possible outcomes in the situation. The sum of the probabilities of the mutually exclusive and collectively exhaustive events must equal 1.

Suppose that the manufacturing firm has just had a machine breakdown due to a defective part and wants to determine the probability that the part came from supplier 1 and the probability that the part came from supplier 2. Thus it is desired to compute $P(A_1|D)$ and $P(A_2|D)$. Since $P(A_1 \cap D)$ and $P(A_2 \cap D)$ are not known, we cannot use the conditional probability rule for these calculations. This is a problem in which Bayes' theorem is needed; we return to make the calculations shortly.

■

Suppose we have two mutually exclusive and collectively exhaustive events, A_1 and A_2, and know the prior probabilities, $P(A_1)$ and $P(A_2)$. If it becomes known that a related event B has occurred, the following formulas, known as Bayes' theorem, can be used to compute the posterior probabilities $P(A_1|B)$ and $P(A_2|B)$.

Bayes' Theorem

$$P(A_1|B) = \frac{P(B|A_1)P(A_1)}{P(B|A_1)P(A_1) + P(B|A_2)P(A_2)} \qquad (4.14)$$

and

$$P(A_2|B) = \frac{P(B|A_2)P(A_2)}{P(B|A_1)P(A_1) + P(B|A_2)P(A_2)} \qquad (4.15)$$

EXAMPLE 4.20
(continued)

Recall that we are interested in finding the posterior probabilities that the defective part came from supplier 1, $P(A_1|B)$, and from supplier 2, $P(A_2|B)$. Using the prior probabilities $P(A_1) = .65$ and $P(A_2) = .35$ and the conditional probabilities $P(D|A_1) = .02$ and $P(D|A_2) = .05$, we can utilize Bayes' theorem to compute the probabilities in question. Using (4.14) and using event D as the related event that has occurred we find

$$P(A_1|D) = \frac{P(D|A_1)P(A_1)}{P(D|A_1)P(A_1) + P(D|A_2)P(A_2)}$$

$$= \frac{(.02)(.65)}{(.02)(.65) + (.05)(.35)} = \frac{.0130}{.0130 + .0175}$$

$$= \frac{.0130}{.0305} = .426$$

Using (4.15) we find $P(A_2|D)$ as follows:

$$P(A_2|D) = \frac{(.05)(.35)}{(.02)(.65) + (.05)(.35)}$$

$$= \frac{.0175}{.0130 + .0175} = \frac{.0175}{.0305} = .574$$

We initially had a probability of .65 that a part selected at random was from supplier 1. However, given the information that the part is defective, the probability that the part is from supplier 1 drops to .426. Thus, if the part is defective, there is a better than 50-50 chance that the part came from supplier 2; that is, $P(A_2|D) = .574$.

■

The Tabular Approach

A tabular approach helpful in organizing and conducting the Bayes' theorem calculations is shown in Table 4.4 for the data presented in Example 4.20. The computations shown in that table are conducted as follows:

STEP 1 Prepare the following three columns.

Column 1: The list of the mutually exclusive and collectively exhaustive events that can occur in the problem.

Column 2: The prior probabilities for the events. Note that since the events are mutually exclusive and collectively exhaustive, the probabilities in column 2 must sum to 1.

Column 3: The conditional probabilities of the new information (event D in example 4.20) *given* each of the mutually exclusive events in column 1.

TABLE 4.4

Tabular Approach to Bayes' Theorem Calculations for the Two-Supplier Problem

COLUMN 1 A_i	COLUMN 2 $P(A_i)$	COLUMN 3 $P(D|A_i)$	COLUMN 4 $P(A_i \cap D)$	COLUMN 5 $P(A_i|D)$
A_1	.65	.02	.0130	$\frac{.0130}{.0305} = .426$
A_2	.35	.05	.0175	$\frac{.0175}{.0305} = .574$
	1.00		$P(D) = .0305$	1.000

STEP 2 In column 4 compute the joint probabilities for each mutually exclusive event and the event providing new information. These joint probabilities are found by multiplying the values in column 2 by the corresponding values in column 3. For Example 4.20, we obtain $P(A_1 \cap D) = P(A_1)P(D|A_1)$ and $P(A_2 \cap D) = P(A_2)P(D|A_2)$.

STEP 3 Add the joint probability column (column 4) to find the probability of the event representing new information. We see that in Example 4.20, there is a .0130 probability of a defective part and supplier 1 and there is a .0175 probability of a defective part and supplier 2. Since these are the only two ways a defective part can be obtained, the sum .0130 + .0175 = .0305 shows there is an overall probability of .0305 of finding a defective part from the combined shipments of both suppliers.

STEP 4 In column 5, compute the posterior probabilities using the basic relationship of conditional probability. For Example 4.20, this is

$$P(A_1|D) = \frac{P(A_1 \cap D)}{P(D)}$$

The joint probabilities $P(A_1 \cap D)$ are found in column 4, whereas the probability $P(D)$ appears as the sum of column 4.

As a final note, we can generalize Bayes' theorem to the case where there are n mutually exclusive and collectively exhaustive events $A_1, A_2, \ldots, A_n$. In such a case, Bayes' theorem for the computation of the posterior probabilities given event B is known to have occurred, $P(A_1|B)$, appears as follows:

$$P(A_i|B) = \frac{P(B|A_i)\,P(A_i)}{P(B|A_1)\,P(A_1) + P(B|A_2)\,P(A_2) + \cdots + P(B|A_n)\,P(A_n)} \quad (4.16)$$

With the prior probabilities $P(A_1), P(A_2), \ldots, P(A_n)$ and the appropriate conditional probabilities $P(B|A_1), P(B|A_2), \ldots, P(B|A_n)$, (4.16) can be used to compute the posterior probability of the events $A_1, A_2, \ldots, A_n$.

EXERCISES

41. Suppose $P(A_1 \cap A_2) = 0$. Suppose further that $P(A_1) = .6$, $P(A_2) = .4$, $P(B|A_1) = .7$, $P(B|A_2) = .20$.
   a. Are A_1 and A_2 mutually exclusive events?
   b. Find $P(A_1 \cup A_2)$.
   c. Find $P(A_1|B)$.
   d. Find $P(A_2|B)$.
   e. Find $P(A_1 \cap B)$.
   f. Find $P(A_2 \cap B)$.
   g. Find $P(B)$.

42. The prior probabilities for the mutually exclusive events A_1, A_2, and A_3 are $P(A_1) = .20$, $P(A_2) = .50$, and $P(A_3) = .30$. The conditional probabilities of event B given A_1, A_2, and A_3 are $P(B|A_1) = .50$, $P(B|A_2) = .40$, and $P(B|A_3) = .30$.

 a. Compute $P(B \cap A_1)$, $P(B \cap A_2)$, and $P(B \cap A_3)$.

 b. Apply Bayes' theorem (4.16), to compute the posterior probability $P(A_2|B)$.

 c. Use the tabular approach to applying Bayes' theorem to compute $P(A_1|B)$, $P(A_2|B)$, and $P(A_3|B)$.

43. A consulting firm has submitted a bid for a large research project. The firm's management initially felt there was a 50-50 chance of getting the bid. However, the agency to which the bid was submitted has subsequently requested additional information on the bid. Past experience indicates that on 75% of the successful bids and 40% of the unsuccessful bids the agency requested additional information.

 a. What is your prior probability the bid will be successful (i.e., prior to receiving the request for additional information)?

 b. What is the conditional probability of a request for additional information given that the bid will ultimately be successful?

 c. Compute a posterior probability that the bid will be successful given that a request for additional information has been received.

44. A local bank is reviewing its credit card policy with a view toward recalling some of its credit cards. In the past approximately 5% of cardholders have defaulted, and the bank has been unable to collect the outstanding balance. Thus management has established a prior probability of .05 that any particular cardholder will default. The bank has further found that the probability of missing one or more monthly payments for those customers who do not default is .20. Of course the probability of missing one or more payments for those who default is 1.

 a. Given that a customer has missed a monthly payment, compute the posterior probability that the customer will default.

 b. The bank would like to recall its card if the probability that a customer will default is greater than .20. Should the bank recall its card if the customer misses a monthly payment? Why or why not?

45. In a major eastern city, 60% of the automobile drivers are 30 years of age or older, and 40% of the drivers are under 30 years of age. Of all drivers 30 years of age or older, 4% will have a traffic violation in a 12-month period. Of all drivers under 30 years of age, 10% will have a traffic violation in a 12-month period. Assume that a driver has just been charged with a traffic violation; what is the probability that the driver is under 30 years of age?

SUMMARY

In this chapter we introduced probability. We described how probability can be interpreted as a numerical measure of the likelihood that an event will occur. In addition, we showed that the probability of an event could be computed directly by summing the probabilities of the experimental outcomes comprising the event or from related events utilizing the rules of probability indirectly. For cases where additional information is available, we demonstrated how Bayes' theorem could be used to obtain revised, or posterior, probabilities.

GLOSSARY

Probability A numerical measure of the likelihood that an event will occur.

Experiment Any process that generates well-defined outcomes.

Sample space The set of all possible sample points (experimental outcomes).

Tree diagram   A graphical device helpful in determining the experimental outcomes for an experiment involving multiple steps.

Basic requirements of probability   Two requirements that restrict the manner in which probability assignments can be made:
1. For each experimental outcome E_i, we must have $0 \leq P(E_i) \leq 1$.
2. If n is the number of experimental outcomes, then $P(E_1) + P(E_2) + \cdots + P(E_n) = 1$.

Classical method   A method of assigning probabilities that assumes the experimental outcomes are equally likely.

Relative frequency method   A method of assigning probabilities based upon experimentation or historical data.

Subjective method   A method of assigning probabilities based upon judgment.

Event   A set consisting of a collection of experimental outcomes.

Complement of event A   The event containing all experimental outcomes that are not in A.

Venn diagram   A graphical device for symbolically representing the sample space and operations involving events.

Union of events A and B   The event containing all experimental outcomes that are in A, in B, or in both.

Intersection of A and B   The event containing all experimental outcomes that are in both A and B.

Mutually exclusive events   Events that have no experimental outcome in common; that is, $A \cap B$ is empty and $P(A \cap B) = 0$.

Addition rule   A probability law used to compute the probability of a union, $P(A \cup B)$. *It is* $P(A \cup B) = P(A) + P(B) - P(A \cap B)$ in general. For mutually exclusive events, since $P(A \cap B) = 0$, it reduces to $P(A \cup B) = P(A) + P(B)$.

Conditional probability   The probability of an event given that another event has occurred. The conditional probability of A given B is $P(A|B) = P(A \cap B)/P(B)$.

Independent events   Two events A and B where $P(A|B) = P(A)$ or $P(B|A) = P(B)$; that is, the events have no influence on each other.

Multiplication rule   A probability rule used to compute the probability of an intersection, $P(A \cap B)$. It is $P(A \cap B) = P(A)P(B|A)$, or $P(A \cap B) = P(B)P(A|B)$. For independent events it reduces to $P(A \cap B) = P(A)P(B)$.

Mutually exclusive and collectively exhaustive events   A set of events that satisfy two properties: (1) If one event occurs, none of the other events can occur; (2) the events represent the only possible outcomes in the situation. The sum of the probabilities for these events must equal 1.

Prior probabilities   Probabilities for a set of mutually exclusive and collectively exhaustive events prior to being updated by Bayes' theorem.

Posterior probabilities   The revised probabilities for events resulting from application of Bayes' theorem.

Bayes' theorem   A formula for revising prior probabilities concerning mutually exclusive and collectively exhaustive events. The revised probabilities are called posterior probabilities.

KEY FORMULAS

COMPUTING PROBABILITY USING THE COMPLEMENT

$$P(A) = 1 - P(A^c). \tag{4.4}$$

ADDITION RULE

$$P(A \cup B) = P(A) + P(B) - P(A \cap B) \tag{4.5}$$

ADDITION RULE FOR MUTUALLY EXCLUSIVE EVENTS

$$P(A \cup B) = P(A) + P(B) \tag{4.6}$$

DEFINITION OF CONDITIONAL PROBABILITY

$$P(A|B) = \frac{P(A \cap B)}{P(B)} \tag{4.7}$$

or

$$P(B|A) = \frac{P(A \cap B)}{P(A)} \tag{4.8}$$

MULTIPLICATION RULE

$$P(A \cap B) = P(B)P(A|B) \tag{4.11}$$

or

$$P(A \cap B) = P(A)P(B|A) \tag{4.12}$$

MULTIPLICATION RULE FOR INDEPENDENT EVENTS

$$P(A \cap B) = P(A)P(B) \tag{4.13}$$

BAYES' THEOREM

$$P(A_i|B) = \frac{P(B|A_i)P(A_i)}{P(B|A_1)P(A_1) + P(B|A_2)P(A_2) + \cdots + P(B|A_n)P(A_n)} \tag{4.16}$$

REVIEW QUIZ

TRUE/FALSE

1. Probabilities can never be greater than 1 or less than 0.

2. The sum of probabilities for all experimental outcomes equals 1.

3. The sum of the probabilities for all experimental outcomes may be any number between 0 and 1, inclusive.

4. The subjective method of assigning probabilities is one of the two methods permitting probabilities less than 0.

5. The sum of the probabilities of the experimental outcomes in an event must equal 1.

6. If we know the probability of an event, then the probability of the complement of the event can also be computed.

7. To use the addition rule to compute the probability of the union of two events, we must know the probability of the intersection of the events.

8. If two events are independent, they must be mutually exclusive.

9. If two events are independent, we need only know each event's probability in order to compute the probability of the intersection of the events.

10. Posterior probabilities must be known before Bayes' theorem can be applied.

MULTIPLE CHOICE

The following event probabilities for a statistical experiment are utilized in questions 11-13.

$$P(A) = .60 \qquad P(B) = .40$$
$$P(A \cap B) = .25$$

11. $P(A \cup B)$ is closest to
    a. .65
    b. .72
    c. .79
    d. .82

12. $P(A|B)$ is closest to
    a. .60
    b. .67
    c. .74
    d. .81

13. $P(A^c)$ is closest to
    a. .50
    b. .60
    c. .70
    d. .80

In questions 14-16, assume events A and B are independent, $P(A|B) = .70$, and $P(A \cap B) = .21$.

14. $P(A)$ is closest to
    a. .30
    b. .50
    c. .70
    d. .90

15. $P(B)$ is closest to
    a. .20
    b. .30
    c. .50
    d. .70

16. $P(A \cup B)$ is closest to
    a. .50
    b. .60
    c. .70
    d. .80

17. If A and B are independent events with $P(A) = .3$ and $P(B) = .5$, then $P(A|B) =$
a. 0
b. .15
c. .20
d. .30

18. If J and K are mutually exclusive events with $P(J) = .4$ and $P(K) = .5$, then $P(J \cap K) =$
a. 0
b. .2
c. .7
d. .9

19. If J and K are mutually exclusive events with $P(J) = .4$ and $P(K) = .5$, then $P(J \cup K) =$
a. 0
b. .2
c. .7
d. .9

<hr>

SUPPLEMENTARY EXERCISES

46. Consider an experiment where eight experimental outcomes exist. We denote the experimental outcomes as $E_1, E_2, \ldots, E_8$. Suppose that the following events are identified.

$$A = \{E_1, E_2, E_3\}$$

$$B = \{E_2, E_4\}$$

$$C = \{E_1, E_7, E_8\}$$

$$D = \{E_5, E_6, E_7, E_8\}$$

Find the following:
a. $A \cup B$
b. $C \cup D$
c. $A \cap B$
d. $C \cap D$
e. $B \cap C$
f. A^c
g. D^c
h. $A \cup D^c$
i. $A \cap D^c$
j. Are A and B mutually exclusive?
k. Are B and C mutually exclusive?

47. Referring to Exercise 46 and assuming that the classical method is an appropriate way of establishing probabilities, find the following probabilities.
a. $P(A), P(B), P(C),$ and $P(D)$
b. $P(A \cap B)$
c. $P(A \cup B)$
d. $P(A|B)$

e. $P(B|A)$
f. $P(B \cap C)$
g. $P(B|C)$
h.   Are B and C independent events?

48.   Suppose that $P(A) = .30$, $P(B) = .25$, and $P(A \cap B) = .20$.
   a.   Find $P(A \cup B), P(A|B)$, and $P(B|A)$.
   b.   Are events A and B independent? Why or why not?

49.   Suppose that $P(A) = .40$, $P(A|B) = .60$, and $P(B|A) = .30$.
   a.   Find $P(A \cap B)$ and $P(B)$.
   b.   Are events A and B independent? Why or why not?

50.   Suppose that $P(A) = .60$, $P(B) = .30$, and events A and B are mutually exclusive.
   a.   Find $P(A \cup B)$ and $P(A \cap B)$.
   b.   Are events A and B independent?
   c.   Can you make a general statement about whether or not mutually exclusive events can be independent?

51.   In September 1988 the House of Representatives voted on an amendment requiring life imprisonment for drug-related murders. Results of the vote were reported as shown below.

	YEA	NAY	DID NOT VOTE	Totals
Democrat	153	83	19	255
Republican	169	0	8	177
Totals	322	83	27	432

   a.   What is the probability that a randomly selected representative voted for the amendment?
   b.   What is the probability that a randomly selected representative is both a Democrat and voted against the amendment?
   c.   What is the probability that a representative known to have voted for the amendment is a Democrat?
   d.   If someone is known not to have voted, is it more likely this person is a Democrat or Republican?
   e.   What is the joint probability of a randomly selected representative being a Republican and voting for the amendment?
   f.   Are the events "voted for the amendment" and "is a Democrat" independent?

52.   A survey of 800 people found the following facts about the ability to recall a television commercial for a particular product and actual purchase of the product.

	COULD RECALL TELEVISION COMMERCIAL	COULD NOT RECALL TELEVISION COMMERCIAL	Totals
Purchased the Product	160	80	240
Had not Purchased the Product	240	320	560
Totals	400	400	800

Let T be the event of the person recalling the television commercial and B the event of buying or purchasing the product.

a. Find $P(T)$, $P(B)$, and $P(T \cap B)$.

b. Are T and B mutually exclusive events? Use probability values to explain.

c. What is the probability that a person who could recall seeing the television commercial has actually purchased the product?

d. Are T and B independent events? Use probability values to explain.

e. Comment on the value of the commercial in terms of its relationship to purchasing the product.

53. Carstab Corporation,* a subsidiary of Morton International, makes an expensive catalyst used in chemical processing. One customer has very stringent requirements for the catalyst and, as a result, has typically returned 40% of Carstab's shipments. In order to cut down on returned shipments, Carstab has developed a test that, it is hoped, will help identify whether or not a lot will be acceptable to the customer.

A sample of production lots was tested using Carstab's new procedure and by the customer. Of the lots tested, 55% passed Carstab's test and 50% passed both the customer's and Carstab's tests.

a. Find the probability a lot will be acceptable to the customer given that it has passed Carstab's new test.

b. Would you recommend that Carstab implement the new testing procedure? Why or why not?

54. A large consumer goods company has been running a television advertisement for one of its soap products. A survey was conducted. On the basis of this survey, probabilities were assigned to the following events.

$$B = \text{individual purchased the product}$$
$$S = \text{individual recalls seeing the advertisement}$$
$$B \cap S = \text{individual purchased the product and recalls seeing the advertisement}$$

The probabilities assigned were $P(B) = .20$, $P(S) = .40$, and $P(B \cap S) = .12$. The following problems relate to this situation.

a. What is the probability of an individual's purchasing the product given that the individual saw the advertisement? Does seeing the advertisement increase the probability the individual will purchase the product? As a decision maker, would you recommend continuing the advertisement (assuming that the cost is reasonable)?

b. Assume that those individuals who do not purchase the company's soap product buy from its competitors. What would be your estimate of the company's market share? Would you expect that continuing the advertisement will increase the company's market share? Why or why not?

c. The company has also tested another advertisement and assigned it values of $P(S) = .30$ and $P(B \cap S) = .10$. What is $P(B \mid S)$ for this other advertisement? Which advertisement seems to have had the bigger effect on customer purchases?

55. Western Airlines has done an analysis of a price promotion it is offering to frequent air travelers in order to increase the number of people on their New York to San Francisco route. Some 20% of the people in a large sample of individuals identified as frequent travelers from New York to San Francisco were aware of the Western promotion and elected to fly with Western on their next trip. It was further found that 80% were aware of the promotion and that prior to the promotion, 25% of these travelers flew with Western.

a. What is the probability that a person will fly with Western on their next trip given that he or she is aware of the price promotion?

b. For a randomly selected traveler, are the events "fly with Western" and "aware of the price promotion" independent? Why or Why not?

*Information provided by Michael Haskell of Morton International's Carstab subsidiary.

c. On the basis of these results, does the promotion appear to be successful in terms of increasing business? Why or why not?

56. Cooper Realty is a small real estate company specializing primarily in residential listings. They have recently become interested in the possibility of determining the likelihood of one of their listings being sold within a certain number of days. An analysis of company sales of 800 homes for the previous years produced the following data.

		DAYS LISTED UNTIL SOLD			
		Under 30	31–90	Over 90	Totals
	Under $50,000	50	40	10	100
Initial	$50,000–75,000	20	150	80	250
Asking Price	$75,000–100,000	20	280	100	400
	Over $100,000	10	30	10	50
	Totals	100	500	200	800

↑
Total Homes Sold

a. If A is defined as the event that a home is listed for over 90 days before being sold, estimate the probability of A.

b. If B is defined as the event that the initial asking price is under $25,000, estimate the probability of B.

c. What is the probability of $A \cap B$?

d. Assuming that a contract has just been signed to list a home that has an initial asking price of less than $25,000, what is the probability the home will take Cooper Realty more than 90 days to sell?

e. Are events A and B independent?

57. In the evalutation of a sales training program, a firm found that of 50 salespersons making a bonus last year, 20 had attended a special sales training program. The firm has 200 salespersons. Let B = the event that a salesperson makes a bonus and S = the event a salesperson attends the sales training program.

a. Find $P(B)$, $P(S|B)$, and $P(S \cap B)$.

b. Assume that 40% of the salespersons have attended the training program. What is the probability that a salesperson makes a bonus given that the salesperson attended the sales training program, $P(B|S)$?

c. If the firm evaluates the training program in terms of the effect it has on the probability of a salesperson's making a bonus, what is your evaluation of the training program? Comment on whether B and S are dependent or independent events.

58. In a study of television-viewing habits among married couples, a researcher found that for a popular Saturday night program 25% of the husbands viewed the program regularly and 30% of the wives viewed the program regularly. The study found that for couples where the husband watches the program regularly 80% of the wives also watch regularly.

a. What is the probability that both the husband and wife watch the program regularly?

b. What is the probability that at least one—husband or wife—watches the program regularly?

c. What percentage of married couples do not have at least one regular viewer of the program?

59. A statistics professor has noted from past experience that students who do the homework for the course have a .90 probability of passing the course. On the other hand, students who do not do the homework for the course have a .25 probability of passing the course. The professor estimates that 75% of the students in the course do the homework. Given a student who passes the course, what is the probability that she or he completed the homework?

60. A salesperson for Business Communication Systems, Inc. sells automatic envelope-addressing equipment to medium- and small-size businesses. The probability of making a sale to a new customer is .10. During the initial contact with a customer, sometimes the salesperson will be asked to call back later. Of the 30 most recent sales, 12 were made to customers who initially told the salesperson to call back later. Of 100 customers who did not make a purchase, 17 had initially asked the salesperson to call back later. If a customer asks the salesperson to call back later, should the salesperson do so? What is the probability of making a sale to a customer who has asked the salesperson to call back later?

61. Migliori Industries, Inc., manufactures a gas-saving device for use on natural gas forced-air residential furnaces. The company is currently trying to determine the probability that sales of this product will exceed 25,000 units during next year's winter sales period. The company believes that sales of the product depend to a large extent on the winter conditions. Management's best estimate is that the probability that sales will exceed 25,000 units if the winter is severe is .8. This probability drops to .5 if the winter conditions are moderate. If the weather forecast is .7 for a severe winter and .3 for moderate conditions, what is Migliori's best estimate that sales will exceed 25,000 units?

62. The Dallas IRS auditing staff is concerned with identifying potential fraudulent tax returns. From past experience they believe that the probability of finding a fraudulent return given that the return contains deductions for contributions exceeding the IRS standard is .20. Given that the deductions for contributions do not exceed the IRS standard, the probability of a fraudulent return decreases to .02. If 8% of all returns exceed the IRS standard for deductions due to contributions, what is the best estimate of the percentage of fraudulent returns?

63. In the January 11, 1988, issue of the *Oil & Gas Journal,* R.A. Baker describes how the Bayesian approach can be used to revise probabilities that a prospect field will produce oil. In one case he describes, geological assessment indicates a 25% chance the field will produce oil. Further, there is an 80% chance that a particular well will strike oil given that oil is present on the prospect field.
   a. Suppose that one well is drilled on the field and it comes up dry. What is the probability the prospect field will produce oil?
   b. If two wells come up dry, what is the probability the field will produce oil?
   c. The oil company would like to keep looking as long as the chances of finding oil are greater than 1%. How many dry wells must be drilled before the field will be abandoned?
   d. If the first well produces oil, what is the chance the field will produce oil?

5

Random Variables and Discrete Probability Distributions

What You Will Learn in This Chapter:

- What random variables are and how they are used
- What is meant by the probability distribution of a random variable
- How to compute and interpret the expected value and variance of a random variable
- The use and interpretation of the binomial probability distribution
- How to use the binomial probability function and binomial tables to obtain probabilities
- The use and interpretation of the Poisson probability distribution
- How to use the Poisson probability function and Poisson probability tables to compute probabilities
- The use and interpretation of the hypergeometric probability distribution

Contents

Pete Rose Versus Joe DiMaggio: Baseball Greats

Many baseball enthusiasts will aggree that the most difficult record for modern baseball players to challenge is Joe DiMaggio's 56-game hitting streak in 1941. In baseball circles, the prevailing view is that Joe's performance set an unbreakable major league standard. However, for 6 weeks of the 1978 season (June 14 to August 1) Pete Rose of the Cincinnati Reds was unstoppable as he knocked down consecutive-game hitting records. In game 38 of the hitting streak, Rose passed Tommy Holmes' 1945 modern National League record. In game 44 he became the National League's all-time consecutive-game hitting record-holder when he matched Willie Keeler's 1897 streak. Joe DiMaggio's 56-game record was next.

However, on August 1, Pete Rose's hitting streak of 44 consecutive games came to an end when he went 0 for 4 in a game with the Atlanta Braves in Atlanta. During the streak Pete hit .376 with 56 singles, 14 doubles, and 13 walks. He provided plenty of excitement when on six occasions he saved the streak in his last at-bat. On four occasions during the streak, Rose's only hit was a bunt. Reds manager Sparky Anderson exclaimed, "Watching Pete break the National League record is the biggest thrill I've had as a manager, but Joe's 56-game record is an impossibility."

© Cincinnati Reds 1985

Pete Rose, baseball's all-time hit leader, challenged DiMaggio's record with a 44 consecutive game hitting streak in 1978.

When Pete set the 44-game record, Las Vegas odds makers were still strong in their belief that he could not continue the hitting streak to DiMaggio's 56-game record. Probability specialists say that a special probability distribution, known as the binomial distribution, can be used to estimate the probability, or odds, of continuing a hitting streak for 12 more games. Using Pete's .376 batting average as the probability of a hit on each at-bat and assuming Pete would come to bat four times in a game, the probability of Pete having at least one hit in the four at-bats can be computed to be .8484. The probability of hitting in 12 successive games can be computed from this to be only .1391. The probability specialists say there was a .1391 probability of Pete reaching Joe's record and a .8609 probability that he would not. Thus the odds were still better than 6 to 1 that Pete could not match DiMaggio's performance.

While Pete Rose's 44-game hitting streak provided much interest and excitement, Joe DiMaggio's 56-game record stands as baseball's "most unbreakable" record.

National Baseball Library, Cooperstown, N.Y.

Joe DiMaggio, the "Yankee Clipper," hit safely in 56 consecutive games during the 1941 season.

Based on "Doing Much," *Sports Illustrated* (August 7, 1978).

In this chapter we introduce the concepts of random variables and probability distributions. We will see that random variables provide a means of assigning numerical values to experimental outcomes. Probability distributions provide a means of determining the probability of the different values of the random variable occurring. Some probability distributions are used so extensively in statistical analysis that special formulas and/or tables have been developed for computing probabilities associated with them. Three of these probability distributions, the binomial, the Poisson, and the hypergeometric are introduced in this chapter.

5.1   Random Variables

In Chapter 4 we studied the role of an experiment and the associated experimental outcomes in statistics. Random variables provide a means of assigning numerical values to experimental outcomes. These numerical values are used in computing means, variances, and other measures used to describe populations of interest. The definition of a random variable is as follows.

> **Random Variable**
> ___
> A *random variable* is a numerical description of the outcome of an experiment.

For any experiment a random variable can be defined such that each possible experimental outcome generates one and only one numerical value for the random variable. The particular numerical value that the random variable takes on depends upon the outcome of the experiment. That is, the value of the random variable is not known until the experimental outcome is observed.

EXAMPLE 5.1

Consider the experiment that consists of tossing a coin twice. With H indicating a head and T indicating a tail, the sample space for this experiment is

$$S = \{(H, H), (H, T), (T, H), (T, T)\}$$

Suppose we let x = number of heads occurring on the two coin tosses. Then x is a random variable (it provides a numerical description of the experimental outcome) that can assume the values 0, 1, and 2.

EXAMPLE 5.2

To receive state certification as medical lab technicians, candidates must pass a series of three examinations. If we define the random variable x as the number of examinations any one candidate passes, then x can assume the values 0, 1, 2, and 3.

EXAMPLE 5.3

The construction of a new library has just gotten underway at Lakeland Community College. If we define a random variable x as the percentage of the project that is completed after 6 months, the possible values of x range from 0 to 100. In other words, $0 \leq x \leq 100$.

■

A random variable is classified as either discrete or continuous depending upon the numerical values it can assume. A random variable that may assume only a finite or countably infinite (e.g., 1, 2, 3, . . .) number of values is referred to as a *discrete random variable*. The number of units sold, the number of defects observed, and the number of customers that enter a bank during one day of operation are examples of discrete random variables. Examples 5.1 and 5.2 involve discrete random variables.

Random variables such as weight, time, and temperature, which may take on all values in a certain interval or collection of intervals, are referred to as *continuous random variables*. For instance, the random variable in Example 5.3 (percentage of project completed after 6 months) is a continuous random variable because it may take on any value in the interval from 0 to 100 (e.g., 56.33 or 64.227). The feature that distinguishes a discrete from a continuous random variable is the separation between successive values it may assume.

Notes and Comments

One way to determine whether a random variable is discrete or continuous is to think first of the values the random variable may assume as points. Then imagine drawing a line segment between every pair of points. If *all* the points on *any* of these line segments represent values the random variable may assume, the random variable is continuous; otherwise, the random variable is discrete.

EXERCISES

1. For each of the following random variables state whether it is discrete or continuous:
   a. Number of heads on 3 tosses of a coin.
   b. Number of heads on 500 tosses of a coin.
   c. Number of people entering a drugstore in a 2-hour period
   d. Time from beginning of year until first baby is born
   e. Weight of a person
   f. Number of customers calling an airline reservation service in 1 minute

2. Listed is a series of experiments and associated random variables. In each case, identify the values that the random variable can take on and state whether the random variable is discrete or continuous.

	EXPERIMENT	RANDOM VARIABLE (x)
a.	Take a 20-question examination	Number of questions answered correctly
b.	Observe cars arriving at a tollbooth for 1 hour	Number of cars arriving at tollbooth
c.	Audit 50 tax returns	Number of returns containing errors

142   CHAPTER 5 RANDOM VARIABLES AND DISCRETE PROBABILITY DISTRIBUTIONS

EXPERIMENT	RANDOM VARIABLE (x)
d. Observe an employee's work	Number of nonproductive hours in an 8-hour workday
e. Weigh a shipment of goods	Number of pounds

3. Three students have interviews scheduled for summer employment at the Brookwood Institute. In each case the result of the interview will either be that a position is offered or not offered.
   a. List the experimental outcomes.
   b. Define a random variable that represents the number of offers made. Is this a discrete or continuous random variable?
   c. Show what value the random variable will assume for each of the experimental outcomes.

4. A new treatment for lower back pain has been developed and is being tested on two patients. The results of the experiment are a rating for each patient of either no improvement, moderate improvement, or strong improvement.
   a. List the experimental outcomes.
   b. Define a random variable that represents the number of patients that show at least some improvement. Show what value the random variable will assume for each of the experimental outcomes.

5.2 Discrete Probability Distributions

For any discrete random variable *x*, the *probability function*, denoted by *f(x)*, gives the probabilities associated with the values the random variable may assume. A *probability distribution* is a table, graph, or mathematical formula that shows all possible values of the random variable, *x*, and the associated probability function, *f(x)*.

EXAMPLE 5.4

As part of a study of 300 households in a village on the coast of Maine, a sociologist collected data showing the number of children in each household. The following data were obtained: 54 of the households had no children, 117 had one child, 72 had two children, 42 had three children, 12 had four children, and 3 had five children.

Suppose we consider the experiment of randomly selecting one of these households to participate in a follow-up study. If we let *x* denote the number of children in the household selected, possible values of *x* are 0, 1, 2, 3, 4, and 5. Thus *f(0)* provides the probability that a randomly selected household has no children, *f(1)* provides the probability that a randomly selected household has one child, and so on. Since 54 of the 300 households have no children, we assign the value $54/300 = .18$ to *f(0)*. Similarly, since 117 of the 300 households have one child, we assign the value $117/300 = .39$ to *f(1)*. Continuing in this fashion for the other values the random variable *x* may assume, we obtain Table 1. This table, showing the values the random variable may assume and the associated probabilities *f(x)*, is the probability distribution for the random variable *x*.

■

We can also present the probability distribution of *x* graphically. In Figure 5.1 the values of the random variable *x* from Example 5.4 are shown on the horizontal axis and the probability that *x* assumes these values is shown on the vertical axis. For many discrete random variables, the probability can also be given as a formula that yields $f(x)$ for every possible value of x.

EXAMPLE 5.5

Consider the random variable x, representing the possible choices on a multiple-choice exam. The probability distribution, as shown by the following table, provides the probability of getting the correct answer by guessing.

x	1	2	3	4
$f(x)$	¼	¼	¼	¼

This probability distribution can also be given by the formula

$$f(x) = \tfrac{1}{4} \quad \text{for } x = 1, 2, 3, 4$$

A discrete probability distribution, such as this, where every value of the random variable has the same probability is called a *uniform probability distribution*.

■

A more complex example of a discrete random variable with its probability distribution given by a formula is the binomial probability distribution; it is introduced in Section 5.4.

In the development of the probability function for a discrete random variable, the following two conditions must always be satisfied:

Required Conditions for a Discrete Probability Function

$$f(x) \geq 0 \tag{5.1}$$

$$\Sigma f(x) = 1 \tag{5.2}$$

The symbol Σ in (5.2) is used to indicate that the summation is over all the values x may assume.

TABLE 5.1

Probability Distribution for the Number of Children per Household

x	$f(x)$
0	.18
1	.39
2	.24
3	.14
4	.04
5	.01
Total	1.00

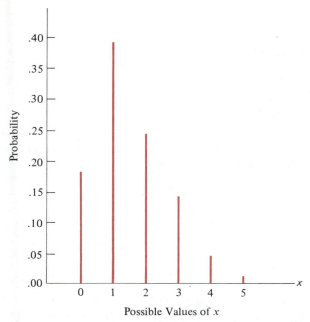

FIGURE 5.1   Graphical Presentation of the Probability Distribution for Number of Children Per Household

EXAMPLE 5.4
(continued)

Table 5.1 shows that the probabilities for the random variable x are all greater than or equal to 0. In addition, we note that

$$\Sigma f(x) = f(0) + f(1) + f(2) + f(3) + f(4) + f(5)$$

$$= .18 + .39 + .24 + .14 + .04 + .01 = 1.00$$

Since (5.1) and (5.2) are satisfied, the probability function developed by the sociologist is a valid discrete probability function.

■

EXERCISES

5. State whether or not each of the following is a valid probability distribution. Why or why not?

    a.

x	f(x)
0	.60
1	.15
2	.20

    b.

x	f(x)
4.5	.05
6.0	.35

x	f(x)
7.33	.40
8.00	.20

c.

x	f(x)
40	.20
80	.70
100	.10

6. The director of admissions at Lakeville Community College has subjectively assessed a probability distribution for x, the number of entering students.

x	f(x)
1000	.15
1100	.20
1200	.30
1300	.25
1400	.10

a. Is this a valid probability distribution?
b. What is the probability there will be 1200 or fewer entering students?

7. The following data were collected by counting the number of operating rooms in use at Tampa General Hospital over a 20-day period: On 3 of the days only 1 operating room was used, on 5 of the days 2 were used, on 8 of the days 3 were used, and on 4 days all 4 of the hospital's operating rooms were used.
a. Use the relative frequency approach to construct a probability distribution for the number of operating rooms in use on any given day.
b. Draw a graph of the probability distribution.
c. Show that your probability distribution satisfies the required conditions for a valid discrete probability distribution.

8. A stockbroker has given the following probability estimates for the price of Mills Corporation stock at the end of next week: $P(\$22) = .10$, $P(\$23) = .40$, $P(\$24) = .30$, $P(\$25) = .20$.
a. Identify an appropriate probability function and specify a probability distribution for the price of the stock at the end of next week.
b. Draw a graph of the probability distribution.
c. Show that the probability distribution satisfies (5.1) and (5.2).

9. QA Properties is considering making an offer to purchase an apartment building. Management has subjectively assessed a probability distribution for x, the purchase price:

x	f(x)
$148,000	.20
$150,000	.40
$152,000	.40

a. Determine if this is a proper probability distribution. (Check (5.1) and (5.2).)
b. What is the probability that the apartment house can be purchased for $150,000 or less?

10. The cleaning and changeover operation for a production system requires from 1 to 4 hours, depending upon the specific product that will begin production. Let x be a random variable indicating the time in hours required to make the changeover. The following probability function can be used to compute the probability associated with any changeover time x:

$$f(x) = \frac{x}{10} \quad \text{for } x = 1, 2, 3, \text{ or } 4$$

a. Show that the probability function meets the required conditions of (5.1) and (5.2).
b. What is the probability that the changeover will take 2 hours?
c. What is the probability that the changeover will take more than 2 hours?
d. Graph the probability distribution for the changeover times.

11. A psychologist has determined that the number of hours required to obtain the trust of a new patient is either 1, 2, or 3. Let x be a random variable indicating the time in hours required to gain the patient's trust. The following probability function has been proposed.

$$f(x) = \frac{x}{6} \quad \text{for } x = 1, 2, \text{ or } 3$$

a. Is this a valid probability function? Explain.
b. What is the probability that it takes exactly 2 hours to gain the patient's trust?
c. What is the probability that it takes at least 2 hours to gain the patient's trust?

12. Shown below is a partial probability distribution for the MRA Company's projected profits (in $1000s) for the first year of operation (the negative value shows a loss).

x	f(x)
−100	.10
0	.20
50	.30
100	.25
150	.10
200	

a. What is the value of $f(200)$? What is your interpretation of this value?
b. What is the probability that MRA will be profitable?
c. What is the probability that MRA will make at least $100,000?

5.3   Expected Value and Variance

Expected Value

The *expected value* of a random variable gives us the average, or central, value for the random variable. The mathematical expression for the expected value of a discrete random variable x is as follows.

Expected Value of a Discrete Random Variable

$$E(x) = \mu = \Sigma \, xf(x) \tag{5.3}$$

Both the notations $E(x)$ and μ are used to refer to the expected value of a random variable.

Equation (5.3) shows that in order to compute the expected value of a discrete random variable, we must multiply each value of the random variable by the corresponding value of its probability function, and then add the resulting products.

EXAMPLE 5.4
(continued)

In Table 5.2 we show the calculation of the expected value of the random variable x, which is the number of children in a randomly selected household. We see that 1.50 is the expected value of the number of children per household. Since it is impossible for any household to have 1.5 children, we see that the expected value of a random variable does not have to be one of the values the random variable can assume. The expected value is thought of as the mean or average value and not necessarily some value we expect the random variable to assume.

■

Variance

While the expected value gives us the mean value for the random variable, we often need a measure of the dispersion, or variability, of the random variable. Just as we used variance in Chapter 3 to summarize the dispersion in a data set, we now use the variance measure to summarize the variability in the values of a random variable about its mean. The mathematical expression for the variance of a discrete random variable is as follows.

Variance of a Discrete Random Variable

$$\text{Var}(x) = \sigma^2 = \Sigma(x - \mu)^2 f(x) \qquad (5.4)$$

As (5.4) shows, an essential part of the variance formula is the deviation, $x - \mu$, which measures how far a particular value of the random variable is from the expected value or mean, μ. In computing the variance of a random variable, the deviations are squared and then weighted by the corresponding value of the probability function. The sum of these weighted squared deviations from the mean for all values of the random variable

TABLE 5.2

Expected Value of Random Variable for Example 5.4

x	f(x)	xf(x)
0	.18	.00
1	.39	.39
2	.24	.48
3	.14	.42
4	.04	.16
5	.01	.05
		1.50

$$E(x) = \mu = \Sigma xf(x)$$

TABLE 5.3

Calculation of Variance for Example 5.4

0	-1.50	2.25	.18	.4050
1	$-.50$	.25	.39	.0975
2	.50	.25	.24	.0600
3	1.50	2.25	.14	.3150
4	2.50	6.25	.04	.2500
5	3.50	12.25	.01	.1225
				1.2500

$$\sigma^2 = \Sigma(x - \mu)^2 f(x)$$

is referred to as the *variance*. The *standard deviation*, σ, is defined as the positive square root of the variance.

**EXAMPLE 5.4
(continued)**

The calculation of the variance for the probability distribution of the number of children per household is summarized in Table 5.3. We see that the variance for the number of children per household is 1.25. The standard deviation of the number of children per household is

$$\sigma = \sqrt{1.25} = 1.118$$

The standard deviation is measured in the same units as the random variable ($\sigma = 1.118$ children per household in Example 5.4); for this reason σ is often preferred in describing the variability of a random variable. The variance (σ^2) is measured in squared units and is thus more difficult to interpret.

An alternate formula for the variance of a random variable is

$$Var(x) = \sigma^2 = \Sigma x^2 f(x) - \mu^2 \tag{5.5}$$

When computing the variance using a calculator, this formula is often preferred because it does not require the computation of the deviations about the mean $(x - \mu)$. To illustrate the use of this formula, we recompute the variance for Example 5.4.

**EXAMPLE 5.4
(continued)**

The calculations necessary to use (5.5) to compute the variance of the number of children per household are summarized below.

x	x^2	$f(x)$	$x^2f(x)$
0	0	.18	.00
1	1	.39	.39
2	4	.24	.96
3	9	.14	1.26
4	16	.04	.64
5	25	.01	.25
15	55		3.50

$$\mu = 1.50$$

$$\text{Var}(x) = \sigma^2 = \Sigma x^2 f(x) - \mu^2 = 3.50 - (1.50)^2 = 1.25$$

As we should expect, this is the same answer we obtained previously using (5.4).

■

EXERCISES

13. Consider the following probability distribution.

x	f(x)
1	.20
2	.50
3	.30
	1.00

  a. Compute the expected value.
  b. Compute the variance.

14. A volunteer ambulance service handles from 0 to 5 service calls on any given day. The following probability distribution for the number of service calls is assumed:

NUMBER OF SERVICE CALLS	PROBABILITY
0	.10
1	.15
2	.30
3	.20
4	.15
5	.10

  a. What is the expected number of service calls?
  b. What is the variance in the number of service calls? What is the standard deviation?

15. Glazer's Winton Woods apartment building has 20 two-bedroom apartments. The number of apartment air-conditioner units that must be replaced during the summer season has the probability distribution shown.

AIR CONDITIONERS REPLACED	PROBABILITY
0	.30
1	.35
2	.20
3	.10
4	.05

  a. What is the expected number of air-conditioner units that will be replaced during a summer season?
  b. What is the variance in the number of air-conditioner replacements?
  c. What is the standard deviation?

16. A roulette wheel at a Las Vegas casino has 18 red numbers, 18 black numbers, and 2 green numbers. Assume that a $5 bet is placed on the black numbers. If a black number comes up, the player wins the bet; otherwise the player loses the $5.

   a. Let x be a random variable indicating the player's net winnings on one bet ($x = 5$ if the player wins and $x = -5$ if the player loses). Show the probability distribution for x.

   b. What is the expected amount won on a bet? What is your interpretation of this value?

   c. What is the variance in the amount won on a bet? What is the standard deviation?

   d. If a player places 100 bets of $5 each, what are the expected winnings? Comment on why casinos like a high volume of betting.

17. The probability distribution for collision insurance claims paid by the Newton Automobile Insurance Company is as follows.

CLAIMS ($)	PROBABILITY
0 (No claims)	.90
200	.04
500	.03
1000	.01
2000	.01
3000	.01

   a. Use the expected collision claim amount to determine the collision insurance premium that would allow the company to break even on the collision portion of the policy.

   b. The insurance company charges an annual rate of $130 for the collision coverage. What is the expected value of the collision policy for the policyholder? Why does the policy holder purchase a collision policy with this expected value?

18. The number of dots up on the roll of a die has the following probability function.

$$f(x) = \frac{1}{6} \quad \text{for } x = 1, 2, 3, 4, 5, 6$$

   a. Show that this probability function possesses the properties necessary for a probability distribution.

   b. Draw a graph of the probability distribution.

   c. What is the expected value? What is the interpretation of this value?

   d. What are the variance and the standard deviation for the number of dots up on the roll of a die?

19. The demand for a product of Carolina Industries varies greatly from month to month. Based on the past 2 years of data, the following probability distribution shows the company's monthly demand.

Unit demand	300	400	500	600
Probability	.20	.30	.35	.15

   a. If the company places monthly orders based on the expected value of the monthly demand, what should Carolina's monthly order quantity be for this product?

   b. Assume that each unit demanded generates $70 in revenue and that each unit ordered costs $50. How much will the company gain or lose in a month if it places an order based on your answer to part (a) and where the actual demand for the item is 300 units?

20. What are the variance and the standard deviation for the number of units demanded in Exercise 19?

21. The J.R. Ryland Computer Company is considering a plant expansion that will enable the company to begin production of a new computer product. The company's president must determine whether to make the expansion a medium- or large-scale project. An uncertainty involves the demand for the new product, which for planning purposes may be low demand, medium demand, or high demand. The probability estimates for the demands are .20, .50 and .30, respectively. Letting x indicate the annual profit in $1000s, the firm's planners have developed profit forecasts for the medium- and large-scale expansion projects.

DEMAND	MEDIUM-SCALE EXPANSION PROFITS		LARGE-SCALE EXPANSION PROFITS	
	x	$f(x)$	x	$f(x)$
Low	50	.20	0	.20
Medium	150	.50	100	.50
High	200	.30	300	.30

a. Compute the expected value for the profit associated with the two expansion alternatives. Which decision is preferred for the objective of maximizing the expected profit?

b. Compute the variance for the profit associated with the two expansion alternatives. Which decision is preferred for the objective of minimizing the risk or uncertainty?

5.4 The Binomial Probability Distribution

The *binomial probability distribution* is a discrete probability distribution that has many applications. It is associated with a multiple-step experiment that we call the binomial experiment.

Binomial Experiment

For a probability experiment to be classified as a *binomial experiment*, it must have the following four properties.

Properties of a Binomial Experiment

---

1. The experiment consists of a sequence of n identical trials.
2. Two outcomes are possible on each trial. We refer to one as a *success* and the other as a *failure*.
3. The probability of a success, denoted by p, does not change from trial to trial. Consequently, the probability of failure, denoted by $1 - p$, does not change from trial to trial. Also, since there are only two possible outcomes on each trial, the probability of a success plus the probability of a failure must equal 1.
4. The trials are independent.

Figure 5.2 depicts a sequence of possible outcomes for a binomial experiment involving eight trials.

Property 1: The experiment consists of
$\qquad$ $n = 8$ identical trials.

$\qquad\qquad\qquad$ *Property 2*: Each trial results
$\qquad\qquad\qquad\qquad$ in either success
$\qquad\qquad\qquad\qquad$ (S) or failure (F).

Trials $\longrightarrow$ 1 $\quad$ 2 $\quad$ 3 $\quad$ 4 $\quad$ 5 $\quad$ 6 $\quad$ 7 $\quad$ 8

Outcomes $\longrightarrow$ S $\quad$ F $\quad$ F $\quad$ S $\quad$ S $\quad$ F $\quad$ S $\quad$ S

FIGURE 5.2 Diagram of Eight-Trial Binomial Experiment

In a binomial experiment, our interest is in the *number of successes occurring in the n trials*. If we let x denote the number of successes occurring in the n trials, we see that x can assume the values of 0, 1, 2, 3, . . ., n. Since the number of values is finite, x is a *discrete* random variable. The probability distribution associated with this random variable is called the *binomial probability distribution*.

EXAMPLE 5.6

Consider the experiment of tossing a coin five times and on each toss observing whether the coin lands with a head or a tail on its upward face. Suppose we are interested in counting the number of heads appearing during the five tosses. Does this experiment have the properties of a binomial experiment? What is the random variable of interest? Note that:

1. The experiment consists of five identical trials, where each trial involves the tossing of one coin.
2. There are two outcomes possible for each trial. The possible outcomes are a head and a tail. We can designate head as success and tail as failure.
3. The probability of a head and the probability of a tail are the same for each trial, with $p = .5$ and $1 - p = .5$.
4. The trials or tosses are independent, since the outcome on any one trial is not affected by what happens on other trials or tosses.

Thus the properties of a binomial experiment are satisfied. The random variable of interest is $x =$ the number of heads appearing in the five trials. In this case, x can assume the values of 0, 1, 2, 3, 4, or 5.

■

Property 2 is not as restrictive as it may first appear. For instance, consider rolling a die and observing whether or not a 5 comes up. Defining a success to be "5 comes up" and a failure to be "5 does not come up" we have defined the experiment in such a fashion that there are exactly two outcomes possible on each trial, 5 or not 5. This is true even though there are actually six numbers that may appear on the upward face of a die.

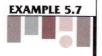

EXAMPLE 5.7

An insurance salesperson pays a visit to 10 randomly selected families. An outcome associated with a visit is classified as a success if the family purchases an insurance policy and a failure if the family does not. From past experience, the salesperson knows the probability that a randomly selected family purchases an insurance policy is .10. Show that the process of the salesperson contacting the ten families and recording the number of families that purchase an insurance policy is a binomial experiment.

Checking the properties of a binomial experiment, we observe the following:

1. The experiment consists of 10 identical experiments, where each experiment, or trial, involves contacting one family.
2. There are two outcomes possible on each trial: the family purchases a policy or the family does not purchase a policy.
3. The probabilities of a purchase and a nonpurchase are assumed to be the same for each family, with $p = .10$ and $1 - p = .90$.
4. The trials are independent since the families are randomly selected.

Since the four assumptions are satisfied, this is a binomial experiment. The random variable of interest is the number of sales obtained in contacting the 10 families. In this case, x can assume the values of 0, 1, 2, 3, 4, 5, 6, 7, 8, 9 and 10.

■

Property 3 of the binomial experiment is often called the *stationarity assumption* and is sometimes confused with Property 4, independence of trials. To see how they differ, consider again the case of the salesperson in Example 5.7 calling on families to sell insurance policies. If, as the day wore on the salesperson got tired and lost enthusiasm, then the probability of success (selling a policy) might drop to .05 by the tenth call. In such a case Property 3 (stationarity) would not be satisfied, and we would not have a binomial experiment. This would be true even if Property 4 held—that is, the purchase decisions of each family were made independently.

In applications involving binomial experiments, a special mathematical formula, called the *binomial probability function,* can be used to compute the probability of x successes in the n trials. We develop the binomial probability function by considering a situation that can be analyzed using the methods from Chapter 4. We then show that if the properties of a binomial experiment are satisified, the binomial probability function can be used to compute the desired probabilities.

EXAMPLE 5.8

A moving target at a policy academy target range can be hit 80% of the time by a particular individual. Suppose this person takes three shots at the target. What is the probability of exactly two hits?

Using a tree diagram, we can see from Figure 5.3 that the experiment of taking three shots at the target has eight possible outcomes. Using S to denote success (hitting the target) and F to denote failure (missing the target), we are interested in outcomes having two successes in the three trials, or shots.

Next, let us verify that the experiment of taking three shots at the target has the properties of a binomial experiment. Check to see if you agree with the following conclusions:

1. The experiment consists of three identical trials or shots at the target.
2. The two outcomes per trial are a hit *(S)* or a miss *(F)*.
3. The probability of a hit and the probability of a miss are the same for each trial, with $p = .80$ and $1 - p = 1 - .80 = .20$.
4. The trials, or shots, are independent.

■

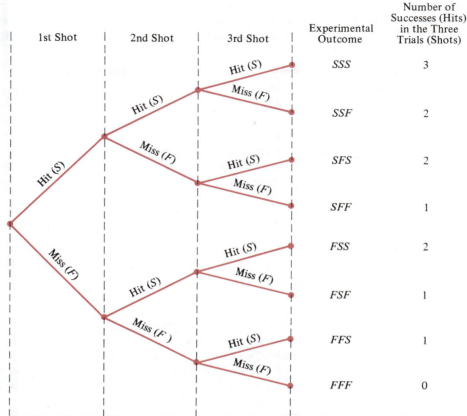

| | | | | Number of Successes (Hits) in the Three |
1st Shot	2nd Shot	3rd Shot	Experimental Outcome	Trials (Shots)
		Hit (S)	SSS	3
	Hit (S)	Miss (F)	SSF	2
Hit (S)	Miss (F)	Hit (S)	SFS	2
		Miss (F)	SFF	1
	Hit (S)	Hit (S)	FSS	2
Miss (F)		Miss (F)	FSF	1
	Miss (F)	Hit (S)	FFS	1
		Miss (F)	FFF	0

FIGURE 5.3   Tree Diagram of Experiment Involving Taking Three Shots at a Target

The number of outcomes of a binomial experiment that result in exactly x successes in n trials can be computed from the following formula.*

$$\text{Number of Experimental Outcomes Providing} \atop \text{Exactly } x \text{ Successes in } n \text{ Trials} = \binom{n}{x} = \frac{n!}{x!(n-x)!} \qquad (5.6)$$

where

$$n! = n(n-1)(n-2) \cdots (2)(1) \qquad (5.7)$$

and

$$0! = 1$$

The term $n!$ is called *n factorial*. For example, $5! = (5)(4)(3)(2)(1) = 120$.

*This formula is commonly used to determine the number of combinations of n objects selected x at a time. For the binomial experiment, this combinatorial formula provides the number of experimental outcomes having x successes in n trials.

**EXAMPLE 5.8
(continued)**

Now let us return to the experiment of taking three shots at a target. Equation (5.6) can be used to determine the number of experimental outcomes involving two hits; that is, the number of ways of obtaining $x = 2$ successes in the $n = 3$ trials. From (5.6) we have

$$\binom{n}{x} = \binom{3}{2} = \frac{3!}{2!(3 - 2)!} = \frac{(3)(2)(1)}{[(2)(1)](1)} = \frac{6}{2} = 3$$

Formula (5.6) shows that three of the outcomes yield two successes. From Figure 5.3 we see these three outcomes are denoted by *SSF, SFS,* and *FSS.*

Using (5.6) to determine how many experimental outcomes have three successes (hits) in the three trials, we obtain:

$$\binom{n}{x} = \binom{3}{3} = \frac{3!}{3!(3 - 3)!} = \frac{3!}{3!0!} = \frac{(3)(2)(1)}{[(3)(2)(1)](1)} = \frac{6}{6} = 1$$

From Figure 5.3 we see that the one experimental outcome with three successes is identified by *SSS.*

We know that (5.6) can be used to determine the number of experimental outcomes that result in x successes. But, if we are to determine the probability of x successes in n trials, we must also know the probability associated with each experimental outcome. Since the trials of a binomial experiment are independent, we can simply multiply the probabilities associated with each trial outcome to find the probability of a particular sequence of outcomes.

**EXAMPLE 5.8
(continued)**

With three shots at a target, the probability of hitting on the first and second shots but missing on the third is given by

$$pp(1 - p)$$

With a .80 probability of hitting the target on any one shot, the probability of hitting the target with the first two shots and missing the target with the third shot is given by

$$(.80)(.80)(.20) = (.80)^2(.20) = .128$$

There are two other sequences of outcomes resulting in two successes and one failure. The probabilities for all three sequences involving 2 successes are as shown.

TRIAL OUTCOMES			SUCCESS-FAILURE NOTATION	PROBABILITY OF OUTCOME
1st Shot	2nd Shot	3rd Shot		
Hit	Hit	Miss	*SSF*	$pp(1 - p) = p^2(1 - p) = (.80)^2(.20) = .128$
Hit	Miss	Hit	*SFS*	$p(1 - p)p = p^2(1 - p) = (.80)^2(.20) = .128$
Miss	Hit	Hit	*FSS*	$(1 - p)pp = p^2(1 - p) = (.80)^2(.20) = .128$

Observe that all three outcomes with two successes have exactly the same probability.

In any binomial experiment, each sequence of trial outcomes yielding x successes in n trials has the *same probability* of occurrence. The probability of each sequence of trials yielding x successes in n trials is as follows:

$$\text{Probability of a Particular Sequence of Trial Outcomes} = p^x(1 - p)^{(n-x)} \tag{5.8}$$
with x Successes in n Trials

For the target practice situation of Example 5.8, this formula shows that any outcome with two successes has a probability of $p^2(1 - p)^{(3-2)} = p^2(1 - p)^1 = (.80)^2(.20)^1 = .128$, as shown.

Since (5.6) shows the number of outcomes in a binomial experiment with x successes and since (5.8) gives the probability for each sequence involving x successes, we combine (5.6) and (5.8) to obtain the following *binomial probability function*.

Binomial Probability Function

$$f(x) = \binom{n}{x} p^x(1 - p)^{(n-x)} \tag{5.9}$$

where

$f(x)$ = the probability of x successes

n = the number of trials

$$\binom{n}{x} = \frac{n!}{x!(n - x)!}$$

p = the probability of a success on any one trial

$(1 - p)$ = the probability of a failure on any one trial

EXAMPLE 5.8 (continued)

In the experiment of taking three shots at the target, what is the probability for each of the following: hits on all three shots, hits on exactly two shots, a hit on exactly one shot, and misses on all three shots?

The binomial probability function can be used to answer these questions.

$$f(3) = \frac{3!}{3!(3 - 3)!} (.80)^3(.20)^0 = \frac{6}{6} (.512)(1) = .512$$

$$f(2) = \frac{3!}{2!(3 - 2)!} (.80)^2(.20)^1 = \frac{6}{2} (.64)(.20) = .384$$

$$f(1) = \frac{3!}{1!(3 - 1)!} (.80)^1(.20)^2 = \frac{6}{2} (.80)(.04) = .096$$

$$f(0) = \frac{3!}{0!(3 - 0)!} (.80)^0(.20)^3 = \frac{6}{6} (1)(.008) = .008$$

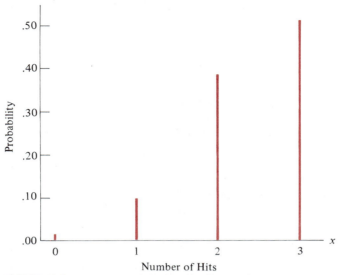

FIGURE 5.4   Binomial Probability Distribution for Number of Hits in Three Trials

Summarizing these calculations in a tabular form provide a tabular presentation of the probability distribution for the number of hits in three shots at the target.

x	f(x)
3	.512
2	.384
1	.096
0	.008
	1.000

A graphical representation of this probability distribution is also shown in Figure 5.4.

■

The binomial probability function can be applied to *any* binomial experiment. If we are satisfied that a situation has the properties of a binomial experiment and if we know the values of n, and p, (5.9) can be used to compute the probability of x successes in the n trials.

EXAMPLE 5.9

For the typical American,* the probability of being hospitalized during a 1-year period is .12. Assume that four randomly selected individuals have been interviewed as part of a health care study.

a. What is the probability that none of the individuals have been hospitalized during the past year?

We need to compute $f(0)$ for the binomial function with $n = 4$ trials, $p = .12$, and $1 - p = .88$. We obtain

*From *How You Rate* by Tom Biracree, New York Dell Publishing Co., 1984.

$$f(0) = \frac{4!}{0!(4-0)!}(.12)^0(.88)^4 = .5997$$

b. What is the probability that two or more of the individuals have been hospitalized during the past year?

We need to determine the probability of 2, 3, or 4 individuals being hospitalized. This is $f(2) + f(3) + f(4)$. But the same probability is given by $1 - f(0) - f(1)$. Since we already know $f(0)$, the second approach involves less computation.

$$f(1) = \frac{4!}{1!(4-3)!}(.12)^1(.88)^3 = .3271$$

Therefore, the probability that two or more of the individuals have been hospitalized is $1 - f(0) - f(1) = 1 - .5997 - .3271 = .0732$.

■

EXAMPLE 5.10

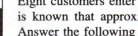

Eight customers enter a clothing store during a 1-hour period. From past experience, it is known that approximately 30% of the people entering the store make a purchase. Answer the following questions.

a. What is the probability exactly three of the eight customers make a purchase?

Assuming the properties of the binomial experiment apply, we have $n = 8$ trials with $p = .30$ and $1 - p = 1 - .30 = .70$:

$$f(3) = \frac{8!}{3!(8-3)!}(.30)^3(.70)^{(8-3)}$$

$$= \frac{40,320}{720}(.027)(.16807) = .2541$$

b. What is the probability at least one customer makes a purchase?

The probability of *at least* one customer purchase is the sum of the probabilities of 1, 2, 3, 4, 5, 6, 7 and 8 customer purchases. Although we could compute each of these probabilities separately and add them together, it is helpful to note that the probability of at least one person making a purchase is 1 minus the probability of no customer purchases. Computing the probability of no successes in the eight trials, we have:

$$f(0) = \frac{8!}{0!(8-0)!}(.30)^0(.70)^{(8-0)}$$

$$= \frac{8!}{1(8!)}(1)(.0576) = .0576$$

Therefore, the probability of at least one success must be:

$$P \text{ (at least one success)} = 1 - f(0)$$
$$= 1 - .0576 = .9424$$

■

Using Tables of Binomial Probabilities

Tables have been developed that give the probability of x successes in n trials for a binomial experiment. These tables are generally easy to use and quicker than Equation (5.9), especially when the number of trials involved is large. A table of binomial probabilities is provided as Table 5 of Appendix B. A portion of this table is given in Table 5.4. In order to use this table it is necessary to specify the values of n, p, and x

TABLE 5.4 **Selected Values of the Binomial Probability Table**
Example: $n = 10$, $x = 3$, $p = .40$; $P(x = 3$ successes$) = .2150$

n	x	.05	.10	.15	.20	.25	.30	.35	.40	.45	.50
						(p)					
9	0	.6302	.3874	.2316	.1342	.0751	.0404	.0207	.0101	.0046	.0020
	1	.2985	.3874	.3679	.3020	.2253	.1556	.1004	.0605	.0339	.0176
	2	.0629	.1722	.2597	.3020	.3003	.2668	.2162	.1612	.1110	.0703
	3	.0077	.0446	.1069	.1762	.2336	.2668	.2716	.2508	.2119	.1641
	4	.0006	.0074	.0283	.0661	.1168	.1715	.2194	.2508	.2600	.2461
	5	.0000	.0008	.0050	.0165	.0389	.0735	.1181	.1672	.2128	.2461
	6	.0000	.0001	.0006	.0028	.0087	.0210	.0424	.0743	.1160	.1641
	7	.0000	.0000	.0000	.0003	.0012	.0039	.0098	.0212	.0407	.0703
	8	.0000	.0000	.0000	.0000	.0001	.0004	.0013	.0035	.0083	.0176
	9	.0000	.0000	.0000	.0000	.0000	.0000	.0001	.0003	.0008	.0020
10	0	.5987	.3487	.1969	.1074	.0563	.0282	.0135	.0060	.0025	.0010
	1	.3151	.3874	.3474	.2684	.1877	.1211	.0725	.0403	.0207	.0098
	2	.0746	.1937	.2759	.3020	.2816	.2335	.1757	.1209	.0763	.0439
	3	.0105	.0574	.1298	.2013	.2503	.2668	.2522	**.2150**	.1665	.1172
	4	.0010	.0112	.0401	.0881	.1460	.2001	.2377	.2508	.2384	.2051
	5	.0001	.0015	.0085	.0264	.0584	.1029	.1536	.2007	.2340	.2461
	6	.0000	.0001	.0012	.0055	.0162	.0368	.0689	.1115	.1596	.2051
	7	.0000	.0000	.0001	.0008	.0031	.0090	.0212	.0425	.0746	.1172
	8	.0000	.0000	.0000	.0001	.0004	.0014	.0043	.0106	.0229	.0439
	9	.0000	.0000	.0000	.0000	.0000	.0001	.0005	.0016	.0042	.0098
	10	.0000	.0000	.0000	.0000	.0000	.0000	.0000	.0001	.0003	.0010
11	0	.5688	.3138	.1673	.0859	.0422	.0198	.0088	.0036	.0014	.0005
	1	.3293	.3835	.3248	.2362	.1549	.0932	.0518	.0266	.0125	.0054
	2	.0867	.2131	.2866	.2953	.2581	.1998	.1395	.0887	.0531	.0269
	3	.0137	.0710	.1517	.2215	.2581	.2568	.2254	.1774	.1259	.0806
	4	.0014	.0158	.0536	.1107	.1721	.2201	.2428	.2365	.2060	.1611
	5	.0001	.0025	.0132	.0388	.0803	.1321	.1830	.2207	.2360	.2256
	6	.0000	.0003	.0023	.0097	.0268	.0566	.0985	.1471	.1931	.2256
	7	.0000	.0000	.0003	.0017	.0064	.0173	.0379	.0701	.1128	.1611
	8	.0000	.0000	.0000	.0002	.0011	.0037	.0102	.0234	.0462	.0806
	9	.0000	.0000	.0000	.0000	.0001	.0005	.0018	.0052	.0126	.0269
	10	.0000	.0000	.0000	.0000	.0000	.0000	.0002	.0007	.0021	.0054
	11	.0000	.0000	.0000	.0000	.0000	.0000	.0000	.0000	.0002	.0005

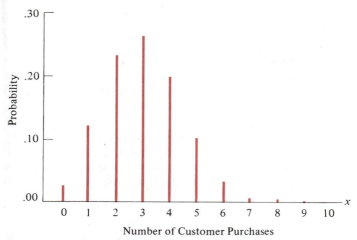

FIGURE 5.5 Binomial Probability Distribution for Number of Customer Purchases in Ten Trials

for the binomial experiment of interest. In the example at the top of Table 5.4, we see that the probability of $x = 3$ successes in a binomial experiment with $n = 10$ and $p = .40$ is .2150. You might want to use (5.9) to verify that this is the answer you would obtain using the binomial probability function directly.

EXAMPLE 5.10 (continued)

Assume that 10 customers enter the clothing store during a 1-hour period and that the probability of a customer purchase is .30. Use the table of binomial probabilities to answer the following questions.

a. What is the probability exactly 3 of the 10 customers make a purchase? From Table 5.4, $f(3) = .2668$

b. What is the probability at least 1 customer makes a purchase? From Table 5.4, P(at least 1 success) $= 1 - f(0) = 1 - .0282 = .9718$

c. What is the probability 3 or fewer customers make a purchase? From Table 5.4, P(3 or fewer successes) $= f(3) + f(2) + f(1) + f(0) = .2668 + .2335 + .1211 + .0282 = .6496$

Using all the probabilities corresponding to $n = 10$ trials with $p = .30$, Figure 5.5 provides a graphical representation of the binomial probability distribution for the number of customer purchases.

■

While Example 5.10 demonstrates the relative ease of using the tables of binomial probabilities, it is impossible to have tables that show all possible values of n and p that might be encountered in a binomial experiment. In cases where the appropriate probabilities are not available, one can interpolate to arrive at an approximation. For instance, suppose we had $p = .275$, $n = 10$, and $x = 3$. Using Table 5.3 for $p = .25$, $n = 10$, and $x = 3$, we find that $f(3) = .2503$. Similarly for $p = .30$, $n = 10$, and $x = 3$ we find that $f(3) = .2668$. Interpolating halfway between these values, we obtain $f(3) = (.2503 + .2668)/2 = .2586$ for the case of $p = .275$. Alternatively, with today's calculators it is not too difficult to calculate the desired probability using (5.9), especially if the number of trials is not too large. In the exercises, you should practice using (5.9) to compute the binomial probabilities unless the problem specifically requests that you use the binomial probability table.

The Expected Value and Variance for the Binomial Probability Distribution

In Section 5.3 we provided formulas for computing the expected value and variance of a discrete random variable. In the special case where the random variable has a binomial probability distribution with a known number of trials *(n)* and known probabilities, the general formulas for the expected value and variance (5.3) and (5.4) can be simplified; the results are as follows.

Expected Value and Variance for the Binomial Probability Distribution

$$E(x) = \mu = np \qquad\qquad (5.10)$$

$$\mathrm{Var}(x) = \sigma^2 = np(1 - p) \qquad\qquad (5.11)$$

EXAMPLE 5.11

A statistics class has 25 students. From past experience it is known that 20% of the students withdraw from the course before the end of the term. What is the expected number of withdrawals, and what is the variance in the number of withdrawals?

Assuming each student makes a decision independently, the process can be viewed as a binomial experiment with $n = 25$, $p = .20$, and $(1 - p) = 1 - .20 = .80$. The random variable of interest is the number of withdrawals during the term.

Using (5.10) and (5.11), the expected value and variance of the number of withdrawals are as follows:

$$\mu = np = (25)(.20) = 5$$

$$\sigma^2 = np(1 - p) = (25)(.20)(.80) = 4$$

The corresponding standard deviation is $\sigma = \sqrt{4} = 2$.

■

EXAMPLE 5.12

A shipment of 500 parts is received from a supplier. From past experience, it is known that the probability that any particular part is defective is .03. What is the expected number of defective parts in the shipment? What are the variance and standard deviation in the number of defective parts?

Viewing this as a binomial experiment with $n = 500$, $p = .03$, and $1 - p = 1 - .03 = .97$, we have:

$$\mu = (500)(.03) = 15$$

$$\sigma^2 = (500)(.03)(.97) = 14.5500$$

$$\sigma = \sqrt{14.5500} = 3.8144$$

■

Notes and Comments

The binomial probability tables can be used only for $0 \le p \le .50$. To use the tables when the probability of success is greater than .50, simply switch what is meant by success and failure. Then the probability of success will be less than .50 and the probability of failure will be greater than or equal to .50.

EXERCISES

22. Suppose in a binomial experiment we have $n = 4$ and $p = .15$.
  a. What is the probability of no successes?
  b. What is the probability of one success?
  c. What is the probability of two, or more successes?

23. For $n = 15$ and $p = .35$, use the binomial probability tables to compute the following:
  a. $f(3)$
  b. $f(5)$
  c. $f(10)$
  d. $1 - f(0)$

24. For $n = 10$ and $p = .125$, use the binomial tables and interpolate to find the following:
  a. $f(2)$
  b. $f(3)$
  c. $f(8)$

25. For $n = 10$ and $p = .60$, use the binomial probability tables to find the following:
  a. $f(4)$
  b. $f(6)$
  c. $f(7)$

26. Suppose that a newly married couple is planning to have three children and that the couple is interested in knowing the probabilities of having no girls, one girl, two girls, and three girls. Assume that the probability of having a girl is .50 on any one birth.
  a. What are the trials of the experiment in this application? How many trials are there?
  b. How many outcomes are possible on each trial and what are they?
  c. What are the probabilities associated with the outcomes for each trial? Are these probabilities the same for each trial?
  d. What additional assumption must be made about the trials in order for this to be a binomial experiment?
  e. What is the random variable of interest in this problem? Is it a discrete or a continuous random variable and what values can it assume?

27. A die is to be rolled five times. Assume that we are interested in the number of times the upward face of the die is a 6. Define success and failure for the trials of this experiment and describe the conditions existing in this example that make it a binomial experiment.

28. A study found that for 60% of the couples who have been married 10 years or less, both spouses work. A sample of 20 couples who have been married 10 years or less will be selected from marital records available at a local courthouse. We will be interested in the number of couples in the sample in which both spouses work. Describe the conditions necessary for this sampling process to be viewed as a binomial experiment.

29. The New York State Bar Examination is the basis for admitting law school graduates into the law profession. Historically, 30% of the individuals taking the examination pass on their first attempt. Suppose a group of 15 individuals will be taking the examination for the first time and that we are interested in the number of individuals in this group who will pass the exam. Describe the conditions necessary for this situation to be a binomial experiment.

30. When a new machine is functioning properly, only 3% of the items produced are defective. Assume that we will randomly select two parts produced on the machine and that we are interested in the number of defective parts found.
   a. Describe the conditions under which this situation would be a binomial experiment.
   b. Draw a tree diagram similar to Figure 5.3 showing this as a two-trial experiment.
   c. How many experimental outcomes result in exactly one defect being found?
   d. Compute the probabilities associated with finding no defects, exactly 1 defect, and 2 defects.

31. Forty-five percent of the residents in a township who are of voting age are not registered to vote.
   a. In a sample of 10 people, what is the probability 5 are not registered to vote?
   b. In a sample of 10 people, what is the probability 2 or fewer are not registered to vote?

32. A baseball player with a batting average of .300 comes to bat four times in a game. What is the probability the player obtains exactly 1 hit? Exactly 2 hits? No hits?

33. Military radar and missile detection systems are designed to warn a country against enemy attacks. A reliability question deals with the ability of the detection system to identify an attack and issue the warning. Assume that a particular detection system has a .90 probability of detecting a missile attack. Answer the following questions using the binomial probability distribution.
   a. What is the probability that a single detection system will detect an attack?
   b. If two detection systems are installed in the same area and operate independently, what is the probability that at least one of the systems will detect the attack?
   c. If three systems are installed, what is the probability that at least one of the systems will detect the attack?
   d. Would you recommend that multiple detection systems be operated? Explain.

34. Assume that the binomial distribution applies for the case of a college basketball player shooting free throws. Late in a basketball game, a team will sometimes foul intentionally in the hope that the player shooting the free throw will miss and the team committing the foul will get the ball. Assume that the best player on the opposing team has a .82 probability of making a free throw and that the worst player has a .56 probability of making a free throw.
   a. What are the probabilities that the best player makes 0, 1, and 2 points if fouled and given two free throws?
   b. What are the probabilities that the worst player makes 0, 1, and 2 points if fouled and given two free throws?

35. A firm estimates the probability of having employee disciplinary problems on any day to be .10.
   a. What is the probability that the company experiences 5 days without a disciplinary problem?
   b. What is the probability of exactly 2 days with disciplinary problems in a 10-day period?
   c. What is the probability of at least 2 days with disciplinary problems in a 20-day period?

36. Consider a binomial experiment with $n = 12$ and $p = .20$.
   a. Find $E(x)$.
   b. Find $\text{Var}(x)$.

37. Suppose a salesperson makes a sale on 20% of customer contacts. A normal work week will enable the salesperson to contact 25 customers. What is the expected number of sales for the week? What is the variance for the number of sales for the week? What is the standard deviation for the number of sales for the week?

38. Eighty-five percent of the next-day express mailings handled by the U.S. Postal Service are actually received by the addressee one day after the mailing. What is the expected value and variance for the number of 1-day deliveries in a group of 250 express mailings?

39. Betting on the color red in the game of roulette has a 18/38 chance of winning. What is the expected value and variance for the number of wins in a series of 100 bets on red?

5.5 The Poisson Probability Distribution

A probability distribution that is often useful when dealing with the number of occurrences of an event over a specified interval of time, space, distance, and so on is the Poisson distribution. For instance, the random variable of interest might be the number of arrivals at a car wash in 1 hour, the number of accidents during rush hour, or the number of leaks in 100 miles of pipeline. If the following two assumptions are satisfied, it can be shown that the number of occurrences is a random variable described by the *Poisson probability distribution*.

1. The probability of an occurrence of the event is the same for any two intervals of equal length.
2. The occurrence or nonoccurrence of the event in any interval is independent of occurrence in any other interval.

The Poisson probability function is given by (5.12).

Poisson Probability Function

$$f(x) = \frac{\mu^x e^{-\mu}}{x!} \qquad \text{for } x = 0, 1, 2, \ldots \qquad (5.12)$$

where

μ = expected value of number of occurrences in the interval

$e = 2.71828$

Note that there is no upper bound for the value that a Poisson random variable may take on. It is a discrete random variable that may assume a countably infinite number of values. Any nonnegative integer value is permissable.

EXAMPLE 5.13

Suppose we are interested in the number of arrivals at the drive-up window of a fast-food restaurant during a 5-minute time period. If we can assume that the probability of a car arriving is the same for any two time periods of equal length and that the arrival or nonarrival of a car in any time period is independent of the arrival or nonarrival in any other time period, the Poisson probability function is applicable. Suppose that we are interested only in the busy lunch hour period, and these assumptions are satisfied. The average number of cars arriving in a 5-minute period of time is 10; thus the following probability function applies:

$$f(x) = \frac{10^x e^{-10}}{x!} \qquad \text{for } x = 0, 1, 2, \ldots$$

The random variable, x, represents the number of cars arriving during a 5-minute time period.

If we wanted to know the probability of exactly 5 arrivals in a 5-minute period, we would set $x = 5$ and obtain*

$$\text{Probability of Exactly} \atop \text{5 Arrivals in 5 Minutes} = f(5) = \frac{10^5 e^{-10}}{5!} = .0378$$

The probability in Example 5.13 was determined by evaluating the Poisson probability function with $\mu = 10$ and $x = 5$. It is often easier to refer to tables to find Poisson probabilities. Poisson probability tables can be used when μ and x are known. Such a table is included as Table 7 of Appendix B. For convenience we have reproduced a portion of this table as Table 5.5. From Table 5.5, we see that the probability of 5 arrivals in a time period with $\mu = 10$ is found by locating the value in the row of the table corresponding to $x = 5$ and the column of the table corresponding to $\mu = 10$. It is $f(5) = .0378$.

Example 5.13 involved computing the probability of 5 arrivals in a 5-minute period. To compute Poisson probabilities for time periods of different length, one must first determine the expected number of occurrences during the time period of interest. Then (5.12), or the Poisson probability tables, can be used.

EXAMPLE 5.13 (continued)

Suppose we want to compute the probability of 3 arrivals during a 2-minute time period. Since 10 is the expected number of arrivals in a 5-minute period, 4 is the expected number of arrivals in a 2-minute time period. Thus, the probability of x arrivals in a 2-minute time period is given by

$$f(x) = \frac{4^x e^{-4}}{x!} \qquad \text{for } x = 0, 1, 2, \ldots$$

To find the probability of 3 arrivals during a 2-minute time period, we can either use the above formula or Table 7 in Appendix B. Using the above formula, we obtain

*Values of $e^{-\mu}$ can be found in Table 6 of Appendix B.

$$f(3) = \frac{4^3 e^{-4}}{3!} = .1954$$

■

The Poisson probability distribution is not limited to computing probabilities for occurrences over time periods. Example 5.14 illustrates another application.

EXAMPLE 5.14

Suppose we are interested in the number of defects in a section of highway shortly after it is resurfaced. It seems reasonable to assume that the probability of a defect is the same for any two intervals of equal length and that the occurrence of a defect in one interval is independent of a defect in any other interval. Assume that after resurfacing, defects can be expected to occur at the mean rate of two per mile.

TABLE 5.5   **Selected Values from the Poisson Probability Tables**
Example: $\mu = 10$, $x = 5$; $f(5) = .0378$

x	9.1	9.2	9.3	9.4	(μ) 9.5	9.6	9.7	9.8	9.9	10
0	.0001	.0001	.0001	.0001	.0001	.0001	.0001	.0001	.0001	.0000
1	.0010	.0009	.0009	.0008	.0007	.0007	.0006	.0005	.0005	.0005
2	.0046	.0043	.0040	.0037	.0034	.0031	.0029	.0027	.0025	.0023
3	.0140	.0131	.0123	.0115	.0107	.0100	.0093	.0087	.0081	.0076
4	.0319	.0302	.0285	.0269	.0254	.0240	.0226	.0213	.0201	.0189
5	.0581	.0555	.0530	.0506	.0483	.0460	.0439	.0418	.0398	**.0378**
6	.0881	.0851	.0822	.0793	.0764	.0736	.0709	.0682	.0656	.0631
7	.1145	.1118	.1091	.1064	.1037	.1010	.0982	.0955	.0928	.0901
8	.1302	.1286	.1269	.1251	.1232	.1212	.1191	.1170	.1148	.1126
9	.1317	.1315	.1311	.1306	.1300	.1293	.1284	.1274	.1263	.1251
10	.1198	.1210	.1219	.1228	.1235	.1241	.1245	.1249	.1250	.1251
11	.0991	.1012	.1031	.1049	.1067	.1083	.1098	.1112	.1125	.1137
12	.0752	.0776	.0799	.0822	.0844	.0866	.0888	.0908	.0928	.0948
13	.0526	.0549	.0572	.0594	.0617	.0640	.0662	.0685	.0707	.0729
14	.0342	.0361	.0380	.0399	.0419	.0439	.0459	.0479	.0500	.0521
15	.0208	.0221	.0235	.0250	.0265	.0281	.0297	.0313	.0330	.0347
16	.0118	.0127	.0137	.0147	.0157	.0168	.0180	.0192	.0204	.0217
17	.0063	.0069	.0075	.0081	.0088	.0095	.0103	.0111	.0119	.0128
18	.0032	.0035	.0039	.0042	.0046	.0051	.0055	.0060	.0065	.0071
19	.0015	.0017	.0019	.0021	.0023	.0026	.0028	.0031	.0034	.0037
20	.0007	.0008	.0009	.0010	.0011	.0012	.0014	.0015	.0017	.0019
21	.0003	.0003	.0004	.0004	.0005	.0006	.0006	.0007	.0008	.0009
22	.0001	.0001	.0002	.0002	.0002	.0002	.0003	.0003	.0004	.0004
23	.0000	.0001	.0001	.0001	.0001	.0001	.0001	.0001	.0002	.0002
24	.0000	.0000	.0000	.0000	.0000	.0000	.0000	.0001	.0001	.0001

Let us determine the probability that there will be no defects in a particular 3-mile stretch of the highway. Since we are intrested in an interval with a length of 3 miles, μ = (2 defects/mile)(3 miles) = 6 represents the expected number of defects over the 3-mile stretch. Using either (5.12) or Table 7 in Appendix B, we find that the probability of no defects is .0025. Thus, it is very unlikely that there will be no defects in the 3-mile section. In fact, there is a $1 - .0025 = .9975$ probability of at least one defect.

■

EXERCISES

40. Consider a Poisson random variable with μ = 4. Compute the following:
   a. $f(0)$
   b. $f(1)$
   c. $f(2)$

41. Suppose we have a Poisson random variable with 3 occurrences per minute.
   a. Find $f(1)$.
   b. What is the probability of 4 or more occurrences in 1 minute?
   c. What is the probability of 4 or more occurrences in 2 minutes?
   d. What is the probability of 10 occurrences in 5 minutes?

42. A certain restaurant has a reputation for good food. Restaurant management boasts that on a Saturday night, groups of customers arrive at the rate of 15 groups every half-hour.
   a. What is the probability that 5 minutes will pass with no groups of customers arriving?
   b. What is the probability that 8 groups of customers will arrive in 10 minutes?
   c. What is the probability that more than 5 groups will arrive in a 10-minute period of time?

43. During rush hours accidents occur in a particular metropolitan area at the rate of 2 per hour. The morning rush period lasts for 1 hour 30 minutes and the evening rush period lasts for 2 hours.
   a. On a particular day what is the probability that there will be no accidents during the morning rush period?
   b. What is the probability of 2 accidents during the evening rush period?
   c. What is the probability of 4 or more accidents during the morning rush period?
   d. On a particular day what is the probability there will be no accidents during both the morning and evening rush periods?

44. Airline passengers arrive randomly and independently at the security checkpoint at a major international airport. The mean arrival rate is 10 passengers per minute.
   a. What is the probability of no arrivals in a 1-minute period?
   b. What is the probability 3 or fewer passengers arrive in a 1-minute period?
   c. What is the probability of no arrivals in a 15-second period?
   d. What is the probability of at least 1 arrival in a 15-second period?

45. For a given model of hand calculator, 3% of the calculators will fail within the first 30 days of operation and be returned to the manufacturer for repair. Assume that there is a batch of 120 calculators.
   a. What is the expected number of calculators that will fail in the first 30 days of operation?
   b. What is the probability that at least 2 will fail?
   c. What is the probability that exactly 3 will fail?

<table>
<tr><td>5.6</td><td></td></tr>
</table>

5.6   The Hypergeometric Probability Distribution

The hypergeometric probability distribution is closely related to the binomial probability distribution. It also provides the probability of obtaining x successes in n trials when there are two possible outcomes (success and failure) on each trial. The key difference between the two probability distributions is that with the hypergeometric distribution, the probability of success changes from trial to trial.

EXAMPLE 5.15

A 5-member committee consists of 3 women and 2 men. Two of the committee members are expected to represent the group at a meeting in Las Vegas. The committee has decided to randomly choose the 2 members that will attend the meeting. What is the probability that both persons chosen will be women?

Since 3 of the 5 committee members are women, the probability that the first person randomly selected is a woman is $\frac{3}{5} = .60$. However, if the first person chosen is a woman, then the probability of randomly selecting another woman from the 4 remaining committee members drops to $\frac{2}{4} = .50$. Therefore, the probability that 2 women will be selected to attend the meeting is $(.60)(.50) = .30$.

■

One of the most important applications of the hypergeometric probability distribution involves sampling without replacement from a finite population. The objective is to choose a random sample of n items out of a population of N items, under the condition that once an item has been selected it is not returned to the population. Thus, on the next selection, the probability of selecting an item of that type goes down.

The usual notation in applications of the hypergeometric probability distribution is to let r denote the number of items in the population that are labeled success and $N - r$ denote the number of items in the population that are labeled failure. The hypergeometric

---

Hypergeometric Probability Function

$$f(x) = \frac{\binom{r}{x}\binom{N-r}{n-x}}{\binom{N}{n}} \qquad x = 0, 1, \cdots, r \qquad (5.13)$$

where

$$
\begin{aligned}
f(x) &= \text{probability of } x \text{ successes} \\
N &= \text{number of items in the population} \\
r &= \text{number of items in the population labeled success} \\
n &= \text{number of items in the sample}
\end{aligned}
$$

probability function is used to compute the probability that in a random sample of n items, selected without replacement, we will obtain x items labeled success and $n - x$ items labeled failure. Note that for this to occur, we must obtain x successes from the r successes in the population and $n - x$ failures from the $N - r$ failures in the population. The hypergeometric probability function provides $f(x)$, the probability of obtaining x successes in a sample of size n.

Note that $\binom{N}{n}$, defined in (5.6), simply represents the number of ways a sample of size n can be selected from a population of size N; $\binom{r}{x}$ represents the number of ways x successes can be selected from a total of r successes in the population; and $\binom{N - r}{n - x}$ represents the number of ways $n - x$ failures can be selected from a total of $N - r$ failures in the population. To illustrate the computations involved in using (5.13), let us reconsider Example 5.15.

EXAMPLE 5.15 (continued)

Recall that the objective is to select 2 members from the 5-member committee consisting of 3 women and 2 men. To determine the probability of obtaining a sample that consists of 2 women, we can use (5.13) with $N = 5$, $r = 3$, and $x = 2$.

$$f(2) = \frac{\binom{3}{2}\binom{2}{0}}{\binom{5}{2}} = \frac{\dfrac{3!}{2!1!}\dfrac{2!}{2!0!}}{\dfrac{5!}{3!2!}} = \frac{3}{10} = .30$$

Note that this is the same answer that we obtained previously. Suppose, however, that we now learn that 3 committee members will be allowed to make the trip. The probability that 2 of the 3 members will be women is

$$f(2) = \frac{\binom{3}{2}\binom{2}{1}}{\binom{5}{3}} = \frac{\dfrac{3!}{1!2!}\dfrac{2!}{1!1!}}{\dfrac{5!}{3!2!}} = \frac{6}{10} = .60$$

∎

EXAMPLE 5.16

Suppose a population consists of 10 items, 6 of which are classified as acceptable and 4 of which are classified as defective. What is the probability that a random sample of size three contains 2 defective items?

For this problem we can think of obtaining a defective item as a "success". Thus, $N = 10$, $r = 4$, $n = 3$, and $x = 2$. Using (5.13) we can compute $f(2)$, as shown below:

$$f(2) = \frac{\binom{4}{2}\binom{6}{1}}{\binom{10}{3}} = \frac{\dfrac{4!}{2!2!}\dfrac{6!}{1!5!}}{\dfrac{10!}{3!7!}} = \frac{36}{120} = .30$$

∎

EXERCISES

46. Suppose $N = 10$, and $r = 3$. Compute the hypergeometric probabilities for the following values of n and x.
   a. $n = 4, x = 1$
   b. $n = 2, x = 2$
   c. $n = 2, x = 0$
   d. $n = 4, x = 2$

47. Suppose $N = 15$ and $r = 4$. What is the probability of $x = 5$ for $n = 1$ to 15?

48. What is the probability of being dealt 3 of a kind in a 7-card poker hand?

49. There are 25 students (14 boys and 11 girls) in the sixth grade class at St. Andrew School. Five students were absent Thursday.
   a. What is the probability 2 were girls?
   b. What is the probability 2 were boys?
   c. What is the probability all were boys?
   d. What is the probability none were boys?

50. Axline Computers manufactures personal computers at two plants; one is in Las Vegas, the other in Hawaii. There are 40 employees at the Las Vegas plant and 20 in Hawaii. A random sample of 10 different employees is to be asked to fill out a benefits' questionnaire.
   a. What is the probability none will be from the plant in Hawaii?
   b. What is the probability 1 will be from the plant in Hawaii?
   c. What is the probability 2 or more will be from the plant in Hawaii?
   d. What is the probability 9 will be from the plant in Las Vegas?

SUMMARY

The concept of a random variable was introduced in order to provide a numerical description of the outcome of an experiment. We saw that the probability distribution for a random variable describes how the probabilities are distributed over the values the random variable can take on. For any discrete random variable x, the probability distribution is defined by a probability function, denoted by $f(x)$, which provides for each value of the random variable its corresponding probability. Once the probability function has been defined, we can then compute the expected value and the variance for the random variable.

The special probability distributions that were discussed in this chapter are the binomial, Poisson, and hypergeometric distributions. The binomial probability distribution can be used to determine the probability of x successes in n trials whenever the experiment has the following properties:

1. The experiment consists of a sequence of n identical trials.
2. Two outcomes are possible on each trial, one called success and the other failure.
3. The probability of a success, p, does not change from trial to trial. Consequently, the probability of failure, $1 - p$, does not change from trial to trial.
4. The trials are independent.

When the above conditions hold, a binomial probability function, or a table of binomial probabilities, can be used to determine the probability of x successes in n trials. Formulas were also presented for the mean and variance of the binomial probability distribution.

The Poisson probability distribution is used when it is desired to determine the probability of obtaining x occurrences over an interval of time or space. The following assumptions were required for the Poisson distribution to be applicable:

1. The probability of an occurrence of the event is the same for any two intervals of equal length.
2. The occurrence or nonoccurrence of the event in any interval is independent of the occurrence or nonoccurrence in any other interval.

Both the binomial and Poisson are discrete probability distributions. The binomial random variable may assume a finite number of values $(0, 1, \ldots, n)$; the Poisson random variable may assume a countably infinite number of values $(0, 1, 2, \ldots)$.

A third discrete probability distribution, the hypergeometric, was introduced in Section 5.6. Like the binomial, it is used to compute the probability of x successes in n trials. But, unlike with the binomial, the probability of success changes from trial to trial.

GLOSSARY

Random variable   A numerical description of the outcome of an experiment

Discrete random variable   A random variable that can assume only a finite or infinite sequence of values.

Continuous random variable   A random variable that may assume all values in an interval or collection of intervals.

Probability function   A function, denoted $f(x)$, that gives the probability that the discrete random variable x assumes a particular value.

Discrete probability distribution   A table, graph, or equation showing the values of a discrete random variable and the associated probabilities.

Expected value   A measure of the average, or central, value of a random variable.

Variance   A measure of the dispersion, or variability, of a random variable.

Standard deviation   The positive square root of the variance.

Binomial experiment   A probability experiment possessing the four properties stated in Section 5.4

Binomial probability function   The function used to compute the probability of x successes in n trials for a binomial experiment.

Poisson probability function   The function used to compute the probability of x occurrences during one interval for a Poisson random variable.

Hypergeometric probability function   The function used to compute the probability of x successes in n trials when the trials are dependent.

EXPECTED VALUE OF A DISCRETE RANDOM VARIABLE

$$E(x) = \mu = \Sigma \, xf(x) \tag{5.3}$$

VARIANCE OF A DISCRETE RANDOM VARIABLE

$$\text{Var}(x) = \sigma^2 = \Sigma(x - \mu)^2 f(x) \tag{5.4}$$

COMPUTATIONAL FORMULA FOR VARIANCE OF A DISCRETE RANDOM VARIABLE

$$\text{Var}(x) = \sigma^2 = \Sigma x^2 f(x) - \mu^2 \tag{5.5}$$

NUMBER OF EXPERIMENTAL OUTCOMES PROVIDING EXACTLY x SUCCESSES IN n TRIALS

$$\binom{n}{x} = \frac{n!}{x!(n-x)!} \tag{5.6}$$

BINOMIAL PROBABILITY FUNCTION

$$f(x) = \binom{n}{x} p^x (1-p)^{(n-x)} \tag{5.9}$$

EXPECTED VALUE FOR THE BINOMIAL PROBABILITY DISTRIBUTION

$$E(x) = \mu = np \tag{5.10}$$

VARIANCE FOR THE BINOMIAL PROBABILITY DISTRIBUTION

$$\text{Var}(x) = \sigma^2 = np(1-p) \tag{5.11}$$

POISSON PROBABILITY FUNCTION

$$f(x) = \frac{\mu^x e^{-\mu}}{x!} \quad \text{for } x = 0, 1, 2, \ldots \tag{5.12}$$

HYPERGEOMETRIC PROBABILITY FUNCTION

$$f(x) = \frac{\binom{r}{x}\binom{N-r}{n-x}}{\binom{N}{n}} \tag{5.13}$$

REVIEW QUIZ

TRUE/FALSE

1. A random variable may assume only numerical values.

2. A random variable that may assume any value between 5 and 6 is a discrete random variable.

3. The sum of the probabilities for all values that a discrete random variable may assume cannot be greater than 1.

4. The expected value for a random variable must be a value the random variable can assume.

5. The variance of a random variable is the sum of the squared deviations from the mean.

6. In a binomial experiment the probability of success is 1 minus the probability of failure.

7. The binomial random variable is a discrete random variable.

8. In a 5-trial binomial experiment, there are 6 possible values for the random variable.

9. In a binomial experiment involving 3 trials with a success probability of .3, the probability of 1 success is .441.

10. In a binomial experiment involving 100 trials with a success probability of .22, the expected number of successes is less than or equal to 20.

11. The Poisson random variable is a continuous random variable.

12. The probability of one success in n trials is the same for the binomial random variable as it is for the hypergeometric random variable.

MULTIPLE CHOICE

For questions 13–15, consider the random variable x, which gives the number of successes in six identical trials, each of which has a probability of success of .3.

13. The probability of one success is
   a. .0000
   b. .1176
   c. .3025
   d. .1780

14. The probability of at least four successes is
   a. .0704
   b. .0595
   c. .0109
   d. not able to be computed from the information given

15. The expected value of x is closest to
   a. 1.50
   b. 1.75
   c. 2.00
   d. 3.00

For questions 16 and 17, consider a Poisson random variable with 8 occurrences per hour.

16. The probability of 3 occurrences in 1 hour is closest to
   a. 0.00
   b. 0.10
   c. 0.20
   d. 0.30

17. The probability of 6 or more occurrences is closest to
- **a.** .20
- **b.** .30
- **c.** .40
- **d.** .50

18. A random variable assumes the values 1, 2, and 3 with probabilities .10, .60, .30, respectively. The expected value of this random variable is
- **a.** 1.80
- **b.** 2.00
- **c.** 2.20
- **d.** 3.00

19. The variance of the random variable in question 18 is
- **a.** 2.00
- **b.** 2.20
- **c.** 8.22
- **d.** .680

20. Consider a binomial random variable with $p = .4$ and $n = 18$. The standard deviation is
- **a.** 7.20
- **b.** 4.32
- **c.** 2.08
- **d.** none of the above

SUPPLEMENTARY EXERCISES

51. Which of the following random variables are discrete and which are continuous?
- **a.** $x =$ amount of time until the first foul is called in a basketball game
- **b.** $x =$ number of leaks in 10 miles of sewer
- **c.** $x =$ distance traveled between meeting oncoming cars
- **d.** $x =$ number of credit hours carried by a randomly selected college student.

52. For the given values of n, x, and p, compute $f(x)$ for a binomial random variable.
- **a.** $n = 10, x = 3, p = .2$
- **b.** $n = 20, x = 14, p = .6$
- **c.** $n = 5, x = 2, p = .7$
- **d.** $n = 8, x = 3, p = .3$

53. Let $\mu =$ mean number of occurrences per interval, $x =$ number of occurrences, and $t =$ number of intervals for a Poisson random variable. Compute $f(x)$ for the cases given.
- **a.** $\mu = 8, t = 1, x = 6$
- **b.** $\mu = 2, t = 3, x = 8$
- **c.** $\mu = 4, t = 1.5, x = 4$
- **d.** $\mu = 1, t = 3, x = 2$

54. Which of the following are and which are not probability distributions? Explain.

x	f(x)	y	f(y)	z	f(z)
0	.20	0	.25	−1	.20
1	.30	2	.05	0	.50
2	.25	4	.10	1	−.10
3	.35	6	.60	2	.40

55. An automobile agency located in Beverly Hills specializes in the rental of luxury automobiles. Assume that the probability distribution of daily demand at their agency is as follows.

x	$f(x)$
0	.15
1	.30
2	.40
3	.10
4	.05
	1.00

a. Compute the expected value of daily demand.
b. If the daily rental cost for an automobile is $75 per day, what is the expected value of daily automobile rental?

56. At a large university, the number of student problems handled by the dean for student affairs varies from semester to semester. Assume that the number of student problems (x) handled by the dean has the following probability distribution.

x	$f(x)$
0	.10
1	.15
2	.30
3	.25
4	.10
5	.10
Total	1.00

What are the mean and variance of the number of student problems handled by the dean each semester?

57. The number of weekly lost-time injuries at a particular plant (x) has the following probability distribution.

x	$f(x)$
0	.05
1	.20
2	.40
3	.20
4	.15

a. Compute the expected value.
b. Compute the variance.

58. Assume that the plant in Exercise 57 initiated a safety training program and that the number of lost-time injuries during the 20 weeks following the training program was as follows.

NUMBER OF INJURIES	NUMBER OF WEEKS
0	2
1	8
2	6
3	3
4	1
Total	20

a. Construct a probability distribution for weekly lost-time injuries based on these data.

b. Compute the expected value and the variance and use both to evaluate the effectiveness of the safety training program.

59. The Hub Real Estate Investment stock is currently selling for $16 per share. An investor plans to buy shares and hold the stock for 1 year. Let x be the random variable indicating the price of the stock after 1 year. The probability distribution for x is shown.

PRICE OF STOCK (x)	$f(x)$
16	.35
17	.25
18	.25
19	.10
20	.05

a. Show that the above probability distribution possesses the properties of all probability distributions.

b. What is the expected price of the stock after 1 year?

c. What is the expected gain per share of the stock over the 1-year period? What percent return on the investment is reflected by this expected value?

d. What is the variance in the price of the stock over the 1-year period?

e. Another stock with a similar expected return has a variance of 3. Which stock appears to be the better investment in terms of minimizing risk or uncertainty associated with the investment? Explain.

60. The budgeting process for a midwestern college resulted in expense forecasts for the coming year (in 1,000,000s) of $9, $10, $11, $12, and $13. Since the actual expenses were unknown, the following respective probabilities were assigned: .3, .2, .25, .05, and .2.

a. Show the probability distribution for the expense forecast.

b. What is the expected value of the expenses for the coming year?

c. What is the variance in the expenses for the coming year?

d. If income projections for the year are estimated at $12 million, comment on the financial position of the college.

61. Exercise 10 provided a probability function for x, the hours required to change over a production system, as follows.

$$f(x) = \frac{x}{10} \quad \text{for } x = 1, 2, 3, \text{ or } 4.$$

a. What is the expected value of the changeover time?

b. What is the variance of the changeover time?

62. The police department of a major midwestern city makes arrests on 40% of its reported robberies. Assume that we are interested in the number of arrests that will be made in the next 20 reported robberies. Describe whether or not you feel the properties of a binomial experiment are satisfied.

63. *Better Homes and Gardens* (October, 1986) published the results of a reader survey. Over 30,000 readers responded. Findings indicated that 34% of the women worked full time outside home and 24% worked part time outside home.

a. For a random sample of 5 women who are *Better Homes and Gardens* readers, estimate the probability that 3 work full time outside the home.

b. For a random sample of 5 women readers, estimate the probability that 2 will be part-time workers.

c. Suppose a random sample of 5 women (not necessarily *Better Homes and Gardens* readers) was taken. Would your answers to parts a and b change? Why or why not?

64. Refer again to the *Better Homes and Gardens* survey in problem 63. In 65% of the two-parent households, the wife does most of the child care; in 1% the husband does most. For a sample of 4 respondents from two-parent households, answer the following:

a. What is the probability that none of the respondents will say the wife does more?

b. What is the probability that none of the respondents will say the husband does more?

c. What is the probability that 3 or more of the respondents will say the wife does more?

d. What is the probability that one of the respondents will say the husband does more?

65. A new clothes-washing compound is found to satisfactorily remove excessive dirt and stains on 88% of the items washed. Assume that 10 items are to be washed with the new compound.

a. What is the probability of satisfactory results on all 10 items?

b. What is the probability at least 2 items are found with unsatisfactory results?

66. In an audit of a company's billings, an auditor randomly selects five bills. If 3% of all bills contain an error, what is the probability that the auditor will find the following?

a. exactly one bill in error

b. at least one bill in error

67. Many companies use a quality control technique referred to as acceptance sampling in order to monitor incoming shipments of parts, raw materials, and so on. In the electronics industry it is common to have component parts shipped from suppliers in large lots. Inspection of a sample of n components can be viewed as the n trials of a binomial experiment. The outcome for each component tested (trial) will be that the component is good or defective.

Reynolds Electronics accepts lots from a particular supplier as long as the percent defective in the lot is not greater than 1%. Suppose a random sample of 5 items from a recent shipment has been tested.

a. Assume that 1% of the shipment is defective. Compute the probability that no items in the sample are defective.

b. Assume that 1% of the shipment is defective. Compute the probability that exactly one item in the sample is defective.

c. What is the probability of observing 1 or more defective items in the sample if 1% of the shipment is defective.

d. Would you feel comfortable accepting the shipment if 1 item was found defective? Why, or why not?

68. On September 13, 1988, *USA Today* reported that 39% of men and 27% of women surveyed have had an extramarital affair.

a. In a sample of 3 men, what is the probability that exactly one has had an extramarital affair?

b. In a sample of 3 women, what is the probability that exactly one has had an extramarital affair?

c. Suppose we randomly sample 1 man and 1 woman. What is the probability both have had an extramarital affair?

69. Cars arrive at a carwash at the average rate of 15 cars per hour. If the number of arrivals per hour follows a Poisson distribution, what is the probability of 20 or more arrivals during any given hour of operation? Use the Poisson probability table.

70. A new automated production process has been experiencing an average of 1.5 breakdowns per day. Because of the cost associated with a breakdown, management is concerned about the possibility of having 3 or more breakdowns during a given day. Assume that the number of breakdowns per day follows a Poisson distribution. What is the probability of observing 3 or more breakdowns?

71. A regional director responsible for business development in Pennsylvania is concerned about the number of businesses that end as failures. If the average number of failures per month is ten, what is the probability that exactly four businesses will fail during a given month? Assume that the number of businesses failing per month follows a Poisson distribution.

72. The arrivals of customers at a bank follow the Poisson distribution. Answer the following questions assuming a mean arrival rate of three per minute.

   a. What is the probability of exactly three arrivals in a 1-minute period?

   b. What is the probability of at least three arrivals in a 1-minute period?

73. During the registration period at a local university, students consult advisors with questions about course selection. A particular advisor noted that during the registration period an average of eight students per hour ask questions, although the exact arrival times of the students were random in nature. Use the Poisson distribution to answer the following questions:

   a. What is the probability that exactly eight students come in for consultation during a particular 1-hour period?

   b. What is the probability that three students come in for consultation during a particular ½-hour period?

6 Continuous Probability Distributions

What You Will Learn in This Chapter:

- The use of the continuous uniform probability distribution
- How to compute probabilities for a continuous probability distribution
- The properties of the normal probability distribution
- How to use the standard normal probability distribution to compute probabilities
- How to use the normal probability distribution to approximate binomial probabilities
- The use of the exponential probability distribution

Contents

IQ Scores: Are You Normal?

Intelligence test scores, referred to as intelligence quotient, or IQ scores, are based on characteristics such as verbal skill, abstract reasoning power, numerical ability, and spatial visualization. An IQ score of 100 is considered average. If plotted on a graph with IQ scores on the horizontal axis, the distribution of intelligence test scores approximates a bell-shaped normal probability curve. This distribution shows that the greatest concentration of scores are near 100 and that the frequency of scores decreases gradually and symmetrically as the extremes of intelligence are approached.

Knowing your IQ score gives you an indication of how you rate or compare to other individuals in terms of intelligence. An IQ score above 115 is considered superior. Studies of intellectually gifted children have generally set the lower limit at an IQ score of 140. Approximately 1% of the population has IQ scores of 140 or more. The average IQ score for a high-school graduate is 110. The aver-

IQ test scores help teachers determine if students are working up to their abilities.

age IQ score of a college graduate is 120, and the average IQ score of a person with a Ph.D. is 130.

Based on "Your Intelligence Quotient," Tom Biracree, in *How You Rate*. New York: Dell Publishing Co., Inc. 1984.

In the previous chapter we introduced the concept of a random variable. Recall that discrete random variables assume a finite or countably infinite number of values and continuous random variables assume an infinite number of values. For each value of a discrete random variable, there is a corresponding probability. For continuous random variables, we cannot specify a probability for each value. Instead, we are able to specify only the probability that the random variable will assume a value within a given interval. That is, we speak of the probability that the random variable lies between given lower and upper limits.

We begin the chapter by introducing the uniform probability distribution; the simple nature of this distribution provides an excellent illustration of how probabilities are computed for continuous random variables. Then, the normal probability distribution is introduced. It is probably the most important probability distribution in statistics and is used extensively in the remainder of the text. Another probability distribution used extensively in waiting-line applications, the exponential distribution, is covered in the last section of the chapter.

6.1   The Uniform Probability Distribution

The continuous uniform probability distribution is used in situations in which all values of the random variable are equally likely.

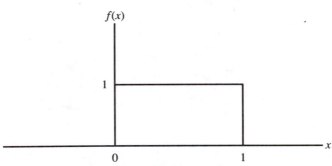

FIGURE 6.1   Uniform Probability Distribution for Random Numbers Generated

EXAMPLE 6.1

Most computer languages have a function that can be used to generate random numbers. In QuickBASIC 4.0,* the RND function can be used to generate random numbers between 0 and 1. If we let x denote the random number generated, then x is a continuous random variable with the following probability density function:

$$f(x) = \begin{cases} 1 & \text{for } 0 \leq x \leq 1 \\ 0 & \text{elsewhere} \end{cases}$$

A graph of this probability density function is shown in Figure 6.1.

In general, the uniform probability density function for a random variable x is

Uniform Probability Density Function

$$f(x) = \begin{cases} \dfrac{1}{b - a} & \text{for } a \leq x \leq b \\ 0 & \text{elsewhere} \end{cases} \tag{6.1}$$

In example 6.1, involving the random number generator, $a = 0$ and $b = 1$.

The graph of the probability density function shows the height, $f(x)$, of the function at any particular value of x. Note that for a uniform probability density function the height, or value of the function, is the same for each value of x. For instance, in Example 6.1, $f(x) = 1$ for all values of x between 0 and 1. It should be clear that the height of the probability density function cannot represent probability; otherwise, the sum of probabilities for all values of the random variable would be infinite. For a continuous random variable, probability statements can be made only in terms of the probability that the random variable takes on a value within a specified interval.

Area as a Measure of Probability

The probability that a continuous random variable will assume a value between a given lower limit, c, and a given upper limit, d, is denoted by $P(c \leq x \leq d)$ and is given by the area under the graph of the probability density function between c and d.

*Microsoft QuickBASIC 4.0 BASIC Language Reference, Microsoft Corporation, 1987.

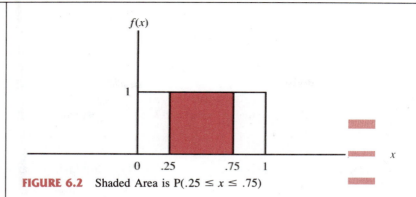

FIGURE 6.2 Shaded Area is P(.25 ≤ x ≤ .75)

**EXAMPLE 6.1
(continued)**

The shaded region in Figure 6.2 shows the area under the graph of the probability density function between .25 and .75 for the uniform random variable corresponding to the random number generator. To compute the probability the random variable assumes a value in the interval, we must find this area. Since this area is a rectangle, we simply multiply the base $(.75 - .25 = .5)$ by the height (1). Thus, $P(.25 \le x \le .75) = (.5)(1) = .50$. We conclude that the probability of generating a random number between .25 and .75 is .50.

For continuous random variables that have a uniform probability distribution, it is easy to compute the probability that the random variable takes on a value in a certain interval. The area under the probability density function will always be a rectangle, and its area is easy to compute.

Note that $P(0 \le x \le 1) = 1$ for the random number generator example. In general, the area under any probability density function is equal to 1. This property holds for all continuous probability distributions and is the analog of the condition that the sum of the probabilities has to equal 1 for a discrete probability function. For a continuous probability density function, we must also require that $f(x) \ge 0$ for all values of x. This is the analog of the requirement that $f(x) \ge 0$ for discrete probability functions.

Summarizing, when we deal with continuous random variables and probability distributions, two major differences stand out as compared to the treatment of their discrete counterparts:

1. We no longer talk about the probability of the random variable taking on a particular value. Instead we talk about the probability of the random variable taking on a value within some given interval.
2. The probability of the random variable taking on a value within some given interval from c to d is defined to be the area under the graph of the probability density function between c and d. *This implies that the probability that a continuous random variable takes on any particular value exactly is zero, since the area under the graph of $f(x)$ at a single point is zero.*

The calculation of the mean and variance for a continuous random variable is analogous to that for a discrete random variable. However, since the computational procedure involves integral calculus, we leave the derivation of the appropriate formulas to more advanced texts.

For the uniform continuous probability distribution introduced in this section, the formulas for the mean and variance are

$$E(x) = \mu = \frac{a + b}{2}$$

$$Var(x) = \sigma^2 = \frac{(b - a)^2}{12}$$

In these formulas a is the smallest value and b is the largest value that the random variable may take on.

EXAMPLE 6.1
(continued)

Applying these formulas to the uniform probability distribution for random numbers, we obtain

$$E(x) = \mu = \frac{(0 + 1)}{2} = \frac{1}{2}$$

$$Var(x) = \frac{(1 - 0)^2}{12} = \frac{1}{12} = .0833$$

The standard deviation can be found by taking the square root of the variance. Thus $\sigma = .2887$.

■

Notes and Comments

1. Since for any continuous random variable the probability of any single value is zero, we have $P(c \leq x \leq d) = P(c < x < d)$. That is, the probability is the same whether or not the endpoints of the interval are included.

2. To remind yourself that the height of a probability density function is not a probability, consider a uniform probability distribution from 0 to .50. The height of the curve, $f(x)$, would be $1/(.50 - 0) = 2$. As we know, probabilities can never be greater than 1.

EXERCISES

1. The random variable x is known to be uniformly distributed between 1.0 and 1.5.
   a. Show the graph of the probability density function.
   b. Find $P(x = 1.25)$
   c. Find $P(1.0 \leq x \leq 1.25)$.
   d. Find $P(1.20 < x < 1.5)$.

2. The random variable x is known to be uniformly distributed between 10 and 20.
   a. Show the graph of the probability density function.
   b. Find $P(x < 15)$.

   c. Find $P(12 \leq x < 18)$.

   d. Find $E(x)$.

   e. Find $Var(x)$.

3. Delta airlines quotes a 1 hour 52 minute flight time for its flights from Cincinnati to Tampa. Suppose that you believe actual flight times are uniformly distributed between the quoted time and 2 hours and 10 minutes.

   a. Show the graph of the probability density function for flight times.

   b. What is the probability the flight will be no more than 5 minutes late?

   c. What is the probability the flight will be more than 10 minutes late?

   d. What is the expected flight time?

4. The travel time for a truck traveling from Davenport to Iowa City is uniformly distributed between 70 and 90 minutes.

   a. Give a mathematical expression for the probability density function.

   b. Compute the probability that the truck will make the trip in 75 minutes or less.

   c. What is the probability that the trip will take longer than 82 minutes?

   d. Find the expected travel time and its standard deviation.

   e. What is the probability that the trip will take exactly 80 minutes?

5. The total time to process a loan application is uniformly distributed between 3 and 7 days.

   a. Give a mathematical expression for the probabiltiy density function.

   b. What is the probability that the loan application will be processed in fewer than 3 days?

   c. Compute the probability that a loan application will be processed in 5 days or less.

   d. Find the expected processing time and its standard deviation.

6. Bus arrival times at a particular location are uniformly distributed between 2:10 and 2:25 P.M., with 2:10 listed as the scheduled arrival time.

   a. Give a mathematical expression for the probability density function.

   b. What is the expected value of the arrival time?

   c. The bus is considered delayed when it arrives more than 5 minutes after the scheduled arrival time. What is the probability that the bus will be considered delayed?

   d. If a person arrives at the bus stop at 2:17, what is the probability that the person will still catch the bus?

7. The label on a bottle of liquid detergent shows contents to be 12 ounces per bottle. The production operation fills the bottle uniformly according to the following probability density function:

$$f(x) = \begin{cases} 8 & \text{for } 11.975 \leq x \leq 12.10 \\ 0 & \text{elsewhere} \end{cases}$$

   a. What is the probability that a bottle will be filled with between 12 and 12.05 ounces?

   b. What is the probability that a bottle will be filled with 12.02 or more ounces?

   c. Quality control accepts production that is within .02 ounces of the number of ounces shown on the container label. What is the probability that a bottle of this liquid detergent will fail to meet the quality control standard?

6.2 The Normal Probability Distribution

The normal probability distribution has been applied in a wide variety of practical applications in which the random variables involved are heights and weights of people, IQ scores, scientific measurements, amounts of rainfall, and so on. In order to use this probability distribution, the random variable must be continuous. However, as we shall see, a continuous normal random variable is often used as an approximation in situations involving discrete random variables. The form, or shape, of the normal probability density function is illustrated by the bell-shaped curve shown in Figure 6.3. The mathematical equation that describes the bell-shaped curve of the normal probability density function is given by

Normal Probability Density Function

$$f(x) = \frac{1}{\sqrt{2\pi}\,\sigma}\, e^{-(x-\mu)^2/2\sigma^2} \tag{6.2}$$

where μ is the mean, σ is the standard deviation, $\pi = 3.14159$ and $e = 2.71828$. The value of $f(x)$ for any choice of x gives the height of the curve (see Figure 6.3).

We make some observations about the characteristics of the normal probability distribution:

1. There is an entire family of normal probability distributions with each specific normal distribution being differentiated by its mean μ and its standard deviation σ.
2. The highest point on the normal curve occurs at the mean, which is also the median and mode of the distribution.
3. The mean of the distribution can be any numerical value: negative, zero, or positive. Three normal curves with the same standard deviation but three different means (-10, 0, and 20) are shown in Figure 6.4.
4. The normal probability distribution is symmetric, with the tails of the curve extending indefinitely in both directions and theoretically never touching the horizontal axis.
5. The standard deviation determines the width of the probability density function. Larger values of the standard deviation result in wider, flatter curves, showing more dispersion in the random variable. Two normal distributions with the same mean but different values for the standard deviation are shown in Figure 6.5.
6. Probabilities for the normal random variable are given by areas under the normal curve. Probabilities for some commonly used intervals are:

FIGURE 6.3 Bell-shaped Curve for the Normal Probability Distribution

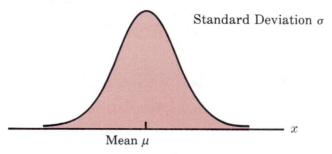

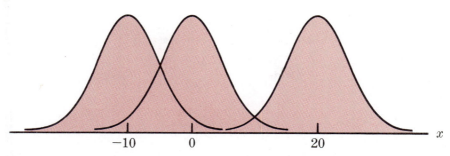

FIGURE 6.4 Three Normal Curves with Different Means but Same Standard Deviation

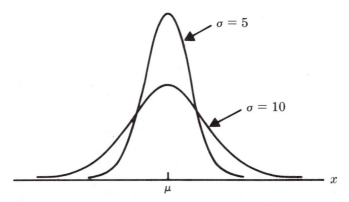

FIGURE 6.5 Two Normal Curves with Same Mean but Different Standard Deviations

  a. 68.26% of the time, a normal random variable assumes a value within plus or minus 1 standard deviation of its mean.
  b. 95.44% of the time, a normal random variable assumes a value within plus or minus 2 standard deviations of its mean.
  c. 99.72% of the time, a normal random variable assumes a value within plus or minus 3 standard deviations of its mean. Figure 6.6 shows properties (a), (b), and (c) graphically.
7. The total area under the curve for the normal probability distribution is 1. (This is true for all continuous probability distributions.)

The Standard Normal Probability Distribution

A random variable that has a normal distribution with a mean of 0 and a standard deviation of 1 is said to have a *standard normal probability distribution*. The letter z is commonly used to designate this particular normal random variable. The graph of the standard normal probability distribution is shown in Figure 6.7. Note that the standard normal probability distribution has the same general appearance as other normal distributions but with the special properties of $\mu = 0$ and $\sigma = 1$.

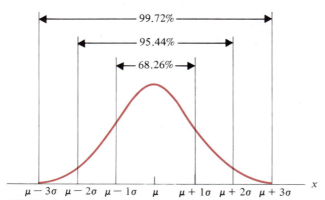

FIGURE 6.6   Areas Under the Curve for any Normal Probability Distribution

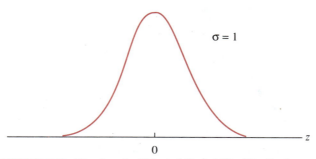

FIGURE 6.7   The Standard Normal Probability Distribution

With a continuous random variable, probability calculations are always concerned with finding the probability that the random variable assumes any value in an interval between two specific points c and d. The probability that a continuous random variable assumes a value between the two points c and d is the area under the graph of the probability density function between c and d. Because of its wide applicability, areas under the standard normal curve have been computed and are available in tables that can be used in computing the probability values for the standard normal probability distribution. Table 6.1 is such a table. This table is also available as Table 1 of Appendix B and inside the back cover of the text.

Let us show how the table of probabilities for the standard normal probability distribution (Table 6.1) can be used by considering some examples. Later we will see how the table for the standard normal distribution can be used to compute probabilities for any normal distribution.

EXAMPLE 6.2

What is the probability that the z value for the standard normal random variable will be between 0.00 and 1.00? That is, what is $P(0.00 \leq z \leq 1.00)$? The shaded region in the following graph shows this area or probability.

The entries in Table 6.1 give the area under the standard normal curve between the mean, $z = 0$, and a specified positive value of z. In this case we are interested in the

TABLE 6.1

Areas, or Probabilities, for the Standard Normal Distribution

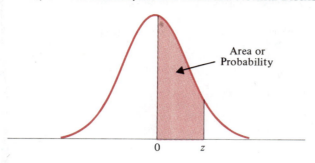

z	.00	.01	.02	.03	.04	.05	.06	.07	.08	.09
.0	.0000	.0040	.0080	.0120	.0160	.0199	.0239	.0279	.0319	.0359
.1	.0398	.0438	.0478	.0517	.0557	.0596	.0636	.0675	.0714	.0753
.2	.0793	.0832	.0871	.0910	.0948	.0987	.1026	.1064	.1103	.1141
.3	.1179	.1217	.1255	.1293	.1331	.1368	.1406	.1443	.1480	.1517
.4	.1554	.1591	.1628	.1664	.1700	.1736	.1772	.1808	.1844	.1879
.5	.1915	.1950	.1985	.2019	.2054	.2088	.2123	.2157	.2190	.2224
.6	.2257	.2291	.2324	.2357	.2389	.2422	.2454	.2486	.2518	.2549
.7	.2580	.2612	.2642	.2673	.2704	.2734	.2764	.2794	.2823	.2852
.8	.2881	.2910	.2939	.2967	.2995	.3023	.3051	.3078	.3106	.3133
.9	.3159	.3186	.3212	.3238	.3264	.3289	.3315	.3340	.3365	.3389
1.0	.3413	.3438	.3461	.3485	.3508	.3531	.3554	.3577	.3599	.3621
1.1	.3643	.3665	.3686	.3708	.3729	.3749	.3770	.3790	.3810	.3830
1.2	.3849	.3869	.3888	.3907	.3925	.3944	.3962	.3980	.3997	.4015
1.3	.4032	.4049	.4066	.4082	.4099	.4115	.4131	.4147	.4162	.4177
1.4	.4192	.4207	.4222	.4236	.4251	.4265	.4279	.4292	.4306	.4319
1.5	.4332	.4345	.4357	.4370	.4382	.4394	.4406	.4418	.4429	.4441
1.6	.4452	.4463	.4474	.4484	.4495	.4505	.4515	.4525	.4535	.4545
1.7	.4554	.4564	.4573	.4582	.4591	.4599	.4608	.4616	.4625	.4633
1.8	.4641	.4649	.4656	.4664	.4671	.4678	.4686	.4693	.4699	.4706
1.9	.4713	.4719	.4726	.4732	.4738	.4744	.4750	.4756	.4761	.4767
2.0	.4772	.4778	.4783	.4788	.4793	.4798	.4803	.4808	.4812	.4817
2.1	.4821	.4826	.4830	.4834	.4838	.4842	.4846	.4850	.4854	.4857
2.2	.4861	.4864	.4868	.4871	.4875	.4878	.4881	.4884	.4887	.4890
2.3	.4893	.4896	.4898	.4901	.4904	.4906	.4909	.4911	.4913	.4916
2.4	.4918	.4920	.4922	.4925	.4927	.4929	.4931	.4932	.4934	.4936
2.5	.4938	.4940	.4941	.4943	.4945	.4946	.4948	.4949	.4951	.4952
2.6	.4953	.4955	.4956	.4957	.4959	.4960	.4961	.4962	.4963	.4964
2.7	.4965	.4966	.4967	.4968	.4969	.4970	.4971	.4972	.4973	.4974
2.8	.4974	.4975	.4976	.4977	.4977	.4978	.4979	.4979	.4980	.4981
2.9	.4981	.4982	.4982	.4983	.4984	.4984	.4985	.4985	.4986	.4986
3.0	.4986	.4987	.4987	.4988	.4988	.4989	.4989	.4989	.4990	.4990

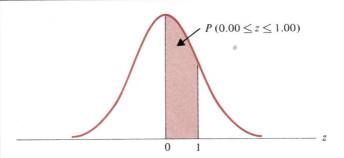
$P(0.00 \leq z \leq 1.00)$

area between $z = 0$ and $z = 1.00$. Thus we must find the entry in the table corresponding to $z = 1.00$. To do this, we first find 1.0 in the left-hand column of the table and then find .00 in the top row of the table. Then by looking in the body of the table we find that the 1.0 row of the table and the .00 column of the table intersect at the value of .3413. We have found the desired probability; $P(0.00 \leq z \leq 1.00) = .3413$. A portion of Table 6.1 showing these steps is shown below.

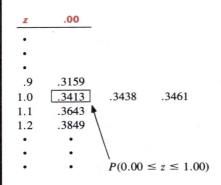

As another example, use Table 6.1 to show that the area between $z = 0.00$ and $z = 1.25$ is .3944. This area or probability value is found by using the the $z = 1.2$ row and the .05 column of the table.

EXAMPLE 6.3

What is the probability of obtaining a z value between $z = -1.00$ and $z = 1.00$? That is, what is $P(-1.00 \leq z \leq 1.00)$?

First note that we have already used Table 6.1 to show that the probability of a z value between $z = 0.00$ and $z = 1.00$ is .3413. Recall now that the normal probability distribution is *symmetric*. That is, the shape of the curve to the left of the mean is the mirror image of the shape of the curve to the right of the mean. Thus the probability of a z value between $z = 0.00$ and $z = -1.00$ is the *same* as the probability of a z value between $z = 0.00$ and $z = +1.00$. Hence the probability of a z value between $z = -1.00$ and $z = +1.00$ is

$$P(-1.00 \leq z \leq 0.00) + P(0.00 \leq z \leq 1.00) = .3413 + .3413 = .6826.$$

This area is shown graphically as follows.

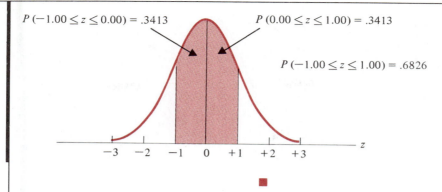

$P(-1.00 \leq z \leq 0.00) = .3413$

$P(0.00 \leq z \leq 1.00) = .3413$

$P(-1.00 \leq z \leq 1.00) = .6826$

In a manner similar to Example 6.3 we can use the values in Table 6.1 to show that the probability of a z value between -2.00 and $+2.00$ is $.4772 + .4772 = .9544$ and that the probability of a z value between -3.00 and $+3.00$ is $.4986 + .4986 = .9972$. Since we know that the total probability or total area under the curve for any continuous random variable must be 1.0000, the probability .9972 tells us that the value of z will almost always fall between -3.00 and $+3.00$.

EXAMPLE 6.4

What is the probability of obtaining a z value of at least 1.58? That is, what is $P(z \geq 1.58)$?

First, we use the $z = 1.5$ row and the .08 column of Table 6.1 to find that $P(0.00 \leq z \leq 1.58) = .4429$. Now, since the normal probability distribution is symmetric and the total area under the curve equals 1, we know that 50% of the area must be above the mean (i.e., $z = 0$) and 50% of the area must be below the mean. Since .4429 is the area between the mean and $z = 1.58$, the area or probability corresponding to $z \geq 1.58$ must be $.5000 - .4429 = .0571$. This probability is shown below.

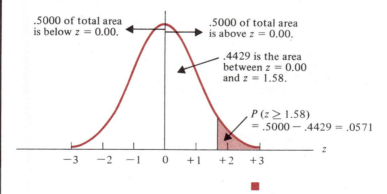

.5000 of total area is below $z = 0.00$.

.5000 of total area is above $z = 0.00$.

.4429 is the area between $z = 0.00$ and $z = 1.58$.

$P(z \geq 1.58) = .5000 - .4429 = .0571$

EXAMPLE 6.5

What is the probability the random variable z assumes a value of $-.50$ or larger? That is, what is $P(z \geq -.50)$.

To make this computation, we note that the probability we are seeking can be written as the sum of two probabilities: $P(z \geq -.50) = P(-.50 \leq z \leq 0.00) + P(z \geq 0.00)$. We have previously seen that $P(z \geq 0.00) = .50$. Also, we know that since the normal distribution is symmetric, $P(-.50 \leq z \leq 0.00) = P(0.00 \leq z \leq .50)$. Referring to Table 6.1 we find that $P(0.00 \leq z \leq .50) = .1915$. Therefore, $P(-.50 \leq z \leq 0.00) = .1915$. Thus $P(z \geq -.50) = P(-.50 \leq z \leq 0.00) + P(z \geq 0.00) = .1915 + .5000 = .6915$. The graph shows this area.

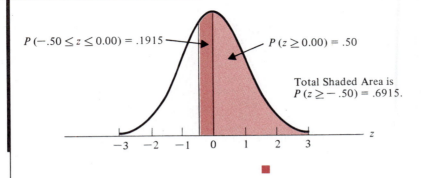

$P(-.50 \le z \le 0.00) = .1915$

$P(z \ge 0.00) = .50$

Total Shaded Area is
$P(z \ge -.50) = .6915$.

EXAMPLE 6.6

What is the probability of obtaining a z value between 1.00 and 1.58. That is, what is $P(1.00 \le z \le 1.58)$?

From Examples 6.2 and 6.4 we know that there is a .3413 probability of a z value between $z = 0.00$ and $z = 1.00$ and that there is a .4429 probability of a z value between $z = 0.00$ and $z = 1.58$. Thus there must be a $.4429 - .3413 = .1016$ probability of a z value between $z = 1.00$ and $z = 1.58$. Thus $P(1.00 \le z \le 1.58) = .1016$. This situation is shown graphically in the following figure.

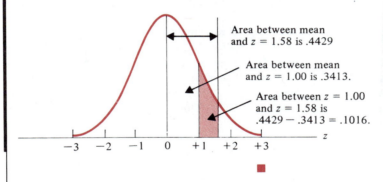

Area between mean and $z = 1.58$ is .4429

Area between mean and $z = 1.00$ is .3413.

Area between $z = 1.00$ and $z = 1.58$ is $.4429 - .3413 = .1016$.

EXAMPLE 6.7

Find a z value such that the probability of obtaining a larger z value is only .10. This situation is shown graphically as follows:

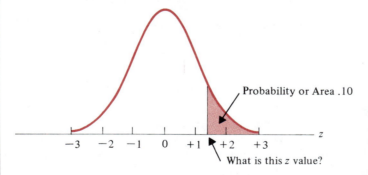

Probability or Area .10

What is this z value?

This problem is somewhat different from the examples we have considered thus far. The difference is that previously we specified the z value of interest and then found the corresponding probability, or *area*. In this example we are given the probability, or area, information and asked to find the corresponding z value. This can be found by using the table of areas for the standard normal probability distribution (Table 6.1) a little differently.

Recall that the body of Table 6.1 gives the area under the curve between the mean and a particular z value. In the above example we are given the information that the area in the upper tail of the curve is .10. Thus we must determine how much of the area is between the mean and the z value of interest. Since we know .5000 of the area is above the mean, $.5000 - .1000 = .4000$ must be the area under the curve *between* the mean and the desired z value. Scanning the body of the table, we find .3997 as the probability value closest to .4000. The section of the table providing this result is shown below:

z ...	.06	.07	.08	.09
·				
·				
·				
1.0			.3599	
1.1			.3810	
1.2	.3962	.3980	.3997	.4015
1.3			.4162	
1.4			.4306	
·				
·				
·				

Reading the z value from the left column and the top row of the table, we find that the corresponding z value is 1.28. Thus there will be an area of approximately .4000 (actually .3997) between the mean and $z = 1.28$. In terms of the question originally asked, there is approximately a .1000 probability of a z value larger than 1.28.

■

The examples illustrate that the table of areas for the standard normal probability distribution can be used to find probabilities associated with values of the standard normal random variable z. Two types of questions can be asked. The first type of question specifies a value, or values, for z and asks us to use the table to determine the corresponding areas, or probabilities. The second type of question provides an area, or probability, and asks us to use the table to determine the corresponding z value. Thus we need to remain flexible in terms of using the standard normal probability table to answer the desired probability question. In most cases, sketching a graph of the standard normal probability distribution and shading the appropriate area helps to visualize the situation and aid in determining the correct answer.

Computing Probabilities for Any Normal Distribution

The reason that we have been discussing the standard normal distribution so extensively is that probabilities for all normal distributions are computed using the standard normal distribution. That is when we have a normal distribution with any mean μ and any standard deviation σ, we answer probability questions about the distribution by first converting to the standard normal distribution. Then we can use Table 6.1 and the appropriate z values to find the desired probabilities. The formula used to convert any normal random variable x with mean μ and standard deviation σ to the standard normal random variable, z, is as follows.

$$z = \frac{x - \mu}{\sigma} \qquad\qquad (6.3)$$

A value of x equal to its mean μ results in $z = (\mu - \mu)/\sigma = 0$. Thus we see that a value of x equal to its mean μ corresponds to a value of z at its mean 0. Now suppose that x is one standard deviation above its mean; that is, $x = \mu + \sigma$. Applying Equation (6.3) we see that the corresponding z value is $z = [(\mu + \sigma) - \mu]/\sigma = \sigma/\sigma = 1$. Thus a value of x that is one standard deviation above its mean yields, $z = 1$. In other words, we can interpret the z value as *the number of standard deviations that the normal random variable, x, is from its mean, μ.*

To see how this conversion enables us to compute probabilities for any normal distribution, suppose we have a normal distribution with $\mu = 10$ and $\sigma = 2$. What is the probability that the random variable, x, is between 10 and 14? Using (6.3) we see that at $x = 10$, $z = (x - \mu)/\sigma = (10 - 10)/2 = 0$ and that at $x = 14$, $z = (14 - 10)/2 = 4/2 = 2$. Thus the answer to our question about the probability of x being between 10 and 14 is given by the equivalent probability that z is between 0 and 2 for the standard normal distribution. In other words, the probability that we are seeking is the probability that the random variable x is between its mean and two standard deviations above the mean. Using $z = 2.00$ and Table 6.1, we see that the probability is .4772. Hence the probability that x is between 10 and 14 is .4772.

EXAMPLE 6.8

IQ scores for a group of sixth graders are normally distributed with a mean of 100 and a standard deviation of 12. What is the probability that a randomly selected student will have an IQ score between 90 and 110?

Letting x be the normally distributed IQ score, we must compute $P(90 \le x \le 110)$ based on the information $\mu = 100$ and $\sigma = 12$. Converting the x values to the corresponding z values we have:

$$\text{For } x = 90, \qquad z = \frac{x - \mu}{\sigma} = \frac{90 - 100}{12} = -.83$$

$$\text{For } x = 110, \qquad z = \frac{x - \mu}{\sigma} = \frac{110 - 100}{12} = +.83$$

Using Table 6.1 for $z = +.83$, we find that the probability of z being between zero and $+.83$ is .2967. Also, since $P(-.83 \le z \le 0.00) = .2967$, we have $P(-.83 \le z \le +.83) = .2967 + .2967 = .5934$. In terms of the IQ scores, we now know the probability that a randomly selected a student will have an IQ score between 90 and 110 is .5934. The graphical representation of this probability with the corresponding z values is shown.

Using the IQ scores with mean $\mu = 100$ and standard deviation $\sigma = 12$, what is the probability of randomly selecting a student with an IQ score of 120 or more?

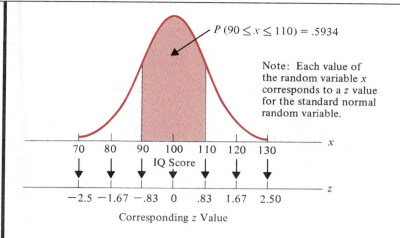

$P(90 \leq x \leq 110) = .5934$

Note: Each value of the random variable x corresponds to a z value for the standard normal random variable.

For $x = 120$, we have

$$z = \frac{x - \mu}{\sigma} = \frac{120 - 100}{12} = 1.67$$

Using Table 6.1, we find an area of .4525 between $z = 0$ and $z = 1.67$. Thus .4525 is the probability that a student's IQ score is between the mean $\mu = 100$ and the IQ score of 120. This is not the answer to the question seeking the probability that a randomly selected student will have an IQ score of 120 or more. However, since .5000 of the area under a normal curve is above the mean, we see the probability that a randomly selected student will have an IQ score of 120 or more must be .5000 − .4525 = .0475. Less than 5% of the students will have an IQ score of 120 or more.

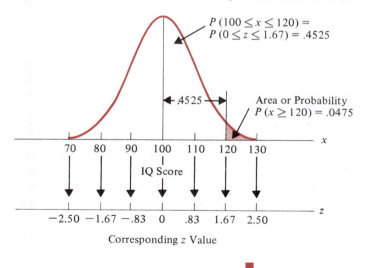

$P(100 \leq x \leq 120) = P(0 \leq z \leq 1.67) = .4525$

Area or Probability
$P(x \geq 120) = .0475$

EXAMPLE 6.9

The Grear Tire Company has just developed a new steel-belted radial tire that will be sold through a national chain of discount stores. From road tests with the tires, it is found that tire mileage is normally distributed with a mean tire mileage of $\mu = 36,500$ miles and a standard deviation of $\sigma = 5000$ miles. What is the probability that a tire will last at least 30,000 miles?

The z value corresponding to $x = 30,000$ miles is

$$z = \frac{x - \mu}{\sigma} = \frac{30,000 - 36,500}{5000} = \frac{-6500}{5000} = -1.30$$

Using Table 6.1, we find that $P(-1.30 \leq z \leq 0.00) = .4032$. Thus the probability a tire will provide at least 30,000 miles of usage is $P(x \geq 30,000) = .4032 + .5000 = .9032$. That is, better than 90% of the tires can be expected to wear for at least 30,000 miles. This situation is shown graphically. Again, values for the corresponding standard normal random variable are also shown.

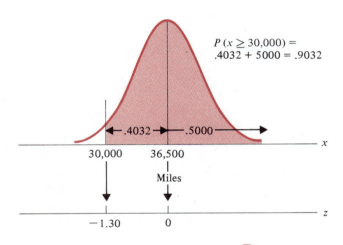

EXAMPLE 6.10

Test scores for a college midterm examination are normally distributed with a mean of $\mu = 72$ and a standard deviation of $\sigma = 13$. Suppose the professor wishes to assign the grade of A to the 15% of the students obtaining the highest scores on the exam. What is the cutoff score for the A grade?

This is a situation in which the probability is known and the question concerns finding a particular value for the random variable, the exam score. With 15% of the area in the upper tail of the normal distribution, we know that the area between the mean score of $\mu = 72$ and the exam score required to obtain the grade of A must be $.5000 - .1500 = .3500$. Using the *body* of Table 6.1, we find the probability value closest to .3500 is .3508. Using the left-hand column and the top row of the table, we find that the z value corresponding to .3508 is $z = 1.04$. This tells us that the midterm exam score required to obtain the grade of A must be at least 1.04 standard deviations above the mean. Computing the corresponding value of x, we find

$$x = \mu + 1.04\sigma = 72 + 1.04(13) = 85.52$$

as the minimum score a student must obtain in order to receive the grade of A for the exam.

Summarizing, the key to answering probability questions about the normal distribution is converting values of the normal random variable x to the corresponding z values and interpreting the table of probabilities for the standard normal probability distribution.

Notes and Comments

Suppose that we know the mean of a normal distribution and the percentage of values greater than some specified value of x. Using (6.3) we can solve for σ.

$$\sigma = \frac{x - \mu}{z}$$

Since in this case x, μ, and z are known, we see that the standard deviation can be computed from the given information. For example, suppose that $\mu = 100$ and 2.5% of the values are greater than 150. Using Table 6.1, we find that $z = 1.96$. Hence

$$\sigma = \frac{150 - 100}{1.96} = 25.51$$

EXERCISES

8. Using Figure 6.6 as a guide, sketch a normal curve for a random variable x that has a mean $\mu = 100$ and a standard deviation $\sigma = 10$. Label the horizontal axis with values of 70, 80, 90, 100, 110, 120 and 130.

9. The length of time required to complete a college examination is normally distributed with a mean of $\mu = 50$ minutes and a standard deviation of $\sigma = 5$ minutes.
   a. Sketch a normal curve for the length of the examination. Label the horizontal axis with values of 35, 40, 45, 50, 55, 60 and 65 minutes. Figure 6.6 shows that the normal curve almost touches the horizontal line at three standard deviations below and at three standard deviations above the mean (in this case at 35 and 65).
   b. What is the probability that a student will take between 45 and 55 minutes to complete the exam?
   c. What is the probability that a student will take between 40 and 60 minutes to complete the exam?

10. Assume that scores on a verbal skills test are normally distributed with a mean of 500 and a standard deviation of 100.
   a. Sketch a normal curve for the test scores. Refer to Figure 6.6 to see that the normal curve almost touches the horizontal axis at three standard deviations below and three standard deviations above the mean.
   b. What is the probability that a randomly selected student who takes the test scores between 400 and 600?
   c. What is the probability that a randomly selected student who takes the test scores between 300 and 700?
   d. What is the probability that a randomly selected student scores 500 or better on the exam?

e. What is the probability that a randomly selected student scores between 500 and 700 on the exam?

f. What is the probability that a randomly selected student who takes the test scores 700 or higher? (*Hint:* Use your answers to (d) and (e) to help answer this question.)

11. The mean annual precipitation in the state of Ohio is 32 inches with a standard deviation of 4 inches.

   a. Sketch a normal curve for the annual precipitation in the state of Ohio. Label the horizontal axis at the mean as well as at 1, 2, and 3 standard deviations above and below the mean.

   b. What is the probability that the precipitation in any one year will be between 24 and 40 inches?

   c. What is the probability that the precipitation in any one year will be between 32 and 40 inches?

   d. What is the probability that the precipitation in any one year will be 32 inches or more?

   e. What is the probability that the precipitation in any one year will be 36 inches or more?

12. Automobile painting times at the Jay Nickerson Auto Body Painting Shop are believed to be normally distributed. It is estimated that 68.26% of the automobiles are painted in from 2.5 to 3.5 hours. In addition, it is estimated that 95.44% of the automobiles are painted in from 2 to 4 hours.

   a. What is the mean painting time for the automobiles?

   b. What is the standard deviation of painting times for the automobiles?

   c. What is the probability that an automobile will be painted in 3 hours or less?

   d. What is the probability that an automobile will require between 2 and 3 hours for painting?

   e. What is the probability that an automobile will be painted in 2 hours or less?

   f. What is the probability that an automobile will require 2 hours or more to be painted?

13. Given that z is the standard normal random variable, sketch the standard normal curve. Label the horizontal axis at values of $-3, -2, -1, 0, 1, 2,$ and 3. Then use the table of probabilities for the standard normal distribution to compute the following probabilities.

   a. $P(0 \leq z \leq 1)$

   b. $P(0 \leq z \leq 1.5)$

   c. $P(0 < z < 2)$

   d. $P(0 < z < 2.5)$

14. Given that z is the standard normal random variable, compute the following probabilities.

   a. $P(-1 \leq z \leq 0)$

   b. $P(-1.5 \leq z \leq 0)$

   c. $P(-2 < z < 0)$

   d. $P(-2.5 \leq z \leq 0)$

   e. $P(-3 < z \leq 0)$

15. Given that z is the standard normal random variable, compute the following probabilities.

   a. $P(0 \leq z \leq .83)$

   b. $P(-1.57 \leq z \leq 0)$

   c. $P(z > .44)$

   d. $P(z \geq -.23)$

   e. $P(z < 1.20)$

   f. $P(z \leq -.71)$

16. Given that z is the standard normal random variable, compute the following probabilities.

   a. $P(-1.98 \leq z \leq .49)$

   b. $P(.52 \leq z \leq 1.22)$

   c. $P(-1.75 \leq z \leq -1.04)$

17. Given that z is the standard normal random variable, find z for each situation.

   a. The area between 0 and z is .4750.

b. The area between 0 and z is .2291.

c. The area to the right of z is .1314.

d. The area to the left of z is .6700.

18. Given that z is the standard normal random variable, find z for each situation.

    a. The area to the left of z is .2119.

    b. The area to between $-z$ and z is .9030.

    c. The area between $-z$ and z is .2052.

    d. The area to the left of z is .9948.

    e. The area to the right of z is .6915.

19. Given that z is the standard normal random variable, find z for each situation.

    a. The area to the right of z is .01.

    b. The area to the right of z is .025.

    c. The area to the right of z is .05.

    d. The area to the right of z is .10.

20. For a normal random variable with $\mu = 50$ and $\sigma = 10$ find the following probabilities.

    a. $P(x \geq 60)$

    b. $P(x < 30)$

    c. $P(42 \leq x < 60)$

    d. $P(65 \leq x < 75)$

21. For a normal random variable with $\mu = 100$ and $\sigma = 8$, answer the following.

    a. Find a value of x such that the probability of exceeding it is .50.

    b. Find a value of x such that the probability of a smaller value occurring is .40.

    c. Find a value of x such that the probability of exceeding it is .72.

22. The demand for a new product is assumed to be normally distributed with $\mu = 200$ and $\sigma = 40$. Letting x be the number of units demanded, find the following.

    a. $P(180 \leq x \leq 220)$

    b. $P(x \geq 250)$

    c. $P(x \leq 100)$

    d. $P(225 \leq x \leq 250)$

23. The Webster National Bank is reviewing its service charge and interest-paying policies on checking accounts. The bank has found that the average daily balance on personal checking accounts is $550.00, with a standard deviation of $150.00. In addition, the average daily balances have been found to be approximately normally distributed.

    a. What percentage of personal checking account customers carry average daily balances in excess of $800.00?

    b. What percentage of the bank's customers carry average daily balances below $200.00?

    c. What percentage of the bank's customers carry average daily balances between $300.00 and $700.00?

    d. The bank is considering paying interest to customers carrying average daily balances in excess of a certain amount. If the bank does not want to pay interest to more than 5% of its customers, what is the minimum average daily balance it should be willing to pay interest on?

24. General Hospital's patient account division has compiled data on the age of accounts receivable. The data collected indicate that the age of the accounts follows a normal distribution, with $\mu = 28$ days and $\sigma = 8$ days.

    a. What percentage of the accounts are between 20 and 40 days old?

    b. The hospital administrator is interested in sending reminder letters to the oldest 15% of accounts. How many days old should an account be before a reminder letter is sent?

    c. The hospital administrator would like to give a 5% discount to those accounts that pay their balance by the 21st day. What percentage of the accounts will receive the discount?

25. The time required to complete a final examination in a particular college course is normally distributed, with a mean of 80 minutes and a standard deviation of 10 minutes. Answer the following questions:

    a. What is the probability of completing the exam in 1 hour or less?

    b. What is the probability a student will complete the exam in more than 60 minutes but less than 75 minutes?

    c. Assume that the class has 60 students and that the examination period is 90 minutes in length. How many students do you expect will be unable to complete the exam in the allotted time?

26. The useful life of a computer terminal at a university computer center is known to be normally distributed, with a mean of 3.25 years and a standard deviation of .5 years.

    a. Historically 22% of the terminals have had a useful life less than the manufacturer's advertised life. What is the manufacturer's advertised life for the computer terminals?

    b. What is the probability that a computer terminal will have a useful life of at least 3 but less than 4 years?

27. From past experience, the management of a well known fast-food restaurant estimates that the number of weekly customers at a particular location is normally distributed, with a mean of 5000 and a standard deviation of 800 customers.

    a. What is the probability that on a given week the number of customers will be between 4760 and 5800?

    b. What is the probability of more than 6500 customers?

    c. For 90% of the weeks the number of customers should exceed what amount?

6.3   Normal Approximation of Binomial Probabilities

In Chapter 5 we introduced the binomial probability distribution as a means for determining the probability of x successes in n trials of a binomial experiment. In cases where the number of trials, n, is large, binomial tables are not usually available and the computations associated with the binomial probability function are not practical. For instance with $n = 120$ and $p = .34$, the probability of 40 successes is given by

$$P(x = 40) = \frac{n!}{x!(n-x)!}p^x(1-p)^{n-x} = \frac{120!}{40!\ 80!}(.34)^{40}(.66)^{80}$$

Evaluating such an expression is laborius and can lead to rounding errors.

    In situations such as this it is often possible to use the normal probability distribution to obtain good approximations of binomial probabilities. In this section we show how the normal probability distribution can be used for this purpose. The normal approximation provides acceptable accuracy whenever the number of trials, n, and the probability of success on each trial, p, have values such that both $np \geq 5$ and $n(1-p) \geq 5$.

EXAMPLE 6.11

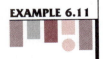

A particular company has a history of making errors on 10% of its invoices. In a sample of 100 invoices, what is the probability that exactly 12 invoices have an error?

    Note that this is a binomial experiment with $n = 100$ trials, $p = .10$ and $x = 12$. Rather than using the binomial probability function directly, we want to show how the normal probability distribution can be used to approximate the desired probability. Checking the requirements for the normal approximation we find that $np = 100(.10) = 10$ and $n(1 - p) = 100(.90) = 90$ are both at least 5. Thus, as previously stated, the normal approximation should provide good results.

Recall from Chapter 5 that the mean of a binomial random variable is $\mu = np$ and standard deviation is $\sigma = \sqrt{np(1 - p)}$. For this example, we have $\mu = np = 100(.10) = 10$ and $\sigma = \sqrt{100(.10) .90} = 3$. A normal distribution with this mean and standard deviation is shown in Figure 6.8. This is the normal distribution that is used to approximate the probabilities in this situation.

Also recall that with a continuous probability distribution, probabilities are computed as areas under the curve. As a result, the probability of any particular value for a continuous random variable is *zero*. Thus to approximate the binomial probability of 12 successes, we compute the area under the corresponding normal curve between 11.5 and 12.5. The .5 that we add to and subtract from 12 to enable the use of a continuous distribution to approximate discrete probabilities is called the *continuity correction*. Thus the interval $11.5 \leq x \leq 12.5$ for the normal random variable is used to approximate the probability that $x = 12$ for the discrete binomial distribution. Thus the binomial probability of $f(12)$ is approximated by the normal probability, $P (11.5 \leq x \leq 12.5)$.

Using the normal distribution shown in Figure 6.8, we use the following z values.

$$\text{At } x = 12.5, \qquad z = \frac{x - \mu}{\sigma} = \frac{12.5 - 10}{3} = .83$$

$$\text{At } x = 11.5, \qquad z = \frac{x - \mu}{\sigma} = \frac{11.5 - 10}{3} = .50$$

Using Table 6.1, we find the area under the curve between 0 and .83 is .2967. Similarly, the area under the normal curve between 0 and .50 is .1915. Therefore, the area between $z = .50$ and $z = .83$ is $.2967 - .1915 = .1052$. Thus the normal approximation of exactly 12 errors in the 100 invoices is .1052. As it turns out, the actual binomial probability is .0988. Thus the error in our approximation is $.1052 - .0988 = .0064$. This is a pretty good approximation.

■

FIGURE 6.8 Normal Approximation of a Binomial Probability with $n = 100$ and $p = .10$: The Probability of 12 Errors in 100 Trials is Approximately .1052

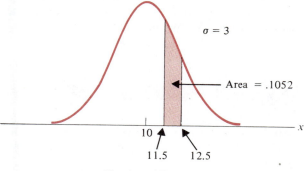

Number of Errors

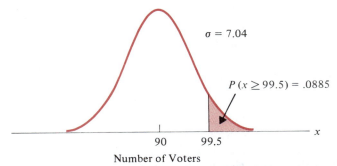

$\sigma = 7.04$

$P(x \geq 99.5) = .0885$

90   99.5

x

Number of Voters

FIGURE 6.9   Normal Approximation to a Binomial Probability Distribution with $n = 200$ and $p = .45$: The Approximation to the Binomial Probability of at least 100 Voters is .0885

EXAMPLE 6.12

It is believed that 45% of a large population of registered voters favor a particular candidate for the state senate. A public opinion poll uses a randomly selected sample of voters and asks each person polled to indicate his or her preference for the candidates. What is the probability that a weekly poll based on the responses of 200 registered voters will show at least 50% of the voters favoring the candidate? That is, what is the probability at least 100 of the 200 voters will favor the candidate?

First note that this is a binomial probability experiment with $n = 200$ voters and $p = .45$. The binomial probability question asks for the probability of at least 100 successes. With $np = 200(.45) = 90$ and $n(1 - p) = 200(.55) = 110$, we see that the normal probability distribution can be used to approximate the desired binomial probability. The mean for the distribution is $\mu = np = 200(.45) = 90$ and the standard deviation is $\sigma = \sqrt{np(1 - p)} = \sqrt{200(.45)(.55)} = 7.04$. Using the continuity correction we see that the interval 99.5 to 100.5 under the normal curve with $\mu = 90$ and $\sigma = 7.04$ provides the approximation of the binomial probability of 100 voters favoring the candidate. Since we are asking for the probability of 100 or more voters, we must compute the corresponding normal distribution probability of $x \geq 99.5$. The graph of the normal probability distribution approximation is shown in Figure 6.9.

Converting to the appropriate z value, we find

$$\text{At } x = 99.5, \qquad z = \frac{x - \mu}{\sigma} = \frac{99.5 - 90}{7.04} = 1.35$$

Using Table 6.1, the area between $z = 0$ and $z = 1.35$ is found to be .4115. Thus we know there must be a $.5000 - .4115 = .0885$ probability of a value of 99.5 or more. We conclude that there is approximately a .0885 probability that the sample of 200 voters will show at least 100 voters favoring the candidate.

■

EXERCISES

For a binomial experiment with $n = 20$ and $p = .4$, compute the following probabilities using the normal approximation and compare the answer with that obtained using the binomial probability tables.

a. $P(x = 3)$
b. $P(x = 4)$

    c. $P(x \geq 8)$

    d. $P(x < 6)$

29. For a binomial experiment with $n = 500$ and $p = .07$, compute the following probabilities using the normal approximation.

    a. $P(x > 40)$

    b. $P(20 \leq x \leq 40)$

    c. $P(x > 50)$

    d. $P(x \leq 30)$

30. In order to obtain cost savings, a company is considering offering an early retirement incentive for its older management personnel. The consulting firm that designed the early retirement program has found that approximately 22% of the employees qualifying for the program will select early retirement during the first year of eligibility. Assume that the company offers the early retirement program to 50 of its management personnel.

    a. What is the expected number of employees who will elect early retirement in the first year?

    b. What is the probability at least 8 but not more than 12 employees will elect early retirement in the first year?

    c. What is the probability that 15 or more employees will select the early retirement option in the first year?

    d. For the program to be judged successful, the company believes that it should entice at least 10 management employees to elect early retirement in the first year. What is the probability that the program is successful?

31. Thirty percent of the students at a particular university attended Catholic high schools. A random sample of 50 of this university's students has been taken. Use the normal approximation to the binomial probability distribution to answer the following questions:

    a. What is the probability that exactly 10 of the students selected attended Catholic high schools?

    b. What is the probability that 20 or more of the students attended Catholic high schools?

    c. What is the probability that the number of students from Catholic high schools is between 10 and 20, inclusive?

32. A Myrtle Beach Resort Hotel has 120 rooms. In the spring months, hotel room occupancy is approximately 75%. Use the normal approximation to the binomial distribution to answer the following questions:

    a. What is the probability that at last half the rooms are occupied on a given day?

    b. What is the probability that 100 or more rooms are occupied on a given day?

    c. What is the probability that 70 or fewer rooms are occupied on a given day?

33. It is known that 30% of all customers of a major national charge card pay their bills in full before any interest charges are incurred. Answer the following questions for a group of 150 credit card holders:

    a. What is the probability that between 40 and 60 customers pay their account charges before any interest charges are incurred? That is, find $P(40 \leq x \leq 60)$.

    b. What is the probability that 30 or fewer customers pay their account charges before any interest charges are incurred?

6.4   The Exponential Probability Distribution

A continuous probability distribution that is often useful in describing the time it takes to complete a task is the *exponential probability distribution*. The exponential random variable can be used to describe such things as the time between arrivals at a carwash, the time required to load a truck, the distance between major defects in a highway, and so on. The exponential probability density function is

---

Exponential Probability Density Function

$$f(x) = \frac{1}{\mu}e^{-x/\mu} \qquad \text{for } x \geq 0, \, \mu > 0 \tag{6.4}$$

where μ is the mean of the probability distribution and $e = 2.71828$.

EXAMPLE 6.13

The time it takes to load a truck at the loading dock for Schips Department Store is described by an exponential probability distribution. The mean, or average, time to load a truck is 15 minutes ($\mu = 15$); thus the appropriate probability density function is

$$f(x) = \frac{1}{15}e^{-x/15} \qquad \text{for } x \geq 0$$

The graph of this density function is shown in Figure 6.10.

■

FIGURE 6.10   Exponential Probability Distribution for the Schips Loading Dock Example

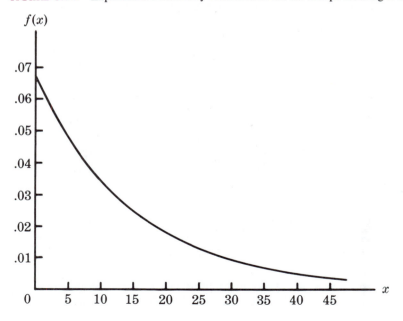

Computing Probabilities for the Exponential Distribution

As with any continuous probability distribution, the area under the curve corresponding to a given interval provides the probability that the random variable takes on a value in that interval.

EXAMPLE 6.13 (continued)

At the Schips loading dock, the probability that it takes 6 *minutes or less* ($x \le 6$) to load a truck is given by the area under the curve in Figure 6.10 from $x = 0$ to $x = 6$. Similarly, the probability that a truck is loaded in 18 *minutes or less* ($x \le 18$) is the area under the curve from $x = 0$ to $x = 18$. Note also that the probability that it takes between 6 minutes and 18 minutes ($6 \le x \le 18$) to load a truck is given by the area from $x = 6$ to $x = 18$.

■

In order to compute exponential probabilities such as those described above, we make use of the following formula, which provides the probability of obtaining a value for the exponential random variable of less than or equal to some specific value of x, denoted by d in the formula:

Computing Exponential Distribution Probabilities

$$P(x \le d) = 1 - e^{-d/\mu} \tag{6.5}$$

EXAMPLE 6.13 (continued)

For the Schips loading dock, formula (6.5) becomes

$$P(\text{loading time} \le d) = 1 - e^{-d/15}$$

Hence, the probability that it takes 6 minutes or less ($x \le 6$) to load a truck is

$$P(x \le 6) = 1 - e^{-6/15} = .3297$$

Note also the probability that it takes 18 minutes or less ($x \le 18$) to load a truck is

$$P(x \le 18) = 1 - e^{-18/15} = .6988$$

Thus we see that the probability that it takes between 6 minutes and 18 minutes to load a truck is equal to $.6988 - .3297 = .3691$. Probabilities for any other interval can be computed in a similar manner.

■

Relationship Between the Poisson and Exponential Distributions

In Section 5.5 we introduced the Poisson distribution as a discrete probability distribution that is often useful when dealing with the number of occurrences of an event over a specified interval of time or space. Recall that the Poisson probability function is

$$f(x) = \frac{\mu^x e^{-\mu}}{x!} \quad \text{for } x = 0, 1, 2, \ldots$$

where

$$\mu = \text{expected value or average number of occurrences in an interval.}$$

The continuous exponential probability distribution is related to the discrete Poisson distribution in that if the Poisson distribution provides an appropriate description of the number of occurrences during an interval, then the exponential distribution provides a description of the length of the interval between occurrences.

EXAMPLE 6.14

The number of cars that arrives at Hatcher's Car Wash during 1 hour of operation is described by a Poisson probability distribution with a mean of 10 cars per hour. Thus the probability function that provides the probability of x arrivals per hour is

$$f(x) = \frac{10^x e^{-10}}{x!} \quad \text{for } x = 0, 1, 2, \ldots$$

Since the average number of arrivals is 10 cars per hour, the average time between cars arriving is

$$\frac{1 \text{ hour}}{10 \text{ cars}} = .1 \text{ hour/car}$$

Thus the corresponding exponential distribution that describes the time between the arrival of cars has a mean of .1 hour per car; the appropriate exponential probability density function is

$$f(x) = \frac{1}{.1} e^{-x/.1} = 10e^{-10x} \text{ for } x \geq 0$$

■

---

34. Answer the following questions concerning the given exponential probability distribution

$$f(x) = \tfrac{1}{4} e^{-x/4} \quad \text{for } x \geq 0$$

a. Find $P(x \leq 4)$.
b. Find $P(x \leq 2)$.
c. Find $P(2 \leq x \leq 4)$.
d. Find $P(1 \leq x \leq 4)$.

35. Answer the following questions concerning the given exponential probability distribution

$$f(x) = \frac{1}{8} e^{-x/8} \qquad \text{for } x \geq 0$$

a. Find $P(x \leq 6)$.
b. Find $P(x \leq 4)$.
c. Find $P(x \geq 6)$.
d. Find $P(4 \leq x \leq 6)$.

36. There were 34 traffic fatalities in Clermont county Ohio (*The Cincinnati Enquirer,* December 8, 1988) during 1987. Assume that an exponential distribution accurately describes the time between fatalities.
a. What is the probability of no fatalities during 1 month?
b. What is the probability of no fatalities during 1 week?

37. The time between arrivals of vehicles at a particular intersection follows an exponential probability distribution with a mean of 12 seconds.
a. Sketch this exponential probability distribution.
b. What is the probability that the arrival time between vehicles is 12 seconds or less?
c. What is the probability that the arrival time between vehicles is 6 seconds or less?
d. What is the probability that there will be 30 or more seconds between arriving vehicles?

38. The lifetime (hours) of an electronic device is a random variable with the following exponential probability density function:

$$f(x) = \frac{1}{50} e^{-x/50} \qquad \text{for } x \geq 0$$

a. What is the mean lifetime of the device?
b. What is the probability the device fails in the first 25 hours of operation?
c. What is the probability the device operates 100 or more hours before failure?

39. A new automated production process has been averaging 2 breakdowns per day, where the number of breakdowns per day follows a Poisson probability distribution.
a. What is the mean time between breakdowns, assuming 8 hours of operation per day?
b. Show the exponential probability density function that can be used for the time between breakdowns.
c. What is the probability the process will run 1 hour or more before another breakdown?
d. What is the probability that the process can run a full 8-hour shift without a breakdown?

40. The time in minutes for a student using a computer terminal at the computer center of a major university follows an exponential probability distribution with a mean of .6 hours. Assume a second student arrives at the terminal just as another student is beginning to work on the terminal.
a. What is the probability the wait for the second student will be 15 minutes or less?
b. What is the probability the wait for the second student will be between 15 minutes and 45 minutes?
c. What is the probability the second student will have to wait an hour or more?

This chapter extended the discussion of probability distributions to the case of continuous random variables. The major conceptual difference between discrete and continuous probability distributions is in the method of computing probabilities. With discrete distributions the probability function, $f(x)$, provided the probability that the random variable, x, assumed various values. With continuous probability distributions, we associate a probability density function, denoted by $f(x)$. The difference is that the probability density function does not provide probability values for a continuous random variable. Probabilities are given by areas under the curve in the graph of the probability density function, $f(x)$. Since the area under the curve above a single point is zero, we observe that the probability of any particular value is zero for any continuous random variable.

Three continuous probability distributions, the uniform, normal and exponential were treated in some detail. Because of its wide range of applicability and because it will be used extensively in the remainder of the text, the normal probability distribution was given greater coverage. It is considered by many to be the most important distribution in probability and statistics. Each normal probability distribution belongs to a family of similar bell-shaped distributions with the specific normal distribution depending upon the value of the mean μ and standard deviation σ.

Tables of probabilities are available for a normal probability distribution with a mean of 0 and a standard deviation of 1. This special normal probability distribution is referred to as the standard normal probability distribution. Common notation is to use z to denote the standard normal random variable. The relationship between z and any other normal random variable x with mean μ and standard deviation σ is given by $z = (x - \mu)/\sigma$. The interpretation of a z value is that it indicates the number of standard deviations the normal random variable x is from its mean μ.

Probability questions about a random variable x with a normal distribution can be answered by first converting the random variable x to its corresponding z value and then using the table of areas for the standard normal probability distribution to determine the appropriate probabilities. We saw that two types of probability questions can be asked: An interval of values of the random variable is given and the question is to determine the probability the random variable assumes a value in the interval; alternatively, a probability value is given and we want to determine the values of the random variable yielding the given probabilities. Both types of questions can be answered by using the table of areas for the standard normal probability distribution. In making the probability calculations, it is recommended that a sketch of the appropriate normal curve be made as an aid to visualizing the probability information desired.

Finally, we noted that binomial probabilities can be difficult to compute whenever the number of trials, n, is large. However, if both $np \geq 5$ and $np(1 - p) \geq 5$, the normal probability distribution with $\mu = np$ and $\sigma = \sqrt{np(1 - p)}$ provides a good approximation to the binomial probabilities. A continuity correction must be utilized to account for the fact that the discrete binomial probability is being approximated by the continuous normal probability distribution.

Uniform probability distribution A continuous probability distribution where the probability that the random variable will assume a value in any interval of equal length is the same for each interval.

Probability density function   The function that defines the probability distribution of a continuous random variable.

Normal probability distribution   A continuous probability distribution. Its probability density function is bell shaped and determined by the mean μ and standard deviation σ.

Standard normal distribution   A normal distribution with a mean of 0 and a standard deviation of 1.

Continuity correction   A value of .5 that is added and subtracted from a value of x when a continuous probability distribution (e.g., the normal) is used to approximate a discrete probability distribution (e.g., the binomial).

Exponential probability distribution   A continuous probability distribution that is useful in describing the time, or space, between occurrences of an event.

KEY FORMULAS

UNIFORM PROBABILITY DENSITY FUNCTION

$$f(x) = \begin{cases} \dfrac{1}{b-a} & \text{for } a \leq x \leq b \\ 0 & \text{elsewhere} \end{cases} \tag{6.1}$$

NORMAL PROBABILITY DENSITY FUNCTION

$$f(x) = \frac{1}{\sqrt{2\pi}\,\sigma}\, e^{-(x-\mu)^2/2\sigma^2} \tag{6.2}$$

CONVERTING TO THE STANDARD NORMAL DISTRIBUTION

$$z = \frac{x-\mu}{\sigma} \tag{6.3}$$

EXPONENTIAL PROBABILITY DENSITY FUNCTION

$$f(x) = \frac{1}{\mu} e^{-x/\mu} \qquad \text{for } x \geq 0,\ \mu > 0 \tag{6.4}$$

COMPUTING EXPONENTIAL DISTRIBUTION PROBABILITIES

$$P(x \leq d) = 1 - e^{-d/\mu} \tag{6.5}$$

REVIEW QUIZ

TRUE/FALSE

1. The uniform probability distribution involves a discrete random variable.

2. For the uniform probability distribution, the height of the probability density function varies but reaches its peak at the mean.

3. The area under the curve to the left of the mean for a uniform probability distribution is .50.

4. As the standard deviation increases, the height of the normal curve increases.

5. The probability the normal random variable assumes a value within one standard deviation of the mean is .50.

6. A standard normal probability distribution has a mean of zero and a variance of 1.

7. In order to compute probabilities for a normal random variable, one must first convert to a standard normal probability distribution.

8. A continuity correction is needed whenever normal probabilities are computed.

9. The normal probability distribution may be used to approximate the binomial, provided that $np \geq 5$ and $n(1 - p) \geq 5$.

10. The probability of a normal random variable assuming a value within two standard deviations of its mean is approximately .95.

MULTIPLE CHOICE

11. The random variable x is uniformly distributed between 12 and 20.
   a. It has a mean of 16.
   b. It has a variance of 5⅓.
   c. The probability $x = 15$ is zero.
   d. All the above are true.

For questions 12–14, consider the normally distributed random variable x, which has a mean of 17 and a standard deviation of 3.

12. The probability that x is less than or equal to 14 is
   a. .1587
   b. .1765
   c. .3414
   d. .4986

13. If a value of x is randomly selected, the probability that it will be between 20 and 22 is
   a. .1112
   b. .3413
   c. .4525
   d. .7938

14. The probability that two randomly selected values of x will both have a value less than 22 is
   a. .2048
   b. .4525
   c. .9073
   d. .9525

15. Assume that a normal probability distribution is being used to approximate binomial probabilities for the case of $n = 500$ and $p = .10$. The value of σ that should be used is closest to
   a. 3
   b. 6
   c. 9
   d. 40

16. Scores on a reading skills test are normally distributed with a mean of $\mu = 500$ and a standard deviation of 50. An agency hires only people whose scores are in the top 5% of individuals taking the test. This company should consider hiring anyone who achieves at least a score of
 a. 500
 b. 538
 c. 550
 d. 583

17. If the average time between fouls in a basketball game is 1 minute, the probability of going 3 minutes with no fouls is closest to
 a. .003
 b. .010
 c. .050
 d. .100

SUPPLEMENTARY EXERCISES

41. The random variable x is described by the following uniform probability distribution:

$$f(x) = \begin{cases} 1/5 & \text{for } 25 \leq x \leq 30 \\ 0 & \text{otherwise} \end{cases}$$

 a. Find $E(x)$.
 b. Find $\text{Var}(x)$.
 c. Find $P(x \leq 28)$.
 d. Find $P(26 \leq x \leq 29)$.

42. Given that z is a standard normal random variable, compute the following probabilities.
 a. $P(-.72 \leq z \leq 0)$
 b. $P(-.35 \leq z \leq .35)$
 c. $P(.22 \leq z \leq .87)$
 d. $P(z \leq -1.02)$

43. Given that z is a standard normal random variable, compute the following probabilities.
 a. $P(z \geq -.88)$
 b. $P(z \geq 1.38)$
 c. $P(-.54 \leq z \leq 2.33)$
 d. $P(-1.96 \leq z \leq 1.96)$

44. Given that z is a standard normal random variable, find z if it is known that
 a. the area between $-z$ and z is .90
 b. the area to the right of z is .20
 c. the area between -1.66 and z is .25
 d. the area to the left of z is .40
 e. the area between z and 1.80 is .20

45. In an office building the waiting time for an elevator is found to be uniformly distributed between 0 minutes and 5 minutes.
 a. What is the probability density function, $f(x)$, for this uniform distribution?
 b. What is the probability of waiting longer than 3.5 minutes?
 c. What is the probability that the elevator arrives in the first 45 seconds?

d. What is the probability of a waiting time between 1 and 3 minutes?

e. What is the expected waiting time?

46. The time required to complete a particular assembly operation is uniformly distributed between 30 and 40 minutes.

a. What is the mathematical expression for the probability density function?

b. Compute the probability that the assembly operation will require more than 38 minutes to complete.

c. If management wants to set a time standard for this operation, what time should be selected such that 70% of the time the operation will be completed within the time specified?

d. Find the expected value and standard deviation for the assembly time.

47. A particular make of automobile is listed as weighing 4000 pounds. Because of weight differences due to the options ordered with the car, the actual weight varies uniformly between 3900 and 4100 pounds.

a. What is the mathematical expression for the probability density function?

b. What is the probability that the car will weigh less than 3950 pounds?

48. A soup company markets eight varieties of homemade soup throughout the Eastern states. The standard-size soup can holds a maximum of 11 ounces, while the label on each can advertises contents of 10¾ ounces. The extra ¼ ounce is to allow for the possibility of the automatic filling machine placing more soup than the company actually wants in a can. Past experience shows that the number of ounces the machine attempts to place in a can is approximately normally distributed, with a mean of 10¾ and a standard deviation of .1 ounces. What is the probability that the machine will attempt to place more than 11 ounces in a can, causing an overflow to occur?

49. The sales of High-Brite Toothpaste are believed to be approximately normally distributed, with a mean of 10,000 tubes per week and a standard deviation of 1500 tubes per week.

a. What is the probability that more than 12,000 tubes will be sold in any given week?

b. In order to have a .95 probability that the company will have sufficient stock to cover the weekly demand, how many tubes should be produced?

50. In the "Statistics in the News" at the beginning of the chapter, it was noted that an IQ score of 100 is considered average, and that approximately 1% of the population have an IQ score of 140 or more. Assume that IQ scores are normally distributed with $\mu = 100$.

a. Determine the standard deviation of IQ scores.

b. What percentage of the population has an IQ score lower than 120?

c. Specify an interval of scores such that 2.5% of the population can be expected to score higher and 2.5% can be expected to score lower.

51. Points scored by the winning team in NCAA college football games are approximately normally distributed, with a mean of 24 and a standard deviation of 6.

a. What is the probability that a winning team in a football game scores between 20 and 30 points; that is, $P(20 \leq x \leq 30)$?

b. How many points does a winning team have to score to be in the highest 20% of scores for college football games?

52. In 1985, the average household income for Americans was $23,618 (*Louis Rukeyser's Business Almanac,* Simon and Schuster, New York, 1988).

a. It was noted that 7.7% of the households earned less than $5000. Assuming that household income is normally distributed, what is the standard deviation of household income?

b. It was also noted that 14.8% of households earned more than $50,000. Does this seem reasonable given the standard deviation computed in part a? Explain.

c. In part a we said to assume that household income is normally distributed. Does this assumption appear to be reasonable? Explain.

53. Ward Doering Auto Sales is considering offering a special service contract that will cover the total cost of any service work required on leased vehicles. From past experience the company manager estimates that yearly service costs are approximately normally distributed, with a mean of $150 and a standard deviation of $25.

    a. If the company offers the service contract to customers for a yearly charge of $200, what is the probability that any one customer's service costs will exceed the contract price of $200?

    b. What is Ward's expected profit per service contract?

54. The attendance at football games at a certain stadium is normally distributed, with a mean of 45,000 and a standard deviation of 3000.

    a. What percentage of the time should attendance be between 44,000 and 48,000?

    b. What is the probability of the attendance exceeding 50,000?

    c. Eighty percent of the time the attendance should be at least how many?

55. The test scores from a college admissions test are normally distributed, with a mean of 450 and a standard deviation of 100.

    a. What percentage of the people taking the test score between 400 and 500?

    b. Suppose that someone receives a score of 630. What percentage of the people taking the test score better? What percentage score worse?

    c. If a particular university will not admit anyone scoring below 480, what percentage of persons taking the test would be acceptable to the university?

56. The Office Products Group of the former Burroughs Corporation manufactures plastic credit cards used in automatic bank teller machines. Any card with a length of less than 3.365 inches is considered defective. One of the dies used in making the credit cards is producing cards with a mean length of 3.367 inches. The lengths are normally distributed with a standard deviation of .001 inches.

    a. What is the probability of obtaining a defective card using this die?

    b. The company does not want to use any die that produces more than 1% defective cards. What should the company do in this instance?

    c. Assuming the standard deviation stays at .001 inches, what is the smallest acceptable mean length for cards manufactured? (*Hint:* For what mean length will no more than 1% of the card be shorter than 3.365 inches?)

57. A machine fills containers with a particular product. The standard deviation of filling weights is known from past data to be .6 ounces. If only 2% of the containers hold less than 18 ounces, what is the mean filling weight for the machine? That is, what must μ equal? Assume the filling weights have a normal distribution.

58. Suppose that 54% of a large population of registered voters favor the Democratic candidate for state senator. A public opinion poll uses randomly selected samples of voters and asks each person in the sample his or her preference; the Democratic candidate or the Republican candidate. The weekly poll is based on the response of 100 voters.

    a. What is the expected number of voters who will favor the Democratic candidate?

    b. What is the standard deviation for the number of voters who will favor the Democratic candidate?

    c. What is the probability that the poll will show that *less than* 50% of the voters favor the Democratic candidate when in fact 54% of the population of registered voters favor the candidate? That is, what is the probability that 49 or fewer individuals in the sample express support for the Democratic candidate?

59. It is estimated that in criminal trials, the jury will reach the correct decision (guilty or not guilty) 90% of the time. Consider a group of 100 cases that are brought to trial before a jury.

    a. What is the expected number of cases where the jury will reach the correct decision?

    b. What is the probability the jury will judge 95 or more cases correctly?

c. What is the probability an incorrect decision is reached in 12 or more cases?

d. Answer the question in (c) if the jury system reaches the correct decision 95% of the time.

60. Consider a multiple-choice examination with 50 questions. Each question has four possible answers. Assume that a student who has done the homework and attended lectures has a .75 probability of answering any question correctly.

a. A student must answer 43 or more questions correctly in order to obtain a grade of A. What percentage of the students who have done their homework and attended lectures will obtain a grade of A on this multiple-choice examination?

b. A student who answers 35 questions to 39 questions correctly will receive a grade of C. What percentage of students who have done their homework and attended lectures will obtain a grade of C on this multiple-choice examination?

c. A student must answer 30 or more questions correctly in order to pass the examination. What percentage of the students who have done their homework and attended lectures will pass the examination?

d. Assume that a student has not attended class and has not done the homework for the course. Furthermore, assume that the student will simply guess at the answer to each question. What is the probability that this student answers 30 or more questions correctly and passes the examination?

61. At the University of Cincinnati computing center, the average runtime for a class G job is 2.68 minutes. The runtimes are known to follow an exponential distribution.

a. What is the probability the runtime for the next class G job will be less than 1 minute?

b. What is the probability the next class G job will have a runtime between 1 and 4 minutes?

c. What is the probability the next class G job will have a runtime greater than 2.68 minutes?

62. Medical research has concluded that individuals experience a common cold roughly two times each year. Assume that the time between colds is exponentially distributed with a mean of 160 days.

a. What is the probability of going 200 or more days between colds?

b. What is the probability of getting a cold within 80 days of a previous cold?

c. What percentage of the population will go 120 to 240 days between colds?

63. The lifetime of a washing machine can be approximated by an exponential probability distribution with a mean of 4 years.

a. What is the probability that the lifetime of the washing machine is between 2.5 and 5.5 years?

b. What is the probability that the lifetime of the washing machine is 5.5 years or more?

c. What is the probability that the washing machine operates 21 months (1.75 years) or less?

64. The lifetime of a color television picture tube is exponentially distributed, with a mean of 7.8 years.

a. What is the probability that a picture tube will last more than 10 years?

b. If the firm guarantees the picture tube for 2 years, what percentage of the television sets sold will have to be replaced because of picture tube failure?

c. If the firm is willing to replace the picture tubes in a maximum of 1% of the television sets sold, what guarantee period can be offered for the television picture tubes?

7

Sampling and Sampling Distributions

What You Will Learn in This Chapter:

- The reasons for sampling

- How to select a simple random sample

- How results from samples can be used to provide estimates of population parameters

- The characteristics and use of the sampling distribution of $\bar{x}$

- What the central limit theorem is and the important role it plays in statistics

- Other sampling techniques, such as stratified sampling, cluster sampling, systematic sampling, convenience sampling, and judgment sampling

Contents

Charitable Contributions: The People Who Give

Independent Sector, a nonprofit organization located in Washington, D.C., promotes philanthropy and nonprofit groups. As part of a campaign designed to encourage Americans to donate 5% of their income and time to charities, Independent Sector commissioned a survey on private giving in America. The survey, conducted by the Gallup Organization, consisted of a sample of 2,775 households throughout the United States.

Households with annual incomes below $10,000 were found to contribute an average of 2.8% of their incomes to charitable causes. Households with annual incomes between $50,000 and $100,000 contributed an average of 1.5%, and households with annual incomes over $100,000 contributed an average of 2.1%. Moreover, households with annual incomes below $30,000 contributed almost half of the total amount given to charity. These results led Brian O'Connell, president of Independent Sector, to conclude that relatively speaking, "People of means cannot be described as particularly caring. For that primary category of humaneness, it is the poor and struggling who generally lead the way."

The survey showed that the average annual contribution for households that donated was $790. Religion, education, and health were the top three categories of giving, with half of all respondents surveyed giving to religious organizations. In addi-

Poster child for the annual United Way campaign

tion, the survey found that individuals who volunteer time give an average of 4.7 hours a week.

Independent Sector will use the results of the survey to help identify potential donors and plan future advertisements, many of which will be targeted at young professionals. Mr. O'Connell thinks that the survey results may remove confusion on the part of most people about the appropriate level of giving. He believes that when people see these results, they may be more willing to be more generous with their time and money.

Based on "Poorer Households Lead In Rate of Charitable Giving" *The Wall Street Journal*, October 19, 1988.

In this chapter we introduce simple random sampling and the process of using a sample mean to provide an estimate of a population mean. In addition we introduce the important concept of a sampling distribution. It is knowledge of the sampling distribution that enables us to make statements concerning the potential errors in using sample results to draw conclusions about a population. In the final section of the chapter, we present a variety of sampling procedures.

7.1 Introductory Sampling Concepts

Populations and Samples

As stated in Chapter 1, the primary purpose of statistics is to provide information about a *population* based on information contained in a *sample*. In the "Statistics in the News" article, the Gallup survey used a sample of 2775 households to learn about charitable giving for the population of all households in the United States. The definitions of a population and a sample are as follows:

Population

A *population* is the set of all elements of interest for a particular study.

Sample

A *sample* is a subset of the population selected to represent the whole population.

EXAMPLE 7.1

Television advertisers paid $1,350,000 per minute for commercials that aired during the 1989 Super Bowl. A total of 49 television commercials were shown during the game. In order to learn about viewer reaction to the ads, *USA Today* gathered 60 randomly selected individuals and asked them to rate each of the commercials. The 60 individuals formed a sample that was used to represent the population of individuals watching the Super Bowl on television. Results from the sample indicated that American Express and Diet Pepsi provided the two best-liked commercials shown during the Super Bowl game (*USA Today*, January 23, 1989).

■

EXAMPLE 7.2

Recent sports psychology studies have investigated the mental-training techniques used by the worlds' greatest athletes (*Peak Performance,* By Charles A. Garfield, 1984). For example assume a particular study is designed to investigate the training techniques of the world's best professional golfers. The previous years Professional Golfers Association (PGA) list of the top 100 money winners could be used to identify the population of world's best golfers. However, since the research project requires lengthy interviews and follow-up studies with each golfer studied, the researcher may not have sufficient time or funds to interview every golfer in the population. Consequently, a sample of 10 golfers can be selected from the population of 100 golfers. Data collected from the sample of 10 golfers will be used to make inferences about the mental training techniques of the population of 100 world's best golfers.

■

Reasons for Sampling

Whenever anyone wants information about a population, two alternatives are available: Collect the information from every item in the population, referred to as a *census,* or collect the information from a subset of the population, referred to as a *sample*. Numerous practical situations indicate that sampling is often the preferred way of collecting the desired information.

The reasons for sampling are largely due to the fact that a sample provides *time* and *cost* savings compared to a census of the entire population. In particular, if the population is large, the time and cost required to conduct a census may be prohibitive. In the case of an extremely large or infinite population, a census of the entire population is impossible. In example 7.1, we described the *USA Today* study of viewer ratings of television commercials shown during the 1989 Super Bowl. A census of the population of all who watched the Super Bowl on television is impractical and essentially impossible. However, the sample of 60 viewers was a relatively inexpensive way to obtain the viewer ratings information. In addition, the use of a sample enabled *USA Today* to report the ratings for each commercial the Monday morning after the Super Bowl. In Example 7.2, a census of the population of 100 top professional golfers would have been possbile. However, the use of the sample of 10 golfers provides substantial time and cost savings.

The use of sampling is also necessary in situations where the information-gathering process results in damaging or destroying the items being studied. For example, an automobile manufacturer obtains information about car safety by crash-testing a sample of its automobiles. Since the testing damages the automobiles, only a small sample can be used to collect the desired safety information. Finally, and perhaps surprisingly to some, a sample can often result in greater accuracy than a census. This is particularly true when a trained interviewer or a trained scientific technician is needed to collect the information. In this case, a census might necessitate the use of some less-skilled interviewers or technicians, which could lead to inaccurate or unreliable data being obtained.

Notes and Comments

The number of elements in a sample, called the *sample size*, can vary greatly depending on the nature of the study. For example, the "Statistics in the News" article at the beginning of this chapter reported a Gallup sample of 2775 households throughout the United States. Generally, sample sizes of 1000 to 3000 are typical for the national polls and surveys conducted by organizations such as Gallup and Harris. However, many other studies use a substantially smaller sample size. For example, *The Wall Street Journal* (November 23, 1988) reported information on operating costs and profits for inns located throughout the United States. Information presented was based on a sample of 72 bed-and-breakfast and full-service inns.

EXERCISES

1. The Federal Bureau of Justice statistics reported a national study of juveniles currently housed in long-term, state-operated correctional institutions *(Democrat and Chronicle.* September 19, 1988). The results of a sample of juveniles in correctional institutions showed that 1835 of 2621 had been raised in homes without both parents being present.
   a. Define the population for this study.
   b. Define the sample for this study.
   c. Why was a sample selected instead of conducting a census?
   d. Use the sample results to estimate the percent of all juveniles in the population who have been raised in homes without both parents present.

2. *The New York Times,* August 14, 1984, reported that a United States General Accounting Office (GAO) sample revealed that many college students who were receiving federal aid were not meeting minimum academic standards. The GAO report, based on an analysis of student records at selected colleges and universities, stated that 10% of the students receiving aid had a grade average of F.

a. Define the population for this study.

b. Define the sample for this study.

c. What were the advantages of sampling in this study?

3. Researchers at Oklahoma State University conducted a study involving a sample of 163 high school juniors *(The School Counselor,* November 1987). The purpose of the study was to explore the relationships among indices of loneliness, perceptions of school, and grade-point average. The 163 high school students who participated in the study were selected from high schools located in three small cities.

a. Define the sample for this study.

b. The researchers would like to use the results of their study to make conclusions about the population of all high school students in the United States. Do you see any problems with such conclusions? Explain.

4. An official for United Airlines reported that out of 104 United flights arriving at Chicago's O'Hare Airport, only 3 arrived more than 15 minutes late *(The Wall Street Journal,* November 7, 1988). Assuming that the 104 flights are a representative sample of all United Airline flights into O'Hare, answer the following questions.

a. Define the population.

b. Define the sample.

c. Use the sample results to estimate the percentage of all United Airline flights into O'Hare that arrive more than 15 minutes late.

d. The source of statistical information should always be considered when accepting and interpreting reported results. The preceding sample information was provided directly by United Airlines. Would you feel better about the results if the study of flight arrivals had been conducted by the Federal Aviation Administration (FAA)? Discuss.

7.2 Simple Random Sampling

There are several methods that can be used to select a sample from a population. One of the most common sampling methods is *simple random sampling*. The definition of a simple random sample and the process of selecting a simple random sample depends upon whether the population is *finite* or *infinite* in size.

Sampling from a Finite Population

Let us assume that the population of interest is finite and consists of N items. We will also assume that it is possible to obtain a list of all N items in the population. A simple random sample of size n from a finite population of size N is defined as follows:

Simple Random Sample (Finite Population)

A *simple random sample* of size n from a finite population of size N is a sample selected such that each possible sample of size n has the same probability of being selected.

EXAMPLE 7.3

A school district uses five buses (identified as A, B, C, D, and E) to transport elementary students to and from school each day. In selecting a simple random sample of size

$n = 2$ buses from the population of size $N = 5$, note that there are 10 different samples of size $n = 2$. These 10 samples consist of buses AB, AC, AD, AE, BC, BD, BE, CD, CE, and DE. If we select a sample in such a way that each of these 10 samples has the same $1/10$ probability of being selected, the sample selected would be a simple random sample. We could find such a sample by writing the two letters corresponding to each of the possible samples on 10 separate, but identical, pieces of paper. Mixing the 10 pieces of paper thoroughly and then randomly selecting one piece of paper would provide a simple random sample of 2 buses from the finite population of 5 buses.

Although listing all possible samples of size n, writing each possible sample on a piece of paper, and then selecting one piece of paper provides a method for identifying a simple random sample, this process becomes cumbersome and impractical as the finite population size increases. Thus, we need a better way to identify a simple random sample from a finite population.

In Example 7.3 the entire sample was selected in one random drawing. Another approach to identifying a simple random sample is to select the elements for the sample in a one-at-a-time fashion. At each selection we make sure that each of the items remaining in the population has the same probability of being selected.

EXAMPLE 7.3 (continued)

The one-at-a-time approach to selecting a simple random sample of two school buses can be accomplished as follows:

1. Write the 5 letters corresponding to each of the buses on a separate piece of paper.
2. Mix the 5 pieces of paper thoroughly and select one piece of paper at random; the letter on this piece of paper is the first bus selected for the simple random sample.
3. Mix the remaining 4 pieces of paper thoroughly and select another piece of paper at random; the letter on this piece of paper is the second bus selected for the simple random sample.

In using the one-at-a-time approach in Example 7.3 we did not replace the first piece of paper after it was selected from the population; this type of sampling is called *sampling without replacement*. If we had replaced the first piece of paper selected prior to choosing the second, we would have been *sampling with replacement*. Although sampling with replacement is a valid way of identifying a simple random sample, sampling without replacement is the sampling procedure used most often. Whenever we refer to simple random sampling from a finite population, we will make the assumption that the sampling is done without replacement.

Although the one-at-a-time approach is a practical way of selecting a simple random sample, the procedure that we used in Example 7.3 can be improved upon. In Example 7.4 we described how random numbers can be used to eliminate the step of writing the name or label of each item in the population on a separate piece of paper.

EXAMPLE 7.4

Suppose a university has received 7000 applications for admission. The director of admissions would like to use a sample of 50 applications in order to obtain information on

Scholastic Aptitude Test (SAT) scores of incoming students. How could a simple random sample of 50 applications be selected?

We begin by numbering the 7000 applications in the population from 1 to 7000. Tables of random numbers are available from a variety of handbooks that contain page after page of random numbers.* One such page of random numbers is shown in Table 8 in Appendix B. A portion of this page is also shown in Table 7.1. The digit appearing in any position in the random number table is a random selection of the digits 0, 1, . . . , 9 with each digit having an equal chance of occurring. By selecting four-digit random numbers that range in value from 0001 to 7000, we have a random number corresponding to each of the numbered applications.

In order to use the random numbers to identify items for the sample, we enter the table at any *arbitrary point* and then select four-digit random numbers by moving systematically down a column or across a row of the table. For example, suppose that we arbitrarily start with the third colmn of random numbers in Table 7.1. Since we need only four-digit numbers, we ignore the first digit in the column. Starting at the top and moving downward, the four-digit random numbers and corresponding application numbers are as follows:

TABLE 7.1   **Random Numbers**

63271	59986	71744	51102	15141	80714	58683	93108	13554	79945
88547	09896	95436	79115	08303	01041	20030	63754	08459	28364
55957	57243	83865	09911	19761	66535	40102	26646	60147	15702
46276	87453	44790	67122	45573	84358	21625	16999	13385	22782
55363	07449	34835	15290	76616	67191	12777	21861	68689	03263
69393	92785	49902	58447	42048	30378	87618	26933	40640	16281
13186	29431	88190	04588	38733	81290	89541	70290	40113	08243
17726	28652	56836	78351	47327	18518	92222	55201	27340	10493
36520	64465	05550	30157	82242	29520	69753	72602	23756	54935
81628	36100	39254	56835	37636	02421	98063	89641	64953	99337
84649	48968	75215	75498	49539	74240	03466	49292	36401	45525
63291	11618	12613	75055	43915	26488	41116	64531	56827	30825
70502	53225	03655	05915	37140	57051	48393	91322	25653	06543
06426	24771	59935	49801	11082	66762	94477	02494	88215	27191
20711	55609	29430	70165	45406	78484	31639	52009	18873	96927
41990	70538	77191	25860	55204	73417	83920	69468	74972	38712
72452	36618	76298	26678	89334	33938	95567	29380	75906	91807
37042	40318	57099	10528	09925	89773	41335	96244	29002	46453
53766	52875	15987	46962	67342	77592	57651	95508	80033	69828
90585	58955	53122	10625	84299	53310	67380	84249	25348	04332
32001	96293	37203	64516	51530	37069	40261	61374	05815	06714
62606	64324	46354	72157	67248	20135	49804	09226	64419	29457
10078	28073	85389	50324	14500	15562	64165	06125	71353	77669
91561	46145	24177	15294	10061	98124	75732	00815	83452	97355
13091	98112	53959	79607	52244	63303	10413	63839	74762	50289

Additional random numbers are provided in Table 8 of Appendix B.

	RANDOM NUMBER	APPLICATION NUMBER TO INCLUDE IN SAMPLE
	1744	1744
	5436	5436
	3865	3865
	4790	4790
	4835	4835
Numbers too high.	9902	—
Cannot use since	8190	—
applications 9902	6836	6836
and 8190 do not	•	•
exist.	•	•
	•	•

Continue until 50 different applications are selected.

*For example, *A Million Random Digits with 100,000 Normal Deviates,* by the Rand Corporation (New York: The Free Press, 1955). Copyright 1955 and 1983 by The Rand Corporation.

Since the numbers selected from the table in Example 7.4 are random, this procedure guarantees that each item in the population has the same probability of being included in the sample and that the sample selected will be a simple random sample. In using random numbers for simple sampling, a random number previously used to identify an item for the sample may reappear in the random number table. In selecting the simple random sample *without replacement*, previously used random numbers are ignored because the corresponding element is already in the sample.

Sampling from an Infinite Population

To this point we have restricted our attention to selecting a simple random sample from a finite population. Although most sampling situations involve finite populations, there are other situations in which the population is either infinite or so large that for practical purposes it must be treated as infinite. In sampling from an infinite population we must give a new definition for a simple random sample. In addition, since the items cannot be listed and numbered, we must use a different process for selecting items for the sample.

Let us consider a situation which can be viewed as requiring a simple random sample from an infinite population. Suppose we want to estimate the average time between placing an order and receiving food for customers arriving at a fast food-restaurant during the 11:30 A.M. to 1:30 P.M. lunch period. If we consider the population as being all possible customer visits, we see that it would be next to impossible to specify a finite limit on the number of possible visits. In fact, if we view the population as being all customer visits that could *conceivably* occur during the lunch period, we can consider the population as being inifnite. Our task is now to select a simple random sample of *n* customers from this population. With this situation in mind, we now state the definition of a simple random sample from an infinite population:

For the example of selecting a simple random sample of customer visits at a fast-food restaurant, we find that the first condition is satisfied by any customer visit occurring during the 11:30 A.M. to 1:30 P.M. lunch period while the restaurant is operating with its regular staff under normal operating conditions. The second condition is satisfied by ensuring that the selection of a particular customer does not influence the selection of any other customer.

A *simple random sample from an infinite population* is a sample selected such that the following conditions are satisfied:

1. Each item selected comes from the same population.
2. Each item is selected independently.

A well-known fast-food restaurant has implemented a simple random sampling procedure for just such a situation. The sampling procedure is based on the fact that some customers present discount coupons, which provide special prices on sandwiches, drinks, french fries, and so on. Whenever a customer presents a discount coupon, the *next* customer is selected for the sample. Since the customers present discount coupons in a random and independent fashion, the firm is satisfied that the sampling plan satisfies the two conditions for a simple random sample from an infinite population.

In other sampling situations, such as the sampling of parts from a production line, the sampling of plants for biological study, the sampling of water for pollution control, and so on, the population can be considered to be infinite in size. In these cases, extra care must be taken to ensure that the sample is representative of the population. Thus selection patterns such as sampling the water supply only at 8:00 A.M. must be avoided. If such precautions can be taken, it is usually reasonable to assume that the properties of a simple random sample have been satisfied.

Notes and Comments

1. A population can be classified as being either finite or infinite in size. Finite populations are often defined by lists such as organization membership rosters, enrolled students, mailing lists, credit card customers, and inventory product numbers. Usually infinite populations are defined by an ongoing process, where a listing of all possible items is impossible, such as all possible parts to be manufactured, all possible customer visits, or all possible bank transactions.

2. The number of different simple random samples of size n that can be selected from a finite population of size N may be determined by using the following formula:

$$\frac{N!}{n!(N - n)!}$$

The $N!$ and $n!$ refer to the factorial computations discussed in Chapter 5.

In Example 7.3, we selected a simple random sample of size $n = 2$ from a finite population of size $N = 5$. Using the preceding formula, we find that the number of different simple random samples possible is

$$\frac{5!}{2!(5 - 2)!} = \frac{(5)(4)(3)(2)(1)}{[(2)(1)][(3)(2)(1)]} = \frac{120}{(2)(6)} = 10$$

With large populations, the number of possible different simple random samples can be quite large.

5. Exercise 1 described a national study of juveniles currently housed in long-term, state-operated correctional institutions. Is the population finite or infinite? Would it be possible to obtain a list of juveniles in these institutions?

6. Exercise 4 described information in *The Wall Street Journal* (November 7, 1988), where United Airlines reported a sample of 104 flights arriving at Chicago's O'Hare Airport. Statistics reported indicated that only 3 of the 104 flights arrived more than 15 minutes late.
   a. If the purpose of the sample is to make inferences about United's flight performance during the preceding month (October 1988), is the population finite or infinite? Explain.
   b. If the purpose of the sample is to make inferences about the ongoing process of all United flights into O'Hare, is the population finite or infinite? Explain.

7. A population consists of four items labeled A, B, C and D.
   a. How many simple random samples of size $n = 2$ can be selected from this population?
   b. List all possible simple random samples.

8. How many simple random samples are possible when samples of size $n = 3$ are to be selected from a finite population of size $N = 6$?

9. Example 7.2 describes a study of the mental-training techniques of the world's best professional golfers. The population was defined to be the top 100 money winners from the Professional Golfers Association. A simple random sample of 10 golfers will be selected for interviews and follow-up studies. How many different simple random samples of 10 golfers are possible?

10. In Example 7.3, a school district used 5 buses to transport elementary students to and from school. The 5 buses were identified by letters *A, B, C, D,* and *E.*
   a. How many samples of size 3 are possible? List the different possible samples.
   b. Using simple random sampling, what is the probability each possible sample will be selected?

11. Consider the following six Midwestern states as a population: Iowa, Illinois, Wisconsin, Michigan, Indiana, and Ohio. Assume that a sample of 4 states will be selected from this population in order to study employment trends in the Midwest.
   a. How many samples of size 4 are possible? List the samples.
   b. How many of the samples contain the state of Illinois?
   c. Using simple random sampling, the possible samples have the same probability of being selected. If this is the case, what is the probability of selecting a sample that contains the state of Illinois?
   d Repeat part (b) for the state of Indiana.
   e. Using the results from parts (b) and (d), what can you say about the probability that a given state will be included in the simple random sample?

12. What is the primary advantage of the one-at-a-time method of simple random sampling as compared to the method of listing all possible samples?

13. A student government organization is interested in estimating the average monthly rent paid by students living in off-campus housing. A list of names and addresses of the 645 students is available from the registar's office. Using every digit in row 10 of Table 7.1 and moving across the row from left to right, identify the first 10 students who would be selected using simple random sampling. Note that when every digit in row 10 is used, the three-digit random numbers begin with 816, 283, and 610.

14. Consider a population of five salespersons selling mobile telephone units to a variety of customers. The individuals in the population are identified by the letters *A, B, C, D,* and *E.* Using the 15th row of random numbers in Table 7.1, use the random digits 1, 2, 3, 4, and

5 to correspond to the five salespersons and select a simple random sample of size 2 from the population.
  a. What sample is selected?
  b. What sample would have been selected if the random numbers in row 20 had been used?

15. The top 10 athletic-footwear manufacturers (based upon sales) are (1) Reebok, (2) Nike, (3) Converse, (4) Avia, (5) Adidas, (6) L.A. Gear, (7) Etonic/Tretorn, (8) New Balance, (9) ASICS Tiger, and (10) British Knights. (*Wall Street Journal,* October 13, 1988) Using the first digits in Table 7.1 and moving down the column, select a simple random sample of 5 manufacturers from this group.

16. Haskel Public Opinion Poll, Inc., conducts telephone surveys concerning a variety of political and general public-interest issues. The households included in the survey are identified by taking a simple random sample from telephone directories in selected metropolitan areas. The telephone directory for a major Midwest area contains 853 pages with 400 lines per page.
  a. Describe a two-stage random selection procedure that could be used to identify a simple random sample of 200 households. The selection process should involve first selecting a page at random and then selecting a line on the sampled page. Use the random numbers in Table 7.1 to illustrate this process. Select your own arbitrary starting point in the table.
  b. What would you do if the line selected in part (a) was clearly inappropriate for the study; that is, the line provided the phone number of a business, restaurant, etc.?

7.3 Sampling Distribution of $\bar{x}$

Once we obtain a simple random sample from a population, we want to summarize the data in order to make estimates or draw conclusions about the population. In general, we use characteristics of the sample, referred to as *sample statistics,* to estimate characteristics of the population, referred to as *population parameters*.

One of the most common statistical procedures is the use of a sample mean $\bar{x}$ to make inferences about an unknown population mean μ. In this case the sample mean $\bar{x}$ is referred to as the sample statistic and the population mean μ is referred to as the population parameter. This statistical process is shown in Figure 7.1.

FIGURE 7.1   The Statistical Process of using a Sample Mean to make Inferences about a Population Mean

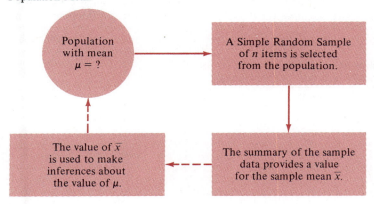

It is important to realize that if we were to repeat the sampling process shown in Figure 7.1, we can anticipate obtaining a different value for the sample mean $\bar{x}$. As we showed in Section 7.2, several different simple random samples of size n are possible. Since each sample consists of different items from the population, we can expect different samples to provide different values for the sample statistic $\bar{x}$.

Since each sample has the same probability of being selected, we can associate a known probability with every possible sample and every possible value of $\bar{x}$. As a result we can identify the probability distribution for the sample mean $\bar{x}$. This probability distribution is called the *sampling distribution of* $\bar{x}$. The name of the probability distribution comes from the fact that the different possible values of $\bar{x}$ are due to the variety of different *samples* that can be selected from the population. Because of the importance of the sampling distribution of $\bar{x}$, we restate its definition.

Sampling Distribution of $\bar{x}$

The *sampling distribution of* $\bar{x}$ is the probability distribution for all possible values of the sample mean $\bar{x}$.

To illustrate the concept of a sampling distribution, let us reconsider the population of five school buses described in Example 7.3.

**EXAMPLE 7.3
(continued)**

The number of students riding on each of the five school buses in the population is shown in Table 7.2. Using the formulas for a population mean and a population standard deviation that were presented in Chapter 3, we can use the data in Table 7.2 to compute the population mean and population standard deviation of the number of students riding on each of the buses as follows:

$$\mu = \frac{\Sigma x_i}{N} = \frac{120}{5} = 24$$

$$\sigma = \sqrt{\frac{\Sigma(x_i - \mu)^2}{N}} = \sqrt{\frac{90}{5}} = \sqrt{18} = 4.24$$

Details of the computations of the values of μ and σ are shown in Table 7.3.

In this situation the population size is small; thus it is easy to compute the mean and standard deviation for the population. However, to illustrate how a sample mean can be

TABLE 7.2

Number of Students per Bus for the Population of Five Buses

BUS	NUMBER OF STUDENTS
A	24
B	30
C	21
D	18
E	27

used to estimate a population mean, *let us assume for the moment that μ is unknown* and that we will have to use a simple random sample of two buses to estimate μ. Recall that there are 10 different samples of size 2 that could be selected. Table 7.4 lists these 10 possible samples and their corresponding sample means.

The column labeled "Sample Mean $(\bar{x})$" in Table 7.4 shows that the value of the sample mean depends upon the sample selected. Since we are using simple random sampling, we know that each possible sample—and, therefore, its corresponding sample mean—has the same $\frac{1}{10}$ probability of being selected. Figure 7.2 is a graph showing each possible value of $\bar{x}$ and the corresponding probability of occurrence. For instance, note that a value of $\bar{x} = 24.0$ has a $\frac{2}{10}$ probability of occurring because two samples, *BD* and *CE*, provide this value. Thus we see that Figure 7.2 is simply a probability distribution that shows all possible values of a random variable and their probabilities of

FIGURE 7.2 The Sampling Distribution of $\bar{x}$ for Example 7.3 based on Samples of Size 2

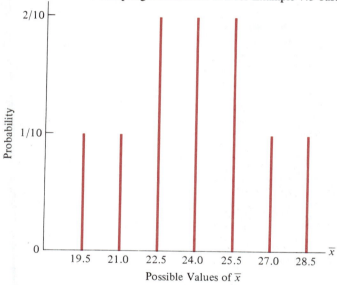

Possible Values of $\bar{x}$

TABLE 7.3

Computation of the Population Mean and Population Standard Deviation for Example 7.3

BUS	NUMBER OF STUDENTS (x_i)	$(x_i - \mu)$		$(x_i - \mu)^2$
A	24	$(24 - 24) =$	0	0
B	30	$(30 - 24) =$	6	36
C	21	$(21 - 24) =$	-3	9
D	18	$(18 - 24) =$	-6	36
E	27	$(27 - 24) =$	3	9
Totals	120			90

$$\mu = \frac{120}{5} = 24 \qquad \sigma = \sqrt{\frac{90}{5}} = \sqrt{18} = 4.24$$

occurrence. The random variable in this case is the sample mean, $\bar{x}$. The probability distribution of $\bar{x}$, as shown in Figure 7.2, is called the sampling distribution of $\bar{x}$.

■

EXAMPLE 7.5

Consider the following population of five families with the data indicating the family size.

FAMILY	FAMILY SIZE
A	2
B	4
C	3
D	5
E	3

TABLE 7.4

Different Possible Simple Random Samples for Example 7.3

BUSES SELECTED IN SAMPLE	PROBABILITY OF SAMPLE	SAMPLE MEAN ($\bar{x}$)
A and B	1/10	$\dfrac{24 + 30}{2} = 27.0$
A and C	1/10	$\dfrac{24 + 21}{2} = 22.5$
A and D	1/10	$\dfrac{24 + 18}{2} = 21.0$
A and E	1/10	$\dfrac{24 + 27}{2} = 25.5$
B and C	1/10	$\dfrac{30 + 21}{2} = 25.5$
B and D	1/10	$\dfrac{30 + 18}{2} = 24.0$
B and E	1/10	$\dfrac{30 + 27}{2} = 28.5$
C and D	1/10	$\dfrac{21 + 18}{2} = 19.5$
C and E	1/10	$\dfrac{21 + 27}{2} = 24.0$
D and E	1/10	$\dfrac{18 + 27}{2} = 22.5$

If a simple random sample of size 3 is used to estimate the mean family size for the population, let us show the sampling distribution of $\bar{x}$.

First, we list all possible samples of size 3. There are 10 such samples consisting of families ABC, ABD, ABE, ACD, ACE, ADE, BCD, BCE, BDE and CDE. The sample mean for each possible sample is shown below.

SAMPLE	DATA VALUES	SAMPLE MEAN ($\bar{x}$)
ABC	2, 4, 3	3.00
ABD	2, 4, 5	3.67
ABE	2, 4, 3	3.00
ACD	2, 3, 5	3.33
ACE	2, 3, 3	2.67
ADE	2, 5, 3	3.33
BCD	4, 3, 5	4.00
BCE	4, 3, 3	3.33
BDE	4, 5, 3	4.00
CDE	3, 5, 3	3.67

With a probability of $\frac{1}{10}$ for each sample, the probabilities for each possible value are $\bar{x}$ are plotted below. This graph shows the sampling distribution of $\bar{x}$.

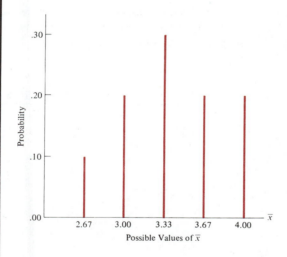

Before continuing the discussion of the sampling distribution of $\bar{x}$, we note that in practice only one sample is actually taken; hence, only one value of $\bar{x}$ is computed and used to make inferences about the population mean μ. The purpose of this section has been to show that there are many possible samples that could be selected. Understanding this provides a much better perspective on the properties and the importance of the sampling distribution of $\bar{x}$. Knowledge of this sampling distribution provides the background to understand the material in Chapter 8; there the focus is on estimating the population mean μ using the information contained in just one sample.

Expected Value of $\bar{x}$

As we have seen, different samples may result in different values for the sample mean. We are often interested in the mean of all possible $\bar{x}$ values that can be generated by the various simple random samples. Let $E(\bar{x})$ denote the expected value of $\bar{x}$, or simply the mean of all possible $\bar{x}$ values. When using simple random sampling from a population with mean μ, the expected value of the sample mean is *equal to μ.*

Expected Value of $\bar{x}$

$$E(\bar{x}) = \mu \tag{7.1}$$

This result states that the expected value of $\bar{x}$—or, stated another way, the mean for all possible $\bar{x}$ values—is the same as the mean of the population from which the samples are taken.

EXAMPLE 7.3 (continued)

Recall that in Example 7.2 we computed the population mean for the number of students on school buses to be $\mu = 24$. Equation (7.1) implies that the mean of the various $\bar{x}$ values must also be 24. Using the 10 values of $\bar{x}$ shown in Table 7.4, we can compute $E(\bar{x})$ as follows:

$$E(\bar{x}) = \frac{27.0 + 22.5 + 21.0 + 25.5 + 25.5 + 24.0 + 28.5 + 19.5 + 24.0 + 22.5}{10}$$

$$= \frac{240.0}{10} = 24$$

Thus we see that equation (8.1) holds with $E(\bar{x}) = \mu$.

■

Standard Deviation of $\bar{x}$

The standard deviation of $\bar{x}$ measures the dispersion in the possible $\bar{x}$ values. To explore what sampling theory says about the standard deviation of $\bar{x}$, we use the following notation:

$$\sigma_{\bar{x}} = \text{standard deviation of the } \bar{x} \text{ values}$$

$$\sigma = \text{standard deviation of the population being sampled}$$

$$n = \text{sample size}$$

$$N = \text{population size}$$

The formula for the standard deviation of $\bar{x}$ is as follows:

$$\sigma_{\bar{x}} = \sqrt{\frac{N - n}{N - 1}} \left(\frac{\sigma}{\sqrt{n}}\right) \qquad (7.2)$$

Later we will see that when only one sample is selected, the value of $\sigma_{\bar{x}}$ is helpful in determining how far the sample mean may be from the population mean. Because of the role that $\sigma_{\bar{x}}$ plays in computing possible estimation errors, $\sigma_{\bar{x}}$ is referred to as the *standard error of the mean*. Thus the standard error of the mean is another name for the standard deviation of $\bar{x}$.

EXAMPLE 7.3 (continued)

Recall that in Example 7.3 the standard deviation for the population of school buses was computed to be $\sigma = 4.24$ students (see Table 7.3). With a population of size $N = 5$ and a sample of size $n = 2$, equation (7.2) shows that the standard deviation of $\bar{x}$ must be

$$\sigma_{\bar{x}} = \sqrt{\frac{N - n}{N - 1}} \left(\frac{\sigma}{\sqrt{n}}\right) = \sqrt{\frac{5 - 2}{5 - 1}} \left(\frac{4.24}{\sqrt{2}}\right) = 2.60$$

■

Table 7.5 shows the computation of the standard deviation of $\bar{x}$ values using the 10 values of $\bar{x}$ generated by the 10 possible samples of two buses. As the computations

TABLE 7.5

Computation of the Standard Deviation of $\bar{x}$ using All Possible Sample Means for the School Bus Example

BUSES SELECTED IN SAMPLE	SAMPLE MEAN ($\bar{x}$)	($\bar{x} - 24$)		($\bar{x} - 24$)²
A and B	27.0	(27.0 − 24) =	3.0	9.00
A and C	22.5	(22.5 − 24) =	− 1.5	2.25
A and D	21.0	(21.0 − 24) =	− 3.0	9.00
A and E	25.5	(25.5 − 24) =	1.5	2.25
B and C	25.5	(25.5 − 24) =	1.5	2.25
B and D	24.0	(24.0 − 24) =	0.0	0.00
B and E	28.5	(28.5 − 24) =	4.5	20.25
C and D	19.5	(19.5 − 24) =	− 4.5	20.25
C and E	24.0	(24.0 − 24) =	0.0	0.00
D and E	22.5	(22.5 − 24) =	− 1.5	2.25
Totals	240.0			67.50

$$E(\bar{x}) = \frac{240}{10} = 24 \qquad \sigma_{\bar{x}} = \sqrt{\frac{\Sigma(\bar{x} - 24)^2}{10}} = \sqrt{\frac{67.50}{10}} = 2.60$$

show, we obtain the same value of $\sigma_{\bar{x}}$ as we found using equation (7.2). However, note that equation (7.2) provides the value of $\sigma_{\bar{x}}$ without having to generate all possible $\bar{x}$ values.

Consider for a moment the factor $\sqrt{(N - n)/(N - 1)}$ that appears in the formula for $\sigma_{\bar{x}}$. This factor is commonly referred to as the *finite population correction factor*. In many practical sampling situations we find that the population being sampled, although finite, is "large," whereas the sample size is relatively "small." In such cases the value of $\sqrt{(N - n)/(N - 1)}$ is close to 1. When this occurs, $\sigma_{\bar{x}} = \sigma/\sqrt{n}$ becomes a very good approximation to the standard deviation of $\bar{x}$. We give the following as a general guideline or rule of thumb for computing the standard deviation of $\bar{x}$.

Guideline for Computing the Standard Deviation of $\bar{x}$:

Whenever the sample size is less than or equal to 5% of the population size (that is, $n/N \leq .05$), use

$$\sigma_{\bar{x}} = \frac{\sigma}{\sqrt{n}} \tag{7.3}$$

In the following chapters, we generally assume the population is large relative to the sample size; that is, $n/N \leq .05$. Thus we use (7.3) to compute the standard deviation of $\bar{x}$. If this assumption is not satisfied in a particular application, we use equation (7.2) to compute $\sigma_{\bar{x}}$.*

EXAMPLE 7.6

Assume that a simple random sample of size 49 is to be taken from a large population with mean $\mu = 100$ and standard deviation $\sigma = 21$. We know that repeating the sampling process will generate different sample means, $\bar{x}$, due to the different samples selected. What are the mean and standard deviation of the values of the sample means?

Using equation (7.1), we see the mean of the $\bar{x}$ values is

$$E(\bar{x}) = \mu = 100$$

Since the population is large relative to the sample size, equation (7.3) provides the standard deviation of the $\bar{x}$ values, as follows.

$$\sigma_{\bar{x}} = \frac{\sigma}{\sqrt{n}} = \frac{21}{\sqrt{49}} = 3$$

---

*This assumption is always satisfied when the population is infinite. If sampling from a finite population is done *with replacement*, (7.3) is used regardless of the sample size because the population size and composition are not changed as the sample is taken.

Note to student: In several of the exercises that follow, the size of the population is not stated. When the population size is *not* provided, you may make the assumption that the population is "large" relative to the sample size and that $n/N \leq .05$. In such cases, equation (7.3) may be used to compute the standard deviation of $\bar{x}$.

17. Four college students are taking the following number of credit hours during the current term:

STUDENT	NUMBER OF CREDIT HOURS
Albert	15
Becky	17
Cindy	19
David	17

Treating this as a population of size 4, answer the following questions.
  a. How many simple random samples of size 2 are possible? List the possible samples.
  b. Compute the sample mean for each of the possible simple random samples of size 2.
  c. Show a graphical representation of the probability distribution of all possible values of the sample mean. This is a graphical representation of the sampling distribution of $\bar{x}$.

18. Using the population of buses as shown in Table 7.2, show the sampling distribution of $\bar{x}$ if a simple random sample of three buses is used to estimate the mean number of students riding a bus.

19. The following data show the number of automobiles owned in a population of five households.

HOUSEHOLD	NUMBER OF AUTOMOBILES
1	2
2	1
3	0
4	2
5	3

  a. If a simple random sample of two households is used to estimate the mean number of automobiles per household, show the sampling distribution of $\bar{x}$.
  b. Repeat part (a) if the simple random sample selected contains three households.

20. Assume we have a population with mean $\mu = 32$ and standard deviation $\sigma = 5$. Furthermore, assume that the population has 500 items and that a simple random sample of 25 items used to obtain information about this population. Let $\bar{x}$ denote the sample mean that will be used to estimate the value of the population mean.
  a. What is the expected value of $\bar{x}$?
  b. What is the standard deviation of $\bar{x}$?

21. The four students mentioned in exercise 17 were taking the following number of credit hours during the current term: Albert, 15; Becky, 17: Cindy, 19: and David, 17.
  a. Treating the four students as the population, use the computational procedure of Table 7.3 to compute the population mean μ and standard deviation σ.
  b. There are six possible simple random samples of size 2 that can be selected from this population. Identify each possible sample and compute its corresponding sample mean.

c. Use the computational procedure of Table 7.5 to compute the mean and standard deviation of the $\bar{x}$ values.

d. Use your results from part (a) and Equations (7.1) and (7.2) to compute the mean and standard deviation of the $\bar{x}$ values. Compare your answers to parts (c) and (d).

22. Five sales representatives sell mobile telephone units to private and commercial customers. Assume that the number of units sold for each sales representative is as follows:

SALESPERSON	UNITS SOLD
Adams *(A)*	14
Baker *(B)*	20
Collins *(C)*	12
Davis *(D)*	8
Edwards *(E)*	16

a. Treating the five sales representatives as the population, use the computational procedure of Table 7.3 to compute the population mean μ and the population standard deviation σ.

b. There are 10 possible simple random samples of size 2 that can be selected from the population. Identify each possible sample and compute its corresponding sample mean, $\bar{x}$.

c. Show the sampling distribution of $\bar{x}$.

d. Use the computational procedure of Table 7.5 to compute the mean and standard deviation of the $\bar{x}$ values.

e. Use Equations (7.1) and (7.2) to determine the expected value and standard deviation of $\bar{x}$. Compare your results with your answer to part (d).

f. If you wish to compute $E(\bar{x})$ and $\sigma_{\bar{x}}$, do you prefer the approach used in part (d) or in part (e)? Explain.

23. The sizes of the 10 offices on the twelfth floor of the new Crosley Tower Bank Building are as follows.

OFFICE	SIZE (SQUARE FEET)	OFFICE	SIZE (SQUARE FEET)
1	150	6	300
2	175	7	140
3	180	8	150
4	180	9	150
5	225	10	200

a. Compute the population mean μ and population standard deviation σ for the population of 10 offices.

b. There are many different possible simple random samples of size 3 that can be selected from the population. What are the values of the mean and standard deviation for the $\bar{x}$ values? That is, compute $E(\bar{x})$ and $\sigma_{\bar{x}}$.

c. Compute $E(\bar{x})$ and $\sigma_{\bar{x}}$ if the sample size is increased to four offices.

24. A statistics class has 80 students. The mean score on the midterm exam was $\mu = 72$ and the standard deviation was $\sigma = 12$. Assume that a simple random sample of 20 students will be selected and the sample mean exam score $\bar{x}$ will be computed. What is the expected value and standard deviation of $\bar{x}$?

25. Weights for males between the ages of 20 and 30 have a mean $\mu = 170$ pounds with a standard deviation of $\sigma = 28$. If a simple random sample of 40 males in this age group is to be selected and the sample mean weight $\bar{x}$ computed, what are the values of $E(\bar{x})$ and $\sigma_{\bar{x}}$?

26. Consider a population of 1000 items. Assume the population standard deviation is $\sigma = 25$. Use (7.2) to compute the standard error of the mean ($\sigma_{\bar{x}}$) for sample sizes of 50, 100, 150, and 200. What can you say about the size of the standard error of the mean as the sample size is increased?

27. In a study of the growth rate of a certain plant, a botanist is planning to use a simple random sample of 25 plants for data collection purposes. After analyzing the data on plant growth rate, the botanist believes that the standard error of the mean is too large. What size simple random sample should the botanist use in order to reduce the standard error to one-half its current value?

28. Simple random samples of size 30 are to be selected from a population of 2000 items. The population standard deviation is $\sigma = 12$.
   a. Use Equation (7.2) to compute the standard error of the mean.
   b. Use Equation (7.3) to compute the standard error of the mean.
   c. Compare your answers to parts (a) and (b) and comment on why it is acceptable to use equation (7.3) whenever $n/N \leq .05$.
   d. What is the value of the finite population correction factor $\sqrt{(N - n)/(N - 1)}$ for this problem?·

29. A simple random sample of size 50 is to be selected from a population with $\sigma = 10$. Find the value of the standard error of the mean in each case.
   a. the population size is infinite
   b. the population size is $N = 50,000$
   c. the population size is $N = 5,000$
   d. the population size is $N = 500$
   e. In which of the above cases is it desirable to use the finite population correction factor and Equation (7.2)?

7.4   The Central Limit Theorem

At this point we know that simple random samples taken from a population will provide different values for the sample statistic $\bar{x}$ due to the fact that different samples consist of different items from the population. The probability distribution showing all possible values of $\bar{x}$ is referred to as the sampling distribution of $\bar{x}$. The final step in identifying the characteristics of the sampling distribution of $\bar{x}$ is to determine the *form* of the sampling distribution of $\bar{x}$. We will consider two cases: one where the population distribution is unknown, and one where the population distribution is known to be a normal probability distribution.

For the situation where the population distribution is unknown, we rely on one of the most important theorems in statistics—the *central limit theorem*. A statement of the central limit theorem as it applies to the sampling distribution of $\bar{x}$ is as follows.*

---

*The theoretical proof of the central limit theorem requires independent observations or items in the sample. This condition exists for infinite populations or for finite populations where sampling is done with replacement. Although the central limit theorem does not directly address sampling without replacement from finite populations, general statistical practice has been to apply the findings of the central limit theorem in this situation provided that the population size is large.

Figure 7.3 shows how the central limit theorem works for three different populations; in each case the population clearly is not normal. However, note what begins to happen to the sampling distribution of $\bar{x}$ as the sample size is increased. When the samples are of size 2, we see that the sampling distribution of $\bar{x}$ begins to take on an appearance different from the population distribution. For samples of size 5, we see all three sampling

FIGURE 7.3   Illustration of the Central Limit Theorem for Three Populations

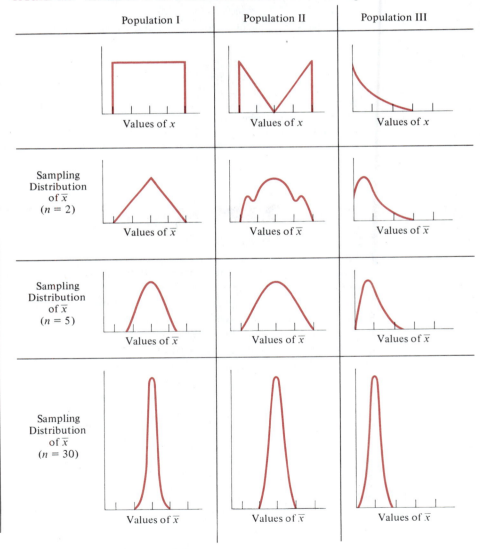

distributions beginning to take on a bell-shaped appearance. Finally, the samples of size 30 show all three sampling distributions to be approximately normal. General statistical practice is to assume that regardless of the population distribution, the sampling distribution of $\bar{x}$ can be approximated by a normal probability distribution whenever the sample size is 30 or more. In effect the sample size of 30 is the rule of thumb that allows us to assume that the large sample condition of the central limit theorem has been satisfied. This observation about the sampling distribution of $\bar{x}$ is so important that we restate it:

> The sampling distribution of $\bar{x}$ can be approximated by a normal probability distribution whenever the sample size is large. The large sample size condition can be assumed for simple random samples of size 30 or more.

The central limit theorem is the key to identifying the form of the sampling distribution of $\bar{x}$ whenever the population distribution is unknown. However, we may encounter some sampling situations where the population is assumed or believed to have a normal probability distribution. When this condition occurs, it is not necessary to rely on the central limit theorem because the following result identifies the form of the sampling distribution of $\bar{x}$.

> Whenever the population being sampled has a normal probability distribution, the sampling distribution of $\bar{x}$ is a normal probability distribution for any sample size.

In summary, whenever we are using a large simple random sample (*rule of thumb: $n \geq 30$*), the central limit theorem enables us to conclude that the sampling distribution of $\bar{x}$ can be approximated by a normal probability distribution. In cases where the simple random sample is small ($n < 30$), the sampling distribution of $\bar{x}$ can be considered to be a normal probability distribution only if we believe that the assumption of a normal probability distribution for the population is appropriate.

EXAMPLE 7.6 (continued)

In Example 7.6, we discussed a situation in which simple random samples of size 49 were to be taken from a population with mean $\mu = 100$ and standard deviation $\sigma = 21$. Show the sampling distribution of $\bar{x}$.

Recall that $E(\bar{x}) = 100$ and $\sigma_{\bar{x}} = 21/\sqrt{49} = 3$. The central limit theorem also tells us that for large samples ($n \geq 30$), the sampling distribution of $\bar{x}$ can be approximated by a normal distribution. In such cases the sampling distribution of $\bar{x}$ is as shown.

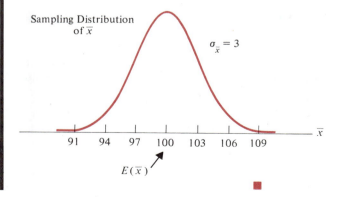

EXAMPLE 7.7

The heights of sixth grade students in a particular school district are *normally distributed* with a mean of $\mu = 58$ inches and a standard deviation of $\sigma = 3.2$ inches. Show the sampling distribution of $\bar{x}$ if simple random samples of 16 students are to be used.

We know that

$$E(\bar{x}) = \mu = 58$$

and

$$\sigma_{\bar{x}} = \frac{\sigma}{\sqrt{n}} = \frac{3.2}{\sqrt{16}} = .80$$

Since $n < 30$, we cannot use the central limit theorem to conclude that the sampling distribution of $\bar{x}$ is approximately normal. However, since the population of heights is described as being normally distributed, the sampling distribution of $\bar{x}$ will be normal for any sample size. Thus we have the distribution shown.

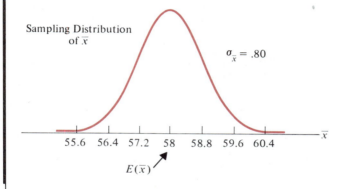

---

30. A sample of size $n = 36$ was selected from a population with a mean of 100 and a standard deviation of 12.
    a. What is the expected value of $\bar{x}$?
    b. What is the standard deviation of $\bar{x}$?
    c. What probability distribution can be used to approximate the sampling distribution of $\bar{x}$.
    d. Sketch a graph of the sampling distribution of $\bar{x}$.

31. A sample of size $n = 9$ was selected from a population that is normally distributed with a mean of 50 and a standard deviation of 15. Sketch the sampling distribution of $\bar{x}$.

32. The mean and standard deviation for the number of calories in a 12-ounce can of light beer are as follows: $\mu = 105$ and $\sigma = 3$. A random sample of 30 cans will be selected and a laboratory test conducted to determine the exact number of calories present in each of the 30 cans. The sample mean $\bar{x}$ will be computed.
    a. What is the expected value of $\bar{x}$?
    b. What is the standard deviation of $\bar{x}$?

c. What probability distribution can be used to approximate the sampling distribution of $\bar{x}$?

d. Sketch a graph of the sampling distribution of $\bar{x}$.

33. The length of time of long-distance telephone calls has a mean $\mu = 18$ minutes and a standard deviation $\sigma = 4$ minutes. Sketch the sampling distribution of $\bar{x}$ if a simple random sample of 50 telephone calls will be used to compute a sample mean length of long-distance telephone calls.

34. A population has a mean $\mu = 400$ and a standard deviation $\sigma = 50$. The probability distribution of the population is unknown.

   a. A research study will use simple random samples of either 10, 20, 30, or 40 items to collect data about the population. In which of these sample-size alternatives will we be able to use a normal probability distribution to describe the sampling distribution of $\bar{x}$? Explain.

   b. Sketch the sampling distribution of $\bar{x}$ for the instances where the normal probability distribution is appropriate.

35. The body length of a certain insect is believed to be *normally distributed* with a mean length of $\mu = 16.5$ mm and a standard deviation of $\sigma = .8$mm. Describe the sampling distribution of the sample mean body length if 20 insects are to be used in the study. Is it necessary to use the central limit theorem to determine the shape of the sampling distribution? Explain.

36. Assume that the number of points scored in basketball games played by a particular college team is normally distributed with $\mu = 68$ and $\sigma = 5$. Show the sampling distribution of $\bar{x}$ for a sample of 10 games played by this team.

7.5   Computing Probabilities Using The Sampling Distribution of $\bar{x}$

As the following examples show, we can use the sampling distribution of $\bar{x}$ to compute the probability of selecting a sample that will provide a value of $\bar{x}$ within any specified distance from the population mean.

**EXAMPLE 7.7
(continued)**

In Example 7.7 we described the sampling distribution of $\bar{x}$ for simple random samples consisting of the heights of 16 sixth-grade students. Specifically, we showed that the sampling distribution of $\bar{x}$ is a normal probability distribution with $E(\bar{x}) = 58$ and $\sigma_{\bar{x}} = .80$. Suppose now we want to compute the probability of obtaining a simple random sample that results in a sample mean ($\bar{x}$) between 57 and 59 inches.

Since the sampling distribution of $\bar{x}$ is or can be approximated by a normal probability distribution, we can use the area under the standard normal, or probability distribution, curve to answer this question. First, we need to compute a z value where

$$z = \frac{\bar{x} - \mu}{\sigma_{\bar{x}}} \tag{7.4}$$

This z value is identical to the z value we used for the normal probability distribution in Chapter 6, with the exception of the notation $\bar{x}$ and $\sigma_{\bar{x}}$. This notation has been used to indicate that the normal random variable here is $\bar{x}$ and its standard deviation is denoted by $\sigma_{\bar{x}}$.

The probability of selecting a simple random sample with $\bar{x}$ between 57 and 59 is given by the shaded area under the normal curve shown.

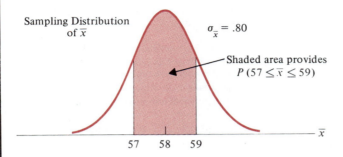

$$\text{At } \bar{x} = 59, \qquad z = \frac{59 - 58}{.80} = 1.25$$

Looking up $z = 1.25$ in the table of areas for the standard normal distribution shows an area of .3944 between 58 and 59.

$$\text{At } \bar{x} = 57, \qquad z = \frac{57 - 58}{.80} = -1.25$$

This z value shows that the area between 57 and 58 must also be .3944. Thus the total probability of selecting a sample with a mean $\bar{x}$ between 57 and 59 inches must be

$$.3944 + .3944 = .7888.$$

Now, let us see how to compute the probability of selecting a simple random sample of 16 students and finding the probability of obtaining a sample mean $\bar{x}$ of 60 inches or more. The graph of the sampling distribution of $\bar{x}$ for this situation is as follows:

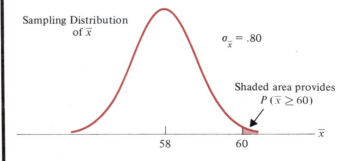

$$\text{At } \bar{x} = 60, \qquad z = \frac{\bar{x} - \mu}{\sigma_{\bar{x}}} = \frac{60 - 58}{.80} = 2.50$$

Using the table of areas for the standard normal distribution and $z = 2.50$, we find that the area between $z = 0.00$ and $z = 2.50$ is .4938. Thus the probability of $z \geq 2.50$ is a very low $.5000 - .4938 = .0062$. Thus the probability of obtaining a sample 16 students from the population with $\bar{x} \geq 60$ is .0062.

■

EXAMPLE 7.8

Suppose that we have a population of high school students with a mean IQ score of $\mu = 100$ and a standard deviation of $\sigma = 10$. If simple random samples of 36 students are to be taken from this population, $\sigma_{\bar{x}} = \sigma/\sqrt{n} = 10/\sqrt{36} = 1.67$. Let us now compute an interval around 100 that include 95% of all possible sample means that could be obtained.

The table of areas for the standard normal distribution shows that 95% of the values in any normal distribution fall between $z = -1.96$ and $z = +1.96$. In other words, 95% of all possible $\bar{x}$ values must be within $-1.96\sigma_{\bar{x}}$ and $+1.96\sigma_{\bar{x}}$ of 100. Thus the range containing 95% of all possible $\bar{x}$ values must be

$$100 - 1.96\sigma_{\bar{x}} = 100 - 1.96(1.67) = 96.7$$

$$100 - 1.96\sigma_{\bar{x}} = 100 + 1.96(1.67) = 103.3$$

That is, there is a .95 probability of selecting a simple random sample having a sample mean IQ score between 96.7 and 103.3. Only 5% of all possible sample means are outside this interval. The situation is shown below.

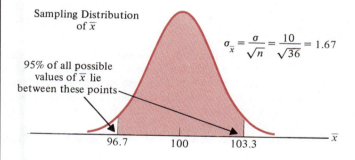

EXERCISES

37. A sample of size $n = 36$ was selected from a population with a mean of 100 and a standard deviation of 12. Compute the probability of finding a sample mean greater than 105.

38. A sample of size $n = 49$ was selected from a population with the following characteristics: $\mu = 500$, $\sigma = 70$. What is the probability that the sample mean will be between 480 and 500?

39. A sample of size $n = 9$ was selected from a population that is normally distributed with a mean of 50 and a standard deviation of 15. What is the probability that the sample mean will be less than 45?

40. An automatic machine used to fill cans of soup has the following characteristics: $\mu = 15.9$ ounces, $\sigma = .5$ ounces.

    a. Show the sampling distribution of $\bar{x}$, where $\bar{x}$ is the sample mean for 40 cans selected randomly by a quality control inspector.

    b. What is the probability of finding a sample of 40 cans with a mean $\bar{x}$ greater than 16 ounces?

41. A library checks out an average of 320 books per day, with a standard deviation of 75 books. Consider a sample of 30 days of operation, with $\bar{x}$ being the sample mean number of books checked out per day.

    a. Show the sampling distribution of $\bar{x}$.

    b. What is the standard error?

    c. What is the probability that the sample mean for the 30 days will be between 300 and 340 books?

    d. What is the probability the sample mean will show more than 325 books checked out per day?

42. A study of the time from computer program submission until program return (i.e., *turnaround time*) was conducted at a university computer center. Assume that under standard operating conditions the population mean is 120 minutes, with a population standard deviation of 40 minutes.

    a. Future studies of turnaround time are to be based on simple random samples of 30 programs. Show the sampling distribution of the sample mean turnaround time.

    b. What is the probability that the sample mean for 30 programs will be less than 100 minutes? Over 125 minutes?

43. An electrical component is designed to provide a mean service life of 3000 hours, with a standard deviation of 800 hours. A customer purchases a batch of 50 components, which can be considered a simple random sample of the population of components. What is the probability that the mean life for the group of 50 components will be at least 2750 hours? At least 3200 hours?

44. The grade-point average for all juniors at Strasser College has a standard deviation of .50.

    a. A random sample of 20 students is to be used to estimate the population mean grade-point average. What assumption is necessary in order to compute the probability of obtaining a sample mean within plus or minus .2 of the population mean?

    b. Provided that this assumption can be made, what is the probability of $\bar{x}$ being within plus or minus .2 of the population mean?

    c. If this assumption cannot be made, what would you recommend doing?

45. A simple random sample of 64 will be used to estimate the mean time required to perform a particular task in a mechanical aptitude test. If the standard deviation in times is $\sigma = 4$ minutes, use the sampling distribution of $\bar{x}$ to comment on the probability of each error shown. The sampling error is defined as the difference between the observed sample mean and the actual population mean μ.

    a. error of 1 minute or less

    b. error of .5 minutes or less

    c. error greater than .25 minutes

46. Three firms have inventories that vary in size. Firm A has an inventory of 2000 items, firm B has an inventory of 5000 items, and firm C has an inventory of 10,000 items. The standard deviation for the cost of the items in inventory is $\sigma = 144$. A statistical consultant recommends that each firm take a sample of 50 items from their respective inventories in order to provide statistically valid estimates of the mean cost per item in inventory. Management of the small firm states that since it has the smallest inventory, it should be able to obtain the data from a smaller sample size than required by the larger firms. However, the consultant states that in order to obtain the same standard error and thus the same precision in the sample results, all firms should take the same sample size regardless of inventory size.

    a. Using the finite population correction factor, compute the standard error for each of the three firms, given a sample of size 50.

    b. For each firm, what is the probability that the sample mean $\bar{x}$ will be within ± 25 of the population mean, μ?

    c. Do you agree with the consultant's statement? Explain.

47. A survey reports its results by stating that the standard error of the mean was 20. The population standard deviation was 500.

    a. How large was the sample used in this survey?

    b. What is the probability that the estimate would be within ± 25 of the population mean?

48. In a study of annual salaries of managers, the population of managers of interest has annual salaries with $\mu = \$31,800$ and $\sigma = \$4000$.

    a. If a sample of 30 managers is selected, what is the probability the sample mean $\bar{x}$ will be within $\pm \$500$ of the population mean of $\$31,800$? That is, what is the probability the sample mean will be between $\$31,300$ and $\$32,300$?

    b. How large a sample should be selected in order for the probability of a sample mean $\bar{x}$ within $\pm \$500$ of μ to be .95?

7.6   Other Sampling Methods

We have described the simple random sampling procedure and discussed the properties of the sampling distribution of $\bar{x}$ when simple random sampling is used. It is important to realize that simple random sampling is not the only sampling method available. Sampling methods such as stratified random sampling, cluster sampling, and systematic sampling offer alternatives that in some situations have advantages over simple random sampling. In this section we briefly describe some of these alternative sampling methods.

Stratified Random Sampling

In *stratified random sampling*, the population is first divided into groups of elements called *strata*, such that each item in the population belongs to one and only one stratum. The basis for forming the various strata, such as department, location, age, industry type, and so on, is up to the discretion of the designer of the sample. Best results, however, are obtained whenever the elements within each stratum are as much alike as possible. Figure 7.4 shows a diagram with the population divided into H strata.

    After the strata are formed, a simple random sample is taken from every stratum. Formulas are available for combining the results for the individual samples into one estimate of the population parameter of interest. The value of stratified random sampling depends upon how homogeneous the elements are within the strata. If units within strata are alike (homogeneity), the strata will have low variances. Thus relatively small sample sizes can be used to obtain good estimates of the strata characteristics. If homogeneous strata exist, the stratified random sampling procedure will provide results similar to simple random sampling but will do so with a smaller total sample size.

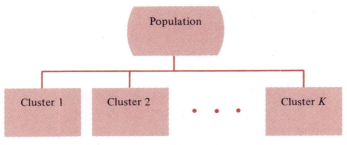

Select a simple random
sample of k of the K clusters.

FIGURE 7.4   Diagram for Stratified Simple Random Sampling

Cluster Sampling

In *cluster sampling,* the population is first divided into separate groups of elements called *clusters*. Each element of the population belongs to one and only one cluster (see Figure 7.5). A simple random sample of the clusters is then taken. All elements within each sampled cluster form the sample. Cluster sampling tends to provide best results whenever the elements within the clusters are heterogeneous (not alike). In the ideal case, each cluster is a representative small-scale version of the entire population. The value of cluster sampling depends upon how representative each cluster is of the entire population. If each cluster is alike in this regard, sampling a small number of clusters will provide good estimates of the population parameters.

One of the primary applications of cluster sampling is area sampling, where clusters are city blocks or other well defined areas. Cluster sampling generally requires a large total sample size than either simple random sampling or stratified random sampling. However, it can result in cost savings because of the fact that when an interviewer is sent to a sampled cluster (or city-block location) many interviews or sample observations can be obtained in a relatively short time. As a result, a larger sample size may be obtainable with a significantly lower cost per element and thus possibly a lower total cost.

Systematic Sampling

In some sampling situations, especially those with large populations, it is often time-consuming to select a simple random sample by first finding a random number and then counting or searching through the list of population items until the corresponging element is found. An alternative to simple random sampling is a *systematic sampling procedure*. For example, if a sample size of 50 is desired from a population containing 5000 elements, we might sample one element for every $5000/50 = 100$ elements in the population. A systematic sample for this case would involve selecting randomly 1 of the first 100 elements from the population list. Other sample elements are identified by starting with the first sampled element and then selecting every 100th element that follows in the population list. In effect the sample of 50 is identified by moving systematically through the population and identifying every 100th element after the first randomly selected element. The sample of 50 usually will be easier to identify in this manner than would

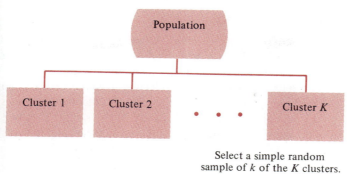

Select a simple random
sample of k of the K clusters.

FIGURE 7.5 Diagram for Cluster Sampling

be the case if simple random sampling were used. Since the first element selected is a random choice, the assumption is usually made that a systematic sample has the properties of a simple random sample. This assumption is expecially applicable when the list of the population elements is believed to be a random ordering of the elements.

Convenience Sampling

The sampling methods discussed thus far are referred to as *probability sampling* techniques. By this we mean that elements selected from the population have a known probability of being included in the sample. The advantage of probability sampling is that the sampling distribution of the appropriate sample statistic generally can be identified. Formulas such as the ones for simple random sampling presented earlier in this chapter can be used to determine the properties of the sampling distribution. Then the sampling distribution can be used to make probability statements about possible sampling errors associated with the sample results.

Convenience sampling is a *nonprobability sampling* technique. As the name implies, the sample is identified primarily by convenience. Items are included in the sample without prespecified or known probabilities of being selected. For example, a professor conducting research at a university may use student volunteers to constitute a sample simply because they are readily available and will often participate as subjects for little or no cost. In another example, a shipment of oranges may be sampled by an inspector who selects oranges haphazardly from among several crates. Labeling each orange and using a probability method of sampling would be impractical. Samples such as wildlife captures and volunteer panels for consumer research are also convenience samples.

Convenience samples have the advantage of relatively easy sample selection and data collection; however, it is impossible to evaluate the "goodness" of the sample in terms of its representativeness of the population. A convenience sample may provide good results, or it may not. However, there is no statistically justified procedure that will allow a probability analysis and inference about the quality of the sample results. Nevertheless, at times you will see statistical methods designed for probability samples applied to a convenience sample. The researcher argues that the convenience sample may well provide a sample which may be treated as if it were a random sample. However, this argument cannot be supported, and we should be very cautious in interpreting the results of convenience samples that are used to make inferences about populations.

Judgment Sampling

One additional nonprobability sampling technique is *judgment sampling*. In this situation the person most knowledgeable on the subject of the study selects individuals or other elements of the population that he or she feels are most representative of the population. Often this is a relatively easy way of selecting a sample. For example, a reporter may sample two or three senators based on the judgment that these senators reflect the general opinion of all senators. However, the quality of the sample results is dependent on the judgment of the person selecting the sample. Again great caution must be used in drawing conclusions based on judgment samples used to make inferences about populations.

SUMMARY

In this chapter we have introduced the important concepts of simple random sampling and the sampling distribution of $\bar{x}$. We introduced the process of using a sample mean $\bar{x}$ (sample statistic) to provide information about a population mean μ (population parameter). Each simple random sample potentially provides a different value for the sample mean $\bar{x}$. The probability distribution for the population of $\bar{x}$ values is called the sampling distribution of $\bar{x}$.

In considering the characteristics of the sampling distribution of $\bar{x}$, we stated that $E(\bar{x}) = \mu$ and that $\sigma_{\bar{x}} = \sigma/\sqrt{n}$, provided $n/N \leq .05$. In addition, the central limit theorem provided the basis for using a normal probability distribution to approximate the sampling distribution of $\bar{x}$. The rule of thumb of $n \geq 30$ provided the large-sample conditions necessary to use the normal probability distribution to approximate the sampling distribution of $\bar{x}$. We also noted that whenever the population being sampled has a normal probability distribution, the sampling distribution of $\bar{x}$ would be normal for any sample size. Knowledge of the sampling distribution of $\bar{x}$ was used to make probability statements about the values of $\bar{x}$ that may be obtained from a simple random sample.

The use of random numbers to select simple random samples was demonstrated. Finally, we concluded the chapter by discussing alternative sampling methods, including stratified random sampling, cluster sampling, systematic sampling, convenience sampling and judgment sampling.

GLOSSARY

Population   The set of all elements of interest for a particular study.

Sample   A subset of the population selected to represent the whole population.

Parameter   A numerical measure that is computed based upon all the elements in the population, such as the population mean μ.

Sample statistic   A numerical measure that is calculated using the observations in a sample, such as the sample mean $\bar{x}$.

Simple random sample (finite population)   A sample selected such that each possible sample of size n has the same probability of being selected.

Simple random sample (infinite population)   A sample selected such that each item comes from the same population and each item is selected independently.

Sampling without replacement   Once an item from the population has been included in the sample, it is removed from further consideration and cannot be selected a second time.

Sampling with replacement   As each item is selected for the sample, it is returned to the population. It is possible that a previously selected item may be selected again and, therefore, appear in the sample more than once.

Sampling distribution   A probability distribution showing all possible values of a sample statistic, such as a sample mean $\bar{x}$.

Standard error of the mean   The standard deviation of $\bar{x}$, denoted by $\sigma_{\bar{x}}$.

Finite population correction factor   The multiplier term $\sqrt{(N - n)/(N - 1)}$ that is used in the formula for $\sigma_{\bar{x}}$; whenever $n/N \leq .05$, the finite population factor is close to 1 and hence $\sigma_{\bar{x}} = \sigma/\sqrt{n}$.

Central limit theorem   A theorem that enables us to use the normal probability distribution to approximate the sampling distribution of $\bar{x}$ whenever the sample size is large. *Rule of thumb:* The central limit theorem applies whenever $n \geq 30$, where n is the sample size.

Probability sampling   Any sampling procedure wherein each element in the population has a known probability of being included in the sample. Simple random sampling, stratified random sampling, cluster sampling, and systematic sampling can be classified as probability sampling techniques.

Nonprobability sample   A sample selected such that the probability of each element being included in the sample is unknown. Convenience and judgment samples are nonprobability samples.

KEY FORMULAS

EXPECTED VALUE OF $\bar{x}$

$$E(\bar{x}) = \mu \tag{7.1}$$

STANDARD DEVIATION OF $\bar{x}$

$$\sigma_{\bar{x}} = \sqrt{\frac{N - n}{N - 1}} \left(\frac{\sigma}{\sqrt{n}} \right) \tag{7.2}$$

If the sample size is less than or equal to 5% of the population size, use

$$\sigma_{\bar{x}} = \frac{\sigma}{\sqrt{n}} \tag{7.3}$$

COMPUTATION OF z VALUE FOR THE SAMPLING DISTRIBUTION $\bar{x}$

$$z = \frac{\bar{x} - \mu}{\sigma_{\bar{x}}} \tag{7.4}$$

REVIEW QUIZ

TRUE/FALSE

1. A simple random sample of size n from a population of size N is a sample selected such that each possible sample of size n has the same probability of being selected.

2. A primary objective of sampling is to choose a sample that is representative of the population being studied.

3. The sample mean and sample standard deviation are not sample statistics.

4. The probability distribution of the sample mean is called the sampling distribution of $\bar{x}$.

5. The standard error of the mean cannot be computed unless the sample size is known.

6. In practice, sampling with replacement is more commonly employed than sampling without replacement.

7. The term $\sqrt{(N - n)/(n - 1)}$ in the formula for the standard diviation of $\bar{x}$ is called the continuity correction factor.

8. When sampling with replacement, we always use the formula $\sigma_{\bar{x}} = \sqrt{(N - n)/(N - 1)} \, (\sigma/\sqrt{n})$ to compute the standard deviation of $\bar{x}$.

9. The central limit theorem ensures that the sampling distribution of $\bar{x}$ is a normal probability distribution regardless of the sample size.

10. When using cluster sampling, we need not be concerned with selecting a sample that is representative of the population.

MULTIPLE CHOICE

11. Consider a population with $\mu = 50$ and $\sigma = 10$. A simple random sample of size $n = 25$ will be used to provide a sample mean $\bar{x}$ to estimate μ. What is the mean value for the sampling distribution of $\bar{x}$.
  a. 2
  b. 10
  c. 50
  d. none of the above

12. Once an item from the population has been included in the sample, it is removed from further consideration and cannot be selected for the sample a second time. This is an example of which of the following?
  a. cluster sampling
  b. probability sampling
  c. sampling without replacement
  d. sampling with replacement

13. In sampling from a large population with $\sigma = 20$, the standard error of the mean is found to be 2. What was the size of the simple random sample used in this situation?
  a. 100
  b. 10
  c. 40
  d. none of the above

14. What condition is required before the central limit theorem justifies approximating the sampling distribution of $\bar{x}$ with a normal probability distribution?
  a. $n/N \geq .05$
  b. $n \geq 30$

    c. $n < 30$

    d. $N \geq 30$

15. Assume a sample of 49 items is taken from a population with $\mu = 16$ and $\sigma = 7$. What is the probability the sample mean, $\bar{x}$, will be within ± 1 of the population mean $\mu = 16$?

    a. .3413

    b. .6826

    c. .1114

    d. cannot be determined from the above information

16. Which of the following is not an example of a probability sample?

    a. simple random sampling

    b. stratified sampling

    c. cluster sampling

    d. convenience sampling

17. As the sample size increases, variability among the sample means

    a. increases

    b. decreases

    c. remains the same

    d. not enough information given

18. Random samples of size 17 are taken from a population that has 200 elements, a mean of 36, and a standard deviation of 8. The distribution of the population is unknown. The mean and standard deviation of the sample mean are

    a. 8.7 and 1.94

    b. 36 and 1.94

    c. 36 and 1.86

    d. 36 and 8

19. Which of the following best describes the form of the sampling distribution of the sample mean for the situation in question 18?

    a. approximately normal because the sample size is small relative to the population size

    b. approximately normal because of the central limit theorem

    c. exactly normal because the population is normally distributed

    d. nothing can be said with the information given

20. Random samples of size 32 are taken from an infinite population whose mean and standard deviation are 40 and 5, respectively. The distribution of the population is unknown. The mean and standard deviation of the sample mean are

    a. 40 and .78

    b. 40 and .88

    c. 24 and .78

    d. 24 and .88

SUPPLEMENTARY EXERCISES

49. Assume that a simple random sample of size 2 is to be taken from the following list of airlines: American, United, TWA, Delta, Eastern, Piedmont, and US Air.

    a. How many simple random samples of size 2 are possible? List the possible samples.

b. What is the probability of each sample being selected?

c. How many samples include United Airlines?

d. What is the probability that United Airlines appears in the simple random sample selected?

50. An apartment complex consists of six buildings. The number of apartments rented in each of the six buildings are as follows: 5, 4, 6, 3, 5, and 4, respectively. Assume that a simple random sample of two buildings will be used to estimate the mean number of apartment rentals per building.

a. Compute the mean μ and standard deviation σ for the population of six apartment buildings.

b. There are 15 possible simple random samples of size 2 that can be selected from this population. Identify each possible sample and compute its corresponding sample mean $\bar{x}$ values.

c. Show the sampling distribution of $\bar{x}$.

d. Use the computational procedure of Table 7.4 to compute the mean and standard deviation of all possible $\bar{x}$ values.

e. Use Equations (7.1) and (7.2) and the results of part (a) to determine the expected value and standard deviation of $\bar{x}$. Compare your results with your answer to part (d).

51. The time it takes a fire department to respond to a request for emergency aid has a mean of $\mu = 14$ minutes with a standard deviation of $\sigma = 4$ minutes. Suppose we randomly sample 50 emergency requests over the past two-month period. Records of aid-request times and arrival times will be used to compute a sample mean response time for the 50 requests.

a. Show the sampling distribution of $\bar{x}$.

b. What role does the central limit theorem play in identifying this sampling distribution?

c. What is the probability that the sample mean will be 15 minutes or less?

d. What is the probability that the sample mean will be within $\pm$ .5 minutes of the mean time for the population?

52. The speed of automobiles on a section of I-75 in northern Florida has a mean of $\mu = 67$ miles per hour with a standard deviation of $\sigma = 6$ miles per hour. Answer the following questions if the population can be assumed to have a *normal distribution* and if a sample of 16 automobiles will be selected to compute a sample mean automobile speed.

a. What is the expected value of $\bar{x}$.

b. What is the value of the standard error of the mean?

c. Show the sampling distribution of $\bar{x}$.

d. What is the probability that the value of the sample mean will be 65 miles per hour or more?

e. What is the probability that the value of the sample mean will be between 66 and 68 miles per hour?

53. Consider a population of size $N = 500$ with a mean $\mu = 200$ and a standard deviation $\sigma = 40$. Assume that a simple random sample of size $n = 100$ will be selected from this population.

a. Should the finite population correction factor be used in computing the standard error of the mean?

b. What is the value of the standard error of the mean for this problem?

c. What is the probability of selecting a simple random sample that provides a value of $\bar{x}$ that is within ± 5 of the population mean μ?

54. In a population of 4000 employees, a simple random sample of 40 employees is selected in order to estimate the mean age for the population.

a. Would you use the finite population correction factor in calculating the standard error of the mean? Explain.

b. If the population standard deviation is $\sigma = 8.2$ years, compute the standard error of the mean, first with and then without the finite population correction factor. What is the rationale for ignoring the finite population correction factor whenever $n/N \leq .05$?

55. In the preceding problem, what is the probability that the sample mean age of the employees will be within ± 2 years of the population mean age?

56. A production process is checked periodically by a quality control inspector. The inspector selects simple random samples of 30 finished products and computes the sample mean product weights, denoted by $\bar{x}$. If test results over a long period of time show that 5% of the $\bar{x}$ values are over 2.1 pounds and 5% are under 1.9 pounds, what are the mean and standard deviation for the population produced with this process?

57. Assume that we wish to identify a simple random sample of 12 of the 372 doctors located in a particular city. The doctors' names are available from a local medical organization. Use the eighth column of five-digit random numbers in Table 7.1 to identify the 12 doctors for the sample. Ignore the first two random digits in each five-digit grouping of the random numbers. This process begins with random number 108 and proceeds down the column of random numbers.

58. Assume that we have a listing of 5500 employees in a large company. Beginning with the four-digit random number 1102 in column 4 of the random numbers of Table 7.1, continue down the column to identify the first 10 employees to be included in the sample.

59. Comment on why each of the samples shown below do not constitute a simple random sample. What kind of samples are they?

    a. To obtain consumer reaction to a new product, a firm contacts women's groups at several local churches and offers to pay the organization for each person participating in the study.

    b. A psychology professor uses a freshman class in Psychology 101 as a sample of subjects for a research project.

    c. After a television debate for presidential candidates, viewers are encouraged to phone the television station to indicate the candidate of their preference.

8 Inferences About a Population Mean

What You Will Learn in This Chapter:

- How to construct and interpret an interval estimate of a population mean
- What is meant by the term confidence level
- The concept of sampling error
- How to determine the sample size when estimating a population mean
- What hypotheses are and how they are formulated
- How to use sample results to test hypotheses about a population mean
- How to interpret the type I and type II errors in hypothesis testing
- What a p-value is and how it is used in hypothesis testing
- The t distribution
- How to use the t distribution to make interval estimates and test hypotheses about a population mean

Contents

Radon Poses Health Hazard

Radon is an odorless gas that is formed in the ground by the natural decay of uranium. As the gas decays, the radioactive particles that are produced travel through the soil and collect in buildings. The problem is that when these radioactive particles are inhaled by humans, the risk of lung cancer is increased. Federal officials have stated that radon is the second-leading cause of lung cancer after smoking. Estimates are that as many as 20,000 deaths can be attributed to this gas each year. The problem is further compounded by the fact that radon is something we cannot see, smell, or feel.

Based upon a Environmental Protection Agency (EPA) survey of 11,000 homes in seven states, the federal government has issued a warning that home contamination by radon is significantly more pervasive than previously indicated. In the EPA study, the measurement of radiation used is called a picocurie. Assistant Surgeon General Vernon Houk stated that exposure to 4 picocuries per liter of air is equivalent to having 200 to 300 chest X rays a year or smoking half a pack of cigarettes a day.

Based on the sample data, the study found North Dakota, with an average home rating of 7 picocuries had the highest level of radon. Arizona, with an average home rating of 1.6 picocuries, had the lowest level of radon. Although there is no federal standard for a "safe" level of radon, the EPA has stated that if a home has a radon level of 4 or more picocuries, the residence should be monitored. If the level of radon persists, the homeowner should take steps to vent the contaminated area.

Based upon the results of the seven states surveyed in 1988 and the results from ten other states

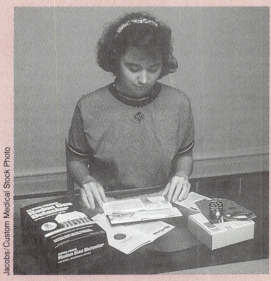

Jacobs/Custom Medical Stock Photo

Easy-to-use test kits can identify the level of radon in your home.

surveyed in 1987, the EPA estimates that more than three million homes in the 17 states are contaminated at a potentially health-threatening level. The statistical evidence collected led government officials to conclude that the potential for radon contamination is so widespread that the only people who do not have to test their homes are those living in apartments above the second floor.

Based on "Federal Health Officials Urge Nationwide Tests For Radon," *Charlotte Observer*, September 13, 1988.

In this chapter we continue the discussion of how a sample mean $\bar{x}$ can be used to make inferences about a population mean μ. First we consider the statistical process known as *estimation*, where the value of $\bar{x}$ is used to estimate the value of μ. This was the process used by the EPA in "Statistics in the News"; the EPA's sample results provided an estimate of the mean radon level for homes in each of the states in the survey.

Next we discuss the statistical process known as *hypothesis testing*. In this process, we hypothesize that the population mean has a specific numerical value; then we use the value of the sample mean $\bar{x}$ to *test the hypothesis* about the value of the population mean.

In the first four sections of the chapter we consider inferences about a population mean based upon a simple random sample which contains at least 30 items. We refer to this situation as the *large-sample* case. In Section 8.5 we discuss inferences about a population mean for the *small-sample* case, where the sample consists of less than 30 items. As you will see, the methodology for the small-sample case requires the use of a new probability distribution known as the *t* distribution.

8.1 Point Estimation of a Population Mean

The statistical procedure called *point estimation* involves the computation of a *sample statistic* in order to estimate a population *parameter*. For example, we use $\bar{x}$, the mean of a simple random sample, to estimate μ, the mean of the population. In such cases we refer to $\bar{x}$ as the *point estimator* of the population mean μ; the actual numerical value obtained for $\bar{x}$ is called the *point estimate* of μ.

In discussing the use of $\bar{x}$ as a point estimator of μ, we will make use of the central limit theorem introduced in Chapter 7. This theorem enables us to conclude that the sampling distribution of $\bar{x}$ can be approximated by a *normal probability distribution* with mean μ and standard deviation $\sigma_{\bar{x}} = \sigma/\sqrt{n}$ whenever the sample size is large. The generally accepted rule of thumb for a large sample is that the sample size n must be at least 30. In this section we show how to compute probability statements that address the question, How good is the estimate? In doing so, we will assume that the large-sample condition exists; in Section 8.5 we discuss the modifications required when the sample size is small.

EXAMPLE 8.1

A simple random sample of 100 sixth grade students was selected from the population of all sixth grade students in a large school district. Each of the students in the sample completed a standardized mathematics achievement test; the average test score for this group was $\bar{x} = 72.5$. The objective of the study was to estimate μ, the mean test score for all sixth grade students in the school district. The mean test score for the sample of 100 students is called the point estimator of μ. The observed value of $\bar{x} = 72.5$ is the point estimate of the population mean μ.

■

Properties of Point Estimators

Intuitively, we feel comfortable in using the sample mean to estimate the population mean. However, there are also several important mathematical reasons why $\bar{x}$ is considered a good point estimator of μ. For example, we know from the discussion of the sampling distribution of $\bar{x}$ in Chapter 7 that the expected value of $\bar{x}$ is equal to μ. Whenever the expected value of a point estimator is equal to the value of the corresponding population parameter, the estimator is said to be an *unbiased estimator*. Thus, the sample mean $\bar{x}$ is an unbiased estimator of μ.

Although unbiasedness is a desirable property for a point estimator, there are other properties that statisticians consider when attempting to determine appropriate point estimators. One such property is called *consistency*. Loosely speaking, an estimator is said to be consistent if values of the point estimator tend to lie closer to the population parameter as the sample size becomes larger. In other words, larger sample sizes tend to provide better point estimates. Recall from Chapter 7 that the standard deviation of the point estimator* $\bar{x}$ is given by $\sigma_{\bar{x}} = \sigma/\sqrt{n}$. Since the standard deviation of $\bar{x}$ decreases as the sample size n increases, we would conclude that larger sample sizes tend to provide estimates that are closer to the population mean. In this sense, we say that $\bar{x}$ is a consistent estimator of μ.

Sampling Error

Any time a sample mean is used to provide a point estimate of a population mean, someone may ask, How good is the estimate? The "how good" question is a way of asking about the error involved when the value of $\bar{x}$ is used as a point estimate of the population mean μ. In general, we shall refer to the magnitude of the difference between an unbiased point estimator and the population parameter as the *sampling error*. When a sample mean is used to estimate a population mean, the sampling error† is defined as follows:

$$\text{Sampling Error} = |\bar{x} - \mu| \tag{8.1}$$

Note, however, that even after a sample is selected and the sample mean is computed, we will not be able to use equation (8.1) to find the value of the sampling error because the population mean μ is unknown. However, the sampling distribution of $\bar{x}$ developed in Chapter 7 can be used to make probability statements about the sampling error. To see how this is done, note that the central limit theorem enables us to conclude that whenever the sample size is large ($n \geq 30$), the sampling distribution of $\bar{x}$ can be approximated by a normal probability distribution with a mean μ and a standard deviation $\sigma_{\bar{x}} = \sigma/\sqrt{n}$.

EXAMPLE 8.1 (continued)

The mean mathematics achievement test score for the sample of 100 students was $\bar{x} = 72.5$. Extensive use of this test throughout the country has also shown that although the mean test scores differ among districts, the standard deviation of test scores has been approximately 12; thus, in the current study it was assumed that the population standard deviation of test scores is $\sigma = 12$. The central limit theorem enables us to conclude that the value of $\bar{x} = 72.5$ is from the sampling distribution shown in Figure 8.1.

■

Although the population mean μ is unknown in Example 8.1, the sampling distribution shown in Figure 8.1 shows how the $\bar{x}$ values are distributed around μ. In effect, this distribution shows the possible differences between $\bar{x}$ and μ and provides a way to make probability statements about the sampling error.

Probability Statements About the Sampling Error

Since the sampling distribution of $\bar{x}$ can be approximated by a normal probability distribution whenever the sample size is 30 or more, we can use the standard normal probability

---

*In this chapter we assume that $n/N \leq .05$. Thus, the finite population correction factor is not needed in the computation of $\sigma_{\bar{x}}$.
†Here, $|\bar{x} - \mu|$ denotes the absolute value of $\bar{x} - \mu$; for example, $|60 - 75| = |-15| = 15$ and $|100 - 75| = |25| = 25$.

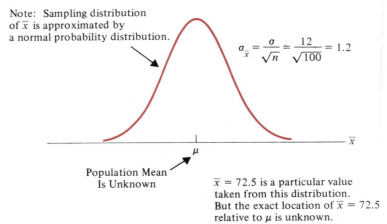

Note: Sampling distribution of $\bar{x}$ is approximated by a normal probability distribution.

$$\sigma_{\bar{x}} = \frac{\sigma}{\sqrt{n}} = \frac{12}{\sqrt{100}} = 1.2$$

Population Mean Is Unknown

$\bar{x} = 72.5$ is a particular value taken from this distribution. But the exact location of $\bar{x} = 72.5$ relative to μ is unknown.

FIGURE 8.1 The Sampling Distribution of x for the Mathematics Achievement Test Scores

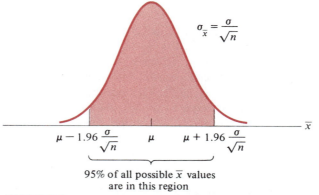

$$\sigma_{\bar{x}} = \frac{\sigma}{\sqrt{n}}$$

$\mu - 1.96 \dfrac{\sigma}{\sqrt{n}}$ μ $\mu + 1.96 \dfrac{\sigma}{\sqrt{n}}$

95% of all possible $\bar{x}$ values are in this region

FIGURE 8.2 The Location of 95% of all possible Sample Means

distribution tables to make probability statements about the sampling error. For example, using the standard normal distribution table (Table 1 of Appendix B), we find that 95% of the values of a normally distributed random variable are within ± 1.96 standard deviations of the mean. Thus, for the sampling distribution of $\bar{x}$, 95% of all possible $\bar{x}$ values are within $\pm 1.96\sigma_{\bar{x}}$ of the mean μ. The location of these sample means is shown in Figure 8.2.

EXAMPLE 8.1 (continued)

In the example of mathematics achievement test scores, we can conclude that 95% of all possible sample means are within $\pm 1.96\sigma_{\bar{x}}$ of the population mean test score μ. Since $1.96\sigma_{\bar{x}} = 1.96(12/\sqrt{100}) = 2.35$, we can state that 95% of all sample means based on simple random samples of 100 students are within ± 2.35 of the population mean μ. Thus, we can make the following probability statement about the sampling error:

> There is a .95 probability that the sample mean mathematics achievement test score based on a sample of 100 students will provide a sampling error of 2.35 or less.

This statement about the sampling error defines the *precision* of the estimate. The value 2.35 is the measure of precision, or closeness, of $\bar{x} = 72.5$ to the population mean

μ. If the school administrators are not satisfied with this degree of precision, a larger sample size should be taken.

■

Let us now generalize the procedure we are using to make probability statements about the sampling error. We will use the Greek letter α (alpha) to indicate the probability that a sampling error is *larger* than the sampling error mentioned in the precision statement. Refer to Figure 8.3. We see that $\alpha/2$ will be the area or probability in each tail of the distribution, and $1 - \alpha$ will be the area or probability that a sample mean will provide a sampling error *less than or equal to* the sampling error in the precision statement.

Using the z notation for the standard normal random variable (Table 1 of Appendix B), we place a subscript on the z value to denote the area in the *upper tail* of the probability distribution. As can be found in the standard normal probability distribution table, $z_{.025} = 1.96$. In order to have a precision statement with probability .99, $\alpha = .01$; in this case we would be interested in an area of $\alpha/2 = .005$ in the upper tail of the distribution, and hence $z_{.005} = 2.575$. We can now provide the following general procedure for making a probability statement about the sampling error whenever $\bar{x}$ is used to estimate μ.

Probability Statement About the Sampling Error

There is a $1 - \alpha$ probability that the value of a sample mean will provide a sampling error of $z_{\alpha/2} \, \sigma_{\bar{x}}$ or less.

FIGURE 8.3   Areas of a Sampling Distribution of $\bar{x}$ used to make Probability Statements about the Sampling Error

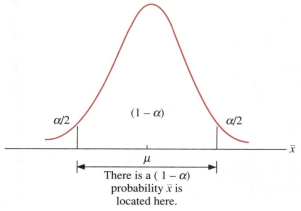

Notes and Comments

Statisticians use the term *bias* to refer to the difference between the expected value of a point estimator and the corresponding population parameter. Thus, for unbiased estimators the amount of bias is 0.   *continued on next page*

---

2. One special class of estimators, referred to as linear estimators, has the property that each estimator can be written as a linear function of the n observations in the sample. To show that $\bar{x}$ is a linear estimator, we can rewrite the formula for $\bar{x}$ as follows:

$$\bar{x} = \frac{x_1 + x_2 + \cdots + x_n}{n}$$

$$= \frac{1}{n}(x_1) + \frac{1}{n}(x_2) + \cdots + \frac{1}{n}(x_n)$$

Since each variable appears in a separate term and is raised to the first power, we see that $\bar{x}$ is a linear function of the sample observations $x_1, x_2, \cdots, x_n$. The importance of this result is that statisticians have shown that $\bar{x}$ is the best linear unbiased estimator of μ; that is, for the class of all possible estimators that are both linear and unbiased, $\bar{x}$ has the smallest variance.

EXERCISES

1. Assume that the population mean is $\mu = 100$ and that a simple random sample of size $n = 64$ results in a value of $\bar{x} = 103$.
  a. What is the point estimator?
  b. What is the point estimate?
  c. What is the value of the sampling error?
  d. If the value of $\bar{x}$ were 96, what would the value of the sampling error be?

2. A simple random sample of size $n = 64$ will be used to estimate a population mean. Answer the following questions assuming that the population mean is $\mu = 80$ and the population standard deviation is $\sigma = 40$.
  a. What is the probability that the sample mean will provide a sampling error of 5 or less?
  b. What is the value of the sampling error if the sample mean is $\bar{x} = 84$?

3. A simple random sample of $n = 100$ was used to estimate μ. If $\mu = 500$ and $\sigma = 50$, determine the sampling error needed to complete the following probability statements.
  a. There is a .95 probability that the sample mean will provide a sampling error of _____ or less.
  b. There is a .99 probability that the sample mean will provide a sampling error of _____ or less

4. A simple random sample of $n = 400$ was used to estimate a population mean μ. The value of $\bar{x}$ that was obtained is 75. Can the value of the sampling error be computed given this information? Explain.

8.2   Interval Estimation of a Population Mean: Large-Sample Case

In this section we show how the point estimate (observed value of $\bar{x}$) can be combined with the probability information about the sampling error to provide an interval estimate of a population mean μ. The procedure shown is appropriate in situations where the large-sample condition ($n \geq 30$) exists.

Calculating an Interval Estimate

We have the ability to make probability statements about the sampling error. We now can combine the point estimate with the probability information about the sampling error to obtain an *interval estimate* of the population mean. The rationale for the interval estimation procedure is as follows: We have already stated that there is a $1 - \alpha$ probability that the value of a sample mean will provide a sampling error of $z_{\alpha/2}\sigma_{\bar{x}}$ or less. This means that there is $1 - \alpha$ probability that the sample mean or point estimate *will not miss* the population mean *by more than* $z_{\alpha/2}\sigma_{\bar{x}}$. Thus if we form an interval by subtracting $z_{\alpha/2}\sigma_{\bar{x}}$ from the sample mean $\bar{x}$ and then adding $z_{\alpha/2}\sigma_{\bar{x}}$ to the sample mean $\bar{x}$, we would have a $1 - \alpha$ probability of obtaining an interval that *includes* the population mean μ. That is, there is a $1 - \alpha$ probability that the interval formed by $\bar{x} \pm z_{\alpha/2}\sigma_{\bar{x}}$ will contain the population mean μ.

EXAMPLE 8.1 (continued)

We previously stated that there is a .95 probability that the value of a sample mean mathematics achievement test score will provide a sampling error of 2.35 or less. Look at the sampling distribution of $\bar{x}$ as shown in Figure 8.4. Let us consider possible values of the sample mean $\bar{x}$ that could be obtained from three different simple random samples,

FIGURE 8.4 The Location of Three Different Sample Means $\bar{x}_1$, $\bar{x}_2$ and $\bar{x}_3$

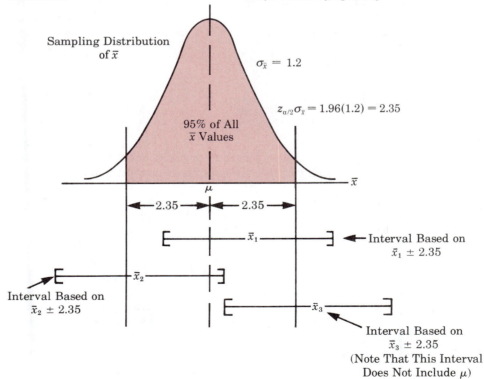

each containing 100 students. Remember, in each case we will form an interval estimate of the population mean by subtracting 2.35 from $\bar{x}$ and also by adding 2.35 to $\bar{x}$.

Consider what happens if the first sample mean turns out to have the value shown in Figure 8.4 as $\bar{x}_1$. Note that in this case the interval formed by subtracting 2.35 from $\bar{x}_1$ and adding 2.35 to $\bar{x}_1$ includes the population mean μ. Now consider what happens if the sample mean turns out to have the value shown in Figure 8.4 as $\bar{x}_2$. Although this next sample mean is different from the first sample mean, we see that the interval based on $\bar{x}_2$ also includes the population mean μ. However, the interval based on the third sample mean, denoted by $\bar{x}_3$, does not include the population mean. The reason for this is that the sample mean $\bar{x}_3$ lies in a tail of the probability distribution at a distance further than 2.35 from μ. Thus subtracting and adding 2.35 to $\bar{x}_3$ forms an interval that does not include μ.

Now think of repeating the sampling process many times, each time computing the value of the sample mean and then forming an interval $\bar{x} - 2.35$ to $\bar{x} + 2.35$. Any sample mean ($\bar{x}$) that falls between the vertical lines in Figure 8.4 will provide an interval that includes the population mean μ. Since 95% of the sample means are in the shaded region, 95% of all intervals that could be formed will include μ. As a result, we say that we are 95% confident that an interval constructed from $\bar{x} - 2.35$ to $\bar{x} + 2.35$ will include the population mean.

■

In common statistical terminology the interval is referred to as a *confidence interval*. With 95% of a sample means leading to a confidence interval including μ, we say the interval is established at the 95% *confidence level*. The value .95 is referred to as the *confidence coefficient* for the interval estimate. Because 95% confidence intervals are commonly used, we restate this confidence interval as follows.

95% Confidence Interval Estimate of a Population Mean

$$\bar{x} \pm 1.96 \frac{\sigma}{\sqrt{n}} \qquad (8.2)$$

EXAMPLE 8.1 (continued)

Using the sample data for the mathematics achievement test scores, we have $\bar{x} = 72.5$, $\mu = 12$, and $n = 100$. Using (8.2), the interval estimate of μ is computed as follows:

$$72.5 \pm 1.96 \frac{12}{\sqrt{100}} = 72.5 \pm \underbrace{2.35}$$

The measure of precision, or closeness, of $\bar{x} = 72.5$ to the population mean μ.

Subtracting 2.35 and adding 2.35 to the value of the sample mean $\bar{x} = 72.5$ provides the interval 70.15 to 74.85 as the 95% confidence interval estimate of μ. In other words, we are 95% confident that the interval 70.15 to 74.85 contains the value of the population mean μ.

■

Other confidence levels are possible for interval estimation of a population mean. The value 1.96 in (8.2) was selected because it is the z value that corresponds to 95% of the values in a standard normal probability distribution. Using a table of areas for the standard normal distribution (Table 1 in Appendix B), other confidence levels can be used by finding the z value that provides the corresponding percentage of values. Such a z value would be substituted for 1.96 in (8.2). Some of the common confidence levels and their corresponding z values are as follows:

---

z Values for Selected Confidence Levels

90% confidence:	$z_{.05} = 1.645$
95% confidence:	$z_{.025} = 1.96$
99% confidence:	$z_{.005} = 2.575$

From these z values, we note that as the confidence level increases, the z value also increases. This means that the confidence intervals become wider as the confidence level increases. This property of interval estimation points out that if we want to increase the confidence that the interval estimate contains the true value of μ, we need a wider interval. Example 8.2 will show this property of interval estimation.

Using the expression $(1 - \alpha)$ to denote the confidence coefficient, the following formula provides the general procedure for interval estimation of a population mean:

Interval Estimation of a Population Mean (Large-Sample Case)

$$\bar{x} \pm z_{\alpha/2} \frac{\sigma}{\sqrt{n}} \tag{8.3}$$

where $(1 - \alpha)$ is the *confidence coefficient* and $z_{\alpha/2}$ is the z value providing an area of $\alpha/2$ in the upper tail of the standard normal probability distribution.

For example, a 95% confidence interval has a confidence coefficient of .95. With $(1 - \alpha) = .95$, we have $\alpha = .05$. This means there must be an area of $\alpha/2 = .025$ in the upper tail of the standard normal probability distribution. From Table 1 of Appendix B, we find that the $z_{\alpha/2}$ value for a 95% confidence interval is $z_{.025} = 1.96$. Note that this is the z value we used in (8.2) when we computed a 95% confidence interval estimate of a population mean.

EXAMPLE 8.2

A large insurance company selected a simple random sample of 64 policyholders in order to estimate the mean age of individuals insured by the company. From past studies, it is assumed that the population standard deviation for age is $\sigma = 7.2$ years. If the sample mean age for the 64 policyholders is $\bar{x} = 39.5$ years, what are the 95% and 99% confidence interval estimates of the mean age for the population of policyholders?

For a 95% confidence interval, $\alpha = .05$ and $z_{\alpha/2} = z_{.025} = 1.96$. Thus, using (8.3), we obtain

$$\bar{x} \pm 1.96 \frac{\sigma}{\sqrt{n}} = 39.5 \pm 1.96 \frac{7.2}{\sqrt{64}}$$

$$= 39.5 \pm 1.76$$

or a confidence interval of 37.74 to 41.26 years.

For a 99% confidence interval, $\alpha = .01$ and $z_{\alpha/2} = z_{.005} = 2.575$. Thus (8.3) provides

$$\bar{x} \pm 2.575 \frac{\sigma}{\sqrt{n}} = 39.5 \pm 2.575 \frac{7.2}{\sqrt{64}}$$

$$= 39.5 \pm 2.32$$

or a confidence interval of 37.18 to 41.82 years.

As we mentioned earlier, these two confidence interval estimates of μ show that if we want to be *more confident* that the interval estimate contains the population mean, we have to use a *wider* interval estimate.

■

Interval Estimation When σ Is Unknown

The general expression for calculating the interval estimate of a population mean is given in (8.3). A difficulty in using (8.3) is that in many sampling situations, the value of the population standard deviation σ is *unknown*. In these instances we simply use the value of the sample standard deviation, *s*, as the point estimate of the population standard deviation σ. The formula for *s* as provided in Chapter 3 is as follows:

Sample Standard Deviation

$$s = \sqrt{\frac{\Sigma(x_i - \bar{x})^2}{n - 1}} \tag{8.4}$$

After computing the value of *s* from the sample data, the interval estimate of the population mean can be computed as follows:

Interval Estimation of a Population Mean—σ Unknown

$$\bar{x} \pm z_{\alpha/2} \frac{s}{\sqrt{n}} \tag{8.5}$$

where $(1 - \alpha)$ is the confidence coefficient, $z_{\alpha/2}$ is the z value providing an area of $\alpha/2$ in the upper tail of the standard normal probability distribution, and s is the sample standard deviation.

EXAMPLE 8.3

A sample of 36 parts was assembled using a proposed production method. Suppose that the sample mean time to assemble a part is $\bar{x} = 15.3$ minutes, and the sample standard deviation is $s = 1.3$ minutes. The objective is to develop a 98% confidence interval estimate for the population mean time to assemble a part.

In this example, the population standard deviation σ is unknown. The sample standard deviation, $s = 1.3$, is used as an estimate of σ. For 98% confidence, $\alpha = .02$; thus we have $z_{\alpha/2} = z_{.01} = 2.33$. The resulting 98% confidence interval estimate of the population mean is calculated using (8.5):

$$\bar{x} \pm 2.33 \frac{s}{\sqrt{n}} = 15.3 \pm 2.33 \frac{1.3}{\sqrt{36}}$$

$$= 15.3 \pm .5$$

Thus we are 98% confident that the interval from 14.8 minutes to 15.8 minutes contains the mean assembly time per part for the population.

■

Determining the Size of the Sample

Recall that in Section 8.1 we were able to make the following probability statement about the sampling error whenever a sample mean was used to provide a point estimate of a population mean:

> There is a $1 - \alpha$ probability that the value of the sample mean will provide a sampling error of $z_{\alpha/2}\sigma_{\bar{x}}$ or less.

Since $\sigma_{\bar{x}} = \sigma/\sqrt{n}$, can rewrite this statement as follows:

> There is a $1 - \alpha$ probability that the value of the sample mean will provide a sampling error of $z_{\alpha/2}(\sigma/\sqrt{n})$ or less.

From this statement we see that the values of $z_{\alpha/2}$, σ, and the sample size n combine to determine the *maximum sampling error*. Once we select a confidence coefficient or probability of $1 - \alpha$, $z_{\alpha/2}$ can be determined. Given values for $z_{\alpha/2}$ and σ, we can adjust the sample size n to provide the maximum sampling error desired. The formula used to compute the required sample size n is developed as follows.

Let E = the maximum sampling error mentioned in the statement about the desired precision. We now have

$$E = z_{\alpha/2} \frac{\sigma}{\sqrt{n}} \tag{8.6}$$

Using (8.6) to solve for $\sqrt{n}$, we have

$$\sqrt{n} = \frac{z_{\alpha/2} \, \sigma}{E}$$

Squaring both sides of this equation, we obtain the following equation for the sample size.

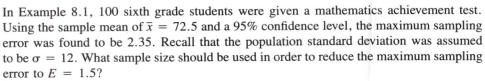

Sample Size for Interval Estimate of a Population Mean

$$n = \frac{(z_{\alpha/2})^2 \sigma^2}{E^2} \qquad (8.7)$$

EXAMPLE 8.1 (continued)

In Example 8.1, 100 sixth grade students were given a mathematics achievement test. Using the sample mean of $\bar{x} = 72.5$ and a 95% confidence level, the maximum sampling error was found to be 2.35. Recall that the population standard deviation was assumed to be $\sigma = 12$. What sample size should be used in order to reduce the maximum sampling error to $E = 1.5$?

Using (8.7), we have

$$n = \frac{(z_{.025})^2 \sigma^2}{E^2} = \frac{(1.96)^2 (12)^2}{(1.5)^2} = 245.9$$

Since we cannot sample a fraction of a student, we round up to a recommended sample size of 246 students. This sample size will provide a 95% confidence level that the sampling error is 1.5 or less.

■

Note that in (8.7) the values of $z_{\alpha/2}$ and E follow directly from the statement about the desired precision. However, the value of the population standard deviation σ may or may not be known. In cases where σ is known we have no problem in determining the sample size. However, in cases where σ is unknown we see that we must at least have a preliminary or *planning value* for σ in order to compute the sample size. In instances where this initial or preliminary sample is unavailable, we may be able to obtain a good approximation of σ from past data on "similar" studies. Without such past data we may have to use a judgment or "best-guess," value for σ. In any case, regardless of the source, we see from (8.7) that we must have a planning value for σ in order to determine a recommended sample size.

EXAMPLE 8.4

The manager of a fast-food restaurant would like to estimate the mean dollar amount spent per customer. Suppose that the manager wants to develop an estimate with a maximum sampling error of $.50. The sample standard deviation for a previous sample of 32 customers is $s = \$2.00$. How many customers should be included in the sample if a 99% confidence interval estimate of the population mean is desired?

For this information, $E = .50$, $z_{.005} = 2.575$, and a planning value for σ is 2. Thus

$$n = \frac{(z_{.005})^2 \sigma^2}{E^2} = \frac{(2.575)^2 (2)^2}{(.50)^2} = 106$$

Hence, for a 99% confidence level, the sampling error will be $.50 or less if we use a simple random sample of $n = 106$ customers.

■

5. A sample of size $n = 100$ was selected from a population with a standard deviation of $\sigma = 50$; the sample mean obtained was $\bar{x} = 200$. Compute the following confidence intervals and comment on what happens to the width of the confidence interval as the confidence level increases.

   a. 90% confidence interval
   b. 95% confidence interval
   c. 99% confidence interval

6. A sample of size $n = 250$ resulted in a sample mean of $\bar{x} = 82.5$ and a sample standard deviation of $s = 20.1$. Develop a 95% confidence interval estimate for the population mean μ.

7. On a final examination for a chemistry course at a large university, 36 randomly selected papers were graded shortly after the exam was over. Based on the previous year, the standard deviation of examination scores for the population was assumed to be $\sigma = 15$. If the sample mean for the 36 papers was $\bar{x} = 72$, provide 90% and 95% confidence intervals for the mean examination score for the population.

8. Data were collected on the golf-ball driving distances by professional golfers in a recent tournament. Using data on the first drives of 30 randomly selected golfers, it was found that the sample mean distance was 250 yards and the sample standard deviation was 10 yards. Develop a 95% confidence interval for the mean driving distance for the population.

9. Nielsen Media Research estimated that the average time that a person spends watching television per day is 6 hours and 59 minutes. (*USA Today,* December 13, 1988) Assume that this estimate was based upon a sample of 1600 viewers and that the sample standard deviation was 130 minutes. Develop a 90% confidence interval estimate of the mean time spent watching television per day.

10. The Stroh Brewery Company estimates that beer drinkers 21-24 years old consume an average of 63 gallons of beer per year. (*The Wall Street Journal,* October 5, 1988.) If their estimate was based upon a sample of 250 beer drinkers 21 to 24 years old, develop a 95% confidence interval estimate of the mean number of gallons consumed for all beer drinkers in this age group; assume that the sample standard deviation was $s = 15$ gallons.

11. E. Lynn and Associates is an energy-research firm that provides estimates of monthly heating costs for new homes based on style of the house, square footage, insulation and so on. The firm's service is used both by builders and potential buyers of new homes who wish advance information on heating costs. For winter months, the standard deviation in the home heating bills for residential homes in a certain area is believed to be $50. Assume that a sample of 30 homes in a particular subdivision will be used to estimate the mean monthly heating bills for all homes in this type of subdivision. If the sample mean is $\bar{x} = \$196.50$, provide a 98% confidence interval for the mean monthly heating bill.

12. A sample of 64 customers at Ron and Ted's Service Station shows a mean number of gallons of gasoline purchased per customer of 13.6 gallons. If the population standard deviation is 3.0 gallons, what is the 95% confidence interval estimate of the mean number of gallons purchased per customer?

13. An exit study collected for the Alaska State Division of Tourism showed that one group of 1,151 tourists spent an average of $3878 on their vacation (*Journal of Travel Research,* vol. XXVI, no. 4, 1988). If the standard deviation for this sample was $625, develop a 95% confidence interval estimate of the mean amount spent.

14. During a water shortage a water company randomly sampled residential water meters in order to monitor daily water consumption. On one particular day, a sample of 50 meters showed

a sample mean of $\bar{x}$ = 240 gallons and a sample standard deviation of s = 45 gallons. Provide a 90% confidence interval estimate of the mean water consumption for the population.

15. The Benson Property Management firm located in St. Louis would like to estimate the mean cost of repairing damages in apartments that are vacated by tenants. A sample of 36 vacated apartments resulted in a sample mean repair cost of $86.00, with a sample standard deviation of $12.25. Develop a 95% confidence interval to estimate the mean repair cost for the population of apartments.

16. A simple random sample of 35 Metro buses shows a sample mean of 225 passengers carried per day per bus. The sample standard deviation is computed to be 60 passengers. Provide a 98% confidence interval estimate of the mean number of passengers carried per bus during a 1-day period.

17. Based upon a sample of size 50, the 90% confidence interval estimate of μ was determined to be 155.5 ± 24.2. How large a sample size would be needed to reduce the maximum sampling error to 12.1?

18. Miles-per-gallon rating tests are being conducted for a particular model of automobile. If it is desired to estimate the mean miles per gallon rating to within ±1 mile per gallon at a 95% level of confidence, how many automobiles should be used in the sample? Assume that preliminary mileage tests indicate a standard deviation in miles per gallon for the automobiles to be 2.9.

19. An educational innovation at the high-school level will enable students to earn credit by completing a self-study workbook on English literature. In order to judge the credit given to the student fairly, educators would like an estimate of the mean number of hours it will take a student to complete the self-study program. In an evaluation study, a sample of students will be monitored during the self-study process. How many students should be in the sample if it is desired to estimate the mean number of hours required to complete the self-study to within ±3 hours at a 95% level of confidence? For planning purposes, it is estimated that the standard deviation of completion times will be approximately 10 hours.

20. Starting annual salaries for college graduates are believed to have a standard deviation of approximately $2000.00. Assume that a 95% confidence interval estimate of the mean annual starting salary is desired. How large a sample size should be taken if the size of the sampling error in the precision statement is to be
    a. $500.00
    b. $200.00
    c. $100.00

21. In developing patient appointment schedules, a medical center desires to estimate the mean time that a staff member spends with each patient. How large a sample should be taken if the precision of the estimate is to be ±2 minutes at a 95% level of confidence? How large a sample for a 99% level of confidence? Use a planning value for the population standard deviation of 8 minutes.

22. A national survey research firm has past data that indicate that the interview time for a consumer opinion study has a standard deviation of 6 minutes.
    a. How large a sample should be taken if the firm desires a .98 probability of estimating the mean interview time to within 2 minutes or less?
    b. Assume that the simple random sample you recommended in (a) is taken and that the mean interview time for the sample is 32 minutes. What is the 98% confidence interval estimate for the mean interview time for the population of interviews?

8.3 Hypothesis Tests About a Population Mean

Statistical inference is the process of drawing conclusions about a population parameter based on information contained in a sample. Point and interval estimation of a population mean, as introduced in Sections 8.1 and 8.2 are forms of statistical inference. Another form of statistical inference is *hypothesis testing*.

In hypothesis tests about a population mean, we begin by making a tentative assumption about the value of the population mean. This tentative assumption, called the *null hypothesis,* is denoted by H_0. We then define another hypothesis, called the *alternative hypothesis,* which is the opposite of what is stated in the null hypothesis. This alternative hypothesis is denoted by H_a. The hypothesis testing procedure involves using data from a sample to test the two competing claims indicated by H_0 and H_a.

The situation encountered in hypothesis testing is similar to the one encountered in a criminal trial. In a criminal trial the assumption is that the defendant is innocent. Thus, the null hypothesis is one of innocence. The opposite of the null hypothesis is the alternative hypothesis—that the defendant is guilty. Thus the hypotheses for a criminal trial would be written

H_0: The defendant is innocent

H_a: The defendant is guilty

To test these competing claims, or hypotheses, a trial is held. The testimony and evidence obtained during the trial provide the sample information. If the sample information is not inconsistent with the assumption of innocence, the null hypothesis that the defendant is innocent cannot be rejected. However, if the sample information is inconsistent with the assumption of innocence, the null hypothesis will be rejected. In this case, action will be taken based upon the alternative hypothesis that the defendant is guilty.

In some hypothesis-testing applications it may not be obvious how the null and alternative hypotheses should be formulated. Care must be taken to be sure that the hypotheses are structured appropriately and that the hypothesis-testing conclusion provides the information that the researcher or decision maker desires. Guidelines for establishing the null and alternative hypotheses will be given for three types of situations that frequently employ hypothesis-testing procedures. A discussion of each of these situations follows.

Testing Research Hypotheses

Consider a particular model automobile that currently obtains an average of 24 miles per gallon. A product-research group has developed a new carburetor designed specifically to increase the miles-per-gallon performance. In order to evaluate the new design, several new carburetors will be manufactured, installed in automobiles, and subjected to research-controlled driving tests. Note that the product-research group is looking for evidence to enable them to conclude that the new design *increases* the mean number of miles per gallon. In this case, the research hypothesis is that the new carburetor will provide a mean miles per gallon exceeding 24; that is, $\mu > 24$. As a general guideline, a research hypothesis such as this should be formulated as the *alternative hypothesis*. Thus, the appropriate null and alternative hypotheses for the study are as follows:

$$H_0: \quad \mu \leq 24$$

$$H_a: \quad \mu > 24$$

If the sample results indicate that H_0 cannot be rejected, we will not be able to conclude that the new carburetor is better. Perhaps more research and subsequent testing should be conducted. However, if the sample results indicate H_0 can be rejected, the inference can be made that $H_a: \mu > 24$ is true. With this conclusion, the researcher has the statistical support necessary to conclude that the new carburetor increases the mean number of miles per gallon.

In research studies such as these, the null and alternative hypotheses should be formulated so that the rejection of H_0, and hence the inference that H_a is true, will provide the researcher with the conclusion and action being sought. Thus, the research hypothesis should be expressed as the alternative hypothesis.

Testing the Validity of Assumption

Examples of testing the validity of assumptions can be found when companies make claims about their products. For example, a manufacturer of soft drinks states that 2-liter containers of its products have an average of at least 67.6 fluid ounces. A sample of 2-liter containers will be selected, and the contents will be measured in order to test the manufacturer's statement. In this type of hypothesis-testing situation, we generally follow the rationale suggested by the criminal trial analogy. That is, the manufacturer's statement should be assumed true (innocent) unless the sample evidence proves otherwise (guilty). Using this approach for the soft-drink example, the null and alternative hypotheses would be stated as follows:

$$H_0: \quad \mu \geq 67.6$$

$$H_a: \quad \mu < 67.6$$

If the sample results indicate H_0 cannot be rejected, the manufacturer's claim cannot be challenged. However, if the sample results indicate H_0 can be rejected, the inference will be made that $H_a: \mu < 67.6$ is true. With this conclusion, statistical evidence indicates that the manufacturer's statement is incorrect and that the soft-drink containers are being filled with a mean less than the claimed 67.6 ounces. Appropriate action against the manufacturer may be considered.

In any situation that involves testing the validity of some assumption, the null hypothesis is generally formulated based on the assumption. The alternative hypotheses is then formulated so that rejection of H_0 will provide the statistical evidence that the stated assumption is incorrect. Action to correct the claim or assumption should be considered whenever H_0 is rejected.

Decision Making

A hypothesis-testing situation involving decision making occurs when a decision maker must choose between two courses of action, one associated with the null hypothesis and another associated with the alternative hypothesis. For example, on the basis of a sample of parts from a shipment that has just been received, a quality-control inspector must decide whether to accept the entire shipment or to return the shipment to the supplier because it does not meet specifications. Assume that specifications for a particular part indicate a mean length of 2 inches per part is required. If the average length of the parts is greater or less than the 2-inch standard, the parts will cause quality problems in the assembly operation. In this case the null and alternative hypotheses would be formulated as follows:

$$H_0: \quad \mu = 2$$

$$H_a: \quad \mu \neq 2$$

If the sample results indicate H_0 cannot be rejected, the quality-control inspector will have no reason to doubt that the shipment meets specifications, and thus the shipment will be accepted. However, if the sample results indicate that H_0 should be rejected, the conclusion can be made that the parts do not meet specifications. In this case, the quality-control inspector has sufficient evidence to return the shipment to the supplier.

A Summary of Forms for Null and Alternative Hypotheses

The three hypothesis-testing examples that we have discussed in this section all concern the value of a population mean. Let μ_0 denote the numerical value being considered in the hypotheses. In general, a hypothesis test concerning the value of a population mean must take one of the following three forms:

$$H_0: \quad \mu \geq \mu_0 \qquad H_0: \quad \mu \leq \mu_0 \qquad H_0: \quad \mu = \mu_0$$

$$H_a: \quad \mu < \mu_0 \qquad H_a: \quad \mu > \mu_0 \qquad H_a: \quad \mu \neq \mu_0$$

Type I and Type II Errors

The null and alternative hypotheses are competing claims about the true state of nature. Either the null hypotheses, H_0, is true or the alternative hypothesis, H_a, is true, but not both. Ideally the hypothesis testing procedure should lead to the acceptance of H_0 when H_0 is the true state of nature and the rejection of H_0 when H_a is the true state of nature. Unfortunately, these results are not always possible. Since hypothesis tests are based upon sample information, we must allow for the possibility of errors. Table 8.1 provides a summary of the correct decisions and the errors in hypothesis testing.

TABLE 8.1

Errors and Correct Decisions in Hypothesis Testing

		STATE OF NATURE	
		H_0 True	H_a True
Decision	Accept H_0	Correct Decision	Type II Error
	Reject H_0	Type I Error	Correct Decision

The first row of Table 8.1 shows what can happen if we make the decision to accept H_0. If H_0 is the true state of nature, that is the correct decision. However, if H_a is the true state of nature, we have made a *type II error;* that is, we have accepted H_0 when it is false.

The second row of Table 8.1 shows what can happen if we make the decision to reject H_0. If the true state of nature is H_0, we have made a *type I error;* that is, we rejected H_0 when it is true. However, if H_a is the true state of nature, then rejecting H_0 is the correct decision.

Although we cannot eliminate the possibility of errors in hypothesis testing, we can consider the probability of their occurrence. Using common statistical notation, we denote the probabilities of making the two errors as follows:

$$\alpha = \text{the probability of making a type I error}$$

$$\beta = \text{the probability of making a type II error}$$

In most applications of hypothesis testing, the probability of making a type I error is controlled, but the probability of making a type II error is not. In practice, the person conducting the hypothesis test specifies the maximum allowable probability of making a type I error, called the *level of significance* for the test. Common choices for the level of significance are .05 and .01. Referring to the second row of Table 8.1, note that the decision to *reject* H_0 indicates that either a type I error or a correct decision has been made. Thus, if the probability of making a type I error is controlled by selecting a low value for the level of significance, we have a high degree of confidence that the decision to reject H_0 is correct. In such cases we have statistical support to conclude that H_0 is false and H_a is true. Any action suggested by the alternative hypothesis, H_a, is appropriate.

As we stated previously, however, although applications of hypothesis testing control the probability of making a type I error, they do not always control for the probability of making a type II error. Thus, whenever we decide to accept H_0 without taking into consideration the probability of making a type II error, we cannot determine how confident we can be with the decision to accept H_0. Because of the uncertainty associated with making a type II error, statisticians often recommend that we use the statement *do not reject* H_0 instead of *accept* H_0. Using the statement *do not reject* H_0 carries the recommendation to withhold both judgment and action. In effect, by never directly concluding that H_0 is true, the statistician avoids the risk of making a type II error. Whenever the probability of making a type II error has not been determined and controlled, we will not make the decision to accept H_0. In such cases, only two conclusions are possible: *do not reject* H_0 or *reject* H_0.

Notes and Comments

As we indicated in this section, the statement *do not reject* H_0 is used to avoid the conclusion that H_0 is true and thus eliminate the possibility of making a type II error. As a result, that statement *do not reject* H_0 means we do not conclude that H_0 is true or that H_a is true. In other words, the statement *do not reject* H_0 indicates that the test results are inconclusive. Since we are unable to conclude H_0 is true or H_a is true, the only action that should be taken is to recommend that further study be conducted in order to clarify the situation.

EXAMPLE 8.5

The United States Golf Association (USGA) has established rules that manufacturers of golf equipment must meet in order to have their products acceptable for use in USGA events. One rule regarding the manufacture of golf balls states that "A brand of golf ball, when tested on apparatus approved by the USGA on the outdoor range at the USGA Headquarters . . . shall not cover an average distance in carry and roll exceeding 280 yards."

Superflight, Inc., has recently developed a high-technology manufacturing method designed to produce golf balls that have an average or mean distance in carry and roll of 280 yards. Superflight realizes that if the new manufacturing process goes out of adjustment the process may produce golf balls with a mean distance less than 280 yards or with a mean distance greater than 280 yards. In the first case, Superflight may experience a downturn in sales as a result of marketing golf balls with a mean distance less than 280 yards. In the second case, Superflight's golf balls may be rejected by the USGA because the mean distance is greater than 280 yards.

As part of a quality-control program, periodically Superflight inspectors will select a random sample of 36 golf balls from the production line and subject the golf balls to tests equivalent to those performed by the USGA. With no reason to doubt the assumption that the manufacturing process is functioning correctly, the "innocent-until-proven-guilty" analogy suggest establishing the following null and alternative hypotheses:

$$H_0: \quad \mu = 280$$

$$H_a: \quad \mu \neq 280$$

Based upon statistical analysis of the sample data, one of the following conclusions will be reached:

Do Not Reject H_0: There is no evidence to suggest that the manufacturing process is out of control. No action should be taken.

Reject H_0: The sample evidence indicates the golf balls are being produced with a mean distance not equal to 280 yards. The conclusion can be made that the manufacturing process is out of control; action should be taken to adjust and correct the process.

Whenever a sample of 36 golf balls is selected and tested, the sample mean distance ($\bar{x}$) will be computed. Under the assumption that $H_0: \mu = 280$ is true, the sample mean comes from the sampling distribution shown in Figure 8.5. As with all sampling distri-

FIGURE 8.5 The Sampling Distribution of $\bar{x}$ for Simple Random Samples of 36 Golf Balls under the Assumption $H_0: \mu = 280$

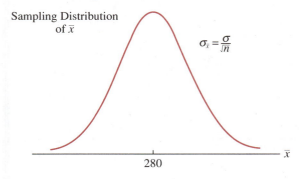

Sampling Distribution of $\bar{x}$

$\sigma_{\bar{x}} = \dfrac{\sigma}{\sqrt{n}}$

280

$\bar{x}$

butions of $\bar{x}$, we cannot expect the value of the sample mean obtained from the test to equal the population mean exactly. That is, even if $\mu = 280$ yards, we cannot expect the sample mean $\bar{x}$ to be exactly equal to 280. However, whenever the null hypothesis is true, we expect $\bar{x}$ to be close to 280.

For a given value of the sample mean $\bar{x}$, we can compute a corresponding z value, which determines how many standard deviations $\bar{x}$ is from the assumed value of the population mean μ. The expression for the z value is as follows:

$$z = \frac{\bar{x} - \mu}{\sigma / \sqrt{n}} \tag{8.8}$$

In this case, z is the standard normal random variable discussed in Chapter 7. Figure 8.6 shows the distribution of z when the null hypothesis is true.

If the z value using (8.8) is close to zero, it indicates that the value of the sample mean $\bar{x}$ must be close to the hypothesized value for μ. While this condition does not prove that the hypothesized value of μ is true, it does indicate that the value of the sample mean $\bar{x}$ is not inconsistent with the hypothesized value of μ. On the other hand, if the z value using (8.8) is substantially different from 0, for instance, $z = 4$, we have an indication that the sample mean is far from the hypothesized value for the population mean. In fact, if $z = 4$, the observed value of $\bar{x}$ is so far from the hypothesized value of μ that we would probably doubt that the sample mean $\bar{x}$ actually came from a sampling distribution with the hypothesized value of $\mu = 280$. In this case, we conclude that the sample results are inconsistent with the null hypothesis. As a result, the null hypothesis would be rejected and any action taken would be based on the alternative hypothesis, $\mu \neq 280$.

We want to reject the claim that $\mu = 280$ when the z value indicates that the sample mean, $\bar{x}$, is significantly less than 280 yards or when the z value indicates that the sample mean is significantly greater than 280 yards. Thus, H_0 should be rejected for values of the test statistic in either the lower tail or the upper tail of the sampling distribution. As a result the test is referred to as a *two-tailed* test.

Following the usual hypothesis-testing procedure, we first specify a level of significance by determining a maximum allowable probability of making a type I error. Suppose we choose $\alpha = .05$ as the level of significance. This means that there will be a

FIGURE 8.6 The Distribution of the Standard Normal Random Variable z when the Null Hypothesis H_0 is True

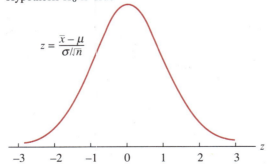

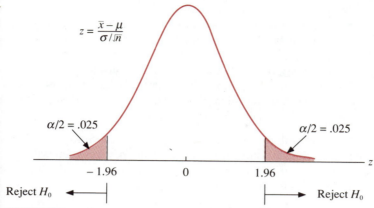

$$z = \frac{\bar{x} - \mu}{\sigma/\sqrt{n}}$$

$\alpha/2 = .025$ $\alpha/2 = .025$

-1.96 0 1.96 z

Reject H_0 Reject H_0

FIGURE 8.7 Rejection Region for the Null Hypothesis H_0: $\mu = 280$ with an $\alpha = .05$ Level of Significance

.05 probability of concluding that the mean distance is not 280 yards when in fact $\mu = 280$.

Figure 8.7 shows the sampling distribution of z with the two-tailed rejection region for $\alpha = .05$. With two-tailed hypothesis tests, we will always determine the rejection region by placing an area or probability of $\alpha/2$ in each tail of the distribution. The values of z that provide an area of .025 in each tail can be found from the standard normal probability distribution table. We see in Figure 8.7 that $-z_{.025} = -1.96$ identifies an area of .025 in the lower tail and $z_{.025} = +1.96$ identifies an area of .025 in the upper tail. Referring to Figure 8.7, we can establish the following rejection rule:

Reject H_0 if $z < -1.96$ or if $z > 1.96$

Suppose that a sample of 36 golf balls provides a sample mean distance of $\bar{x} = 278.5$ yards and a sample standard deviation of $s = 12$ yards. Using the value of μ from the null hypothesis and the sample standard deviation of $s = 12$ as an estimate of the population standard deviation, σ, the value of the test statistic is

$$z = \frac{\bar{x} - \mu}{\sigma/\sqrt{n}} = \frac{278.5 - 280}{12/\sqrt{36}} = -0.75$$

According to the rejection rule, H_0 cannot be rejected. The sample results do not indicate that the quality-control manager has reason to doubt that the manufacturing process is producing golf balls with a population mean distance of 280 yards.

EXAMPLE 8.6

A school administrator has developed an individualized reading-comprehension program for eighth grade students. To evaluate this new program, a random sample of 45 eighth grade students was selected; these students participated in the new reading program for one semester and then took a standard reading-comprehension examination. The mean test score for the population of students who had taken this test in the past was 76.

a. Determine the null and alternative or research hypotheses that will enable the administrator to determine whether or not the mean test score for students who take the individualized reading program is *greater than* the historical mean test score of 76.

b. Develop a rejection rule for the hypothesis test using a .05 level of significance.

c. The sample results for the 45 students provided a sample mean of $\bar{x} = 79$ and a sample standard deviation of $s = 8$; what is the appropriate conclusion about the proposed individualized reading program?

We begin with the null hypothesis H_0: $\mu \leq 76$. Since the administrator is interested in the research question of determining if the new program will provide an *increase* in the mean test score, the alternative hypothesis is H_a: $\mu > 76$. Thus we have a *one-tailed hypothesis test* with the hypotheses as shown:

$$H_0: \quad \mu \leq 76$$

$$H_a: \quad \mu > 76$$

In this example the rejection of the null hypothesis is not necessarily a "bad" conclusion. In fact, the administrator would probably be delighted to learn that H_0 can be rejected, leading to the conclusion that the new program is better.

The hypothesis test for Example 8.6 can be made by referring to the rejection region shown in Figure 8.8. Note that with a .05 level of significance, the rejection region with an area or probability of .05 is located in the upper tail of the distribution. Using the table of areas for the standard normal probability distribution, we find that the critical z value for the test is 1.645. The rejection rule for this one-tailed test is written as follows:

$$\text{Reject } H_0 \text{ if } z > 1.645$$

From the information in (c) we know that $n = 45$ and $\bar{x} = 79$. The sample standard deviation $s = 8$ can be used as the estimate of the population standard deviation σ. With the null hypothesis H_0: $\mu \leq 76$, the value of the test statistic z is as follows:

FIGURE 8.8 Rejection Region for the Null Hypothesis in Example 8.6 using an $\alpha = .05$ Level of Significance

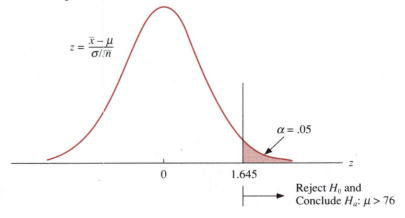

$$z = \frac{\bar{x} - \mu}{\sigma/\sqrt{n}} = \frac{79 - 76}{8/\sqrt{45}} = +2.52$$

Using the rejection rule, we reject H_0 and conclude that the proposed reading program provides a significant improvement in the mean test scores for the students.

EXAMPLE 8.7

The Federal Trade Commission (FTC) has decided to select a simple random sample of 36 cans of coffee in order to check a manufacturer's claim regarding the amount of coffee in the cans. If the population mean weight is 3 or more pounds of coffee per can, no action will be taken by the FTC. However, if the sample data leads the FTC to believe that the company produces coffee with a mean weight of less than 3 pounds per can, the FTC will claim a violation exists and take action against the manufacturer.

a. Develop the null and alternative hypotheses for this test.
b. Provide the rejection rule if the FTC is willing to tolerate a .01 probability of making the error of claiming the manufacturer is in violation when the manufacturer is actually filling cans with an acceptable mean of 3 pounds per can.
c. From past data, the population standard deviation of weights is $\sigma = .18$ pounds. If the sample of 36 items shows a sample mean of $\bar{x} = 2.96$ pounds, what is your conclusion and what action should the FTC take?

a. The hypotheses test can be written as follows:

$$H_0: \quad \mu \geq 3$$
$$H_a: \quad \mu < 3$$

This is a one-tailed test with the rejection of H_0 occurring only if the sample results indicate that the population mean filling weight is less than 3 pounds per can.

b. The probability of the type I error, or level of significance, is specified to be $\alpha = .01$. With the rejection region in the lower tail of the distribution, the table of areas for the standard normal probability distribution shows the rejection rule is written:

Reject H_0 if $z < -2.33$

c. The sample mean of $\bar{x} = 2.96$ and the hypothesized value of $\mu = 3$ provide the following value for the test statistic:

$$z = \frac{\bar{x} - \mu}{\sigma/\sqrt{n}} = \frac{2.96 - 3.00}{.18/\sqrt{36}} = -1.33$$

Since $z = -1.33$, we cannot reject H_0. Thus, although we have not proven that the mean weight of coffee per can is $\mu \geq 3$, the sample evidence does not indicate the null hypothesis can be rejected. As a result, the FTC should not take any action against the manufacturer.

A Summary of Rejection Rules

Using the notation

$$\alpha = \text{the level of significance}$$

$$z_\alpha = \text{the } z \text{ value with an area of } \alpha \text{ in the upper tail}$$
$$\text{of the standard normal probability distribution}$$

the three forms of hypothesis tests about a population mean and the corresponding rejection rules are summarized in Figure 8.9.

The Relationship Between Interval Estimation and Hypothesis Testing

In this chapter we have discussed statistical procedures that can be used to make inferences about the value of a population mean. In Section 8.2 we discussed interval estimation, and in Section 8.3 we focused on hypothesis testing. In the case of interval estimation, the population mean μ was unknown. Once the sample was selected and the sample mean $\bar{x}$ computed, we developed an interval around the value of $\bar{x}$ that had a good chance of including the value of the parameter μ. The interval estimate computed was referred to as a confidence interval with $1 - \alpha$ defined as the confidence coefficient. In the large-sample case, the formula for interval estimation of a population mean was given by

$$\bar{x} \pm z_{\alpha/2} \frac{\sigma}{\sqrt{n}} \tag{8.9}$$

Conducting a hypothesis test requires us first to make an assumption about the value of the population mean. The two-tailed hypothesis test has the form

$$H_0: \quad \mu = \mu_0$$

$$H_a: \quad \mu \neq \mu_0$$

where μ_0 is the hypothesized value for the population mean. Using the rejection rule for a two-tailed test, we see that the region over which we do not reject includes all values of the sample mean $\bar{x}$ that are within $-z_{\alpha/2}$ and $+z_{\alpha/2}$ standard errors of μ_0. Thus the following expression provides the do-not-reject region for the sample mean $\bar{x}$ in a two-tailed hypothesis test with a level of significance of α:

$$\mu_0 \pm z_{\alpha/2} \frac{\sigma}{\sqrt{n}} \tag{8.10}$$

A close look at (8.9) and (8.10) will provide insight into the relationship between the estimation and hypothesis-testing approaches to statistical inference. Note in particular that both procedures require the computation of the values $z_{\alpha/2}$ and $\sigma/\sqrt{n}$. Focusing on α, we see that a confidence coefficient of $(1 - \alpha)$ for interval estimation corresponds to a level of significance of α in hypothesis testing. For example, a 95% confidence interval

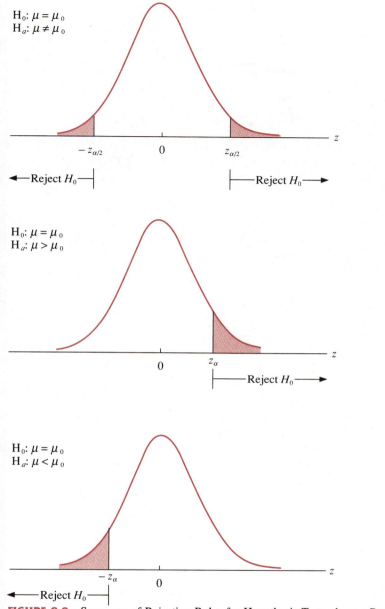

$H_0: \mu = \mu_0$
$H_a: \mu \neq \mu_0$

$-z_{\alpha/2}$ 0 $z_{\alpha/2}$ z

◄—Reject H_0—| |—Reject H_0—►

$H_0: \mu = \mu_0$
$H_a: \mu > \mu_0$

0   z_α z

|—Reject H_0—►

$H_0: \mu = \mu_0$
$H_a: \mu < \mu_0$

$-z_\alpha$ 0 z

◄—Reject H_0—|

FIGURE 8.9   Summary of Rejection Rules for Hypothesis Tests about a Population Mean

for estimation corresponds to a .05 level of significance for hypothesis testing. Furthermore, (8.9) and (8.10) show that since $z_{\alpha/2}$ ($\sigma/\sqrt{n}$) is the plus or minus value for both expressions, if $\bar{x}$ falls in the do-not-reject region defined by (8.10), the hypothesized value μ_0 will be in the confidence interval defined by (8.9). Conversely, if the hypothesized value μ_0 falls in the confidence interval defined by (8.9), the sample mean $\bar{x}$ will be in the do-not-reject region for the hypothesis $H_0: \mu = \mu_0$. These observations lead to the following procedure for using confidence interval results to draw hypothesis-testing conclusions:

EXAMPLE 8.5 (continued)

Let us return to the Superflight golf ball study discussed earlier to demonstrate the use of a confidence interval for hypothesis testing. The Superflight golf ball study resulted in the following two-tailed test:

$$H_0: \quad \mu = 280$$

$$H_a: \quad \mu \neq 280$$

In order to test this hypothesis with a level of significance of $\alpha = .05$, we sampled 36 golf balls and found a sample mean distance of $\bar{x} = 278.5$ yards and a sample standard deviation of $s = 12$ yards. Using these results with $z_{.025} = 1.96$, the 95% confidence interval estimate of the population mean becomes

$$\bar{x} \pm z_{.025} \frac{\sigma}{\sqrt{n}}$$

$$278.5 \pm (1.96) \left(\frac{12}{\sqrt{36}} \right)$$

$$278.5 \pm 3.92$$

or

$$274.58 \text{ to } 282.42$$

This finding enables the quality-control manager to conclude with 95% confidence that the mean distance for the population of golf balls is between 274.58 yards and 282.42 yards. Since the hypothesized value for the population mean, $\mu_0 = 280$, is in this interval, the hypothesis-testing conclusion is that the null hypothesis, $H_0: \mu = 280$, cannot be rejected.

Notes and Comments

Whenever a hypothesis test results in the rejection of H_0, we say that the sample results are *statistically significant*. While statistical significance is of interest to the manager or decision maker, additional judgment may be needed to determine the practical value of the significance information. For example, assume that a research study provides statistical significant results rejecting H_0: $\mu \leq 40{,}000$ and concluding H_a: $\mu > 40{,}000$. If the statistical significance is based on a sample mean of $\bar{x} = 40{,}100$, the decision maker may show little concern for immediate action. However, if the statistical significance is based on a sample mean of $\bar{x} = 48{,}000$, the decision maker may desire immediate action due to the large difference between $\bar{x}$ and $\mu \leq 40{,}000$. Thus, we see that the practical value of a statistical test may be based on more than simply the conclusion of statistically significant results.

EXERCISES

23. Consider the following two-tailed test:

$$H_0: \quad \mu = 500$$

$$H_a: \quad \mu \neq 500$$

Assume that $\sigma = 120$ and that a sample of size 100 resulted in a sample mean of $\bar{x} = 480$.
   a. What is the value of the test statistic z?
   b. At the .05 level of significance, what conclusion should be reached?

24. The hypotheses for a particular study are as follows:

$$H_0: \quad \mu \leq 125$$

$$H_a: \quad \mu > 125$$

Assume that the sample mean is $\bar{x} = 120$.
   a. Could the null hypothesis be rejected in this case? Explain.
   b. If the null hypothesis is accepted, what type of error could we be making? Explain.

25. An experiment resulted in the following one-tailed test:

$$H_0: \quad \mu \geq 25$$

$$H_a: \quad \mu < 25$$

Assume that a sample of size 49 resulted in a sample mean of 23 and a sample standard deviation of 10.5.
   a. Compute the value of the test statistic z.
   b. At the .10 level of significance, what conclusion should be reached?

26. In any test of hypotheses, is it possible to make a type I error *and* a type II error at the same time? Explain.

27. A quality control inspector at Morgan Manufacturing Company tests part dimensions for a machining operation. The desired mean part diameter is two inches. A sample of parts is periodically selected and the size of each part is checked. If the sample leads the quality control inspector to believe that the mean part diameter for the population is either too large or too small, the machine will be shut down and readjusted.

a. Which of the following hypothesis tests should be used to determine whether or not the machine should be shutdown?

A	B	C
$H_0: \quad \mu = 2$	$H_0: \quad \mu \geq 2$	$H_0: \quad \mu \leq 2$
$H_a: \quad \mu \neq 2$	$H_a: \quad \mu < 2$	$H_a: \quad \mu > 2$

b. What are the type I and type II errors in hypothesis testing? Define the consequences of making these errors in the context of this problem.

28. The manager of an automobile dealership is considering a new bonus plan that may increase sales volume for the dealership. Historically, the mean sales volume per sales employee has been four automobiles per month. A sample of 32 sales employees will be allowed to sell under the new bonus plan for a trial period. The manager is interested in implementing the new plan throughout the company if the sample evidence indicates the plan increases the mean sales volume. What hypotheses are appropriate for determining whether or not the new bonus plan should be implemented?

A	B	C
$H_0: \quad \mu = 4$	$H_0: \quad \mu \geq 4$	$H_0: \quad \mu \leq 4$
$H_a: \quad \mu \neq 4$	$H_a: \quad \mu < 4$	$H_a: \quad \mu > 4$

29. Spread Easy paint is labeled as having a mean coverage of 400 square feet per gallon. A mean coverage of more than 400 square feet is unsatisfactory and indicates that the paint is too thin. A mean coverage of less than 400 square feet is unsatisfactory and indicates that the paint is too thick. Assume $\sigma = 25$ square feet per gallon. A sample of 30 gallons of paint will be selected and used to test the coverage of the paint.

a. What are the null and alternative hypotheses for this test?
b. If the sample mean is $\bar{x} = 380$ square feet, what is your conclusion about the paint coverage? Use a .05 level of significance.

30. At Western University the mean scholarship examination score for entering students has been 900, with a standard deviation of 80. Each year a sample of applications is taken to see if the mean examination scores are at the same level or are changing. The null and alternative hypotheses are $H_0: \mu = 900$ and $H_a: \mu \neq 900$. A random sample of 60 students in this year's class provides a sample mean examination score of $\bar{x} = 924$. Using a .05 level of significance, what conclusion should be made about any change in the examination scores?

31. An automobile assembly-line operation has a scheduled mean completion time of 12.2 minutes. Because of the effect of completion time on both earlier and later assembly operations, it is important to maintain the 12.2-minute standard.

a. Define the null and alternative hypotheses and determine the decision rule for conducting the test at a .02 level of significance.
b. If a random sample of 45 completion times show $\bar{x} = 12.39$ and $s = 1.20$ minutes, what is your conclusion?

32. A study of the operation of a city-owned parking garage shows a historical mean parking time of 220 minutes per car. The garage area has recently been remodeled and the parking

charges have been increased. The city manager would like to know if these changes have had any effect on the mean parking time of the garage customers. What is your conclusion if a sample of 50 cars showed $\bar{x} = 208$ and $s = 80$?

33. Fowle Marketing Research, Inc., estimates costs for a client on the assumption that a telephone interview can be completed with a mean time of 15 minutes or less. If a greater mean time is required, the client is charged a premium rate. Does a sample of 35 surveys that results in a sample mean of 17 minutes and a sample standard deviation of 4 minutes justify the premium rate? Test at a .01 level of significance.

34. The manager of the Danvers-Hilton Hotel believes that the mean guest bill is $250 or more. Assume that we wish to test the manager's claim with the following hypotheses: $H_0: \mu \geq 250$, $H_a: \mu < 250$. In this case, the manager's claim will be accepted unless H_0 is rejected. If $\sigma = 50$ and if a simple random sample of 60 billings shows a sample mean of $\bar{x} = \$235$, what conclusion would you draw? Use a .05 level of significance.

35. The manager of the Keeton Department Store believes that the mean annual income of the stores credit-card customers is at least $18,000 per year. A sample of 58 credit-card customers shows a sample mean of $17,200 and a sample standard deviation of $3000. At the .05 level of significance, can the manager's claim be rejected?

36. A federal funding program is available to low-income neighborhoods. To qualify for the funding, a neighborhood must have a mean household income of less than $7000 per year. Funding decisions are based on a sample of residents in a neighborhood. The hypothesis test involves $H_0: \mu \geq 7000$ and $H_a: \mu < 7000$. Federal funding is granted only if H_0 is rejected.
   a. Assume that a sample of 36 households in a particular neighborhood shows $\bar{x} = \$6600$ and $s = \$1200$. At a .05 level of significance, does this neighborhood qualify for federal funding?
   b. In a sample of 36 households, what is the maximum sample mean income level that will enable the neighborhood to still qualify for federal funding?

8.4 The Use of *p*-values

We now consider another approach that can be used to establish decision rules for a hypothesis test. This approach is based upon what is called a *p-value*. We will show how the *p*-value for a sample can be computed based on the value of z. The *p*-value can then be used to make the conclusion of whether or not to reject H_0. The *p*-value is defined as follows:

> **p-value**
>
> The *p-value* is the probability, assuming the null hypothesis is true, of obtaining a sample result that is more unlikely than the observed sample result.

In order to show how *p*-values are computed, we will reconsider the one-tailed test introduced in Example 8.6.

**EXAMPLE 8.6
(continued)**

In Example 8.6, a school administrator selected a random sample of 45 eighth grade students in order to test the effectiveness of a new reading program. The performance of the program was measured by having students who participated in the program for one semester take a standardized reading-comprehension examination. The sample results for n = 45 students provided a sample mean of $\bar{x}$ = 79 and a sample standard deviation of s = 8.

Since the mean test score for the population of students who had taken the test prior to the development of the new program was 76, the following null and alternative hypotheses were developed:

$$H_0: \quad \mu \leq 76$$

$$H_a: \quad \mu > 76$$

Thus, if the null hypothesis is rejected, we would conclude that the mean test score for the new reading program is greater than for the previous reading program. The school administrator had specified that the test should be conducted using a .05 level of significance.

In this example the value of z is computed as follows:

$$z = \frac{\bar{x} - \mu}{s/\sqrt{n}} = \frac{79 - 76}{8/\sqrt{45}} = +2.52$$

To compute the p-value for this one-sided hypothesis test, we must now determine the probability of obtaining a sample result that is more extreme than $\bar{x}$ = 79. In other words, we must determine the probability of obtaining a value for z greater than +2.52. Using the table of areas for the standard normal probability distribution, we see that the area in the right-hand tail corresponding to z = +2.52 is .5000 − .4941 = .0059; thus, the p-value is .0059.

The p-value of .0059 can now be used to make hypothesis-test conclusion. First, note that if the *p-value is greater than or equal to α,* the value of z must be in the do-not-reject region. Similarly, if the *p-value is less than α,* the value of z must be in the rejection region. Thus, in this example, the fact that the p-value of .0059 is less than the stated level of significance, α = .01, indicates that z is in the rejection region and the null hypothesis can be rejected.

The computation of the p-value for a two-sided hypothesis test is somewhat more involved. To illustrate the approach, let us reconsider Example 8.5.

**EXAMPLE 8.5
(continued)**

In Example 8.5, Superflight, Inc., used a hypothesis test for the mean driving distance of new golf balls. The null and alternative hypotheses were stated as follows:

$$H_0: \quad \mu = 280$$

$$H_a: \quad \mu \neq 280$$

A sample of 36 golf balls resulted in a mean driving distance of $\bar{x}$ = 278.5 yards and a standard deviation of s = 12 yards. A .05 level of significance was specified.

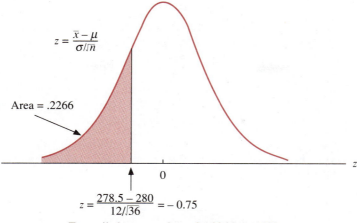

$$z = \frac{\bar{x} - \mu}{\sigma/\sqrt{n}}$$

Area = .2266

0

z

$$z = \frac{278.5 - 280}{12/\sqrt{36}} = -0.75$$

Two-tailed test: p-value = 2 (.2266) = .4532

FIGURE 8.10 Computation of the p-value for Example 8.5

To compute the p-value for a two-tailed test we must first calculate the area in the tail of the sampling distribution corresponding to the observed value of $\bar{x}$. The p-value for the two-tailed test is then computed by *doubling* this area or probability. For instance, for the Superflight golf ball example, $\bar{x} = 278.5$; the corresponding z value was -0.75. The table for the standard normal probability distribution shows that the area between the mean and $z = -0.75$ is .2734. Thus, the area in the lower tail is .5000 − .2734 = .2266. The p-value for this two-tailed test is double this value: p-value = 2(.2266) = .4532. Since the p-value is greater than the level of significance ($\alpha = .05$), the p-value indicates that the null hypothesis cannot be rejected. The p-value calculation for Superflight is summarized in Figure 8.10.

■

Regardless of whether the hypothesis test is one-tailed or two-tailed, the p-value approach to hypothesis testing will always provide the same conclusion as the approach using the test statistic z. However, the p-value approach carries the interpretation of being the probability of obtaining a difference between $\bar{x}$ and the hypothesized value for μ that is more extreme than the value actually observed. Under this interpretation, a relatively large p-value is an indication that the observed difference is not unusual. Thus the sample result would be judged not inconsistent with the null hypothesis and H_0 cannot be rejected. However, a small p-value indicates that the difference between $\bar{x}$ and the hypothesized value for μ is so unusual that the null hypothesis should be rejected. For a given level of significance α, the decision rule can be defined in terms of the p-value as follows:

p-value Criterion for Hypothesis Testing

Reject H_0 if the p-value $< \alpha$

1. The *p*-value is called the observed level of significance; it depends only on the sample outcome. One does not need to know the level of significance to compute the *p*-value. But, it is necessary to know whether the hypothesis test being investigated is one-tailed or two-tailed. Given the value of $\bar{x}$ in a sample, the *p*-value for a two-tailed test will always be *twice* the area in the tail of the sampling distribution for the observed value of $\bar{x}$.

2. Computer programs that provide hypothesis testing outputs almost always provide the *p*-value for the sample results. The *p*-values on computer outputs can be referred to under a variety of headings including P, PROB, and Probability.

EXERCISES

37. Consider the following hypothesis test:

$$H_0: \quad \mu = 100$$

$$H_a: \quad \mu \neq 100$$

If the experimenter wanted to perform the test using a level of significance of .05, what conclusion should be reached if the *p*-value is 0.65? Explain.

38. Consider the following hypothesis test:

$$H_0: \quad \mu \leq 25$$

$$H_a: \quad \mu > 25$$

If the experimenter wanted to perform the test using a level of significance of .05, what conclusion should be reached if the *p*-value is .025? Explain.

39. To test the null hypothesis $H_0: \mu \geq 250$ versus the alternative hypothesis $H_a: \mu < 250$, a random sample of size $n = 100$ was obtained. The sample results are $\bar{x} = 245$ and $s = 60$. Assume that the experimenter wanted to use a level of significance of .10.
    a. Compute the *p*-value.
    b. What conclusion should be reached given the *p*-value computed in (a)? Explain:

40. Statistics have been compiled for the selling prices of condominiums in Dade County, Florida. During the previous year the mean selling price of condominiums was $91,000. A sample of 50 current condominium sales showed a sample mean price of $\bar{x} = \$89,500$, with a sample standard deviation of $s = \$5000$.
    a. Formulate null and alternative hypotheses that can be used to test for *any change* in the mean selling price of condominiums during the 1-year period.
    b. What is the *p*-value associated with the sample mean of $89,500?
    c. What is the hypothesis-testing conclusion? Used $\alpha = .05$.

41. In order to test whether or not the mean driving speeds on the Pennsylvania Turnpike are within the posted speed limit of 65 miles per hour, the speeds for a random sample of 40 vehicles were recorded; the sample results show a mean of $\bar{x} = 67.6$ miles per hour and a standard deviation of $s = 8$ miles per hour. Using a .10 level of significance, calculate the *p*-value that can be used to determine whether or not the mean highway driving speeds are within the 65-mile-per-hour speed limit. What conclusion should be reached?

42. The president of Fightmasster and Associates Real Estate, Inc., claims that the mean selling time of a residential home is 40 days or less after it is listed with the company. A sample of 50 recently sold residential homes shows a sample mean selling time of 45 days and a sample standard deviation of 20 days. Test the president's claim at the .02 level of significance using the p-value approach.

43. An automobile company currently pays an average wage of $20 per hour for its production employees. The company is planning to build a new factory, and several locations are being considered. The availability of labor at a rate less than $20.00 per hour is a major factor in the location decision. For one location, a sample of 40 workers showed a current mean hourly wage of $\bar{x} = \$19.00$ and a sample standard deviation of $s = \$2.60$ Using the p-value and a .10 level of significance, does the sample data indicate that the location has a mean wage rate significantly below the $20.00 per hour rate?

44. Medical Economics magazine reported that the average physician made $115,440 (*The Wall Street Journal*, October 11, 1988). To test this claim, suppose that a simple random sample of 100 physicians was selected from the American Medical Association's directory; the average salary reported for this group was $113,650, with a standard deviation of $10,000. At the .05 level of significance, use the p-value approach to test this claim.

8.5   Inferences Using Small Samples

The methods of estimation and hypothesis testing that we have discussed thus far have required sample sizes of at least 30 items. The reason for this is that in the large-sample situation ($n \geq 30$), the central limit theorem can be used to approximate the sampling distribution of $\bar{x}$ with a normal probability distribution. Thus the random variable

$$z = \frac{\bar{x} - \mu}{\sigma/\sqrt{n}} \tag{8.11}$$

is a standard normal random variable. Recall that z values were used in both interval-estimation and hypothesis-testing procedures.

    If the sample size is small (that is, $n < 30$), the central limit theorem can no longer be used to specify the sampling distribution of $\bar{x}$; in this case we have to consider other methods of estimation and hypothesis testing.

The *t* Distribution

W. S. Gosset, who published his writings under the pen name Student, found that if the population being sampled has a *normal probability distribution* and the sample standard deviation s is used as an estimate of the population standard deviation σ, statistical inferences about a population mean can be based on the random variable

$$t = \frac{\bar{x} - \mu}{s/\sqrt{n}} \tag{8.12}$$

The sampling distribution of t is called the Student t probability distribution, or simply the *t distribution*.

    It is important to realize that if the population has a normal probability distribution and if s is used as an estimator of σ, the t distribution is applicable for *any sample size*.

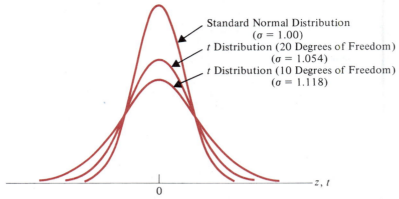

FIGURE 8.11 Comparison of the Standard Normal Distribution with *t* Distributions having 10 and 20 Degrees of Freedom

In fact, as the sample size increases, the *t* distribution approaches the standard normal distribution. As a result, in the large-sample case, interval estimation and hypothesis tests based on *z* values and *t* values provide similar conclusions. However, in Sections 8.2 and 8.3 we already have presented methods for interval estimation and hypothesis tests about a population mean with large samples. Thus we do not need to draw upon the *t* distribution and its required assumption of a normally distributed population until we encounter a small-sample case.

The *t* distribution is actually a family of similar probability distributions. Each specific *t* distribution depends upon a parameter known as its *degrees of freedom;** there is a unique *t* distribution with 1 degree of freedom, a unique *t* distribution with 2 degrees of freedom, a unique *t* distribution with 3 degrees of freedom, and so on. As the number of degrees of freedom increases, the difference between the *t* distribution and the standard normal distribution becomes smaller and smaller. Figure 8.11 shows *t* distributions with 10 and 20 degrees of freedom and their relationships to the standard normal probability distribution.

We use a subscript for *t* to indicate the area in the *upper tail* of the *t* distribution. For example, we use $t_{.025}$ to indicate a *t* value with .025 area in the upper tail of the distribution. A table for the *t* distribution is provided in Table 2 of Appendix B. This table is also shown in Table 8.2. Note, for example, that for a *t* distribution with 10 degrees of freedom, $t_{.025} = 2.228$. Similarly, for a *t* distribution with 20 degrees of freedom, $t_{.025} = 2.086$. Now that we have an idea of what the *t* distribution is, let us show how it is used in estimating a population mean.

Estimation

Assuming that the population has a normal probability distribution and that the sample standard deviation *s* is used as an estimate of the population standard deviation σ, interval estimation of a population mean is similar to the approach we used in Section 8.2; the exception is that a *t* value is used instead of a *z* value. Using $1 - \alpha$ to denote the confidence coefficient, the formula on page 288 provides the general procedure for an interval estimate of a population mean.

---

*The reason the terminology degrees of freedom is used is that the computation of *s* is based on the sum of the squared deviations $\Sigma(x_i - \bar{x})^2$ and only $n - 1$ of the deviations, $(x_i - \bar{x})$, are independent. That is, if we know $n - 1$ of the $(x_i - \bar{x})$ values, the remaining deviation can be determined, since for any sample $\Sigma(x_i - \bar{x}) = 0$. In this case, we have $n - 1$ degrees of freedom.

TABLE 8.2

t Distribution Tables for Areas in the Upper Tail.
Example: With 10 Degrees of Freedom $t_{.025} = 2.228$

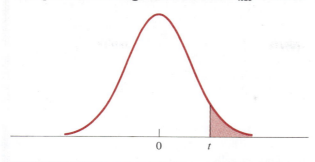

DEGREES OF FREEDOM	UPPER-TAIL AREAS (SHADED)				
	.10	.05	.025	.01	.005
1	3.078	6.314	12.706	31.821	63.657
2	1.886	2.920	4.303	6.965	9.925
3	1.638	2.353	3.182	4.541	5.841
4	1.533	2.132	2.776	3.747	4.604
5	1.476	2.015	2.571	3.365	4.032
6	1.440	1.943	2.447	3.143	3.707
7	1.415	1.895	2.365	2.998	3.499
8	1.397	1.860	2.306	2.896	3.355
9	1.383	1.833	2.262	2.821	3.250
10	1.372	1.812	2.228	2.764	3.169
11	1.363	1.796	2.201	2.718	3.106
12	1.356	1.782	2.179	2.681	3.055
13	1.350	1.771	2.160	2.650	3.012
14	1.345	1.761	2.145	2.624	2.977
15	1.341	1.753	2.131	2.602	2.947
16	1.337	1.746	2.120	2.583	2.921
17	1.333	1.740	2.110	2.567	2.898
18	1.330	1.734	2.101	2.552	2.878
19	1.328	1.729	2.093	2.539	2.861
20	1.325	1.725	2.086	2.528	2.845
21	1.323	1.721	2.080	2.518	2.831
22	1.321	1.717	2.074	2.508	2.819
23	1.319	1.714	2.069	2.500	2.807
24	1.318	1.711	2.064	2.492	2.797
25	1.316	1.708	2.060	2.485	2.787
26	1.315	1.706	2.056	2.479	2.779
27	1.314	1.703	2.052	2.473	2.771
28	1.313	1.701	2.048	2.467	2.763
29	1.311	1.699	2.045	2.462	2.756
30	1.310	1.697	2.042	2.457	2.750
40	1.303	1.684	2.021	2.423	2.704
60	1.296	1.671	2.000	2.390	2.660
120	1.289	1.658	1.980	2.358	2.617
∞	1.282	1.645	1.960	2.326	2.576

$$\bar{x} \pm t_{\alpha/2} \frac{s}{\sqrt{n}} \tag{8.13}$$

where $1 - \alpha$ is the confidence coefficient and $t_{\alpha/2}$ is the t value providing an area of $\alpha/2$ in the upper tail of a t distribution with $n - 1$ *degrees of freedom*.

EXAMPLE 8.8

Assume that the duration time of long-distance telephone calls is normally distributed. A sample of 20 telephone calls resulted in a sample mean of $\bar{x} = 12.5$ minutes and a sample standard deviation of $s = 4.1$ minutes. Develop 95% and 98% confidence interval estimates of the mean duration time for the population of long-distance telephone calls.

At 95% confidence, $1 - \alpha = .95$ and $\alpha = .05$. Using (8.13), we obtain

$$\bar{x} \pm t_{.025} \frac{s}{\sqrt{n}}$$

The $t_{.025}$ value is based on $n - 1 = 20 - 1 = 19$ degrees of freedom. From Table 8.2, $t_{.025} = 2.093$. Thus we have

$$12.5 \pm 2.093 \frac{4.1}{\sqrt{20}} = 12.5 \pm 1.92$$

Thus the 95% confidence interval for the population mean is 10.58 to 14.42 minutes per telephone call.

At 98% confidence, $1 - \alpha = .98$ and $\alpha = .02$. Using (8.13), we have

$$\bar{x} \pm t_{.01} \frac{s}{\sqrt{n}}$$

or

$$12.5 \pm 2.539 \frac{4.1}{\sqrt{20}} = 12.5 \pm 2.33$$

This result provides a 98% confidence interval for the population mean from 10.17 to 14.83 minutes per telephone call.

Hypothesis Testing

The small-sample procedure for hypothesis tests about a population mean follows the general hypothesis-testing procedure of Section 8.3; the exception is that the t distribution is used to determine the critical value(s) for the test.

EXAMPLE 8.9

The heights of a particular plant are normally distributed with a mean of 28 inches. A new plant food is tested on a sample of 12 plants. Results of the sample show a sample mean height of 29.4 inches and a sample standard deviation of 3 inches. Using a .10 level of significance, is there reason to believe that the new plant food *increases* plant growth?

The hypothesis test is as follows:

$$H_0: \quad \mu \leq 28$$

$$H_a: \quad \mu > 28$$

The rejection region is located in the upper tail of the sampling distribution. With $n - 1 = 12 - 1 = 11$ degrees of freedom, we have $t_{.10} = 1.363$ as the critical value for the test. Thus if the value of the test statistic

$$t = \frac{\bar{x} - \mu}{s/\sqrt{n}}$$

is greater than 1.363, reject H_0; otherwise do not reject H_0. Using the sample results, we have the following value for t:

$$t = \frac{\bar{x} - \mu}{s/\sqrt{n}} = \frac{29.4 - 28}{3/\sqrt{12}} = 1.62$$

Since 1.62 is greater than 1.363, the null hypothesis is rejected. At a .10 level of significance it can be concluded that the mean plant height exceeds 28 inches when the new plant food is used.

The Use of *p*-values

The *p*-value for Example 8.9 can be determined by using the observed t value of 1.62. However, due to the format of the t distribution table, p-values are slightly more difficult to determine than they were in Section 8.3. For example, the t distribution in Example 8.9 has 19 degrees of freedom. Referring to the row for 19 degrees of freedom, in Table 8.2, we see that 1.62 is between 1.328, occurring at a *p*-value of .10, and 1.729, occurring at a *p*-value of .05. Interpolation shows that the value 1.62 corresponds to a *p*-value of approximately .064. With a *p*-value of .064, which is less than $\alpha = .10$, we reject H_0 and conclude that the new plant food will increase the mean plant height.

EXAMPLE 8.10

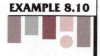

A production process is designed to fill containers with a mean filling weight of $\mu = 16$ ounces. An undesirable condition exists if the process is underfilling containers and the consumer is not receiving the amount of product indicated on the container label. In addition, an equally undesirable condition exists if the process is overfilling containers; in this case the firm is losing money since the process is placing more product in the container than is required. Suppose that quality control personnel periodically select a simple random sample of eight containers in order to make the following two-tailed hypothesis test:

$$H_0: \quad \mu = 16$$

$$H_a: \quad \mu \neq 16$$

If H_0 is rejected, the production manager will request that the production process be stopped and that the mechanism for regulating filling weights be readjusted to provide a mean filling weight of 16 ounces. If the sample provides data values of 16.02, 16.22, 15.82, 15.92, 16.22, 16.32, 16.12, and 15.92 ounces, what action should be taken at a .05 level of significance? Assume that the population of filling weights is normally distributed.

Since the data have not been summarized, we must first compute the sample mean and sample standard deviation. Doing so provides the following results:

$$\bar{x} = \frac{\Sigma x_i}{n} = \frac{128.56}{8} = 16.07 \text{ ounces}$$

and

$$s = \sqrt{\frac{\Sigma(x_i - \bar{x})^2}{n-1}} = \sqrt{\frac{.22}{7}} = .18 \text{ ounces}$$

With a two-tailed test and $\alpha = .05$, the critical values are $-t_{.025}$ and $+t_{.025}$ from a t distribution with $n - 1 = 8 - 1 = 7$ degrees of freedom. Using Table 8.2, we obtain -2.365 and $+2.365$ as the critical values. Thus H_0 will be rejected if the value of t is less than -2.365 or greater than $+2.365$.

Using $\bar{x} = 16.07$ and $s = .18$, we have

$$t = \frac{\bar{x} - \mu}{s/\sqrt{n}} = \frac{16.07 - 16.00}{.18/\sqrt{8}} = 1.10$$

Thus, since t is not in the rejection region, the null hypothesis $\mu = 16$ ounces cannot be rejected.

Using Table 8.2 and the row for 7 degrees of freedom, we see that the computed t value of 1.10 has an upper tail area of more than .10. While the format of the table prevents us from being more specific, we can at least conclude that the two-tailed p-value is greater than .20. Since this is greater than .05 level of significance, we see that the p-value also leads to the conclusion that H_0 cannot be rejected.

EXERCISES

45. For a t distribution with 12 degrees of freedom, find the area, or probability, that lies in each region.

a. to the left of 1.782
b. to the right of -1.356
c. to the right of 2.681
d. to the left of -1.782
e. between -2.179 and $+2.179$
f. between -1.356 and $+1.782$

46. Find the t values for each of the following.
a. upper tail area of .05 with 18 degrees of freedom
b. lower tail area of .10 with 22 degrees of freedom
c. upper tail area of .01 with 5 degrees of freedom
d. 90% of the area is between these two t values with 14 degrees of freedom
e. 95% of the area is between these two t values with 28 degrees of freedom

47. A test is made on the breaking strength of a new synthetic fishing line. The breaking strength of six sections of the line showed a sample mean of $\bar{x} = 20$ pounds and a sample standard deviation of $s = 2.3$ pounds. Develop a 95% confidence interval for the mean breaking strength of the new line under the assumption that the population of breaking strengths is normally distributed.

48. A sample of 12 cab fares in New York City shows a sample mean of $\bar{x} = \$8.50$ and a sample standard deviation of $s = \$2.40$. Develop a 90% confidence interval estimate of the mean cab fares in New York City.

49. A simple random sample of five people provided the following data on ages: 21, 25, 20, 18, and 21. Develop a 95% confidence interval for the mean age of the population being sampled.

50. In the testing of a new production method, 18 employees were randomly selected and asked to try the new method. The sample mean production rate for the 18 employees was 80 parts per hour. The sample standard deviation was 10 parts per hour. Provide 90% and 95% confidence interval estimates for the mean production rate for the new method.

51. The directors of a university computer center have been studying how many of the center's 30 computer terminals are in use at 9:00 P.M. on Friday evenings. A sample of 5 weeks resulted in the following data.

Week	1	2	3	4	5
Number in use	12	18	21	15	9

Treat the data as being from a simple random sample and develop a 95% confidence interval estimate for the mean number of terminals in use on Friday evenings at 9:00 P.M.

52. Following are the duration times in minutes for a sample of 20 telephone flight reservations:

2.1	4.8	5.5	10.4
3.3	3.5	4.8	5.8
5.3	5.5	2.8	3.6
5.9	6.6	7.8	10.5
7.5	6.0	4.5	4.8

a. What is the point estimate of the population mean time for flight-reservation phone calls?
b. Develop a 95% confidence interval estimate of the population mean time.

53. A bath soap manufacturing process is designed with the expectation that each batch prepared in the mixture department will produce a mean of 120 bars of soap per batch. A mean over or under this standard is undersirable. A sample of ten batches shows the following numbers of bars of soap.

$$108, 118, 120, 122, 119, 113, 124, 122, 120, 123$$

a. Using a .05 level of significance, should the null hypothesis $H_0: \mu = 120$ be rejected?

b. What is the p-value?

54. It is estimated that a housewife with a husband and two children works an average of 55 hours or less per week on household related activities. Shown below are the hours worked during a week for a sample of eight housewives.

$$58, 52, 64, 63, 59, 62, 62, 55$$

a. Use $\alpha = .05$ to test the hypotheses $H_0: \mu \leq 55$, $H_a: \mu > 55$. What is your conclusion about the mean number of hours worked per week?

b. What is the p-value?

55. A study of a drug designed to reduce blood pressure used a sample of 25 men between the ages of 45 and 55. With μ indicating the mean change in blood pressure for the population of men receiving the drug, the hypotheses in the study were written: $H_0: \mu \geq 0$ and $H_a: \mu < 0$. Rejection of H_0 shows that the mean change is negative, indicating that the drug is effective in lowering blood pressure.

a. At a .05 level of significance, what conclusion should be made if $\bar{x} = -10$ and $s = 15$?

b. What is the p-value?

56. Last year the number of lunches served at an elementary school cafeteria was normally distributed with a mean of 300 lunches per day. At the beginning of the current year, the price of a lunch was raised by 25¢. A sample of 6 days during the months of September, October and November provided the following number of children being served lunches: 290, 275, 305, 260, 270, and 275. Do these data indicate that the mean number of lunches per day has dropped compared to last year? Test the hypothesis $H_0: \mu \geq 300$ against the alternative hypothesis $H_a: \mu < 300$ at a .10 level of significance.

8.6   The Role of the Computer in Statistical Inference

In this section we describe the role of computers and computer software packages in making inferences about a population mean μ. To begin with, we will show how the Minitab statistical computing system can be used to develop interval estimates using the t distribution approach introduced in Section 8.5.

EXAMPLE 8.8 (continued)

A sample of 20 long-distance calls showed a sample mean of 12.5 minutes and a sample standard deviation of 4.1 minutes. The data that resulted in these values are as follows:

12.6	11.0	21.5	6.4	18.8
9.1	14.0	13.6	13.7	15.3
16.1	10.6	13.0	14.9	8.6
16.4	4.8	10.2	8.7	10.7

```
MTB > tinterval 95 cl

            N      MEAN    STDEV   SE MEAN    95.0 PERCENT C.I.
C1         20      12.50    4.10     0.92   (  10.58,   14.42)

MTB > tinterval 98 cl

            N      MEAN    STDEV   SE MEAN    95.0 PERCENT C.I.
C1         20      12.50    4.10     0.92   (  10.17,   14.83)
```

FIGURE 8.12   Confidence Intervals for Example 8.8 using Minitab

To develop a confidence interval for these data using Minitab, we must first enter the data into the Minitab worksheet; assume that we have done this and that these data are in the column labeled C1. To develop a confidence interval corresponding to any specified confidence coefficient, we us the Minitab command named TINTERVAL. Figure 8.12 shows the results obtained for both a 95% and 98% confidence interval estimate. Note that these are the same results that we obtained following the procedure described in Section 8.5.

■

Although the TINTERVAL command can be used to develop confidence intervals for any sample size, we know from the discussion in Section 8.5 that this approach is appropriate only if the population from which we are sampling is normally distributed. In cases where this may not be appropriate but where the sample size is large ($n \geq 30$), the Minitab command named ZINTERVAL can be applied.

In addition to the confidence coefficient to be used and the column in which the data is to be found, the ZINTERVAL command requires the user to specify the known or assumed value of the population standard deviation; for example, if we had entered a sample of size 50 into C2 of the worksheet, the command ZINTERVAL 95 22.5 C2 would produce a 95% confidence interval corresponding to an assumed standard deviation of 22.5. The use of the ZINTERVAL command corresponds to the large-sample interval estimation case that we described in Section 8.2. Note also that if the user does not specify a confidence coefficient for either command, a 95% confidence interval estimate is computed.

Minitab can also be used to perform the small-sample procedure for hypothesis tests that we described in the previous section. Recall that this procedure also requires that the population we are sampling from be normally distributed.

EXAMPLE 8.8
(continued)

Assume that there was some concern that the mean duration time for the population of long-distance telephone calls had exceeded the 10-minute suggested guideline. To test this, the following hypotheses were proposed:

$$H_0: \quad \mu \leq 10$$

$$H_a: \quad \mu > 10$$

To perform this hypothesis test, we can use the Minitab command name TTEST. For a two-tailed test the user must provide the value of μ stated in the null hypothesis and the column in which the data are located. For a one-tailed test, a subcommand must also be entered in order to specify whether the test is a lower-tailed or upper-tailed test. As shown in Figure 8.13, this is done by first ending the TTEST command with a semicolon; the

```
MTB > TTEST 10 C1;
SUBC> ALTERNATIVE = 1.

TEST OF MU = 10.0 VS MU G.T. 10.0

             N      MEAN    STDEV   SE MEAN      T    P VALUE
C1          20     12.50     4.10      0.92    2.73    0.0067
```

FIGURE 8.13 Hypothesis Test for Example 8.8 using Minitab

SUBC prompt is then followed by ALTERNATIVE = 1 or ALTERNATIVE = −1. If the value of ALTERNATIVE is 1, Minitab assumes an upper-tailed test; if the value is −1, Minitab assumes a lower-tailed test. Note that this command produces both the t value for the test and the p-value. If the desired level of significance is .05, a p-value of .0067 indicates that we should reject H_0; the conclusion is that the mean duration time is greater than the 10-minute guideline.

For the large-sample case, we can use the Minitab command named ZTEST. The use of the ZTEST command is similar to the TTEST command with the exception that the user must also specify the assumed value for the population standard deviation.

■

SUMMARY

In this chapter we discussed how interval estimation and hypothesis testing can be used for making statistical inferences about a population mean. In the large-sample case ($n \geq 30$), the central limit theorem enables us to approximate the sampling distribution of $\bar{x}$ with a normal probability distribution. As a result, the standard normal random variable z was used in both the large-sample interval-estimation and hypothesis-testing procedures.

We indicated that an interval estimate is developed for a stated confidence level. For a given sample size, higher confidence levels require wider intervals. We showed that given information regarding the maximum acceptable sampling error, the desired level of confidence, and a planning value for the population standard deviation σ, a formula can be used for determining the sample size that would meet the desired level of precision.

We introduced the t distribution to make inferences about a population mean whenever the population is normal and the sample standard deviation s is used as an estimate of σ. The t distribution is particularly helpful in making inferences about a population mean whenever the sample size is small ($n < 30$). The specific steps of using the t distribution parallel the steps for the large-sample case.

Figure 8.14 summarizes the interval estimation procedures for a population mean. The figure shows that the expression used to compute an interval estimate depends on whether or not the population standard deviation σ is known, whether the sample size is large ($n \geq 30$) or small ($n < 30$), and in some cases, whether or not the population has a normal probability distribution. If the sample size is large, no assumption is required about the distribution of the population and $z_{\alpha/2}$ is always used in the computation of the interval estimate. If the sample size is small, the population must be normally distributed in order to compute an interval estimate of μ. If the sample size is small and the assumption of a normally distributed population is inappropriate, the sample size must be increased to $n \geq 30$ in order to develop an interval estimate of μ. Finally, note that the use of the t distribution is reserved for the case where the population standard deviation σ is unknown, the sample size is small, and the population has a normal probability distribution.

We then showed how to conduct hypothesis tests about the value of a population mean. The purpose of hypothesis testing is to make a statistical conclusion about a hypothesized value of the population mean. A null hypothesis H_0 and an alternative hypothesis H_a must be formulated for the test. The type I error is the error of rejecting H_0 when it is true, and the type II error is the error of accepting H_0 when it is false. By specifying a maximum allowable probability of making a type I error, called the level of significance, a rejection rule can be established for determining whether or not the null hypothesis should be rejected.

The p-value was introduced as another criterion for determining whether or not to reject the null hypothesis. The p-value indicates the probability of obtaining a difference between the observed sample mean and the hypothesized value of the population mean that is larger than the difference actually observed. The decision rule indicates the null hypothesis should be rejected only if the p-value is less than the level of significance for the test.

FIGURE 8.14   Summary of Interval Estimation Procedures for a Population Mean

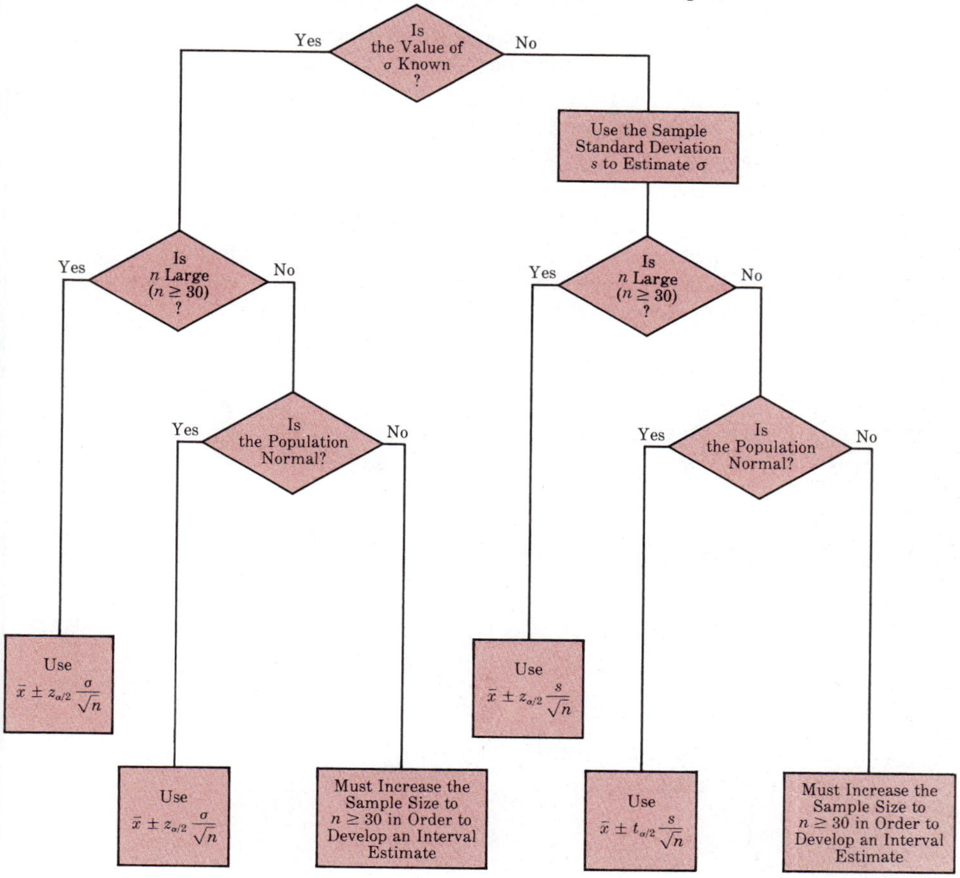

Point estimate   A single numerical value used as the estimate of the value of a population parameter. In this chapter, the value of the sample mean $\bar{x}$ provided the point estimate of the population mean μ.

Sampling error   The difference between an unbiased point estimate and the value of the population parameter. In the case of the mean, the sampling error can be written $|\bar{x} - \mu|$.

Interval estimate   An estimate of a population parameter that provides an interval of values believed to contain the value of the parameter.

Confidence coefficient   The confidence that an interval estimate contains the value of the parameter of interest. For example, if an interval-estimation procedure provides intervals such that 95% of the time the value of the population mean μ is included in the interval, the interval estimate is said to be constructed at the 95% confidence level; .95 is referred to as the confidence coefficient.

Null hypothesis   The hypothesis such as H_0: $\mu = \mu_0$. This hypothesis is assumed true, with the sample results indicating whether or not the null hypothesis is rejected.

Alternative hypothesis   The hypothesis concluded if the null hypothesis is rejected.

Type I error   The error of rejecting H_0 when it is true.

Type II error   The error of accepting H_0 when it is false.

Level of significance   The maximum probability of making a type I error that the user will tolerate in the hypothesis-testing procedure.

Two-tailed test   A hypothesis test in which rejection of the null hypothesis occurs in either tail of the sampling distribution.

One-tailed test   A hypothesis test in which rejection of the null hypothesis occurs in only one tail of the sampling distribution.

***p*-value**   The probability, assuming the null hypothesis is true, of obtaining a sample result that is more unlikely than the observed sample result. If the *p*-value is less than the level of significance for the test, the null hypothesis should be rejected.

***t* distribution**   A family of probability distributions that can be used to make inferences about a population mean whenever the population is normal and the sample standard deviation s is used as an estimate of the population standard deviation σ.

Degrees of freedom   A parameter of the t distribution that specifies the t distribution of interest. Whenever the t distribution is used to make inferences about a population mean, the appropriate t distribution has $n - 1$ degrees of freedom, where n is the size of the simple random sample.

INTERVAL ESTIMATION OF A POPULATION MEAN (LARGE-SAMPLE CASE)

$$\bar{x} \pm z_{\alpha/2}\frac{\sigma}{\sqrt{n}} \tag{8.3}$$

SAMPLE STANDARD DEVIATION

$$s = \sqrt{\frac{\Sigma(x_i - \bar{x})^2}{n - 1}} \tag{8.4}$$

INTERVAL ESTIMATION OF A POPULATION MEAN—σ UNKNOWN

$$\bar{x} \pm z_{\alpha/2} \frac{s}{\sqrt{n}} \qquad (8.5)$$

SAMPLE SIZE FOR INTERVAL ESTIMATION OF A POPULATION MEAN

$$n = \frac{(z_{\alpha/2})^2 \sigma^2}{E^2} \qquad (8.7)$$

STANDARD NORMAL RANDOM VARIABLE

$$z = \frac{\bar{x} - \mu}{\sigma/\sqrt{n}} \qquad (8.8)$$

t RANDOM VARIABLE

$$t = \frac{\bar{x} - \mu}{s/\sqrt{n}} \qquad (8.12)$$

INTERVAL ESTIMATION OF A POPULATION MEAN (SMALL-SAMPLE CASE)

$$\bar{x} \pm t_{\alpha/2} \frac{s}{\sqrt{n}} \qquad (8.13)$$

REVIEW QUIZ

TRUE/FALSE

1. Whenever the central limit theorem is applicable, the sampling error will be zero when using $\bar{x}$ as a point estimator of μ.

2. Interval estimates provide information on how close the value of the sample mean $\bar{x}$ is to the population mean μ.

3. A point estimate does not contain information about the size of the sampling error.

4. An interval estimate contains information about the size of the sampling error.

5. In order to determine the appropriate sample size, the user must specify a maximum allowable sampling error.

6. In hypothesis testing we do not assume that the null hypothesis is true until after the sample results have been analyzed.

7. In hypothesis testing the z value can be interpreted as the number of standard deviations the sample mean is from the hypothesized value of μ.

8. The type I error is the error of accepting H_0 when it is false.

9. If the level of significance is made smaller, the rejection region becomes larger.

10. Decision rules based on critical z values and p-values will always provide the same hypothesis-testing conclusions.

11. In order to use the t distribution, we must be willing to assume that the population has a normal probability distribution.

12. When using the t distribution to form a 95% confidence interval, the interval will be smaller than if a z value from a standard normal probability distribution is used.

13. The t distribution can never be used if the sample size is 30 or more.

MULTIPLE CHOICE

Use the following information for questions 14–16. A random sample of 81 automobile tires has a mean tread life of 36,000 miles. It is known that the standard deviation of tread life of tires is $\sigma = 4500$ miles.

14. A 95% confidence interval for the population mean is
   a. 35,500 to 36,500
   b. 35,177.5 to 36,822.5
   c. 35,020 to 36,980
   d. none of the above

15. If the sample mean of 36,000 had been from a random sample size 50, the 95% confidence interval would have been
   a. the same
   b. a wider interval
   c. a narrower interval
   d. none of the above

16. A 90% confidence interval for the population mean is
   a. 35,000 to 37,000
   b. 35,177.5 to 36,822.5
   c. 35,020 to 36,980
   d. none of the above

17. The useful life of a certain type of light bulb is known to have a standard deviation of $\sigma = 40$ hours. How large a sample should be taken if it is desired to have a sampling error of 10 hours or less at a 95% level of confidence?
   a. 62
   b. 44
   c. 37
   d. 8

18. If a hypothesis test leads to the rejection of the null hypothesis
   a. a type I error is always committed
   b. a type II error is always committed
   c. a type I error may have been committed
   d. a type II error may have been committed

Use the following information for questions 19–21. The ABC Electronics Company claims that the batteries it produces have a useful life of at least 100 hours. It is known that the standard deviation is 20 hours. A test is undertaken to check the validity of this claim.

19. With the level of significance set at .05, the critical value or values for the test based on a sample of 49 batteries is
   a. $z = -1.645$
   b. $z = -1.435$

c. $z = -1.96$ and $+1.96$

d. $z = +1.645$

20. If the random sample of 49 batteries resulted in an average life of 96 hours, can the manufacturer's claim be rejected at the .05 level of significance?

    a. Yes, the null hypothesis can be rejected.

    b. No, do not reject the null hypothesis.

    c. Not enough information is given to answer this question.

21. What is the p-value associated with the sample mean of 96 hours?

    a. .042

    b. .081

    c. .419

    d. .96

22. If the level of significance of a hypothesis test is increased from .01 to .05, the probability of a type II error

    a. will also be increased from .01 to .05

    b. will not be changed

    c. will be decreased

    d. Not enough information is given to answer this question.

SUPPLEMENTARY EXERCISES

57. Sales personnel for Skillings Distributors are required to submit weekly reports listing the customer contacts made during the week. A sample of 60 weekly contact reports shows a mean of 22.4 customer contacts per week for the sales personnel. The sample standard deviation was 5 contacts. Compute a 95% confidence interval for the mean number of weekly customer contacts for the population of sales personnel.

58. The Gordon S. Black Corporation surveyed 439 working parents with children younger than 12; these parents used an average of 37.4 hours of child care a week at an average cost of $56 weekly. Develop a 95% confidence interval estimate for the mean weekly hours of child care if the sample standard deviation is $s = 4.2$ hours.

59. The North Carolina Savings and Loan Association would like to develop an estimate of the mean size of home improvement loans granted by its member institutions. A sample of 100 loans granted by member institutions resulted in a sample mean of $3400 and a sample standard deviation of $650. With these data develop a 98% confidence interval for the mean dollar amount of home improvement loans.

60. In a test of phone utilization, a firm recorded the length of time for phone calls handled by its main switchboard. A sample of 50 phone calls provided a sample mean of 8.9 minutes and a sample standard deviation of 5 minutes.

    a. What is a 90% confidence interval estimate for the mean length of time for phones?

    b. What is a 99% confidence interval estimate?

61. A utility company found that a sample of 100 delinquent accounts yielded an average amount owed of $131.44, with a sample standard deviation of $16.19. Develop a 90% confidence interval for the population mean amount owed.

62. In Exercise 61 a utility company sampled 100 delinquent accounts in order to estimate the mean amount owed by these accounts. The sample standard deviation was $16.19. How large a sample should be taken if the company wants to be 90% confident that the estimate of the population mean will have a sampling error of $1.50 or less?

63. The owner of a pay-fishing lake wishes to determine the mean weight of the fish caught by the patrons. A preliminary sample of 10 fish caught shows a sample standard deviation of 1.6 pounds. How many fish should be included in the sample if we would like to estimate the mean weight of the fish to within $\pm .25$ pounds at a 95% level of confidence?

64. A gasoline service station shows a standard deviation of $6.25 for the charges made by the credit-card customers. Assume that the station's management would like an estimate of the mean gasoline bill for its credit-card customers to within $\pm \$1.00$ of the actual population mean. For a 95% confidence level, how large a sample would be necessary?

65. Because of production changeover time and costs, a director of manufacturing must convince management that a proposed manufacturing method reduces costs before the new method can be implemented. The current production line operates with a mean cost of $220 per hour. A new production line has been proposed and a sample production period specified. What hypothesis test should be formulated in order to test whether or not the company should convert to the new production line?

66. The monthly rent for a two-bedroom apartment in a particular city is reported to average $350. Assume that we would like to test the hypothesis $H_0: \mu = 350$ versus $H_a: \mu \neq 350$. A sample of 36 two-bedroom apartments is selected. The sample mean turns out to be $\bar{x} = \$338$, with a sample standard deviation of $s = \$40$.
   a. Conduct this hypothesis test with a .05 level of significance.
   b. Use the sample results to construct a 95% confidence interval for the population mean.

67. A long-distance trucking firm believes that its mean weekly loss due to damaged shipments is $2000 or less. A sample of 35 weeks of operations shows a sample mean weekly loss of $2200, with a sample standard deviation of $500. Using a .05 level of significance, test the trucking firm's claim that the mean weekly loss is $2,000 or less. (Note: $H_0: \mu \leq 2000$ and $H_a: \mu > 2000$.)

68. New tires manufactured by a company in Findlay, Ohio, are designed to provide a mean of at least 28,000 miles. Tests with 40 tires show a sample mean of 27,200 miles with a sample standard deviation of 1000 miles. Using a .01 level of significance, test for whether or not there is sufficient evidence to reject the claim of a mean of at least 28,000 miles. What is the p-value?

69. In making bids on building projects, Sonneborn Builders, Inc. assumes construction workers are idle on the average no more than 15% of the time. For a normal 8-hour shift, the mean idle time per worker should be 72 minutes or less per day. A sample of 30 construction workers found a mean idle time of 80 minutes per day. The sample standard deviation was 20 minutes.
   a. Formulate the null and alternative hypotheses so that the claim of 72 minutes or less per day will be rejected if H_0 is rejected.
   b. What is the p-value associated with the sample result?
   c. Using a .05 level of significance and the p-value, what is your conclusion?

70. Stout Electric Company operates a fleet of trucks for its electrical service to the construction industry. Monthly mean maintenance costs have been $75 per truck. A random sample of 40 trucks shows a sample mean maintenance cost of $82.50 per month, with a sample standard deviation of $30. Management would like a test to determine whether or not the mean monthly maintenance cost has increased.
   a. What is your conclusion based on the sample mean of $82.50? Use a .05 level of significance.
   b. What is the p-value associated with this sample result? What is your conclusion based on the p-value?

71. The lifetimes of a certain battery are normally distributed. In a random sample of 25 batteries, the sample results showed that $\bar{x} = 60$ hours and $s = 10$ hours. Provide a 90% confidence interval for the mean lifetime of this type of battery.

72. Sample assembly times for a particular manufactured part were 8, 10, 10, 12, 15, and 17 minutes. If the mean of the sample is to be used to estimate the mean of the population of assembly times, provide a point estimate and a 90% confidence interval estimate of the population mean.

73. Dailey Paints, Inc. implemented a long-term painting test study designed to check the wear resistance of its major brand of paint. The test consisted of painting 8 houses in various parts of the United States and observing the number of months until signs of peeling were observed. The following data were obtained.

House	1	2	3	4	5	6	7	8
Time Until Signs of Peeling (months)	60	51	64	45	48	62	54	56

a. What is a point estimate of the mean number of months until signs of peeling are observed?

b. Develop a 95% confidence interval to estimate the population mean number of months until signs of peeling are observed.

c. Develop a 99% confidence interval for the population mean.

74. The hemoglobin level for a sample of seven heart-lung transplant recipients is shown below (*New England Journal of Medicine,* November 3, 1988, 1186–1192)

Patient	1	2	3	4	5	6	7
Hemoglobin Level	10.0	10.1	10.2	11.8	13.4	10.2	11.3

Assuming that the distribution of hemoglobin level for all heart-lung transplant recipients is approximately normally distributed, develop a 95% confidence interval of the mean hemoglobin level.

75. Joan's Nursery specializes in custom designed landscaping for residential areas. The labor cost associated with a particular landscaping proposal is estimated based on the number of plantings of trees, shrubs, and so on associated with the project. For labor cost estimating purposes, management allows a maximum of 2 hours of labor time for the planting of a medium-size tree. Actual times from a sample of 10 plantings during the past month are as follows (times in hours).

$$1.9, 2.1, 2.8, 3.0, 2.6, 2.5, 2.8, 3.2, 2.2, 2.5$$

Using a .05 level of significance, test the claim that the mean tree planting time is 2 hours or less. What is your conclusion, and what recommendations would you consider making to management? What is the *p*-value?

COMPUTER EXERCISE

A consumer research organization has been studying the repair history of automobiles produced by a major Detroit manufacturer. Of particular concern has been the performance of engines used in the manufacturer's full-sized cars. Preliminary evidence shows that owners of these cars have experienced relatively early failures in the car's transmission system. Part of the consumer research organization's study has uncovered the fact that the transmission used by the manufacturer may be too small for the engines. To aid in the investigation, the research organization sent questionnaires to owners of the automobiles. Data were collected from 50 owners who had experienced a transmission failure. The following data shows the number of miles the vehicle had been driven at the time of the failure.

85,092	32,609	59,465	77,437	32,534	64,090	32,464	59,902
39,323	89,641	94,219	116,803	92,857	63,436	65,605	85,861
64,342	61,978	67,998	59,817	101,769	95,774	121,352	69,568
74,276	66,998	40,001	72,069	25,066	77,098	69,922	35,662
74,425	67,202	118,444	53,550	79,294	64,544	86,813	116,269
37,831	89,341	73,341	85,288	138,114	53,402	85,586	82,256
77,539	88,798						

QUESTIONS

1. Use appropriate descriptive statistics to summarize this data.

2. Develop a 95% confidence interval for the mean number of miles driven until transmission failure for the population of vehicles that have experienced transmission failure.

3. Discuss the implications of your statistical findings in terms of the claim that some owners of the automobiles have been experiencing early transmission failures.

9

Inferences About a Population Proportion

What You Will Learn in This Chapter:

- The process of using a sample proportion to make inferences about a population proportion

- The characteristics of the sampling distribution of the sample proportion

- How to construct and interpret an interval estimate of a population proportion

- How to conduct hypothesis tests about a population proportion

Contents

Cost is the Major Reason for Not Going to College

The Council for Advancement and Support of Education used a Gallup survey to learn about the attitudes of 13- to 21-year-olds toward college. A sample of 1001 people between the ages of 13 and 21 found that 48% believed that the major reason more students do not go to college is that colleges are too expensive. Although 59% of the high school juniors and seniors in the survey indicated that they had savings available for their college education, 41% did not have savings for college. Nonetheless, 92% of the juniors and seniors plan to go to college or attend some type of vocational or trade school.

An interesting result of the study was that 53% of the students surveyed who were between the ages of 13 and 15 agreed with the statement that "the higher the tuition costs of a college, the better the quality of education a student will receive." Although 41% of the 16- or 17-year-old students also agreed with this statement, only 27% of current college students and college graduates felt that way. In addition, 70% of those surveyed believed that public colleges provide just as good an education as private colleges, and 60% felt that 2-year schools are as good as 4-year schools.

When asked which factors were "extremely im-

Stock, Boston

A proud family on graduation day.

portant" in selecting a college, 67% selected availability of particular courses and curricula, whereas 44% cited the academic reputation of the college. In addition, 45% said that the expense of attending a particular college was crucial in the selection decision. In contrast, only 20% of the students said that the social life and athletic reputation of the college were "extremely important."

Based on "Cost Seen as Key Reason for not Going to College," The Cincinnati Enquirer, October 10, 1988.

In Chapters 7 and 8 we showed how a sample mean can be used to estimate and conduct hypothesis tests about a population mean. In some statistical applications, however, we find that a population proportion is of more interest than a population mean. The "Statistics in the News" article described a survey that was conducted in order to estimate the *proportion* of people between the ages of 13 and 21 who support statements involving a variety of college issues. Several proportions were reported in the article, including the proportion who believe that the major reason for not going to college is that college is too expensive, the proportion that believe that higher tuition costs are associated with a better quality of education, and so on. Other situations where the interest is on a population proportion include the proportion of voters who prefer a particular political candidate, the proportion of adults who have a particular disease, the proportion of college students who are in-state residents, and so on. In situations such as these, we use the sample proportion $\bar{p}$ to make statistical inferences about the population proportion p. This statistical process is depicted in Figure 9.1.

EXAMPLE 9.1

A sample of 200 married couples selected from throughout the United States showed that for 84 of the couples, both the husband and the wife held full-time jobs. Use the sample results to estimate the proportion of all married couples in the United States for which both the husband and wife hold full-time jobs. Let

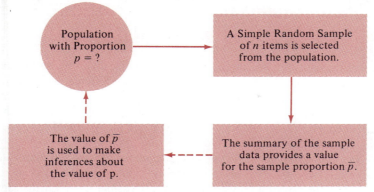

FIGURE 9.1 The Statistical Process of Using a Sample Proportion to Make Inferences about a Population Proportion

$$x = \text{the number of couples in the sample for which both the husband and the wife hold full-time jobs}$$

$$n = \text{the number of couples in the sample}$$

The sample proportion is computed as follows:

$$\bar{p} = \frac{x}{n} \qquad (9.1)$$

Using the given data, we have

$$\bar{p} = \frac{x}{n} = \frac{84}{200} = .42$$

This sample proportion $\bar{p} = .42$ is the *point estimate* of the population proportion p. The value $\bar{p} = .42$ provides the estimate that for 42% of married couples in the United States both the husband and the wife hold full-time jobs.

■

Notes and Comments

The notation $\bar{p}$ is used to denote the sample proportion because it is analogous to the sample mean $\bar{x}$. For example, in Example 9.1 the sample proportion $\bar{p}$ represented the proportion of couples in a sample of 200 in which both the husband and the wife hold full-time jobs. If we define the random variable

$$x_i = \begin{cases} 1 & \text{if the } i\text{th couple in the sample both} \\ & \text{have full-time jobs} \\ 0 & \text{otherwise} \end{cases}$$

then the sample proportion $\bar{p} = \Sigma x_i/n$ is also a sample mean.

The Sampling Distribution of $\bar{p}$

As was the case with the sample mean $\bar{x}$, we realize that if we repeat the sampling process shown in Figure 9.1 several times, we expect the different samples to provide different values for the sample statistic $\bar{p}$. This situation would be observed in Example 9.1 if we randomly selected a new sample of 200 married couples. Perhaps this time we would find 96 couples for which both the husband and the wife hold full-time jobs. The sample proportion for this sample would be $\bar{p} = 96/200 = .48$.

Because different simple random samples can provide different values for the sample proportion, whenever we make inferences about a population proportion p, we are interested in the probability distribution of all possible values of the sample proportion $\bar{p}$. This probability distribution is referred to as the *sampling distribution of $\bar{p}$*. If we can identify the properties of the sampling distribution of $\bar{p}$, we can use this distribution to make inferences about a population proportion p in *exactly* the same way that we used the sampling distribution of $\bar{x}$ to make inferences about the population mean μ. Let us begin by defining the properties of the sampling distribution of $\bar{p}$.

Expected Value of $\bar{p}$

The *expected value of $\bar{p}$* is the mean value of all possible $\bar{p}$ values that can be observed when simple random samples are selected from a population with a proportion denoted by p. Letting $E(\bar{p})$ denote the expected value of $\bar{p}$, or simply the mean of all possible $\bar{p}$ values, it can be shown that

Expected Value of $\bar{p}$

$$E(\bar{p}) = p \qquad (9.2)$$

Since the expected value of $\bar{p}$ is equal to the population proportion p, $\bar{p}$, is an unbiased estimator of p.

Standard Deviation of $\bar{p}$

The *standard deviation* of $\bar{p}$ is a measure of dispersion in the possible $\bar{p}$ values. When simple random samples of size n are selected from a population with proportion p, the expression for the standard deviation of $\bar{p}$ shown as Equation (9.3).

---

*In this chapter we assume that the population is *large* relative to the sample size, with $n/N \leq .05$. In this situation for finite population correction factor, $\sqrt{(N - n)/(N - 1)}$ is approximately 1 and is not needed to the formula for the standard deviation or standard error. When the populations are considered large (9.3) may be used to compute $\sigma_{\bar{p}}$. If the population size N and the sample size n are values such that $n/N > .05$, the finite population correction factor should appear in (9.3) with $\sigma_{\bar{p}} = \sqrt{(N - n)/(N - 1)} \sqrt{p(1 - p)/n}$.

$$\sigma_{\bar{p}} = \sqrt{\frac{p(1 - p)}{n}} \qquad\qquad (9.3)$$

The standard deviation of $\bar{p}$ is called the *standard error of the proportion*.

Form of the Sampling Distribution of $\bar{p}$

In Chapter 7 we saw that the central limit theorem enabled us to conclude that the sampling distribution of $\bar{x}$ can be approximated by a normal probability distribution whenever the sample size is large. Applying the same central limit theorem as it relates to the sample proportion $\bar{p}$, we have the following statement about the form of the sampling distribution of $\bar{p}$.

> In selecting simple random samples of size n from a population with proportion p, the sampling distribution of the sample proportion $\bar{p}$ approaches a normal distribution with mean p and standard deviation $\sqrt{p(1 - p)/n}$ as the sample size becomes large.

Thus we can use a normal probability distribution to approximate the sampling distribution of $\bar{p}$ provided the sample size is large. With proportions, the sample size can be considered large whenever the following two conditions are satisfied:

$$np \geq 5$$

$$n(1 - p) \geq 5$$

In the following examples, we determine the sampling distribution of $\bar{p}$ and then show how the sampling distribution can be used to make probability statements about the value of the sample proportion $\bar{p}$ that will be found from a simple random sample.

EXAMPLE 9.2

Consider the population of fourth-grade students in the Cleveland, Ohio, public school system. Assume that the proportion of students in this population who wear eyeglasses is $p = .25$. If a random sample of 50 students is selected, define the characteristics of the sampling distribution of $\bar{p}$, where $\bar{p}$ is the proportion of students in the sample who wear eyeglasses.

From (9.2), we know the mean of the sampling distribution is

$$E(\bar{p}) = p = .25$$

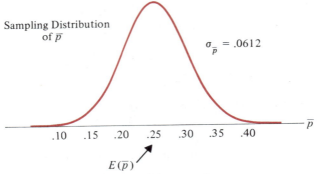

FIGURE 9.2   Sampling Distribution of $\bar{p}$ for the Sample Proportion of Students Wearing Eyeglasses

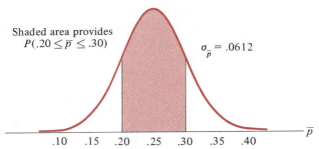

FIGURE 9.3   Sampling Distribution of $\bar{p}$ Showing $P(.20 \leq \bar{p} \leq .30)$ for Example 9.3

From (9.3), we know that the standard deviation is

$$\sigma_{\bar{p}} = \sqrt{\frac{p(1-p)}{n}} = \sqrt{\frac{.25(1-.25)}{50}} = .0612$$

Finally, for the sample size of $n = 50$, we have $np = 50(.25) = 12.5$ and $n(1-p) = 50(1-.25) = 37.5$; thus the large sample conditions are satisfied and the sampling distribution of $\bar{p}$ can be approximated by a normal probability distribution. The complete sampling distribution of $\bar{p}$ for this example is shown in Figure 9.2.

■

EXAMPLE 9.3

As an extension of Example 9.2, use the sampling distribution of $\bar{p}$ in Figure 9.2 to compute the probability that the sample proportion of students who wear eyeglasses is between .20 and .30. The sampling distribution of $\bar{p}$ and the area showing the probability of $.20 \leq \bar{p} \leq .30$ is shown in Figure 9.3.

Since the sampling distribution of $\bar{p}$ can be approximated by a normal distribution, we can use the areas under the curve of the standard normal distribution to answer the probability question. With the normal distribution, we need to compute a z value, where

$$z = \frac{\bar{p} - p}{\sigma_{\bar{p}}} \tag{9.4}$$

The preceding value is identical to the z value we have always used, with the exception of the notation $\bar{p}$ and $\sigma_{\bar{p}}$, which is used to indicate we are using the sampling distribution of $\bar{p}$.

$$\text{At } \bar{p} = .30, \qquad z = \frac{\bar{p} - p}{\sigma_{\bar{p}}} = \frac{.30 - .25}{.0612} = +.82$$

Using $z = .82$ in the table of areas for the standard normal distribution, we find an area between .25 and .30 of .2939.

$$\text{At } \bar{p} = .20, \qquad z = \frac{\bar{p} - p}{\sigma_{\bar{p}}} = \frac{.20 - .25}{.0612} = -.82$$

This z value shows that the area between .20 and .25 must also be .2939. Thus the total probability of selecting a sample with a sample proportion $\bar{p}$ between .20 and .30 must be $.2939 + .2939 = .5878$.

■

EXAMPLE 9.4

According to the Census Bureau's 1988 population survey (*The Wall Street Journal*, October 20, 1988), 28% of all households are non-family households where people live alone or with unrelated people. If a simple random sample of 100 households is selected, what is the probability that the proportion of non-family households in the sample will be .37 or more?

Letting $\bar{p}$ denote the proportion of nonfamily households in the sample and noting that $p = .28$, the sampling distribution of $\bar{p}$ is shown in Figure 9.4.

$$\text{At } \bar{p} = .37, \qquad z = \frac{\bar{p} - p}{\sigma_{\bar{p}}} = \frac{.37 - .28}{.0449} = 2.00$$

Using the table of areas for the standard normal distribution, we find that at $z = 2.00$, the area between .28 and .37 must be .4772. Thus the probability of a sample proportion $\bar{p} = .37$ or more is $.5000 - .4772 = .0228$.

■

FIGURE 9.4 Sampling Distribution of $\bar{p}$ for the Proportion of Non-Family Households in a Sample of 100 Households

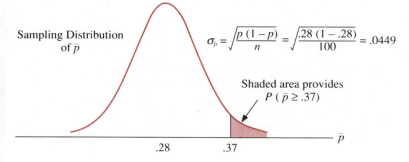

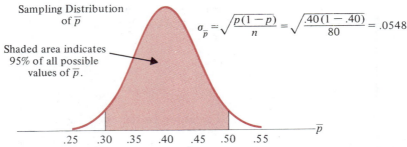

Sampling Distribution of $\bar{p}$

$$\sigma_{\bar{p}} = \sqrt{\frac{p(1-p)}{n}} = \sqrt{\frac{.40(1-.40)}{80}} = .0548$$

Shaded area indicates 95% of all possible values of $\bar{p}$.

$\bar{p}$

.25 .30 .35 .40 .45 .50 .55

FIGURE 9.5 Sampling Distribution of $\bar{p}$ for the Proportion of Women Students Belonging to a Sorority

EXAMPLE 9.5

At a large university, 40% of all undergraduate women students are members of a sorority. Thus, the population proportion of undergraduate women students who are members of a sorority is $p = .40$. If simple random samples of size 80 are to be taken from this population, we expect different samples to consist of different students and, therefore, provide a variety of values for the sample proportion $\bar{p}$. Find a range of values for $\bar{p}$ that will include 95% of all possible sample proportions that can be observed.

First we show the sampling distribution for the sample proportion of women students belonging to a sorority. Using $p = .40$, this sampling distribution is shown in Figure 9.5.

Using the standard normal probability distribution, we know that 95% of the area under the curve is between $z = -1.96$ and $z = 1.96$. Thus the range containing 95% of all possible $\bar{p}$ values is

$$p - 1.96\sigma_{\bar{p}} = .40 - 1.96\sigma_{\bar{p}} = .40 - 1.96(.0548) = .2926$$
$$p + 1.96\sigma_{\bar{p}} = .40 + 1.96\sigma_{\bar{p}} = .40 + 1.96(.0548) = .5074$$

Thus, there is a .95 probability of selecting a simple random sample of 80 women students and obtaining a sample proportion between .2926 and .5074.

■

Notes and Comments

To understand the rationale behind the requirement that $np \geq 5$ and $n(1 - p) \geq 5$ for a large sample size, first note that the population proportion p is equivalent to the probability of success associated with the binomial probability distribution. In fact, the exact sampling distribution of $\bar{p}$ can be determined by using the binomial probability distribution. However, as we saw in Chapter 6, whenever n is large it is computationally convenient to use the normal distribution to approximate the binomial distribution. The rule of thumb for a large sample indicates when the normal approximation of the binomial distribution is appropriate and thus also when the normal approximation is appropriate for the sampling distribution of $\bar{p}$.

1. Nine nurses selected at random from a large hospital were asked if they believed that the hospital nursing department is understaffed. The responses are as shown.

Nurse	1	2	3	4	5	6	7	8	9
Response	No	Yes	Yes	No	No	Yes	No	No	No

Develop a point estimate of the proportion of all hospital nurses that believe that the hospital nursing department is understaffed.

2. A survey of 400 women college students was conducted to determine future plans concerning career, marriage, and family. The following results were recorded:

310 women answered yes to the question, Do you plan to begin a full-time career immediately following graduation?

225 women answered yes to the question, Do you plan to marry before the age of 30?

175 women answered yes to the question, Do you plan to have children?

Use the survey results to provide point estimates of each of the following.
   a. the proportion of college women planning to begin full-time careers immediately following graduation
   b. the proportion of college women planning to marry before the age of 30
   c. the proportion of college women planning to have children

3. Develop a point estimate of the proportion of the pages in this text that contain a figure or a table. That is, if a randomly selected page has at least one figure or at least one table, the response is yes; otherwise the response is no. Use the random numbers in Table 6 of Appendix B to select a simple random sample of 20 pages. Compare your point estimate with those of others in the class.

4. Consider the following population of 25-year-old males, where Yes indicates that the individual has a life insurance policy and No indicates that the individual does not have such a policy.

INDIVIDUAL	RESPONSE
1	Yes
2	No
3	No
4	Yes
5	No
6	Yes

   a. Selecting simple random samples of size 4 provides a total of 15 possible samples. List the 15 samples.
   b. Compute the proportion of Yes responses for each sample, and show a graph of the sampling distribution of $\bar{p}$.

5. A family has five children, three girls and two boys. A random sample of three children will be selected in order to estimate the proportion of girls in the family.
   a. List the 10 possible samples.
   b. Compute the sample proportion $\bar{p}$ for each possible sample.
   c. Show a graphical representation of the sampling distribution of $\bar{p}$.
   d. Compute the mean of the 10 values of $\bar{p}$ and compare this to $E(\bar{p})$ given in Equation (9.2).

6. A research finding published in the *Journal of Clinical Gastroenterology* indicates that one-third of all adults experience gastrointestinal problems after consuming as little as 10 grams of sorbitol, a common artificial sweetener. (*Journal of Clinical Gastroenterology, 1988*). A sample of 50 individuals will be taken to verify this result.
   a. Assuming $p = .33$ is true, state as much as you can about the characteristics of the probability distribution of the various values of $\bar{p}$ that can be observed.
   b. Assuming $p = .33$ is true and that the sample size is increased to $n = 100$, state as much as you can about the characteristics of the sampling distribution of $\bar{p}$.

7. The Bureau of Labor Statistics reports that 26% of the labor force has had at least 4 years of college education (*Wall Street Journal*, October 20, 1988).
   a. Explain how the sampling distribution of $\bar{p}$ results from random samples of size 80 being used to estimate the proportion of the labor force that have at least 4 years of college education.
   b. Show the sampling distribution for $\bar{p}$ in this case.
   c. If the sample size is increased to 200, what happens to the sampling distribution of $\bar{p}$? Compare the standard error for the $n = 80$ and $n = 200$ alternatives.

8. The president of Doerman Distribution, Inc. believes that 30% of the firm's orders come from new customers. A simple random sample of 100 orders will be used to estimate the proportion of new customers. The results of the sample will be used to verify the president's claim of $p = .30$.
   a. Assume that the president is correct ($p = .30$). What is the sampling distribution of $\bar{p}$ for this study?
   b. What is the probability that the sample proportion $\bar{p}$ will be between .20 and .40?
   c. What is the probability that the sample proportion will be within plus or minus .05 of the population proportion $p = .30$?

9. A particular county in West Virginia has a 9% unemployment rate. A monthly survey of 800 individuals is conducted by a state agency. This study provides the basis for monitoring the unemployement rate of the county.
   a. Assume that $p = .09$. What is the sampling distribution of $\bar{p}$ when a sample of size 800 is used?
   b. What is the probability that a sample proportion $\bar{p}$ of at least .08 will be observed?

10. A doctor believes that 80% of all patients having a particular disease will be fully recovered within 3 days after receiving a new drug.
   a. A simple random sample of 20 medical records will be used to develop an estimate of the proportion of patients who were fully recovered within 3 days after receiving the drug. If a data analyst suggests using the normal probability distribution to approximate the sampling distribution of $\bar{p}$, what would you say? Explain.
   b. If the sample of patient records is increased to 60, what is the probability that the sample proportion will be within $\pm .10$ of the population proportion? (Assume that the population proportion is .80.)

11. The proportion of individuals insured by the All-Driver Automobile Insurance Company who have received at least one traffic ticket during a 5-year period is .15.
   a. Show the sampling distribution of $\bar{p}$ if a random sample of 150 insured individuals is used to estimate the proportion having received at least one ticket.
   b. What is the probability that the sample proportion will be within $\pm .03$ of the population proportion?

12. Historical records show that .50 of all orders placed at Big Burger Fast Food restaurants include a soft drink. With a simple random sample of 40 orders, what is the probability that between .45 and .55 of the sampled orders will include a soft drink?

13. Lori Jeffrey is a successful sales representative for a major publisher of college textbooks. Historically, Lori obtains a book adoption on 25% of her sales calls. Viewing her sales calls

for one month as a sample of all possible sales calls, a statistical analysis of the data yields a standard error of the proportion of $\sigma_{\bar{p}} = .0625$.

a. How large was the sample used in this analysis? That is, how many sales calls did Lori make during the month?

b. Let $\bar{p}$ indicate the sample proportion of book adoptions obtained during the month. Show the sampling distribution $\bar{p}$.

c. Use the sampling distribution of $\bar{p}$. What is the probability that Lori will obtain book adoptions on 30% or more of her sales calls during the one month period?

9.2 Estimation of a Population Proportion

In Chapter 8 we used the value of the sample mean $\bar{x}$ to develop an interval estimate of the population mean μ. The procedure that we used was based upon knowledge of the characteristics of the sampling distribuion of $\bar{x}$. In this section we show how the value of the population proportion $\bar{p}$ can be used to develop an interval estimate of the population proportion p. Since the characteristics of the sampling distribution of $\bar{p}$ are known, we can use the logic we used to estimate μ in order to develop an interval estimate of p.

For the case of estimating the population proportion p using the sample proportion $\bar{p}$, the sampling error is defined as the absolute value of the difference between $\bar{p}$ and p; that is,

$$\text{Sampling Error} = |\bar{p} - p| \qquad (9.5)$$

The probability statements that we can make about the sampling error for the proportion take the following form:

There is a $1 - \alpha$ probability that the value of the sample proportion will provide a sampling error of $z_{\alpha/2}\sigma_{\bar{p}}$ or less. The rationale for the preceding statement is the same as we used when the value of a sample mean was used as an estimate of a population mean. Specifically, since we know that the sampling distribution of $\bar{p}$ can be approximated by a normal probability distribution, we can use the value of $z_{\alpha/2}$ and the value of the standard error of the proportion, $\sigma_{\bar{p}}$, to make the probability statement about the sampling error.

Once we see that the probability statement concerning the sampling error is based on $z_{\alpha/2}\sigma_{\bar{p}}$, we can subtract this value from $\bar{p}$ and add it $\bar{p}$ in order to obtain an interval estimate of the population proportion. Such as interval estimate is given by

$$\bar{p} \pm z_{\alpha/2}\sigma_{\bar{p}} \qquad (9.6)$$

where $1 - \alpha$ is the confidence coefficient. Since $\sigma_{\bar{p}} = \sqrt{p(1 - p)/n}$, we can rewrite (9.6) as follows:

$$\bar{p} \pm z_{\alpha/2}\sqrt{\frac{p(1 - p)}{n}} \qquad (9.7)$$

However, note that in using (9.7) to develop an interval estimate of a population proportion p, the value of p would have to be *known*. Since the value of p is what we are trying to estimate, and is thus *unknown*, we substitute the sample proportion $\bar{p}$ for p. As a result, the general expression for a confidence interval estimate of a population proportion is as follows.*

Interval Estimation of a Population Proportion

$$\bar{p} \pm z_{\alpha/2} \sqrt{\frac{\bar{p}(1 - \bar{p})}{n}} \qquad\qquad (9.8)$$

where $1 - \alpha$ is the confidence coefficient and $z_{\alpha/2}$ is the z value providing an area of $\alpha/2$ in the upper tail of the standard normal probability distribution.

By choosing different values of $z_{\alpha/2}$, we can obtain interval estimates with different levels of confidence. Some of the common confidence levels and their corresponding z values are as follows:

---

z Values for Selected Confidence Levels

90% Confidence:	$z_{.05} = 1.645$
95% Confidence:	$z_{.025} = 1.96$
99% Confidence:	$z_{.005} = 2.575$

EXAMPLE 9.6

Based upon a sample of 250 credit-card holders, a department store found that 185 card holders incurred a monthly interest charge on an unpaid balance. Develop 95% and 90% confidence interval estimates of the population of credit-card holders who incur a monthly interest charge.

The point estimate of the population p is $\bar{p} = 185/250 = .74$. For a 95% confidence interval, we have $\alpha = .05$ and $z_{\alpha/2} = z_{.025} = 1.96$. Using (9.8) provides

$$\bar{p} \pm z_{.025} \sqrt{\frac{\bar{p}(1 - \bar{p})}{n}} = .74 \pm 1.96 \sqrt{\frac{.74(1 - .74)}{250}} = .74 \pm .0544$$

Thus the 95% confidence interval estimate of the population proportion is .6856 to .7944.

---

*An unbiased estimate of the standard error of the proportion is given by $\sqrt{\bar{p}(1 - \bar{p})/(n - 1)}$. The bias introduced by using n in the denominator does not cause any difficulty because large samples are used in making estimates concerning population proportions and the numerical difference between the results using n and $n - 1$ is negligible.

At 90% confidence, $z_{\alpha/2} = z_{.05} = 1.645$. Using (9.8) provides

$$.74 \pm 1.645 \sqrt{\frac{.74(1 - .74)}{250}} = .74 \pm .0456$$

Thus we are 90% confident that the population proportion p is between .6944 and .7856.

■

EXAMPLE 9.7

A survey conducted by the United States Department of Labor found that 48 out of 500 heads of households were unemployed. Develop a 98% confidence interval estimate of the proportion of unemployment heads of households in the population.

The point estimate of the population proportion is $\bar{p} = 48/500 = .096$. At a 98% confidence level, $z_{.01} = 2.33$. Using (9.8), the confidence interval becomes

$$.096 \pm 2.33 \sqrt{\frac{.096(1 - .096)}{500}} = .096 \pm .0307$$

Thus we are 98% confident that the population proportion of unemployed heads of households is between .0653 and .1267.

■

Determining the Size of the Sample

Let us consider the question of how large the sample size should be in order to obtain an estimate of a population proportion at a specified level of precision. The rationale for the sample-size determination in developing an interval estimate of p is very similar to the rationale used in Section 8.2 to determine the sample size for developing an interval estimate of a population mean μ.

Earlier in this section we provided the following probability statement about the sampling error:

> There is a $1 - \alpha$ probability that the value of the sample proportion will provide a sampling error of $z_{\alpha/2}\sigma_{\bar{p}}$ or less.

With $\sigma_{\bar{p}} = \sqrt{p(1 - p)/n}$, the sampling error in the above statement is based on the values of $z_{\alpha/2}$, the population proportion p, and the sample size n. For a given confidence coefficient $1 - \alpha$, $z_{\alpha/2}$ can be determined. Then, since the value of the population proportion is fixed, the sampling error mentioned in the precision statement is determined by the sample size n. Larger sample sizes again provide better precision.

Let E = the sampling error mentioned in the statement about the desired precision. Thus we have

$$E = z_{\alpha/2} \sqrt{\frac{p(1 - p)}{n}} \tag{9.9}$$

Solving (9.9) for n provides the following formula for the sample size.

$$n = \frac{z_{\alpha/2}^2 \, p(1 - p)}{E^2}$$

(9.10)

This expression shows that in order to determine the sample size, we must have a preliminary idea of the value of the population proportion, p. Past data, a preliminary sample, or an educated guess are suggested ways for obtaining the necessary planning value for p. However, in some cases it may be difficult or impossible to specify the planning value. In order to handle these situations, note that the numerator of (9.10) shows that the sample size is proportional to the quantity $p(1 - p)$. In Table 9.1 we show some possible values for this quantity. To be on the conservative side, we need to consider the largest possible value for $p(1 - p)$, since this will provide the largest sample size. As the values in Table 9.1 suggest, if an appropriate planning value for p cannot be obtained, use $p = .50$. This planning value will provide the largest possible recommended sample size. If the population proportion is different than the .50 planning value, the sample proportion will have more precision than necessary. However, in using $p = .50$ to determine n, we have guaranteed that the required precision will be obtained.

EXAMPLE 9.8

A medical experiment is being conducted to determine the recovery rate of patients given a new drug. In particular, the researcher would like to estimate the proportion of patients who fully recover within two weeks from when they begin taking the drug. The desired precision of the estimate is expressed as a 98% confidence level that the maximum sampling error will be .04 or less. For planning purposes, the researcher anticipates that .80 of the patients receiving the new drug will fully recover within the 2-week period. How large should the sample size be in this study?

Using (9.10) with $z = 2.33$ for a 98% confidence level, a planning value of $p = .80$, and a maximum sampling error $E = .04$, we have

$$n = \frac{z_{\alpha/2}^2 \, p(1 - p)}{E^2} = \frac{(2.33)^2(.80)(1 - .80)}{(.04)^2} = 543$$

■

TABLE 9.1

Possible Values for the Quantity $p(1 - p)$

Value of p	Value of $p(1 - p)$	
.10	$(.10)(.90) = .09$	
.30	$(.30)(.70) = .21$	
.40	$(.40)(.60) = .24$	
.50	$(.50)(.50) = .25$	← Largest value
.60	$(.60)(.40) = .24$	
.70	$(.70)(.30) = .21$	
.90	$(.90)(.10) = .09$	

EXAMPLE 9.9

A national survey of registered voters is being conducted to determine the proportion of voters who favor a particular candidate. Assume that the desired confidence level is 95% and that the desired maximum sampling error is .02.

a. How large a sample is needed if it is believed that approximately 35% of the population currently support the candidate?

b. How large a sample is needed if no information is available on the proportion of voters currently supporting the candidate?

In part (a) the planning value for p is .35. With $z_{.025} = 1.96$ and $E = .02$, the necessary sample size is

$$n = \frac{z_{\alpha/2}^2 \, p(1 - p)}{E^2} = \frac{(1.96)^2(.35)(1 - .35)}{(.02)^2} = 2185$$

In part (b) no planning value is available for p. Using the conservative approach discussed earlier, we base the sample size on a planning value of $p = .50$. Doing so provides

$$n = \frac{z_{\alpha/2}^2 \, p(1 - p)}{E^2} = \frac{(1.96)^2(.50)(1 - .50)}{(.02)^2} = 2401$$

Note that the recommended sample size of 2401 voters is larger than the sample size of 2185 voters recommended in (a). This larger sample size should have been anticipated due to the fact that $p = .50$ is a conservative planning value and guarantees that the sample size will be large enough to satisfy the precision requirement regardless of the actual value of p.

■

Notes and Comments

The sample size needed to generate an interval estimate for a population proportion is generally large. With a confidence coefficient of .95 and a conservative planning value of $p = .50$, the sample size required to keep the sampling error at .10 or less is 96. In order to reduce the sampling error to .05 or less, the sample size must be increased to almost 400. Thus, the large sample conditions of $np \geq 5$ and $n(1 - p) \geq 5$ are almost always satisfied in practice.

EXERCISES

14. Researchers who studied 2000 prescriptions found that most doctors failed to adjust dosages for elderly people or weight. (*Archives of Internal Medicine*, 1987). Assuming that 1600 of the 2000 doctors studied failed to adjust dosages, develop a 95% confidence interval estimate of the population proportion.

15. A simple random sample of 100 residents of Watkins Glen, New York, resulted in 65 individuals stating that they would support a newly proposed water-treatment facility. Develop a 95% confidence interval estimate of the proportion of all Watkins Glen residents that would support the new water-treatment facility.

16. In a telephone follow-up survey of a new advertising campaign, 45 of 150 individuals contacted could recall the new advertising slogan associated with the product. Devlop a 90% confidence interval estimate of the proportion in the population that will recall the advertising slogan.

17. A sample of 90 students at a particular college showed that 27 students favor pass-fail grades for elective courses.
   a. What is the point estimate of the proportion of all students who would favor pass-fail grades for elective courses?
   b. Provide a 90% confidence interval estimate of the proportion of the population of students who would favor pass-fail grades for elective courses.

18. The New Orleans Beverage Company has been experiencing problems with the automatic machine that places labels on bottles. The company wants to estimate the percentage of bottles that have improperly applied labels. A simple random sample of 400 bottles resulted in 18 bottles with improperly applied labels. Using these data, develop a 90% confidence interval estimate of the population proportion of bottles with improperly applied labels.

19. When Leo J. Shapiro & Associates, a Chicago market-research firm, surveyed consumers about their family's holiday spending plans, 29% said they planned to spend more than last year, 36% said they planned to spend less, and 35% said they planned to spend the same (*Wall Street Journal*, October 17, 1988). If 400 shoppers were sampled, develop a 90% confidence interval estimate of the percentage of shoppers that planned to spend more than last year.

20. A sample of 200 people were asked to identify their major source of news information; 110 stated that their major source was television news coverage.
   a. Construct a 95% confidence interval for the proportion of the people in the population that consider television news their major source.
   b. How large a sample would be necessary to estimate the population proportion with a sampling error of .05 or less at a 95% confidence level?

21. A survey is to be taken to estimate the proportion of high school graduates in a particular school district that plan to attend college. How large a sample of students should be selected if the survey is to provide a 95% confidence of reporting a sample proportion that is within $\pm$ .025 of the population proportion? Use $p = .35$ as a planning value for the proportion of high school students who plan to attend college.

22. A new cheese product is to be test-marketed by giving a free sample to randomly selected customers and asking them to state whether or not they like the product. With a 98% confidence level and a target sampling error of .05 or less, what sample size would you recommend in each of the following cases?
   a. Preliminary estimates are that approximately 35% of the individuals in the population will like the product.
   b. No information is available about the proportion in the population that will like the product.

23. In an election campaign, a campaign manager requests that a sample of voters be polled to determine the support for the candidate. From a sample of 120 voters, 64 express plans to support the candidate.
   a. What is the point estimate of the proportion of the voters in the population who will support the candidate?
   b. Develop and interpret the 95% confidence interval for the proportion of voters in the population who will support the candidate.

c. From the result obtained in (b), is the campaign manager justified in feeling confident that the candidate has the support of at least 50% of the voters? Explain.

d. How many voters should be sampled if we desired to estimate the population proportion with a sampling error of 5% or less? Continue to use the 95% confidence level.

9.3   Hypothesis Tests About a Population Proportion

Hypothesis tests about a population proportion p are based on knowledge of the sampling distribution of $\bar{p}$. The mechanics of conducting the test follow the procedure used in Chapter 8 to make hypothesis tests about a population mean μ. The only difference is that in this section we use the sample proportion $\bar{p}$ and the sampling distribution of $\bar{p}$ in the analysis.

We begin by formulating a null hypothesis and an alternative hypothesis about the value of the population proportion. We then consider the sampling distribution of $\bar{p}$ under the assumption that the null hypothesis is true. Based on whether the hypotheses are one-tailed or two-tailed and the level of significance, a critical value(s) is selected for z. Then, using the value of the sample proportion $\bar{p}$, we compute a value for the test statistic z. The comparison of this value of z to the critical value enables us to determine whether or not the null hypothesis should be rejected.

EXAMPLE 9.10

A newspaper article contains the statement that, nationwide, 60% of all college seniors have a job prior to graduation. The director of a college placement office at a large university is interested in testing this claim for her university. The hypotheses about the proportion of the students having a job prior to graduation are as follows:

$$H_0: p = .60$$

$$H_a: p \neq .60$$

If a random sample shows that 40 of 75 recent graduates had a job prior to graduation, what conclusion can be drawn? Use a .05 level of significance.

Under the assumption that the null hypothesis is true, we have $np = 75(.60) = 45$ and $n(1 - p) = 75(.40) = 30$. Since both of these values exceed 5, the sampling distribution of $\bar{p}$ can be approximated by a normal probability distribution. Again assuming $p = .60$, the standard error of the proportion is computed to be

$$\sigma_{\bar{p}} = \sqrt{\frac{p(1 - p)}{n}} = \sqrt{\frac{.60(1 - .60)}{75}} = .0566$$

The quantity

$$z = \frac{\bar{p} - p}{\sigma_{\bar{p}}} \tag{9.11}$$

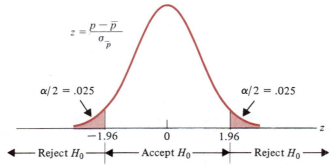

$$z = \frac{p - \bar{p}}{\sigma_{\bar{p}}}$$

$\alpha/2 = .025$ $\alpha/2 = .025$

-1.96 0 1.96

Reject H_0 ← | ← Accept H_0 → | → Reject H_0 →

FIGURE 9.6   Rejection Region for a Two-Tailed Test with $\alpha = .05$

is a standard normal random variable that can be used to determine the number of standard deviations (or standard errors) an observed value of a sample proportion $\bar{p}$ is from the hypothesized value for the population proportion p. Using a .05 level of significance, the rejection regions for the two-tailed hypothesis test are shown in Figure 9.6. The rejection rule is

$$\text{Reject } H_0 \text{ if } z < -1.96 \text{ or } z > +1.96$$

With the sample proportion $\bar{p} = 40/75 = .5333$ and a hypothesized value for the population proportion $p = .60$, using (9.11) we have the following value for z:

$$z = \frac{\bar{p} - p}{\sigma_{\bar{p}}} = \frac{.5333 - .60}{.0566} = -1.18$$

This value of z is not in the rejection region of Figure 9.6. Thus, the hypothesis that 60% of the graduating seniors have a job prior to graduation cannot be rejected.

Using the table of areas for the standard normal probability distribution, the area in the tail of distribution at $z = 1.18$ is $.5000 - .3810 = .1190$. For a two-tailed test, we have a p-value of $2 \times .1190 = .2380$. With $.2380 > \alpha$, we see that the p-value criterion also indicates that the null hypothesis cannot be rejected. The interpretation of p-values for tests about a population proportion is the same as we used for tests about a population mean.

■

EXAMPLE 9.11

The manager of an Italian restaurant is considering opening a carryout food service. However, the manager is concerned that not all individuals placing orders by phone actually pick up the order. If 90% or less of the phone orders will be picked up, the restaurant will not have a profitable carryout operation. However, if it can be concluded that more that 90% of the phone orders will be picked up, the carryout operation will be a worthwhile addition for the restaurant. The hypotheses test of interest are

$$H_0: p \leq .90$$

$$H_a: p > .90$$

This one-tailed test indicates that the restaurant should implement the carryout operation if H_0 can be rejected. During a 2-week test period, 234 orders in a sample of 250 phone-in orders were picked up. Using a .05 level of significance, what conclusion should be made?

With the one-tailed test and $\alpha = .05$, rejection of the null hypothesis will occur in the upper tail of the distribution with a critical value of $z = 1.645$. Thus, if the data show $z > 1.645$, the null hypothesis can be rejected.

Given the hypothesized value of $p = .90$ and the sample proportion $\bar{p} = 234/250 = .936$, the value of z is as follows:

$$z = \frac{\bar{p} - p}{\sigma_{\bar{p}}} = \frac{\bar{p} - p}{\sqrt{p(1 - p)/n}} = \frac{.936 - .900}{\sqrt{.90(1 - .90)/250}} = 1.90$$

Since this value is greater than 1.645, the null hypothesis is rejected. The manager of the restaurant should be safe in concluding that the proportion of carryout orders picked up for the population exceeds .90. Thus it is recommended the restaurant begin the carryout food service. With $z = 1.90$, the corresponding p-value is $.5000 - .4713 = .0287$. Since the p-value is less than $\alpha = .05$, the p-value criterion also indicates H_0 should be rejected.

■

EXAMPLE 9.12

During a water shortage in Florida, restaurants were asked not to serve water with meals unless requested to do so by the customers. In the initial 3-month period, 45% of the customers served at a particular restaurant requested water with their meal. Recently the restaurant placed a card a each table describing the water-shortage problem and pointing out that considering the drinking water, the ice, and the water to wash the glass, it requires 24 ounces of water every 8 ounces of water served. The restaurant would like to use a statistical test to determine if placing their cards at each table significantly decreases the porportion of customers requesting water with their meal. Use a .02 level of significance. If a sample of 150 customers showed 53 customers ordering water, what is your conclusion?

The hypothesis test is as follows:

$$H_0\colon p \geq .45$$

$$H_a\colon p < .45$$

Rejection of H_0 will occur in the lower tail of the distribution. With an area of $\alpha = .02$ in the lower tail, the table of areas for the standard normal distribution shows the critical z value to be -2.05. Thus, the rejection rule is to reject H_0 if $z < -2.05$.

Using the hypothesized value $p = .45$ and the sample proportion $\bar{p} = 53/150 = .3533$, the value of z is

$$z = \frac{\bar{p} - p}{\sigma_{\bar{p}}} = \frac{\bar{p} - p}{\sqrt{p(1 - p)/n}} = \frac{.3533 - .450}{\sqrt{.45(1 - .45)/150}} = -2.38$$

Since $z = -2.38$ is in the rejection region, the null hypothesis H_0: $p = .45$ is rejected. The restaurant can conclude that the cards have helped reduce the proportion of customers who order water with their meal.

∎

EXERCISES

24. A magazine claims that 25% of its readers are college students. A random sample of 200 readers is taken. It is found that 42 of these readers are college students.
   a. Use a .10 level of significance to test H_0: $p = .25$ versus H_a: $p \neq .25$.
   b. Using the sample results, develop a 90% confidence interval for the proportion of the population of readers that are college students.

25. A new television series must prove that it has more than 25% of the viewing audience after its initial 13-week run in order to be judged successful. Assume that in a sample of 400 households, 112 were watching the series.
   a. At a .10 level of significance, can the series be judged successful based on the sample information?
   b. What is the p-value for the sample results? What is your hypothesis-testing conclusion?

26. An accountant believes that the company's cash flow problems are a direct result of the slow collection of accounts receivable. The accountant claims that at least 70% of the current accounts receivable are over 2 months old. A sample of 120 accounts receivable yielded 78 over 2 months old. Test the accountant's claim at the .05 level of significance. What is the p-value?

27. A supplier claims that at least 96% of the parts it supplies meet the product specifications. In a sample of 500 parts received over the past 6 months, 36 were defective. Use the p-value to test the supplier's claim at a .05 level of significance.

28. If 20% or more of the population have a negative reaction to a new drug, the drug will not be marketed.
   a. Define the hypotheses such that the drug will be marketed only if H_0 is rejected.
   b. In a sample of 80 patients, 14 patients experienced a negative reaction to the drug. Test the hypotheses in part (a) at a .02 level of significance. What is your conclusion?

29. The manager of K-Mark Supermarkets estimates that at least 30% of the Saturday customers purchase the price-reduced special advertised in the Friday newspaper. Using $\alpha = .05$ test the manager's claim if the results of a sample of 250 Saturday customers showed that 60 purchased the advertised special. What is the p-value?

30. The Gordon S. Black Corporation surveyed 439 working parents with a total of 736 children younger than 12; 73% of those surveyed said that one parent would stay at home if money were not a factor. (*Democrat and Chronicle*, November 28, 1988). Is there sufficient evidence to conclude that p, the population proportion that would stay at home, is greater than .70? Use $\alpha = .02$.

31. A radio station in a major resort area announced that at least 90% of the hotels and motels would be full for the Memorial Day weekend. The station went on to advise listeners to make reservations in advance if they planned to be in the resort over the weekend. On Saturday night a sample of 58 hotels and motels showed 49 with a no-vacancy sign and 9 with vacancies. What is your reaction to the radio station's claim based on the sample evidence? Use $\alpha = .05$ in making this statistical test. What is the p-value for the sample results?

In this chapter we discussed the procedures for making statistical inferences about a population proportion. After describing the properties of the sampling distribution of $\bar{p}$, we presented the methods of interval estimation and hypothesis testing for a population proportion. In the large-sample case where both $np \geq 5$ and $n(1 - p) \geq 5$, the central limit theorem enables us to use the normal probability distribution to approximate the sampling distribution of $\bar{p}$.

The procedure for developing a confidence interval estimate of a population proportion uses the same logic as the procedure for developing a confidence interval estimate of a population mean. The essential difference is that the sampling distribution of $\bar{p}$ is used as the basis for developing the confidence interval instead of the sampling distribution of $\bar{x}$.

A procedure was presented for determining the sample size that will meet a desired precision when estimating a population proportion. This procedure requires the person conducting the test to specify a planning value for the population proportion p. If p is unknown, using a planning value of $p = .50$ provides a sample size that will satisfy the precision requirements regardless of the actual value of the population proportion.

The procedure for testing hypotheses about a population proportion follows the logic used for testing hypotheses about a population mean. Examples of both one-tailed and two-tailed hypothesis tests about a population proportion were presented.

GLOSSARY

Sampling distribution of $\bar{p}$ The probability distribution showing all possible values of the sample proportion $\bar{p}$.

Standard error of the proportion   The standard deviation of all possible values of the sample proportion $\bar{p}$.

KEY FORMULAS

SAMPLE PROPORTION

$$\bar{p} = \frac{x}{n} \tag{9.1}$$

EXPECTED VALUE OF $\bar{p}$

$$E(\bar{p}) = p \tag{9.2}$$

STANDARD DEVIATION OF $\bar{p}$

$$\sigma_{\bar{p}} = \sqrt{\frac{p(1 - p)}{n}} \tag{9.3}$$

STANDARD NORMAL RANDOM VARIABLE

$$z = \frac{p - \bar{p}}{\sigma_{\bar{p}}} \tag{9.4}$$

INTERVAL ESTIMATION OF A POPULATION PROPORTION

$$\bar{p} \pm z_{\alpha/2} \sqrt{\frac{\bar{p}(1 - \bar{p})}{n}} \tag{9.8}$$

SAMPLE SIZE FOR INTERVAL ESTIMATION OF A POPULATION PROPORTION

$$n = \frac{z_{\alpha/2}^2\, p(1 - p)}{E^2} \tag{9.10}$$

REVIEW QUIZ

TRUE/FALSE

1. The sample proportion $\bar{p}$ is not a point estimator of the population proportion p.

2. The central limit theorem cannot be applied to the sampling distribution of $\bar{p}$.

3. The standard error of the proportion, $\sigma_{\bar{p}}$, depends upon the value of the population proportion p.

4. In computing confidence intervals for a population proportion, the sample proportion $\bar{p}$ can be used to obtain an estimate of the standard error of the proportion, $\sigma_{\bar{p}}$.

5. In determining the sample size, the larger the planning value for p, the larger the sample size.

6. In conducting hypothesis tests about a population proportion, the value of p specified in the null hypothesis is used to compute the standard error of the proportion, $\sigma_{\bar{p}}$.

MULTIPLE CHOICE

Use the following for questions 7-9. A random sample of 300 voters showed .47 in favor of a certain ballot proposal.

7. Which of the following best describes the form of the sampling distribution of the sample proportion?
   a. When standardized, it is exactly the standard normal distribution.
   b. When standardized, it is the t distribution.
   c. It is approximately normal because $n > 30$.
   d. It is approximately normal because $n\bar{p} \geq 5$ and $n(1 - \bar{p}) \geq 5$.

8. An estimate of the standard deviation of the sampling distribution of the sample proportion is
   a. .016
   b. .025
   c. .029
   d. .035

9. A 90% confidence interval estimate for the true proportion of voters favoring the proposal is
  a. .38 to .56
  b. .401 to .539
  c. .413 to .527
  d. none of the above

10. In choosing a sample size for a public-opinion survey, what hypothesized value of the population proportion will lead to the largest sample size when the confidence level and error allowance are given?
  a. $p = .1$
  b. $p = .5$
  c. $p = .99$
  d. The confidence level must be known before an answer can be given.

SUPPLEMENTARY EXERCISES

32. Consider a population of six people, four of whom are college graduates.
  a. Identify the 15 different possible samples of four people that can be selected from this population.
  b. Compute the sample proportion of college graduates for each sample.
  c. Show the sampling distribution of $\bar{p}$ by showing the probability distribution of the values of $\bar{p}$ found in (b).

33. Assume that 60% of the management staff in a large corporation has completed the company's special management-training program on improving communication skills. What is the probability that a sample of managers will provide a sample proportion $\bar{p}$ that is within $\pm$ .05 of the population proportion of $p = .60$? Answer this question for sample sizes of 30, 60, and 100. What is the advantage of the larger sample size?

34. Assume that 15% of the items produced in an assembly-line operation are defective but that the firm's production manager is not aware of this situation. Assume further that 50 parts are tested by the quality assurance department in order to determine the quality of the assembly operation. Let $\bar{p}$ be the sample proportion defective found by the quality assurance test.
  a. Show the sampling distribution for $\bar{p}$.
  b. What is the probability that the sample proportion will be within $\pm$ .03 of the population proportion defective?

35. A market research firm conducts telephone surveys with a 40% historical response rate. What is the probability that in a new sample of 400 telephone numbers at least 150 individuals will cooperate and respond to the questions? In other words, what is the probability of a sample proportion $\bar{p} \geq 150/400 = .375$?

36. A production run is not acceptable for shipment to customers if a sample of 100 items contains 5% or more defective items. If a production run has a population proportion defective of $p = .10$, what is the probability that $\bar{p}$ will be at least .05?

37. A Boston University survey of 1600 employees in the Northeast found that 11% of the respondents were part of the "traditional work force;" that is, married males with wives at home full time. (Boston University: Balancing Job and Homelife Study, 1987, as reported in *The Wall Street Journal*, November 1, 1988). Assuming that this is a representative sample of all employees in the Northeast, develop a 95% confidence interval estimate of the population proportion.

38. H.G. Forester and Company is a distributor of lumber supplies throughout the southwest United States. Management of H.G. Forester would like to check a shipment of over one million pine boards in order to determine if excessive warpage exists for the boards. A sample of 50 boards resulted in the identification of 7 boards with excessive warpage. With these data, develop a 95% confidence interval estimate of the proportion of boards defective in the whole shipment.

39. Consider the H.G. Forester and Company problem represented in Exercise 38. How large a sample would be required to estimate the proportion of boards with warpage to within $\pm .01$ at a 95% confidence level?

40. The Tourism Institute for the State of Florida plans to sample visitors at major beaches throughout the state in order to estimate the proportion of beach visitors that are residents of states other than Florida. Preliminary estimates are that 55% of the beach visitors are not residents of Florida.
   a. How large a sample should be taken to estimate the proportion of out-of-state visitors to within $\pm 3\%$ of the actual value at a 95% confidence level?
   b. How large a sample should be taken if the acceptable error is increased to $\pm 6\%$?

41. A fast-food restaurant plans to initiate a special offer, which will enable customers to purchase specially designed drink glasses featuring well-known cartoon characters. If 15% or fewer of the customers will purchase the glasses, the special offer will not be initiated. If a preliminary test is set up at several locations and 88 of 500 customers purchase the glasses, should the special glass offer be made? Test the hypothesis that will support your decision. Use a .01 level of significance. What is your recommendation?

42. It has been claimed that 5% of the students at one college make blood donations during a given year. If, in a random sample, 10 of 250 students have given blood during the past year, test $H_0: p = .05$ versus $H_a: p \neq .05$. Use a .05 level of significance in reaching your conclusion.

43. It has been claimed that at least 90% of juvenile first-time criminals are given probation upon the admission of guilt. The results of a sample of 92 juvenile criminal convictions showed that 78 juveniles received probation. Test the claim at a .02 level of significance.

COMPUTER EXERCISE

One of the critical issues facing the Congress of the United States deals with whether or not budget cuts should be made in the area of federal aid to higher education. In order to learn about the views of constituents, a congressperson from an eastern state requested a survey be conducted to obtain a more accurate picture of public opinion on the issue. Part of the data collected during the survey showed whether or not the respondent's household had college-age children and whether or not the respondent opposed budget cuts in federal aid to higher education. Results for the first 70 responses received during the survey are as follows:

HOUSEHOLD WITH CHILDREN	OPPOSES BUDGET CUTS	HOUSEHOLD WITH CHILDREN	OPPOSES BUDGET CUTS
Yes	Yes	No	No
Yes	Yes	No	Yes
No	No	Yes	Yes
No	Yes	No	Yes
Yes	No	No	No

HOUSEHOLD WITH CHILDREN	OPPOSES BUDGET CUTS	HOUSEHOLD WITH CHILDREN	OPPOSES BUDGET CUTS
Yes	No	Yes	Yes
No	No	No	No
No	Yes	Yes	No
No	No	Yes	Yes
No	Yes	No	Yes
No	No	No	No
Yes	Yes	Yes	Yes
Yes	No	Yes	Yes
No	Yes	Yes	Yes
Yes	Yes	Yes	No
No	Yes	No	No
Yes	No	Yes	No
No	No	Yes	No
No	No	No	Yes
No	Yes	Yes	No
Yes	Yes	Yes	Yes
Yes	Yes	No	No
No	Yes	Yes	No
No	No	No	Yes
No	No	No	Yes
Yes	Yes	Yes	Yes
No	No	No	No
Yes	No	No	No
Yes	No	No	No
No	No	Yes	Yes
No	Yes	Yes	Yes
No	No	Yes	Yes
No	No	No	No
Yes	Yes	Yes	Yes
No	Yes	Yes	No

QUESTIONS

1. Use descriptive statistics to summarize the data.

2. Provide a 95% confidence interval for the proportion in the population that have households with college-age children.

3. Provide a 95% confidence interval for the proportion in the population that oppose budget cuts in the federal-aid program for higher education.

4. Provide 95% confidence intervals for the proportion opposing budget cuts for the population of households with college-age children and the proportion opposing budget cuts for the population of households without college-age children.

10

Inferences About Means and Proportions with Two Populations

What You Will Learn in This Chapter:

■ How to construct interval estimates and conduct hypothesis tests about the difference between the means of two populations

■ When and how to use the *t* distribution to conduct inferences about the difference between the means of two populations

■ The difference between independent and matched samples

■ How to compute a pooled variance estimate

■ How to construct interval estimates and conduct hypothesis tests about the difference between the proportions of two populations

Contents

Nicotine Chewing Gum Helps Stop Smoking

A group of physicians from Copenhagen, Denmark, recently reported the results of a study on the effectiveness of nicotine chewing gum in helping people stop smoking. In a sample of 173 smokers, 60 were classified as having a high dependence on nicotine and 113 were classified as having a medium or low dependence. Of the 60 subjects classified as highly dependent on nicotine, 27 were given gum containing 4 mg of nicotine and the remaining 33 were given gum containing 2 mg of nicotine. For the 113 smokers with medium or low dependence, 60 were given gum containing 2 mg of nicotine, and 53 were given a placebo gum.

In carrying out the experiment, a double-blind study was used. That is, neither the subjects nor the individuals who monitored their performance knew what type of gum each smoker had been given. At the end of 6 weeks, 1 year, and 2 years of treatment, each smoker was chemically tested to determine if he or she had abstained from smoking. The results obtained are summarized in the following two tables. The entries in each table show the percentage of the group that were abstinent after each of the three follow-up periods.

With the exception of the 2-mg nicotine gum versus the placebo gum at 1 year, each of the differences in outcomes can be shown to be statistically significant at the 5% level. Thus, the study

UPI/Bettmann

Even with good intentions, many have difficulty quitting the smoking habit.

indicates that the effectiveness of nicotine gum is not due to random effects and that it is related to the amount of nicotine in the gum. The researchers concluded that, "The use of nicotine gum in appropriate doses should be helpful to persons who are attempting to stop smoking."

Based on "Effect of Nicotine Chewing Gum in Combination with Group Counseling on the Succession of Smoking," *New England Journal of Medicine*, January, 1988.

TYPE OF GUM	SAMPLE SIZE	PERCENTAGE OF HIGHLY DEPENDENT SMOKERS ABSTINENT AFTER		
		6 Weeks	1 Year	2 Years
4 mg gum	27	81.5%	44.4%	33.3%
2 mg gum	33	54.5%	12.1%	6.1%

TYPE OF GUM	SAMPLE SIZE	PERCENTAGE OF MEDIUM- OR LOW-DEPENDENCE SMOKERS ABSTINENT AFTER		
		6 Weeks	1 Year	2 Years
2 mg gum	60	73.3%	38.3%	28.3%
Placebo gum	53	41.5%	22.6%	9.4%

In the preceding three chapters we have developed statistical methodology for interval estimation and hypothesis tests for population means and population proportions. However, the statistical procedures we have discussed thus far have considered only single-population situations. In this chapter we expand our discussion to cases where *two populations* are involved. Specifically, we will be selecting random samples and performing statistical analyses that will enable us to draw conclusions about the difference between the means and/or the proportions for two populations. An example of this type of situation appeared in "Statistics in the News," where differences between the percentage of smokers who remain abstinent were reported for smokers given either nicotine gum or a placebo. We begin the study of two-population situations by showing how to estimate the difference between the means of two populations.

10.1 Estimation of the Difference Between the Means of Two Populations—Independent Samples

In some statistical applications we are faced with two populations where the difference between the means of the two populations is of prime importance. We know from Chapter 8 that we can take a simple random sample from a single population and use the sample mean $\bar{x}$ to estimate the population mean μ. In the two-population case we will select two independent simple random samples, one from population 1 and another from population 2. Let

μ_1 = mean of population 1

μ_2 = mean of population 2

$\bar{x}_1$ = sample mean for a simple random sample selected from population 1 (i.e., the point estimator of μ_1)

$\bar{x}_2$ = sample mean for a simple random sample selected from population 2 (i.e., the point estimator of μ_2)

The difference between the two population means is given by $\mu_1 - \mu_2$. A point estimator of this difference is given by $\bar{x}_1 - \bar{x}_2$. Thus we see that an estimate of the difference between two population means is given by the difference between the two sample means. The situation is depicted in Figure 10.1.

Sampling Distribution of $\bar{x}_1 - \bar{x}_2$

In Chapter 7 we showed that different simple random samples taken from the same population provide a variety of different values for the sample mean. Accordingly, we can anticipate that both $\bar{x}_1$ and $\bar{x}_2$ will take on different values for different simple random samples. Thus $\bar{x}_1 - \bar{x}_2$, just like other sample statistics, is subject to variability. The probability distribution showing all possible values of $\bar{x}_1 - \bar{x}_2$ is called the *sampling distribution of $\bar{x}_1 - \bar{x}_2$*.

If we can identify the properties of the sampling distribution of $\bar{x}_1 - \bar{x}_2$, we can develop interval estimates and conduct hypothesis tests about $\mu_1 - \mu_2$ in much the same way we used the sampling distribution of $\bar{x}$ to make inferences about the single population mean μ. The properties of the sampling distribution of $\bar{x}_1 - \bar{x}_2$ are as follows:

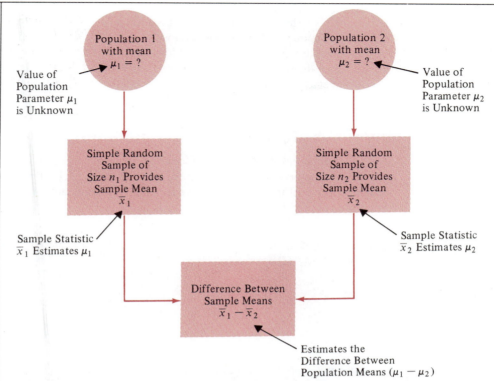

FIGURE 10.1 Statistical Process of Using the Difference Between Sample Means to Estimate the Difference Between Population Means

---

Sampling Distribution of $\bar{x}_1 - \bar{x}_2$

$$\text{Expected Value:} \quad E(\bar{x}_1 - \bar{x}_2) = \mu_1 - \mu_2 \qquad (10.1)$$

$$\text{Standard Deviation:} \quad \sigma_{\bar{x}_1 - \bar{x}_2} = \sqrt{\frac{\sigma_1^2}{n_1} + \frac{\sigma_2^2}{n_2}} \qquad (10.2)$$

where

σ_1 = standard deviation of population 1

σ_2 = standard deviation of population 2

n_1 = size of the simple random sample selected from population 1

n_2 = size of the simple random sample selected from population 2

Distribution Form: Based on the central limit theorem, if both sample sizes are large ($n_1 \geq 30$ and $n_2 \geq 30$), the sampling distribution of $\bar{x}_1 - \bar{x}_2$ can be approximated by a normal probability distribution.

Figure 10.2 shows the sampling distribution of $\bar{x}_1 - \bar{x}_2$ and its relationship to the individual sampling distributions of $\bar{x}_1$ and $\bar{x}_2$. We will use the sampling distribution of $\bar{x}_1 - \bar{x}_2$ to make inferences about the difference between the means of two populations. We first show the method for large sample sizes (both $n_1 \geq 30$ and $n_2 \geq 30$). Then we show the method for situations where one or both samples are small ($n_1 < 30$ and/or $n_2 < 30$).

Interval Estimation of $\mu_1 - \mu_2$: Large-Sample Case

Using the properties of the sampling distribution of $\bar{x}_1 - \bar{x}_2$, we can follow the same procedure for interval estimation that we used in previous chapters. Using $(1 - \alpha)$ as the confidence coefficient, the general statement for a confidence interval estimate of the difference between two population means is as follows:

Interval Estimation of the Difference Between the Means of Two Populations—(Large-Sample Case with $n_1 \geq 30$ and $n_2 \geq 30$)

$$\bar{x}_1 - \bar{x}_2 \pm z_{\alpha/2} \sqrt{\frac{\sigma_1^2}{n_1} + \frac{\sigma_2^2}{n_2}} \qquad (10.3)$$

If the two population variances σ_1^2 and σ_2^2 are unknown, the sample variances s_1^2 and s_2^2 can be substituted for the population variances to provide the interval estimate.

EXAMPLE 10.1

The Educational Testing Service conducted a study to investigate possible differences between the scores of males and females on the Scholastic Aptitude Test (SAT) (*Journal of Educational Measurement*, Spring 1987). A random sample of 562 females and 852 males provided a sample mean SAT verbal score of $\bar{x}_1 = 547$ for the females and $\bar{x}_2 = 525$ for the males. The sample standard deviations were $s_1 = 83$ for the females and $s_2 = 78$ for the males. Using a 95% confidence level, estimate the difference between the mean SAT verbal scores for the population of females and the population of males.

The point estimate of the difference between mean scores is $\bar{x}_1 - \bar{x}_2 = 547 - 525 = 22$ points. Thus, we estimate that females score an average of 22 points higher on the verbal test than the males. Using $s_1 = 83$ and $s_2 = 78$ to estimate σ_1 and σ_2, respectively, the 95% confidence interval estimate of the difference between mean verbal scores can be computed using (10.3):

$$\bar{x}_1 - \bar{x}_2 \pm z_{\alpha/2} \sqrt{\frac{\sigma_1^2}{n_1} + \frac{\sigma_2^2}{n_2}} = 547 - 525 \pm 1.96 \sqrt{\frac{(83)^2}{562} + \frac{(78)^2}{852}}$$

$$= 22 \pm 1.96(4.404)$$

$$= 22 \pm 8.63$$

Thus an interval estimate of 22 ± 8.63, or approximately 22 ± 9, tells us that for this group of students we can be 95% confident that the mean verbal score of females is approximately 13 to 31 points higher than the mean verbal score for males.

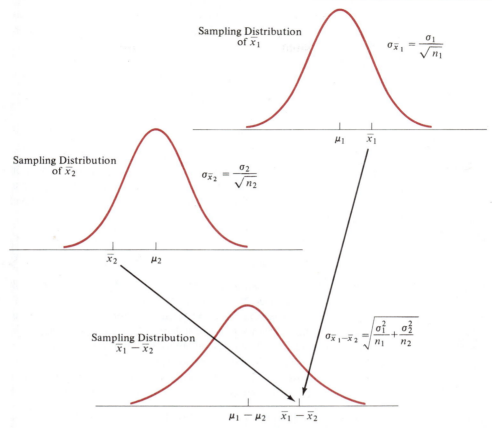

FIGURE 10.2 Sampling Distribution of $\bar{x}_1 - \bar{x}_2$ and its Relationship to the Individual Sampling Distributions of $\bar{x}_1$ and $\bar{x}_2$

EXAMPLE 10.2

A firm with department stores in Atlanta, Georgia has some stores located in the inner city and some stores located in suburban shopping centers. In a study designed to learn about the different characteristics of the inner-city and suburban customer populations, a sample of 60 inner-city customers and a sample of 80 suburban customers was taken. The data obtained on customer ages is summarized below:

STORE TYPE	SAMPLE SIZE	SAMPLE MEAN AGE	SAMPLE STANDARD DEVIATION
Inner city	60	$\bar{x}_2 = 40$ years	$s_1 = 9$ years
Suburban	80	$\bar{x}_2 = 35$ years	$s_2 = 10$ years

Using 90% and 99% confidence intervals, estimate the difference between the mean ages of the two populations of customers.

At 90%, $z_{\alpha/2} = z_{.05} = 1.645$. Using (10.3), we have

$$40 - 35 \pm 1.645 \sqrt{\frac{(9)^2}{60} + \frac{(10)^2}{80}} = 5 \pm 1.645(1.61)$$

$$= 5 \pm 2.65$$

Thus, we are 90% confident that the difference between the mean ages of the two populations is between 2.35 and 7.65 years with the inner-city stores, on the average, having the older customers.

At 99%, $z_{\alpha/2} = z_{.005} = 2.575$. Using (10.3), we have

$$40 - 35 \pm 2.575 \sqrt{\frac{(9)^2}{60} + \frac{(10)^2}{80}} = 5 \pm 2.575(1.61)$$

$$= 5 \pm 4.15$$

Hence, the 99% confidence interval for the difference between the mean ages is .85 to 9.15 years. As expected, the interval estimate with a 99% confidence level results in a wider interval than the interval estimate with a 90% confidence level.

■

Interval Estimation of $\mu_1 - \mu_2$: Small-Sample Case

If either or both of the sample sizes are small ($n_1 < 30$ and/or $n_2 < 30$), the central limit theorem *cannot* be used to conclude that the sampling distribution of $\bar{x}_1 - \bar{x}_2$ can be approximated by a normal probability distribution. However, as we saw in Chapter 8, whenever the population has a normal probability distribution and the sample standard deviation is used to estimate the population standard deviation, the t distribution can be used to make inferences about a population mean.

In order to use the t distribution to develop interval estimates of the difference between the means of two populations, the following two assumptions must be satisfied.

1. Both populations must have normal probability distributions.
2. The variances of the two populations must be equal; i.e., $\sigma_1^2 = \sigma_2^2 = \sigma^2$, where σ^2 is the variance for both populations.

Because of the equal variance assumption, the expression for the standard deviation of $\bar{x}_1 - \bar{x}_2$ as given by (10.2) can be written

$$\sigma_{\bar{x}_1 - \bar{x}_2} = \sqrt{\frac{\sigma^2}{n_1} + \frac{\sigma^2}{n_2}} = \sqrt{\sigma^2 \left(\frac{1}{n_1} + \frac{1}{n_2} \right)} \qquad (10.4)$$

Generally, the value of σ^2 will be unknown. However, since the two sample variances (s_1^2 and s_2^2) both provide estimates of σ^2, we can *combine* s_1^2 and s_2^2 to obtain an estimate of σ^2. The process of combining the results of the two independent samples to provide one estimate of σ^2 is referred to as *pooling*. The *pooled variance estimate* of σ^2 is as follows.

Pooled Variance Estimate

$$s_p^2 = \frac{(n_1 - 1)s_1^2 + (n_2 - 1)s_2^2}{(n_1 + n_2 - 2)} \qquad (10.5)$$

Substituting s_p^2 for σ^2 in (10.4), we obtain the following estimate of the standard deviation of $\bar{x}_1 - \bar{x}_2$:

$$s_{\bar{x}_1 - \bar{x}_2} = \sqrt{s_p^2 \left(\frac{1}{n_1} + \frac{1}{n_2} \right)} \qquad (10.6)$$

The notation $s_{\bar{x}_1 - \bar{x}_2}$ is used to indicate that (10.6) provides an estimate of $\sigma_{\bar{x}_1 - \bar{x}_2}$.

With normally distributed populations and with the common variance σ^2, estimated by s_p^2, the t distribution can be used to compute a confidence interval for the difference between the means of two populations. The confidence interval for this difference is given by the following expression.

Interval Estimation of the Difference Between the Means of Two Populations—(Small-Sample Case with $n_1 < 30$ and/or $n_2 < 30$)

$$\bar{x}_1 - \bar{x}_2 \pm t_{\alpha/2} \sqrt{s_p^2 \left(\frac{1}{n_1} + \frac{1}{n_2} \right)} \qquad (10.7)$$

where the t value is based on a t distribution with $n_1 + n_2 - 2$ degrees of freedom.

The $t_{\alpha/2}$ values in (10.7) are found in the table for the t distribution (see Table 2 in Appendix B). As we saw in Chapter 8, $t_{\alpha/2}$ provides an area of $\alpha/2$ in the upper tail of the t distribution and thus corresponds to an interval with a confidence coefficient of $1 - \alpha$.

EXAMPLE 10.3

An urban-planning group is interested in estimating the difference between the mean household incomes for two neighborhoods in a large metropolitan area. Independent random samples of households in the neighborhoods provided the following results:

	NEIGHBORHOOD 1	NEIGHBORHOOD 2
Sample size	$n_1 = 8$ households	$n_2 = 12$ households
Sample mean	$\bar{x}_1 = \$15,700$	$\bar{x}_2 = \$13,500$
Sample standard deviation	$s_1 = \$700$	$s_2 = \$850$

The point estimate of the difference between the mean household incomes for the two neighborhoods is

$$\bar{x}_1 - \bar{x}_2 = \$15,700 - \$13,500 = \$2,200$$

Making the assumptions that the incomes are normally distributed in both neighborhoods and that the population variances are equal, (10.7) can be used to develop a 95% confidence interval for the difference between the mean incomes in the two neighborhoods. First, we use (10.5) to compute the pooled estimate of σ^2. This computation is as follows:

$$s_p^2 = \frac{(n_1 - 1)s_1^2 + (n_2 - 1)s_2^2}{n_1 + n_2 - 2}$$

$$= \frac{(8 - 1)(700)^2 + (12 - 1)(850)^2}{8 + 12 - 2}$$

$$= \frac{7(490,000) + 11(722,500)}{18} = 632,083$$

With $n_1 + n_2 - 2 = 18$, we use the t distribution with 18 degrees of freedom to find $t_{.025} = 2.101$. (See Table 2 in Appendix B.) Thus, using (10.7), we obtain the 95% confidence interval of

$$\bar{x}_1 - \bar{x}_2 \pm t_{.025} \sqrt{s_p^2 \left(\frac{1}{n_1} + \frac{1}{n_2} \right)} = 2200 \pm (2.101) \sqrt{(632,083) \left(\frac{1}{8} + \frac{1}{12} \right)}$$

$$= 2200 \pm (2.101)(362.883)$$

$$= 2200 \pm 762.42$$

Subtracting and adding 762.42 to the estimate of 2200 provides the 95% confidence interval of \$1437.58 to \$2962.42. Thus we are 95% confident that the difference between the mean household incomes for the two neighborhoods is in this interval.

■

EXAMPLE 10.4

A sociology class project involves the study of dating practices on a major college campus. One aspect of the study compared the frequency of dating by freshman women and freshman men. A sample of 15 freshman women showed a mean of 8.2 dates per month with a standard deviation of 2.5. A sample of 10 freshman men showed a mean of 6.2 dates per month with a standard deviation of 2.2. What is the 90% confidence interval estimate of the difference between the mean number of dates per month for freshman women and freshman men?

With the small-sample case, we make the assumptions that the populations are normally distributed with equal variances. The pooled estimate of the variance is

$$s_p^2 = \frac{(n_1 - 1)s_1^2 + (n_2 - 1)s_2^2}{n_1 + n_2 - 2}$$

$$= \frac{(14)(2.5)^2 + 9(2.2)^2}{15 + 10 - 2} = 5.698$$

At 90% confidence with $n_1 + n_2 - 2 = 15 + 10 - 2 = 23$ degrees of freedom, the t distribution table shows $t_{.05} = 1.714$. Thus we have

$$\bar{x}_1 - \bar{x}_2 \pm t_{.05} \sqrt{s_p^2 \left(\frac{1}{n_1} + \frac{1}{n_2} \right)} = 8.2 - 6.2 \pm (1.714) \sqrt{5.698 \left(\frac{1}{15} + \frac{1}{10} \right)}$$

$$= 2.0 \pm 1.714(.975)$$

$$= 2.0 \pm 1.67$$

Thus we are 90% confident that freshman women students average from .33 to 3.67 more dates per month than do freshman men.

■

As a final comment, in Chapter 8 we pointed out that the t distribution is not restricted to the small-sample situation. Anytime we are interested in the difference between two population means and the populations are normally distributed with equal variances, the t distribution can be used to develop the appropriate confidence interval. However, (10.3) shows how to compute confidence intervals when the sample sizes are large. In this situation the use of the t distribution and its corresponding assumptions are not required. Thus, we do not need to refer to the t distribution unless we have a small-sample-size case.

Notes and Comments

1. In developing a confidence interval estimate of the difference between two population means, we can refer to either of the populations as population 1. For example, assume that two independent samples were selected from the population of service stations in Miami and in Boston. The purpose of the sample is to estimate the difference between the mean price for unleaded gasoline in the two cities. Assume that the Miami sample resulted in a mean of $1.04 per gallon and the Boston sample resulted in a mean of $1.09 per gallon. If we let μ_1 denote the mean price in Miami and μ_2 denote the mean price in Boston, then the point estimate of the difference between the population means would be $1.04 - $1.09 = -$.05. Alternatively, we could have defined μ_1 as the mean price in Boston and μ_2 as the mean price in Miami. In this case, the point estimate of the difference between the means of the two populations, $1.09 - $1.04 = $.05, would be positive. Many experimenters prefer to use the second approach in order to avoid the use of a negative sign in presenting the results.

2. In the small-sample case, we present a procedure based on the assumption that the variances of the two populations are equal. In Chapter 11 we will show how to test for equal population variances. In situations where this assumption is not appropriate, however, several alternate methods can be used. One method, which we describe in Section 10.3, applies to random samples that are not independent.

EXERCISES

1. Independent simple random samples were selected from two populations. Assume that $\mu = 500$, $\sigma_2 = 485$, $\sigma_1 = 89$, $\sigma_2 = 60$, and $n_1 = n_2 = 100$.
   a. Show the sampling distribution of $\bar{x}_1 - \bar{x}_2$.
   b. What is the probability that the sample mean from population 1 will exceed the sample mean from population 2?
   c. What is the probability that $\bar{x}_1 - \bar{x}_2$ will be negative? What situation does this correspond to?

d. What is the probability that the sample from population 1 will show a sample mean between 10 and 20 points higher than the sample mean from population 2?

2. In the evaluation of an eighth-grade reading comprehension program, a standardized test is to be given to a sample of eighth graders and to a sample of seventh graders from the same school. Let the population of eighth graders be denoted as population 1, with test scores having characteristics $\mu_1 = 77$ and $\sigma_1 = 12$. Let the population of seventh graders be denoted as population 2, with test scores having characteristics $\mu_2 = 72$ and $\sigma_2 = 14$. Answer the following questions if a random sample of 40 eighth graders and a random sample of 40 seventh graders will be used in the study.

a. Show the sampling distribution of $\bar{x}_1 - \bar{x}_2$.

b. What is the probability that the eighth graders in the study will show a mean test score between 3 and 7 points greater than the seventh graders?

c. What is the probability of $\bar{x}_1 - \bar{x}_2$ being negative, which would mean that the seventh graders did better on the test than the eighth graders?

3. A college admissions board is interested in estimating the difference between the mean grade point averages of students from two high schools. Independent simple random samples of students at the two high schools provide the following results.

MT. WASHINGTON	COUNTRY DAY
$n_1 = 46$	$n_2 = 33$
$\bar{x}_1 = 3.02$	$\bar{x}_2 = 2.72$
$s_1 = .38$	$s_2 = .45$

a. What is the point estimate of the difference between the means of the two populations?

b. Develop an interval estimate of the difference between the two population means with a confidence coefficient of .90.

c. Answer (b) using a .95 confidence coefficient.

4. The Butler County Bank and Trust Company is interested in estimating the difference between the mean credit card balances at two of its branch banks. Independent samples of credit card customers provide the following results.

BRANCH 1	BRANCH 2
$n_1 = 32$	$n_2 = 36$
$\bar{x}_1 = \$500$	$\bar{x}_2 = \$375$
$s_1 = \$150$	$s_2 = \$130$

a. Develop a point estimate of the difference between the mean balances at the two branches.

b. Develop an interval estimate of the difference between the mean balances. Use a confidence coefficient of .99.

5. Starting annual salaries for individuals with master's and bachelor's degrees were obtained from two independent random samples. Use the data shown below to provide a 90% confidence interval estimate of the increase in mean starting salary that can be expected upon completion of the master's degree.

MASTER'S DEGREE	BACHELOR'S DEGREE
$n_1 = 60$	$n_2 = 80$
$\bar{x}_1 = \$24,000$	$\bar{x}_2 = \$22,000$
$s_1 = \$2,500$	$s_2 = \$2,000$

6. A sample of 15 recently released prisoners who had convictions for armed robbery showed that the time in prison had a sample mean of 5.2 years, with a sample standard deviation of 1.4 years. A sample of 12 recently released prisoners who had convictions for assault showed a sample mean time in prison of 2.7 years, with a sample standard deviation of 1.1 years.

    a. What is the pooled estimate of the population variance σ^2?

    b. Using a 95% confidence interval, estimate the difference in mean time in prison for armed robbery and assault convictions.

10.2   Hypothesis Tests About the Difference Between the Means of Two Populations—Independent Samples

In this section we show how to conduct hypothesis tests about the difference between the means of two populations. These hypothesis tests use the same logic as the hypothesis tests in Chapters 8 and 9, except that the tests are based on the sample statistic $\bar{x}_1 - \bar{x}_2$. Thus we use the sampling distribution of $\bar{x}_1 - \bar{x}_2$ to conduct the test. In the large-sample case, the sampling distribution can be approximated by a normal probability distribution. Thus the following quantity is a standard normal random variable z which is used for the hypothesis test.

Observed Value of $\bar{x}_1 - \bar{x}_2$ Hypothesized Value of $\mu_1 - \mu_2$

$$z = \frac{(\bar{x}_1 - \bar{x}_2) - (\mu_1 - \mu_2)}{\sqrt{\dfrac{\sigma_1^2}{n_1} + \dfrac{\sigma_2^2}{n_2}}} \qquad (10.8)$$

Standard deviation of $\bar{x}_1 - \bar{x}_2$

In this large-sample case, if the population variances σ_1^2 and σ_2^2 are unknown, (10.8) can be used with the sample variances s_1^2 and s_2^2 substituted for σ_1^2 and σ_2^2.

EXAMPLE 10.5

A medical research study was conducted to determine if there is a difference between the effectiveness of two pain-relief medicines used for headaches. Over a 6-month period, a sample of individuals used one of the medicines, whereas another sample of individuals used the other medicine. Data collected during the study showed the time required to receive pain relief. Letting

$$\mu_1 = \text{mean pain-relief time for medicine 1}$$
$$\mu_2 = \text{mean pain-relief time for medicine 2}$$

the hypothesis test is expressed as follows:

$$H_0 \colon \mu_1 - \mu_2 = 0$$
$$H_a \colon \mu_1 - \mu_2 \neq 0$$

Note that these hypotheses are equivalent to H_0: $\mu_1 = \mu_2$ and H_a: $\mu_1 \neq \mu_2$. In either case, if the null hypothesis is rejected, the test will have shown that the two medicines differ in terms of pain-relief speed.

Using the following data, conduct the test and draw a conclusion comparing the two medicines. Use $\alpha = .05$.

	INDIVIDUALS USING MEDICINE 1	INDIVIDUALS USING MEDICINE 2
Sample size	$n_1 = 248$	$n_2 = 225$
Sample mean	$\bar{x}_1 = 24.8$ minutes	$\bar{x}_2 = 26.1$ minutes
Sample standard deviation	$s_1 = 3.3$ minutes	$s_2 = 4.2$ minutes

With $\alpha = .05$ and a two-tailed test, the critical z values are -1.96 and $+1.96$. The rejection rule can be stated as follows:

$$\text{Reject } H_0 \text{ if } z < -1.96 \text{ or } z > +1.96$$

Using (10.8), the value of z is

$$z = \frac{(\bar{x}_1 - \bar{x}_2) - (\mu_1 - \mu_2)}{\sqrt{\dfrac{\sigma_1^2}{n_1} + \dfrac{\sigma_2^2}{n_2}}} = \frac{(24.8 - 26.1) - 0}{\sqrt{\dfrac{(3.3)^2}{248} + \dfrac{(4.2)^2}{225}}}$$

$$= \frac{-1.3}{.35} = -3.71$$

With this value of z, we reject H_0 and conclude that there is a significant difference between the mean pain-relief times for the two medicines.

Using the p-value approach the table of areas in the standard normal probability distribution shows that $z = 3.71$ has an area less than .001 in the tail of the distribution. For the two-tailed hypothesis test, the p-value would be less than $.001 \times 2 = .002$. Thus using this p-value, we would reject H_0 at the .05 level of significance.

■

EXAMPLE 10.6

It has been suggested that college students learn more and obtain higher grades in small classes (40 students or less) when compared to large classes (150 students or more). To test this claim, a university assigned a professor to teach a small class and a large class of the same course. At the end of the course students from the two classes were given the same final exam. Final grade differences for the two classes would provide a basis for testing the difference between the small-class and large-class situations.

Letting μ_1 denote the mean exam score for the population of students taking a small class and μ_2 denote the mean exam score for the population of students taking a large class, the hypothesis test is as follows:

$$H_0: \mu_1 - \mu_2 \leq 0$$

$$H_a: \mu_1 - \mu_2 > 0$$

Rejecting H_0 will lead to the conclusion that the mean exam score is greater in the small class. However, if the test is unable to reject H_0 the conclusion will be that the small class does not show a statistically significant higher grade performance.

Viewing the students actually taking the courses as samples from the populations of students in small and large classes, the following data were obtained.

	INDIVIDUALS TAKING SMALL CLASS	INDIVIDUALS TAKING LARGE CLASS
Sample size	$n_1 = 35$	$n_2 = 170$
Sample mean exam score	$\bar{x}_1 = 74.2$	$\bar{x}_2 = 71.7$
Sample standard deviation	$s_1 = 14$	$s_2 = 13$

Using $\alpha = .05$, test the hypothesis and draw a conclusion about the mean exam scores for the small and large classes.

For the one-tailed test, the critical z value is 1.645. The corresponding rejection rule becomes

$$\text{Reject } H_0 \text{ if } z > +1.645$$

The point estimate of $\mu_1 - \mu_2$ is given by $\bar{x}_1 - \bar{x}_2 = 74.2 - 71.7 = 2.5$; thus the small class group has a sample mean that is 2.5 points greater than the sample mean for the large class. Using (10.8), the value of z becomes

$$z = \frac{(\bar{x}_1 - \bar{x}_2) - (\mu_1 - \mu_2)}{\sqrt{\dfrac{\sigma_1^2}{n_1} + \dfrac{\sigma_2^2}{n_2}}} = \frac{(2.5) - 0}{\sqrt{\dfrac{(14)^2}{35} + \dfrac{(13)^2}{170}}}$$

$$= \frac{2.5}{2.57} = .97$$

Since $z < 1.645$, we are unable to reject H_0. As a result, we are unable to conclude that small classes enable students to obtain statistically significant higher grades. With $z = .97$, the p-value for the test is $.5000 - .3340 = .1660$.

Our analysis of the data indicates H_0 could not be rejected. However, we must note that these results were based upon an experiment with only one professor. While the evidence is not sufficient to conclude that grade performance improves in smaller classes, further testing should be done before arriving at a final conclusion.

In the preceding examples we have shown how hypothesis tests can be conducted for situations involving the means of two populations using *large* samples. Similar hypothesis tests can be made for the small-sample case ($n_1 < 30$ and/or $n_2 < 30$) by assuming that the populations have normal probability distributions with equal variances. In such cases the t distribution is used; the test statistic t is given by

$$t = \frac{(\bar{x}_1 - \bar{x}_2) - (\mu_1 - \mu_2)}{\sqrt{s_p^2 \left(\dfrac{1}{n_1} + \dfrac{1}{n_2} \right)}} \tag{10.9}$$

The t distribution corresponding to (10.9) has $n_1 + n_2 - 2$ degrees of freedom. The denominator in (10.9) provides an estimate of the standard deviation of $\bar{x}_1 - \bar{x}_2$ based on the pooled variance estimate of σ^2 denoted by s_p^2, where

$$s_p^2 = \frac{(n_1 - 1)s_1^2 + (n_2 - 1)s_2^2}{n_1 + n_2 - 2}$$

EXAMPLE 10.7

Automobile gasoline mileage tests were conducted for similar-sized foreign and domestic automobiles. Test the hypothesis that the mean number of miles per gallon is the same for foreign and domestic automobiles based on the following sample results. Use $\alpha = .05$.

	FOREIGN AUTOMOBILES	DOMESTIC AUTOMOBILES
Sample size	$n_1 = 8$	$n_2 = 10$
Sample mean	$\bar{x}_1 = 36.5$	$\bar{x}_2 = 32.4$
Sample standard deviation	$s_1 = 2.3$	$s_2 = 2.8$

The null and alternative hypotheses for this test are

$$H_0: \mu_1 - \mu_2 = 0$$
$$H_a: \mu_1 - \mu_2 \neq 0$$

With $n_1 + n_2 - 2 = 8 + 10 - 2 = 16$ degrees of freedom, the table for the t distribution shows $t_{.025} = 2.12$. The rejection rule for the hypothesis test is

Reject H_0 if $t < -2.12$ or $t > 2.12$

The calculations made with the sample data are as follows:

$$s_p^2 = \frac{(n_1 - 1)s_1^2 + (n_2 - 1)s_2^2}{n_1 + n_2 - 2} = \frac{7(2.3)^2 + 9(2.8)^2}{8 + 10 - 2} = 6.72$$

Then using (10.9) we have

$$t = \frac{(36.5 - 32.4) - 0}{\sqrt{6.72\left(\frac{1}{8} + \frac{1}{10}\right)}} = \frac{4.1}{1.23} = 3.33$$

Since $3.33 > 2.12$, we reject H_0 and conclude that there is a significant difference between the mean number of miles per gallon achieved by foreign and domestic automobiles.

1. In some hypothesis tests, the concern is not the equality of the two population means, but instead how much the two population means differ. For example, in Example 10.6, suppose that because of the increased cost of small classes, the administration had decided to use large classes unless it can be concluded that the mean score for small classes (μ_1) is at least 5 points higher than the mean score for large classes (μ_2). In this case the relevant hypotheses would be

$$H_0: \quad \mu_1 - \mu_2 \leq 5$$
$$H_a: \quad \mu_1 - \mu_2 > 5$$

and the value of z would be

$$z = \frac{(\bar{x}_1 - \bar{x}_2) - 5}{\sqrt{\dfrac{\sigma_1^2}{n_1} + \dfrac{\sigma_2^2}{n_2}}}$$

In general, if we let D_0 denote the hypothesized difference between the population means, z can be written as follows:

$$z = \frac{(\bar{x}_1 - \bar{x}_2) - D_0}{\sqrt{\dfrac{\sigma_1^2}{n_1} + \dfrac{\sigma_2^2}{n_2}}}$$

2. In Section 8.3 we showed that confidence interval estimation was equivalent to a two-tailed hypothesis test about a population mean. The same conclusion is valid for conducting two-tailed hypothesis tests for two-population situations. For example, if we let D_0 denote the hypothesized difference between the means, the hypotheses for a two-tailed test can be written

$$H_0: \quad \mu_1 - \mu_2 = D_0$$
$$H_a: \quad \mu_1 - \mu_2 \neq D_0$$

To use a confidence interval approach to test these hypotheses we first use equation (10.3) to develop an interval estimate of the difference between the two population means. If the value of D_0 is contained within this interval, we cannot reject H_0; if the value of D_0 is *not* contained within the confidence interval, we can reject H_0 in favor of H_a.

EXERCISES

7. A sample of the weights of babies (in pounds) born in two different countries show the following:

COUNTRY 1	COUNTRY 2
$\bar{x}_1 = 7.1$	$\bar{x}_2 = 6.5$
$s_1 = .7$	$s_2 = .4$
$n_1 = 125$	$n_2 = 100$

Using a .05 level of significance, do these data support the conclusion that there is a difference in the weights of babies born in the two countries? What is the p-value?

8. In a wage discrimination case involving male and female employees, independent samples of male and female employees with 5 years or more experience show the following hourly wage results:

MALE EMPLOYEES	FEMALE EMPLOYEES
$n_1 = 44$	$n_2 = 32$
$\bar{x}_1 = \$6.25$	$\bar{x}_2 = \$5.70$
$s_1 = \$1.00$	$s_2 = \$.80$

The null hypothesis is stated such that male employees have a mean hourly wage less than or equal to that of the female employees. Rejection of H_0 leads to the conclusion that male employees have a mean hourly wage exceeding the female employee wages. Test the hypothesis using $\alpha = .01$. Does wage discrimination appear to exist in this case?

9. Safegate Foods, Inc. is redesigning the checkout lanes in its supermarkets throughout the country. Two designs have been suggested. Tests on customer checkout times (in minutes) have been collected at two stores where the two new systems have been installed. The sample data are as follows:

TIMES FOR CHECKOUT SYSTEM A	TIMES FOR CHECKOUT SYSTEM B
$n_1 = 120$	$n_2 = 100$
$\bar{x}_1 = 4.1$	$\bar{x}_2 = 3.3$
$s_1 = 2.2$	$s_2 = 1.5$

Test at the .05 level of significance to determine if there is a difference in the mean checkout times for the two systems. Which system is preferred? What is the p-value?

10. Samples of final examination scores for two statistics classes with different instructors showed the following results:

INSTRUCTOR A'S CLASS	INSTRUCTOR B'S CLASS
$n_1 = 12$	$n_2 = 15$
$\bar{x}_1 = 72$	$\bar{x}_2 = 78$
$s_1 = 8$	$s_2 = 10$

With $\alpha = .05$, use the p-value to test whether or not these data are sufficient to conclude that the mean grades differ for the two classes.

11. A firm is studying the delivery times (in days) for two raw material suppliers. The firm is basically satisfied with its current supplier, referred to as supplier A, and will stay with this supplier provided that the mean delivery times are the same as or less than those of supplier B. However, if the firm finds that the mean delivery times from supplier B are less than those of supplier A, it will begin making raw material purchases from supplier B.

a. What are the null and alternative hypotheses for this situation?

b. Assume that independent samples show the following delivery time characteristics for the two suppliers:

SUPPLIER A	SUPPLIER B
$n_1 = 50$	$n_2 = 30$
$\bar{x}_1 = 14$	$\bar{x}_2 = 12.5$
$s_1 = 3$	$s_2 = 2$

Show the sampling distribution of $\bar{x}_1 - \bar{x}_2$ for the hypothesis test.

c. For $\alpha = .05$, what is the rejection rule for the test?

d. What is your conclusion for the hypotheses from part (a)? What action do you recommend in terms of supplier selection?

10.3   Inferences About the Difference Between the Means of Two Populations—Matched Samples

Suppose that a manufacturing company has two methods available for employees to perform a certain production task. In order to maximize production output, the company would like to identify the method with the smallest mean completion time per unit. If a difference between mean completion times exists, the company will consider implementing the method with the smaller mean completion time. If no difference between means can be detected, the choice between the two production methods will be based on a criterion other than completion time. The hypotheses to be tested are stated as follows:

HYPOTHESIS	CONCLUSION
$H_0: \mu_1 - \mu_2 = 0$	Unable to conclude that a difference exists between the mean completion times for the two methods
$H_a: \mu_1 - \mu_2 \neq 0$	A difference exists between the mean completion times for the two methods

In designing the sampling procedure that will be used to collect production time data and test the above hypotheses, we consider two alternative designs. One is based on *independent samples,* and the other is based on *matched,* or *paired, samples*. The designs are described as follows:

Independent-sample design—A random sample of workers is selected; each worker in this sample uses method 1. A second *independent* random sample of workers is selected; each worker in this sample uses method 2. The test of the difference between means is based on the procedures of Section 10.2.

Matched-sample design—One random sample of workers is selected; each worker in the sample first uses one method and then uses the other method. The order of the two methods is assigned randomly to the workers, with some workers performing method 1 first and others performing method 2 first. Each worker provides a pair of data values, one value for method 1 and another value for method 2.

In the matched-sample design both production methods are tested under similar conditions (i.e., same workers); thus, this design often leads to a smaller sampling error than the independent sample design. The primary reason for this is that each worker in a matched sample design provides data first under one method and then under the other method. Thus variation between workers is eliminated as a source of the sampling error. This variation between workers cannot be eliminated when the independent-sample design is used.

EXAMPLE 10.8

A matched-sample design was used for two production methods. A sample of six workers was taken; each worker performed the task for both production methods. The data on completion times for the six workers are as follows.

WORKER	COMPLETION TIME, METHOD 1 (MINUTES)	COMPLETION TIME, METHOD 2 (MINUTES)	DIFFERENCE IN COMPLETION TIMES (d_i)
1	6.0	5.4	.6
2	5.0	5.2	−.2
3	7.0	6.5	.5
4	6.2	5.9	.3
5	6.0	6.0	.0
6	6.4	5.8	.6

Note that each worker provides a pair of data values, one for each production method. Also note that the last column contains the difference (d_i) in completion times for each worker in the sample. For example, $d_i = .6$ shows that worker 1 required .6 minutes more time for method 1. A negative d_i value indicates the worker required more time for method 2.

The key to analyzing a matched-sample design is to use the *difference data* only. Doing so converts the situation to a single sample containing six values. Then the procedures introduced in Chapter 8 can be used to make an inference about the mean for the population of all possible differences.

Let μ_d = the mean of the *difference* values for the population of workers. With this notation the null and alternative hypotheses are rewritten as follows:

HYPOTHESIS	CONCLUSION
H_0: $\mu_d = 0$	Unable to conclude that a difference exists between the mean completion times for the two methods
H_a: $\mu_d \neq 0$	A difference exists between the mean completion times

The sample mean and sample standard deviation for the six difference values are as follows:

$$\bar{d} = \frac{\Sigma d_i}{n} = \frac{[.6 + (-.2) + .5 + .3 + .0 + .6]}{6} = \frac{1.8}{6} = .3$$

$$s_d = \sqrt{\frac{\Sigma(d_i - \bar{d})^2}{n - 1}}$$

$$= \sqrt{\frac{(.3)^2 + (-.5)^2 + (.2)^2 + (0)^2 + (-.3)^2 + (.3)^2}{6 - 1}}$$

$$= \sqrt{\frac{.56}{5}} = .335$$

In Chapter 8, we found that if the population has a normal probability distribution and if s is used as an estimate of σ, the t distribution with $n - 1$ degrees of freedom can be used to test a hypothesis about the population mean. The test statistic t was given by

$$t = \frac{\bar{x} - \mu}{s/\sqrt{n}}$$

Assuming the population of difference values (d_i) has a normal probability distribution and using the sample standard deviation of differences, s_d, to estimate the population standard deviation of differences, this same formula can be used to test the hypothesis H_0: $\mu_d = 0$. To indicate that we are using the d_i values of the matched-sample design, we write the test statistic t as follows:

$$t = \frac{\bar{d} - \mu_d}{s_d/\sqrt{n}} \tag{10.10}$$

With a sample of six workers, we have $n - 1 = 5$ degrees of freedom. Using $\alpha = .05$, we find that $t_{.025} = 2.571$. The rejection rule is to reject H_0 if $t < -2.571$ or $t > +2.571$. Using the sample results, we have

$$t = \frac{\bar{d} - \mu_d}{s_d/\sqrt{n}} = \frac{.3 - 0}{.335/\sqrt{6}} = 2.19$$

Since t is between -2.571 and 2.571, H_0 cannot be rejected. The sample data do not provide sufficient evidence to reject the assumption of no difference between the mean completion times of the two methods.

Using the sample results, we could also develop an interval estimate of the difference between the means of the two populations using the methodology of interval estimation for one population, as introduced in Chapter 8. Using this approach, we have

$$\bar{d} \pm t_{.025} \frac{s_d}{\sqrt{n}} \tag{10.11}$$

or

$$.3 \pm 2.571 \frac{.335}{\sqrt{6}} = .3 \pm .35$$

Thus the 95% confidence interval estimate of the difference in the means of the two production methods is $-.05$ to $.65$ minutes.

◼

In Example 10.8, workers performed the production task using first one method and then the other method. This is an example of a matched-sample design, where each sampled item (worker) provides a pair of data values. Although this is often the procedure used in the matched-samples analysis, it is possible to use different but similar items to provide the pair of data values. In this sense, a worker at one location could be matched with a similar worker at another location (similarity based on age, education, sex, experience, etc.). The pairs of workers would provide the difference data that could be used in the matched-sample analysis.

Since a matched-sample procedure for inferences about two population means generally provides a better estimate than the two-independent-samples approach, it is the recommended design. However, in some applications the matching cannot be achieved, or the time and cost associated with matching is excessive. In these cases the independent-sample design should be used.

Example 10.8 employed a sample size of six workers. As such, the small-sample case existed, and the t distribution was used in both the test of hypothesis and interval estimation computations. If the sample size is large ($n \geq 30$), the statistical computations can be based on the z values of the standard normal probability distribution.

EXERCISES

12. A manufacturer produces both a deluxe and a standard model automatic sander designed for home use. Selling prices obtained from a sample of retail outlets are as follows.

RETAIL OUTLET	PRICE, DELUXE MODEL	PRICE, STANDARD MODEL
1	$39	$27
2	39	28
3	45	35
4	38	30
5	40	30
6	39	34
7	35	29

The manufacturer's suggested retail prices for the two models show a $10 difference in prices. Using a .05 level of significance, test to see if the mean difference between the prices of the two models is $10. What is the 95% confidence interval estimate of the difference between the mean prices for the two models?

13. Figure Perfect, Inc. is a women's figure salon that specializes in weight-reduction programs. Weights (pounds) for a sample of clients before and after a 6-week introductory program are as follows.

CLIENT	WEIGHT BEFORE	WEIGHT AFTER
1	140	132
2	160	158
3	210	195

CLIENT	WEIGHT BEFORE	WEIGHT AFTER
4	148	152
5	190	180
6	170	164

Using $\alpha = .05$, determine if the introductory program provides a weight loss. What is the p-value?

14. The pulse rates of patients before and after being given a certain tranquilizer are as follows.

BEFORE	AFTER
81	77
80	79
82	75
79	80
84	78
80	74

a. Using a .05 level of significance, test for the ability of the tranquilizer to reduce the pulse rate of the patients.
b. Provide a 95% confidence interval estimate of the mean decrease in pulse rate attributable to the tranquilizer.

15. Word processing systems are often justified on the basis of improved efficiencies for a secretarial staff. Given are typing rates in words per minute for seven secretaries who previously used electronic typewriters and who are now using computer-based word processors. Test at the .05 level of significance to see if there has been any change in the mean typing rate due to the word processor system.

SECRETARY	ELECTRONIC TYPEWRITER	WORD PROCESSOR
1	72	75
2	68	66
3	55	60
4	58	64
5	52	55
6	55	57
7	64	64

10.4

Inferences About the Difference Between the Proportions of Two Populations

We now consider the case where two populations are involved and we are interested in making inferences about the difference between the proportions of the two populations based upon two *independent* simple random samples. Let

p_1 = proportion for population 1

p_2 = proportion for population 2

$$\bar{p}_1 = \text{sample proportion for a simple random sample}$$
$$\text{selected from population 1 (i.e., the point estimator of } p_1)$$

$$\bar{p}_2 = \text{sample proportion for a simple random sample}$$
$$\text{selected from population 2 (i.e., the point estimator of } p_2)$$

The difference between the two population proportions is given by $p_1 - p_2$. The point estimator of this difference is:

$$\bar{p}_1 - \bar{p}_2 \qquad (10.12)$$

Thus we see that the estimator of the difference between two population proportions is the difference between the two sample proportions.

Sampling Distribution of $\bar{p}_1 - \bar{p}_2$

In the study of the difference between two population proportions, $\bar{p}_1 - \bar{p}_2$ is the sample statistic of interest. As we have seen in several previous cases, the sampling distribution of the sample statistic is a key factor in developing confidence interval estimates and in testing hypotheses about the parameters of interest. The properties of the sampling distribution of $\bar{p}_1 - \bar{p}_2$ are as follows:

Sampling Distribution of $\bar{p}_1 - \bar{p}_2$

Expected Value: $\quad E(\bar{p}_1 - \bar{p}_2) = p_1 - p_2 \qquad (10.13)$

Standard Deviation: $\quad \sigma_{\bar{p}_1 - \bar{p}_2} = \sqrt{\dfrac{p_1(1 - p_1)}{n_1} + \dfrac{p_2(1 - p_2)}{n_2}} \qquad (10.14)$

where

$\quad n_1 = \text{size of the simple random sample selected from population 1}$

$\quad n_2 = \text{size of the simple random sample selected from population 2}$

Distribution Form: Provided that the sample sizes are large (that is, $n_1 p_1$, $n_1(1 - p_1)$, $n_2 p_2$, and $n_2(1 - p_2)$ are all greater than or equal to 5), the sampling distribution of $\bar{p}_1 - \bar{p}_2$ can be approximated by a normal probability distribution.

Figure 10.3 shows the sampling distribution of $\bar{p}_1 - \bar{p}_2$.

FIGURE 10.3 Sampling Distribution of $\bar{p}_1 - \bar{p}_2$

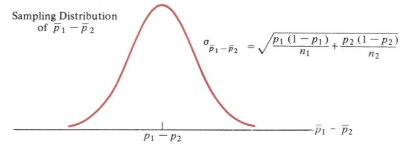

Sampling Distribution
of $\bar{p}_1 - \bar{p}_2$

$$\sigma_{\bar{p}_1 - \bar{p}_2} = \sqrt{\frac{p_1(1 - p_1)}{n_1} + \frac{p_2(1 - p_2)}{n_2}}$$

$p_1 - p_2$

$\bar{p}_1 - \bar{p}_2$

Note that the formula for the standard deviation of $\bar{p}_1 - \bar{p}_2$ as given by (10.14) requires that we know the actual values for the population proportions p_1 and p_2. Since p_1 and p_2 are unknown, we cannot use (10.14) to compute $\sigma_{\bar{p}_1 - \bar{p}_2}$. However, using $\bar{p}_1$ as the estimate of p_1 and $\bar{p}_2$ as the estimate of p_2, we can estimate $\sigma_{\bar{p}_1 - \bar{p}_2}$ as follows:

Estimate of $\sigma_{\bar{p}_1 - \bar{p}_2}$

$$s_{\bar{p}_1 - \bar{p}_2} = \sqrt{\frac{\bar{p}_1(1 - \bar{p}_1)}{n_1} + \frac{\bar{p}_2(1 - \bar{p}_2)}{n_2}} \qquad (10.15)$$

Interval Estimation of $p_1 - p_2$

With the sampling distribution of $\bar{p}_1 - \bar{p}_2$ known and with $s_{\bar{p}_1 - \bar{p}_2}$ providing an estimate of $\sigma_{\bar{p}_1 - \bar{p}_2}$, we follow the same procedure for interval estimation that we have used previously. We have the following expression for a confidence interval:

Interval Estimation of the Difference Between the Proportions of Two Populations—(Applies with $n_1 p_1$, $n_1(1 - p_1)$, $n_2 p_2$, $n_2(1 - p_2)$ all greater than or equal 5)

$$\bar{p}_1 - \bar{p}_2 \pm z_{\alpha/2} \sqrt{\frac{\bar{p}_1(1 - \bar{p}_1)}{n_1} + \frac{\bar{p}_2(1 - \bar{p}_2)}{n_2}} \qquad (10.16)$$

Note that as we have done with previous interval estimations, we can change the $z_{\alpha/2}$ value to reflect different confidence levels.

EXAMPLE 10.9

A firm that specializes in preparing tax returns for its clients is interested in comparing the quality of work at two of its regional offices. A random sample of tax returns prepared at each office is selected and verified for accuracy. Of concern to the firm is the proportion of erroneous returns prepared at each office. In particular, the firm would like an estimate of the difference between the proportions of erroneous returns prepared at the two offices.

Sample results show that of 250 returns verified at office 1, 35 were in error and that of 300 returns verified at office 2, 27 were in error. Develop 95% and 90% confidence interval estimates for the difference between the two population proportions.

The sample proportions at the two offices are as follows.

$$\bar{p}_1 = \frac{35}{250} = .14$$

$$\bar{p}_2 = \frac{27}{300} = .09$$

Using (10.12), the point estimate of the difference between the proportion of erroneous tax returns for the two populations is $\bar{p}_1 - \bar{p}_2 = .14 - .09 = .05$. Specifically, we are led to believe that office 1 possesses a 5% greater error rate than office 2. Using (10.16), the 95% confidence interval for the difference between the two population proportions becomes

$$\bar{p}_1 - \bar{p}_2 \pm z_{.025} \sqrt{\frac{\bar{p}_1(1 - \bar{p}_1)}{n_1} + \frac{\bar{p}_2(1 - \bar{p}_2)}{n_2}}$$

$$.14 - .09 \pm 1.96 \sqrt{\frac{.14(1 - .14)}{250} + \frac{.09(1 - .09)}{300}}$$

$$.05 \pm 1.96(.0275) = .05 \pm .0539$$

Thus the 95% confidence interval estimate of the difference in error rates at the two offices is $-.0039$ to $.1039$.

The 90% confidence interval uses $z_{.05} = 1.645$. Thus, replacing the 1.96 with 1.645 in (10.16) provides the 90% confidence interval estimate. This substitution will result in a confidence interval of $.05 \pm 1.645(.0275)$, or $.05 \pm .0452$. Thus we can be 90% confident that the difference in the error rates at the two offices is in the interval $.0048$ to $.0952$.

■

EXAMPLE 10.10

In a study of household television-viewing habits, individuals in sampled households were asked to participate by keeping a daily diary of television viewing. The letters requesting participation in the study were sent to two groups. The first group received a letter requesting participation and $1.00 as a token of appreciation for their participation. The second group received only the letter requesting participation in the study. Of 300 letters sent to the first group, 141 households participated in the study. Of 500 letters sent to the second group, 150 households participated in the study. Provide a 95% confidence interval estimate for the difference in participation rate that exists if $1.00 is included with the request for participation.

Using the sample results, we have

$$\bar{p}_1 = \frac{141}{300} = .47$$

$$\bar{p}_2 = \frac{150}{500} = .30$$

Using $z_{.025} = 1.96$ and (10.16) provides the confidence interval estimate of

$$.47 - .30 \pm 1.96 \sqrt{\frac{.47(1 - .47)}{300} + \frac{.30(1 - .30)}{500}} = .17 \pm 1.96(.0354)$$

$$= .17 \pm .0694$$

Thus, we can be 95% confident that including $1.00 with the letter increases the participation rate between .1006 and .2394 or approximately be 10% and 24%.

■

Hypothesis Tests About $p_1 - p_2$

Let us now consider statistical inferences involving hypothesis tests about the difference between the proportions of two populations. As with other hypothesis-testing situations, we use the standardized normal random variable z to compare the value of the sample statistic with the hypothesized value for the population parameter. For hypothesis tests concerning the proportions of two populations, z is as follows:

$$z = \frac{(\bar{p}_1 - \bar{p}_2) - (p_1 - p_2)}{\sigma_{\bar{p}_1 - \bar{p}_2}} \tag{10.17}$$

In this expression, $\sigma_{\bar{p}_1 - \bar{p}_2}$ is the standard deviation of the $\bar{p}_1 - \bar{p}_2$ statistic and is defined in (10.14).

We may be tempted to use (10.15) to compute $s_{\bar{p}_1 - \bar{p}_2}$ and use this value as an estimate of $\sigma_{\bar{p}_1 - \bar{p}_2}$. However, whenever the null hypothesis for a hypothesis test about the difference between two population proportions is $H_0: p_1 - p_2 = 0$, the hypothesis-testing procedure requires us to assume H_0 is true and $p_1 - p_2 = 0$. Whenever we make this assumption there is no need to use the individual values of $\bar{p}$ and $\bar{p}_2$ to estimate $\sigma_{\bar{p}_1 - \bar{p}_2}$. Using $p_1 = p_2 = p$, (10.14) can be rewritten

$$\sigma_{\bar{p}_1 - \bar{p}_2} = \sqrt{\frac{p(1 - p)}{n_1} + \frac{p(1 - p)}{n_2}} = \sqrt{p(1 - p)\left(\frac{1}{n_1} + \frac{1}{n_2}\right)} \tag{10.18}$$

With (10.18), we see that we need an estimate of p in order to estimate $\sigma_{\bar{p}_1 - \bar{p}_2}$. The estimate of p is provided by

$$\bar{p} = \frac{n_1 \bar{p}_1 + n_2 \bar{p}_2}{n_1 + n_2} \tag{10.19}$$

In effect, (10.19) provides a combined, or pooled estimate of the population proportion p. The following expression for $s_{\bar{p}_1 - \bar{p}_2}$ can then be used to estimate $\sigma_{\bar{p}_1 - \bar{p}_2}$ in (10.17).

Estimate of $\sigma_{\bar{p}_1 - \bar{p}_2}$ when $H_0: p_1 - p_2 = 0$

$$s_{\bar{p}_1 - \bar{p}_2} = \sqrt{\bar{p}(1 - \bar{p})\left(\frac{1}{n_1} + \frac{1}{n_2}\right)} \tag{10.20}$$

EXAMPLE 10.11

A sample of driving records over a 2-year period show that 16 of 400 adult drivers had received traffic citations, whereas 24 of 300 teen-age drivers had received traffic citations. Test the hypothesis that there is no difference between the traffic citation rate for adult and teenage drivers.

Letting p_1 = the population proportion of adult drivers with citations and p_2 = the population proportion of teenage drivers with citations, we want to test the hypotheses

$$H_0: p_1 - p_2 = 0$$
$$H_a: p_1 - p_2 \neq 0$$

With $\alpha = .05$, we reject H_0 if $z < -1.96$ or $z > 1.96$.

Sample results show the following values for the sample proportions:

$$\text{Adults:} \quad \bar{p}_1 = \frac{16}{400} = .04$$

$$\text{Teenagers:} \quad \bar{p}_2 = \frac{24}{300} = .08$$

Using (10.19), we have

$$\bar{p} = \frac{n_1\bar{p}_1 + n_2\bar{p}_2}{n_1 + n_2} = \frac{400(.04) + 300(.08)}{400 + 300} = .0571$$

as the pooled estimate of p. The value of z is then computed as follows:

$$z = \frac{(\bar{p}_1 - \bar{p}_2) - (p_1 - p_2)}{\sqrt{\bar{p}(1 - \bar{p})\left(\frac{1}{n_1} + \frac{1}{n_2}\right)}} = \frac{(.04 - .08) - 0}{\sqrt{.0571(1 - .0571)\left(\frac{1}{400} + \frac{1}{300}\right)}}$$

$$= \frac{-.04}{.0177} = -2.26$$

With this value for z, we reject H_0 and conclude that there is a significant difference between the traffic-citation rates for adult and teenage drivers.

EXAMPLE 10.12

In the validation of examination questions used in a physics course at a large university, an instructor would like to compare the proportions of A students and F students answering the questions correctly. A particular examination question is judged to be a good discriminator if the proportion of A students (p_1) answering the question correctly is greater than the proportion of F students (p_2) answering the question correctly. That is, rejecting H_0 in the following hypothesis test will support the conclusion that the examination question is a good discriminator between A and F students.

$$H_0: p_1 - p_2 \leq 0$$
$$H_a: p_1 - p_2 > 0$$

Using $\alpha = .05$, H_0 will be rejected if $z > 1.645$.

At the end of the term, sample results showed that 85 of the 110 students who received an A in the course had answered the examination question correctly, whereas 50 of the 95 students who received an F in the course answered the examination question correctly. The computations required to conduct the test of hypothesis are as follows:

$$\text{A Students:} \quad \bar{p}_1 = \frac{85}{110} = .7727$$

$$\text{F Students:} \quad \bar{p}_2 = \frac{50}{95} = .5263$$

$$\text{Overall:} \quad \bar{p} = \frac{110(.7727) + 95(.5263)}{110 + 95} = .6585$$

$$z = \frac{(\bar{p}_1 - \bar{p}_2) - (p_1 - p_2)}{\sqrt{\bar{p}(1 - \bar{p})\left(\dfrac{1}{n_1} + \dfrac{1}{n_2}\right)}} = \frac{(.7727 - .5263) - 0}{\sqrt{.6585(1 - .6585)\left(\dfrac{1}{110} + \dfrac{1}{95}\right)}}$$

$$= \frac{-.2464}{.0664} = 3.71$$

Thus we reject H_0 and conclude that A students do significantly better than F students on the examination question.

Alternatively, we can test these hypotheses using the p-value approach. First, to compute the p-value for this one-sided hypothesis test, we must determine the probability of getting a value for z that is greater than 3.71. Since this probability is approximately 0 and clearly less than $\alpha = .05$, the null hypothesis can be rejected. Thus, we see that the use of p-values in testing hypotheses regarding the difference in population proportions is analogous to the one-population case.

■

EXERCISES

16. A sample of 400 items produced by supplier A contained 70 defective items. A sample of 300 items produced by supplier B contained 40 defective items. Compute a 90% confidence interval estimate of the difference between the proportion defective for the two suppliers.

17. During the primary elections a particular presidential candidate showed the following pre-election voter support in Wisconsin and Illinois:

STATE	VOTERS SURVEYED	VOTERS FAVORING THE CANDIDATE
Wisconsin	500	270
Illinois	360	162

Compute a 95% confidence interval estimate for the difference between the proportion of voters favoring the candidate in the two states.

18. In a study of the eating habits of teenagers, it was found that more girls than boys did not eat breakfast on a regular basis. Provide a 90% confidence interval estimate of the difference between the proportion of girls and proportion of boys that do not eat breakfast on a regular basis if sample data show that 85 of 210 girls and 48 of 200 boys do not eat breakfast on a regular basis.

19. In a study of coffee-drinking habits, 50 of 240 men and 55 of 180 women expressed a preference for decaffeinated coffee. Using a .05 level of significance, test for a difference between the proportion of men and the proportion of women who prefer decaffeinated coffee. What is your conclusion? What is the p-value?

20. A survey firm conducts door-to-door surveys on a variety of issues. Some individuals cooperate with the interviewer and complete the interview questionnaire, while others do not. The following sample data are available (showing the response data for men and women).

	SAMPLE SIZE	NUMBER COMPLETING THE SURVEY
Men	200	110
Women	300	210

   a. Use the p-value and $\alpha = .05$ to test the null hypothesis that the response rate is the same for both men and women.

   b. Compute the 95% confidence interval estimate for the difference between the proportion of men and the proportion of women that cooperate with the survey.

21. In a test of the quality of two television commercials, each commercial was shown in a separate test area six times over a one-week period. The following week a telephone survey was conducted to identify individuals who had seen the commercials. The individuals who had seen the commercials were asked to state the primary message in the commercial. The following results were recorded.

	NUMBER REPORTING HAVING SEEN THE COMMERCIAL	NUMBER RECALLING PRIMARY MESSAGE
Commercial A	150	63
Commercial B	200	60

   a. Using $\alpha = .05$, test the claim that there is no difference in the recall proportions for the two commercials.

   b. Compute a 95% confidence interval estimate the difference between the recall proportions for the two populations.

22. A political opinion survey shows that of 200 Republicans surveyed, 80 opposed the building of power plants using fission processes (nuclear energy). Similar results for a sample of 300 Democrats showed that 150 opposed building nuclear power plants. Do these results indicate that there is a significant difference in the attitudes of Republicans and Democrats on this issue? Use $\alpha = .05$.

23. In the "Statistics in the News" article at the beginning of this chapter, 27 of the 60 subjects classified as highly dependent on nicotine were given gum containing 4 mg of nicotine and 33 were given gum containing 2 mg of nicotine. Let p_1 denote the percentage of smokers who received the 4-mg gum and remained abstinent and p_2 denote the percentage of smokers who received the 2-mg gum and remained abstinent. Using the data provided, test the following hypotheses with $\alpha = .10$.

$$H_0: \quad p_1 - p_2 = 0$$

$$H_a: \quad p_1 - p_2 \neq 0$$

a. What conclusion can be reached at then end of 6 weeks?
b. What conclusion can be reached at the end of 1 year?
c. What conclusion can be reached at the end of 2 years?
d. Overall, what conclusion can be reached regarding the effectiveness of nicotine chewing gum?

SUMMARY

In this chapter we have discussed procedures for interval estimation and hypothesis testing involving two populations. Specifically, we showed how to make inferences about the difference between the means of two populations when independent simple random samples are selected. Two cases were considered. In the first case, where the sample sizes were large, the z values from the standard normal probability distribution are used for inferences about the difference between two population means. In the second case, where the populations are normally distributed with equal variances the t distribution is used. This case also is used when the sample sizes are small.

Inferences about the difference between the means of two populations were discussed for the matched-sample design. In the matched-sample design each data value from one sample is matched with a data value in the other sample. The difference in the pair of data values is then used in the statistical analysis. The matched-sample design is generally preferred over the independent-sample design because the matched-sample procedure tends to reduce the sampling error and thus improves the precision of the estimate.

Finally, interval estimation and hypothesis testing involving the difference between two population proportions were discussed. Statistical procedures for analyzing the difference between two population proportions are similar to the procedures for analyzing the difference between two population means.

GLOSSARY

Pooled variance estimate   An estimate of the variance of a population based on the combination of two (or more) samples. The pooled variance estimate is appropriate whenever the variances of two (or more) populations are assumed equal.

Independent samples   Samples selected from two (or more) populations where the elements making up one sample are chosen independently of the elements making up the other sample(s).

Matched samples   Samples where each data value is a difference between matched or paired observations.

KEY FORMULAS

EXPECTED VALUE OF $\bar{x}_1 - \bar{x}_2$

$$E(\bar{x}_1 - \bar{x}_2) = \mu_1 - \mu_2 \tag{10.1}$$

STANDARD DEVIATION OF $\bar{x}_1 - \bar{x}_2$

$$\sigma_{\bar{x}_1-\bar{x}_2} = \sqrt{\frac{\sigma_1^2}{n_1} + \frac{\sigma_2^2}{n_2}}$$

(10.2)

INTERVAL ESTIMATION OF THE DIFFERENCE BETWEEN THE MEANS OF TWO POPULATIONS (LARGE SAMPLE CASE WITH $n_1 \geq 30$ AND $n_2 \geq 30$)

$$\bar{x}_1 - \bar{x}_2 \pm z_{\alpha/2} \sqrt{\frac{\sigma_1^2}{n_1} + \frac{\sigma_2^2}{n_2}}$$

(10.3)

POOLED VARIANCE ESTIMATE

$$s_p^2 = \frac{(n_1 - 1)s_1^2 + (n_1 - 1)s_2^2}{(n_1 + n_2 - 2)}$$

(10.5)

INTERVAL ESTIMATION OF THE DIFFERENCE BETWEEN THE MEANS OF TWO POPULATIONS (SMALL SAMPLE CASE WITH $n_1 < 30$ AND/OR $n_2 < 30$)

$$\bar{x}_1 - \bar{x}_2 \pm t_{\alpha/2} \sqrt{s_p^2 \left(\frac{1}{n_1} + \frac{1}{n_2}\right)}$$

(10.7)

STANDARD RANDOM VARIABLE z

$$z = \frac{(\bar{x}_1 - \bar{x}_2) - (\mu_1 - \mu_2)}{\sqrt{\frac{\sigma_1^2}{n_1} + \frac{\sigma_2^2}{n_2}}}$$

(10.8)

RANDOM VARIABLE t

$$t = \frac{(\bar{x}_1 - \bar{x}_2) - (\mu_1 - \mu_2)}{\sqrt{s_p^2\left(\frac{1}{n_1} + \frac{1}{n_2}\right)}}$$

(10.9)

RANDOM VARIABLE t FOR MATCHED SAMPLES

$$t = \frac{\bar{d} - \mu_d}{s_d/\sqrt{n}}$$

(10.10)

EXPECTED VALUE OF $\bar{p}_1 - \bar{p}_2$

$$E(\bar{p}_1 - \bar{p}_2) = p_1 - p_2$$

(10.13)

STANDARD DEVIATION OF $\bar{p}_1 - \bar{p}_2$

$$\sigma_{\bar{p}_1-\bar{p}_2} = \sqrt{\frac{p_1(1 - p_1)}{n_1} + \frac{p_2(1 - p_2)}{n_2}}$$

(10.14)

INTERVAL ESTIMATION OF THE DIFFERENCE BETWEEN THE PROPORTIONS OF TWO POPULATIONS

$$\bar{p}_1 - \bar{p}_2 \pm z_{\alpha/2} \sqrt{\frac{\bar{p}_1(1 - \bar{p}_1)}{n_1} + \frac{\bar{p}_2(1 - \bar{p}_2)}{n_2}}$$

(10.16)

ESTIMATE OF $\sigma_{\bar{p}_1 - \bar{p}_2}$ WHEN H_0: $p_1 - p_2 = 0$

$$\bar{p} = \frac{n_1\bar{p}_1 + n_2\bar{p}_2}{n_1 + n_2} \tag{10.19}$$

and

$$s_{\bar{p}_1 - \bar{p}_2} = \sqrt{\bar{p}(1 - \bar{p})\left(\frac{1}{n_1} + \frac{1}{n_2}\right)} \tag{10.20}$$

REVIEW QUIZ

TRUE/FALSE

1. A point estimator of the difference between two population means is the corresponding difference between the two sample means.

2. The central limit theorem cannot be applied to the sampling distribution of $\bar{x}_1 - \bar{x}_2$.

3. The pooled estimate of σ^2 is a weighted average of the two sample variances.

4. The only assumption necessary in using the t distribution for interval estimation of the difference between population means is that the variances of the two populations are equal.

5. The normal probability distribution cannot be used for large-sample hypothesis tests concerning the difference between population means.

6. If sampling n_1 items from one population and n_2 items from a second population, the large-sample case is applicable as long as $n_1 + n_2 \geq 60$.

7. When the matched-sample approach is used for inferences about the difference between population means each item in the sample provides two data values.

8. The advantage of the matched-sample design is that it allows the experimenter to control some factors in order to obtain a sharper measure of others.

9. The sampling distribution of $\bar{p}_1 - \bar{p}_2$ can be approximated by a normal distribution, provided that one of the sample sizes is large.

10. When conducting the hypothesis test H_0: $p_1 - p_2 = 0$, a pooled estimate of the population proportion is computed from the two sample proportions.

MULTIPLE CHOICE

11. Independent samples are obtained from two normal populations with equal variances in order to construct a confidence interval estimate for the difference between the population means. If the first sample contains 16 items and the second sample contain 21 items, the correct form to use for the sampling distribution is:

a. normal distribution
b. t distribution with 15 degrees of freedom
c. t distribution with 37 degrees of freedom
d. t distribution with 35 degrees of freedom

12. The t distribution can be used in the estimation of $\mu_1 - \mu_2$.
   a. if σ_1^2 and σ_2^2 are equal
   b. if either $n_1 \geq 30$ or $n_2 \geq 30$
   c. always
   d. only when both populations are normal

Use the following for questions 13–16. A testing company is checking to see if there is any significant difference in the coverage of two different brands of paint for a hardware store chain. The results are summarized below.

	AMAZON PAINT	COVERUP PAINT
Mean coverage (in square feet)	305	295
Standard deviation	20	25
Sample size	31	41

13. A point estimate of the difference between the population mean is
   a. −5
   b. 0
   c. 5
   d. 10

14. A point estimate of the standard deviation of the difference between the sample means is
   a. −5
   b. 5.3
   c. 28.1
   d. 32

15. The form of the sampling distribution of the difference between the sample means is the
   a. normal distribution, approximately
   b. t distribution with 30 degrees of freedom
   c. t distribution with 35 degrees of freedom
   d. t distribution with 40 degrees of freedom

16. If a two-tailed test is used with a .05 level of significance, the critical z values are
   a. −1.96 and +1.96
   b. −1.645 and +1.645
   c. −10.4 and +10.4
   d. none of the above

SUPPLEMENTARY EXERCISES

24. Samples of dinner and luncheon receipts at a major downtown restaurant show the following results.

DINNER RECEIPTS	LUNCHEON RECEIPTS
$n_1 = 70$	$n_2 = 55$
$\bar{x}_1 = \$32.65$	$\bar{x}_2 = \$12.80$
$s_1 = \$7.20$	$s_2 = \$3.60$

Provide 90% and 98% confidence interval estimates of the difference between the mean receipt amounts for the two meals.

25. Sociologist John P. Robinson surveyed 2500 men and 2500 women in order to study how they handle household chores such as cooking, laundering, housecleaning, and so on. (*American Demographics,* December 1988). Robinson found that men averaged 9.8 hours of housework weekly, while women averaged 19.5 hours. Assume that the sample standard deviations were 2.8 hours for the men and 3.4 hours for the women. Develop a 95% confidence interval for the difference between the two population means. What conclusions can be reached?

26. In a study of job attitudes and job satisfaction, a sample of 50 men and 50 women were asked to rate their overall job satisfaction on a 1 to 10 scale. A high rating indicates a higher degree of job satisfaction. Using the sample results shown, does there appear to be a significant difference in the level of job satisfaction of men and women? Use $\alpha = .05$. What is the *p*-value?

MEN	WOMEN
$\bar{x}_1 = 7.2$	$\bar{x}_2 = 6.8$
$s_1^2 = 2.8$	$s_2^2 = 1.8$

27. A production line is designed on the assumption that the difference in mean assembly times for two operations is 5 minutes. Independent tests for the two assembly operations show the following results.

OPERATION A	OPERATION B
$n_1 = 100$	$n_2 = 50$
$\bar{x}_1 = 14.8$ minues	$\bar{x}_2 = 10.4$ minutes
$s_1 = .8$ minutes	$s_2 = .6$ minutes

Using $\alpha = .02$, test the hypothesis that the difference between the mean assembly times is 5 minutes.

28. A realtor is interested in estimating the difference between the mean selling prices of new homes in two sections of the city. Assume that the standard deviations of the selling prices are approximately $12,000 for both areas. How large a sample should be taken in each area to have a 95% confidence that the sampling error for the difference between mean prices will be $5000 or less? Use the same sample size for both sections of the city.

29. Production quantities for two assembly-line workers are shown. Each data value indicates the amount produced during a randomly selected 1-hour period.

WORKER 1	WORKER 2
20	22
18	18
21	20
22	23
20	24

a. Develop a point estimate of the difference between the mean hourly production rates of the two workers. Which worker appears to have the higher mean production rate?

b. Develop a 90% confidence interval estimate for the difference between the mean production rates of the two workers. Consider the confidence interval estimate. Does the result provide conclusive evidence that the worker having the higher sample mean production rate is actually the worker with the overall higher production rate? Explain.

30. Salary surveys of chemistry and physics majors show the following starting annual salary data:

CHEMISTRY MAJORS	PHYSICS MAJORS
$n_1 = 14$	$n_2 = 16$
$\bar{x}_1 = \$19,800$	$\bar{x}_2 = \$19,300$
$s_1 = \$1,000$	$s_2 = \$1,400$

Consider the test of the hypothesis that the mean annual salaries are the same for both majors.

a. What assumptions must be made in order to test the hypothesis?

b. Assume that these assumptions are appropriate. What is the pooled estimate of the population variance?

c. Using $\alpha = .05$, can you conclude that a difference exists in the mean annual salary for the two majors?

d. What is the p-value?

31. A market research firm used a sample of individuals to rate their potential to purchase a particular product before and after they saw a new television commercial about the product. The potential-to-purchase ratings were based on a 0 to 10 scale, with higher values indicating a higher potential-to-purchase. The null hypothesis stated that the mean rating after seeing the commercial would be less than or equal to the mean rating before. Rejection of this hyothesis would provide the conclusion that the commercial improved the mean potential-to-purchase rating. Using $\alpha = .05$ and the following data, test the hypotheses. Comment on the value of the commercial.

INDIVIDUAL	PURCHASE RATING BEFORE SEEING COMMERCIAL	PURCHASE RATING AFTER SEEING COMMERCIAL
1	5	6
2	4	6
3	7	7
4	3	4
5	5	3
6	8	9
7	5	7
8	6	6

32. A company wants to evaluate the potential for a new bonus plan by having a random sample of five salespersons use the bonus plan for a trial period. The weekly sales volumes (units) before and after implementing the bonus plan are shown below:

SALESPERSON	WEEKLY SALES BEFORE	WEEKLY SALES AFTER
1	15	18
2	12	14
3	18	19
4	15	18
5	16	18

a. Using $\alpha = .05$, test to see if the bonus plan will result in an increase in the mean weekly sales.

b. Provide a 90% confidence interval estimate for the mean *increase* in weekly sales that can be expected if a new bonus plan is implemented.

33. A cable television firm is considering submitting bids for rights to operate in two regions of the state of Florida. Surveys of the two regions provide the following data on customer acceptance of the cable television service.

REGION I	REGION II
$n_1 = 500$	$n_2 = 800$
Number indicating likely	Number indicating likely
to purchase $= 175$	to purchase $= 360$

Develop a 99% confidence interval estimate of the difference between the population proportions of likely-to-purchase customers in the two regions.

34. When Leo J. Shapiro & Associates, a Chicago market-research firm, surveyed consumers about their family's holiday spending plans, 36% said they planned to spend less this year (*Wall Street Journal*, October 17, 1988). In contrast, when consumers were asked this question last year, 28% said they planned to spend less. If 400 shoppers were sampled each year, develop a 95% confidence interval estimate of the difference between these two percentages. What value would the results of such a study be to retailers?

35. A large automobile insurance company selected samples of single and married male policyholders and recorded the number who had made an insurance claim over the past 3-year period:

SINGLE POLICYHOLDERS	MARRIED POLICYHOLDERS
$n_1 = 400$	$n_2 = 900$
Number making claims $= 76$	Number making claims $= 90$

a. Using $\alpha = .05$, determine if the claim rates differ between single and married male policyholders. What is the *p*-value?

b. Provide a 95% confidence interval estimate of the difference between the claim proportions for the two populations.

36. In studying food service and tourism in Hawaii, researchers asked tourists from the United States and Canada if they prepared some of their own meals during their stay in Hawaii. (*Journal of Foodservice Systems*, 1988). Of the 332 tourists from the United States, 149 indicated they prepared some of their own meals; in contrast, 89 of 160 tourists from Canada said they prepared some of their own meals. Test whether or not there is any difference between the population proportions for U.S. and Canadian tourists. Use $\alpha = .10$.

37. Two loan officers at the North Ridge National Bank show the following data for defaults on loans that they have approved (the data are based on samples of loans granted over the past 5 years).

	NUMBER OF LOANS REVIEWED	NUMBER OF DEFAULTED LOANS
Loan officer A	60	9
Loan officer B	80	6

Using $\alpha = .05$, test the hypothesis that the default rates are the same for the two loan officers. What is the *p*-value?

38. In the "Statistics in the News" article, 60 of the 113 subjects classified as having medium or low dependence on nicotine were given gum containing 2 mg of nicotine and 53 were given a placebo gum. Let p_1 denote the percentage of smokers who received the 2-mg gum and remained abstinent and p_2 denote the percentage of smokers who received the placebo gum and remained abstinent. Using $\alpha = .10$ and the data provided, test the following hypotheses:

$$H_0: \quad p_1 - p_2 = 0$$

$$H_a: \quad p_1 - p_2 \neq 0$$

a. What conclusion can be reached at the end of 6 weeks?
b. What conclusion can be reached at the end of 1 year?
c. What conclusion can be reached at the end of 2 years?
d. Overall, what conclusion can be reached regarding the effectiveness of nicotine chewing gum for medium- or low-dependence smokers?

COMPUTER EXERCISE

Par, Inc. is a major manufacturer of golf equipment. The research group at Par has been investigating a new golf ball designed to resist cuts and yet still offer good driving distances. In tests with the new golf balls, 40 balls of the new model and 40 balls of the current model were subjected to distance tests. The testing was performed with a mechanical hitting machine in ideal weather conditions, so that if a difference existed between the mean distance of the two models it could be attributed to a real difference in design performance. The results of the test are shown. The distance data are measured to the nearest yard.

CURRENT MODEL	NEW MODEL	CURRENT MODEL	NEW MODEL
242	274	248	269
239	266	265	256
245	260	267	261
250	263	258	277
236	259	250	271
261	248	253	278
236	259	243	273
244	286	238	266
237	283	256	265
248	261	253	259
241	271	259	280
242	263	252	247
262	259	251	250
241	268	241	257
238	257	253	267
261	278	245	260
233	247	257	258
250	260	252	252
244	275	254	260
246	261	240	276

QUESTIONS

1. Develop numerical and tabular measures to summarize the given data.
2. Develop confidence interval estimates for the mean distance traveled for both types of balls.
3. What statistical conclusion can you reach regarding the mean distances for the two models? What are your recommendations?

11 Inferences About Population Variances

What You Will Learn in This Chapter:

■ Situations where inferences about population variances are needed

■ How the chi-square distribution is used to make an inference about a population variance

■ How the F distribution is used for hypothesis tests concerning the equality of two population variances

Contents

Product Quality: The Challenge from Japan

The Japanese emphasis on quality has pushed many Japanese products to the top of the competitive ladder. In recent years, U.S. companies have been striving to improve product quality and recapture their market leadership. In many cases, the quality of U.S. products is still lagging behind Japanese products, and indications are that U.S. companies still have a long way to go to meet the Japanese challenge. One wonders how the Japanese obtained the knowledge that enabled them to leap past the United States in product quality.

In tracing the history of the Japanese emphasis on quality, many point to the year 1950, when an American statistician by the name of W. Edwards Deming began to talk with Japanese managers about how to build quality products. Deming's efforts were part of Japan's postwar recovery program. Eventually his contributions to Japanese product quality and productivity resulted in the Emperor giving him a medal and naming a prestigious prize after him.

The heart of Deming's method for achieving high quality is statistical in nature. Every process, whether it be on the factory floor or in the office, has variations from the ideal. These variations can be measured statistically with variances and standard deviations. Deming shows clients systematic methods for identifying the causes of variations and then reducing them. The result is a steady improvement in the production process and, ultimately, product quality.

Thirty years after the Japanese started listening to Deming, American companies began to listen.

Adding the grille and making final engine adjustments, these workers send a Cadillac on its way at the General Motors Orion assembly line.

Firms such as Ford and A.T.&T. pay consulting fees as high as $5000 per day for Deming's keys to product quality. Deming's message emphasizes improving the product by reducing the variance. Ford has adopted Deming's statistical process controls throughout the company.

At the age of 83, Deming still feels compelled to deliver his message wherever possible. He believes the United States is still falling behind Japan, and—as a result—the American standard of living is getting lower. Deming estimates that it will take another 30 years for the United States to catch up with Japan.

Based on the "The Curmudgeon Who Talks Tough on Quality," *Fortune* (June 25, 1984).

In the previous four chapters we have discussed methods of statistical inference involving means and proportions. In this chapter we continue the discussion of statistical inference by considering methods for making inferences about population variances and standard deviations. As an example of where variance is important, consider the process of filling containers with a liquid detergent. The filling mechanism is adjusted so that the mean filling weight is 16 ounces per container. Although a mean of 16 ounces is desired, the variance of the filling weights is also critical. That is, even with the filling mechanism properly adjusted for the mean of 16 ounces, we cannot expect every container to have exactly 16 ounces. By selecting a sample of containers, we can compute a sample variance for the filling weights. This value serves as an estimate of the variance for the population of containers being filled by the production process. If the sample variance is modest, the production process is continued. However, if the sample variance is excessive, overfilling and underfilling can create problems even though the mean filling weight is correct (16 ounces). In this case, the filling mechanism must be readjusted to reduce the filling variance for the containers.

In the following section we consider methods for making inferences about the variance of a single population. Later we discuss procedures for making inferences about the variances of two populations. Since the standard deviation is the square root of the variance, the methods introduced in this chapter can also be used to make inferences about population standard deviations.

11.1 Inferences About a Population Variance

Consider a population with an unknown variance σ^2. The sample variance s^2 will be used to make inferences about the population variance σ^2. The formula for computing a sample variance is restated.

Sample Variance

$$s^2 = \frac{\Sigma(x_i - \bar{x})^2}{n - 1} \tag{11.1}$$

where

$$\bar{x} = \text{sample mean}$$
$$n = \text{sample size}$$

The process of using s^2 to make an inference about σ^2 is shown in Figure 11.1.

FIGURE 11.1   The Statistical Process of Using a Sample Variance to Make Inferences About a Population Variance

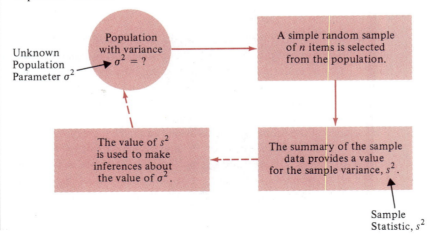

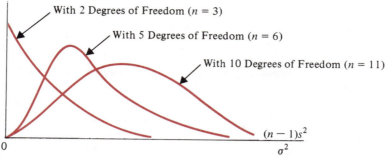

FIGURE 11.2 Examples of the Sampling Distribution of $(n - 1)s^2/\sigma^2$ with 2, 5, and 10 Degrees of Freedom

The Sampling Distribution of $(n - 1)s^2/\sigma^2$

In previous chapters we showed that knowledge of a sampling distribution is essential for computing interval estimates and conducting hypothesis tests about a population mean or a population proportion. Thus it should not be surprising that in order to make inferences about a population variance, we again work with a sampling distribution. Specifically, it can be shown that for normally distributed populations, the sampling distribution of $(n - 1)s^2/\sigma^2$ is a special probability distribution referred to as a *chi-square distribution*. Since tables of probabilities are available for the chi-square distribution, it is relatively easy to use the chi-square distribution to make interval estimates and test hypotheses about the value of a population variance.

The sampling distribution of $(n - 1)s^2/\sigma^2$ is described as follows:

Sampling Distribution of $(n - 1)s^2/\sigma^2$

Whenever a simple random sample of size n is selected from a *normally distributed population*,

$$\frac{(n - 1)s^2}{\sigma^2} \tag{11.2}$$

has a *chi-square distribution* with $n - 1$ degrees of freedom, where s^2 is the sample variance and σ^2 is the population variance.

Typical graphs of the sampling distributions of $(n - 1)s^2/\sigma^2$ are shown in Figure 11.2.

The Chi-Square Distribution

Like the t distribution, the chi-square distribution is a family of similar probability distributions. Each specific chi-square distribution depends upon its *degrees of freedom* parameter. That is, there is a chi-square distribution with 1 degree of freedom, a chi-square distribution with 2 degrees of freedom and so on. Figure 11.2 is thus a graph of the chi-square distributions for 2, 5, and 10 degrees of freedom. In using a sample of size n to make inferences about a population variance, we find that the appropriate chi-square distribution has $n - 1$ degrees of freedom.

To see how tables of chi-square values can be used to make an inference about a population variance, consider a situation where the sample size is 20. The degrees of freedom for this case are $n - 1 = 20 - 1 = 19$. A graph of the chi-square distribution with 19 degrees of freedom is shown in Figure 11.3. Using the symbol χ^2 to refer to the chi-square value, note that the distribution shows that 95% of the possible χ^2 values are between 8.90655 and 32.8523. That is, when a random sample is selected from a normal population, there is a .95 probability that the chi-square value (i.e., $\chi^2 = (n - 1)s^2/\sigma^2$) will be between 8.90655 and 32.8523.

Table 11.1 contains a table of values for selected chi-square distributions. In referring to the chi-square distribution table, we use a subscript on χ^2 to denote the area or probability under the curve to the *right* of the stated χ^2 value. For example, the are under the curve to the right of $\chi^2_{.025}$ is .025. Thus $\chi^2_{.025}$ corresponds to a chi-square value in the upper tail of the distribution. Similarly, $\chi^2_{.975}$ (97.5% of the chi-square values are to the right of $\chi^2_{.975}$) corresponds to a chi-square value in the lower tail of the distribution.

Now let us use the chi-square distribution information in Table 11.1 to show how the values of χ^2 in Figure 11.3 were obtained. Since the chi-square distribution in Figure 11.3 is based on 19 degrees of freedom, we see that $\chi^2_{.975} = 8.90655$ and $\chi^2_{.025} = 32.8523$ are found in row 19 and the .975 and .025 columns of Table 11.1 Thus there is a .95 probability that a chi-square value will be between $\chi^2_{.975}$ and $\chi^2_{.025}$. This statement holds for all chi-square distributions. However, the numerical values of $\chi^2_{.975}$ and $\chi^2_{.025}$ change depending upon the number of degrees of freedom. If we had wanted an interval containing 90% of the chi-square values, we could have used the interval from $\chi^2_{.95}$ to $\chi^2_{.05}$. With 19 degrees of freedom, we see from Table 11.1 that this interval is from 10.1170 to 30.1435.

EXAMPLE 11.1

Consider a chi-square distribution with 15 degrees of freedom. Find the value of χ^2 that provides an area of .01 in the upper tail of the distribution.

Referring to Table 11.1, we find that with 15 degrees of freedom $\chi^2_{.01} = 30.5779$ has an area of .01 in the upper tail.

■

EXAMPLE 11.2

Consider a chi-square distribution with 9 degrees of freedom. Find the value of χ^2 that provides an area of .05 in the lower tail of the distribution.

Referring to Table 11.1, we find that with 9 degrees of freedom $\chi^2_{.95} = 3.32511$ has an area of .05 in the lower tail of the distribution.

■

FIGURE 11.3   A Chi-Square Distribution with 19 Degrees of Freedom

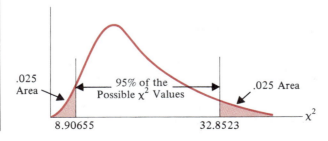

TABLE 11.1

Chi-Square Distribution Table

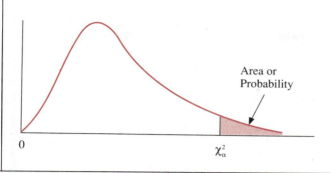

Degrees of Freedom	AREAS TO RIGHT OF χ_α^2					
	.99	.975	.95	.05	.025	.01
1	$157,088 \times 10^{-9}$	$982,069 \times 10^{-9}$	$393,214 \times 10^{-8}$	3.84146	5.02389	6.63490
2	.0201007	.0506356	.102587	5.99147	7.37776	9.21034
3	.114832	.215795	.351846	7.81473	9.34840	11.3449
4	.297110	.484419	.710721	9.48773	11.1433	13.2767
5	.554300	.831211	1.145476	11.0705	12.8325	15.0863
6	.872085	1.237347	1.63539	12.5916	14.4494	16.8119
7	1.239043	1.68987	2.16735	14.0671	16.0128	18.4753
8	1.646482	2.17973	2.73264	15.5073	17.5346	20.0902
9	2.087912	2.70039	3.32511	16.9190	19.0228	21.6660
10	2.55821	3.24697	3.94030	18.3070	20.4831	23.2093
11	3.05347	3.81575	4.57481	19.6751	21.9200	24.7250
12	3.57056	4.40379	5.22603	21.0261	23.3367	26.2170
13	4.10691	5.00874	5.89186	22.3621	24.7356	27.6883
14	4.66043	5.62872	6.57063	23.6848	26.1190	29.1413
15	5.22935	6.26214	7.26094	24.9958	27.4884	30.5779
16	5.81221	6.90766	7.96164	26.2962	28.8454	31.9999
17	6.40776	7.56418	8.67176	27.5871	30.1910	33.4087
18	7.01491	8.23075	9.39046	28.8693	31.5264	34.8053
19	7.63273	8.90655	10.1170	30.1435	32.8523	36.1908
20	8.26040	9.59083	10.8508	31.4104	34.1696	37.5662
21	8.89720	10.28293	11.5913	32.6705	35.4789	38.9321
22	9.54249	10.9823	12.3380	33.9244	36.7807	40.2894
23	10.19567	11.6885	13.0905	35.1725	38.0757	41.6384
24	10.8564	12.4011	13.8484	36.4151	39.3641	42.9798
25	11.5240	13.1197	14.6114	37.6525	40.6465	44.3141
26	12.1981	13.8439	15.3791	38.8852	41.9232	45.6417
27	12.8786	14.5733	16.1513	40.1133	43.1944	46.9630
28	13.5648	15.3079	16.9279	41.3372	44.4607	48.2782
29	14.2565	16.0471	17.7083	42.5569	45.7222	49.5879
30	14.9535	16.7908	18.4926	43.7729	46.9792	50.8922
40	22.1643	24.4331	26.5093	55.7585	59.3417	63.6907
50	29.7067	32.3574	34.7642	67.5048	71.4202	76.1539
60	37.4848	40.4817	43.1879	79.0819	83.2976	88.3794
70	45.4418	48.7576	51.7393	90.5312	95.0231	100.425
80	53.5400	57.1532	60.3915	101.879	106.629	112.329
90	61.7541	65.6466	69.1260	113.145	118.136	124.116
100	70.0648	74.2219	77.9295	124.342	129.561	135.807

Additional values of chi-square can be found in Table 3 of Appendix B.

EXAMPLE 11.3

Consider a chi-square distribution with 50 degrees of freedom. Find values of χ^2 that provide an area of .01 in each tail of the distribution.

Referring to Table 11.1, we find that with 50 degrees of freedom $\chi^2_{.99} = 29.7067$ provides an area of .01 in the lower tail and $\chi^2_{.01} = 76.1539$ has an area of .01 in the upper tail of the distribution.

■

Interval Estimation of σ^2

Based upon the previous discussion of the chi-square distribution, we can conclude that there is a $1 - \alpha$ probability that χ^2 will be between $\chi^2_{(1-\alpha/2)}$ and $\chi^2_{\alpha/2}$. That is, there is a $1 - \alpha$ probability $\chi^2_{(1-\alpha/2)} \leq \chi^2 \leq \chi^2_{\alpha/2}$. From (11.2) we also know the quantity $(n - 1)s^2/\sigma^2$ follows the chi-square distribution. Therefore, there must be a $1 - \alpha$ probability that

$$\chi^2_{(1-\alpha/2)} \leq \frac{(n - 1)s^2}{\sigma^2} \leq \chi^2_{\alpha/2} \tag{11.3}$$

Working with the right-hand side of (11.3), we have

$$\frac{(n - 1)s^2}{\sigma^2} \leq \chi^2_{\alpha/2} \tag{11.4}$$

After taking a sample, the sample size n and the sample variance will be known. In addition, the value of $\chi^2_{\alpha/2}$ can be found by using the chi-square distribution table. Thus the population variance σ^2 is the only unknown in (11.4). Multiplying (11.4) by σ^2 provides the following inequality:

$$(n - 1)s^2 \leq \sigma^2 \chi^2_{\alpha/2}$$

Dividing both sides of this inequality by $\chi^2_{\alpha/2}$, we obtain (11.5):

$$\frac{(n - 1)s^2}{\chi^2_{\alpha/2}} \leq \sigma^2 \tag{11.5}$$

With n, s^2, and $\chi^2_{\alpha/2}$ known, we can use (11.5) to compute a lower limit for the value of the population variance σ^2. Using a similar approach with the left-hand inequality in (11.3), we obtain an upper limit for the value of the σ^2 expressed in terms of n, s^2, and $\chi^2_{(1-\alpha/2)}$.

Based on these results the procedure shown in (11.6) can be used to find a confidence interval estimate of a population variance. By altering the values of α in (11.6), we can obtain the desired level of confidence. For example, $\chi^2_{.025}$ and $\chi^2_{.975}$ will provide a 95% confidence interval, $\chi^2_{.05}$ and $\chi^2_{.95}$ will provide a 90% confidence interval, and $\chi^2_{.005}$ and $\chi^2_{.995}$ will provide a 99% confidence interval.

Interval Estimation for a Population Variance

$$\frac{(n-1)s^2}{\chi^2_{\alpha/2}} \le \sigma^2 \le \frac{(n-1)s^2}{\chi^2_{(1-\alpha/2)}}$$ (11.6)

where the values of χ^2 are based on the chi-square distribution with $n-1$ degrees of freedom and where $(1-\alpha)$ is the confidence coefficient.

EXAMPLE 11.4

A production process is designed to fill 16-ounce containers with liquid detergent. The production manager is concerned about the variance in filling weights. A sample of 20 containers provides a sample variance of $s^2 = .0025$. Develop a 95% confidence interval for the population variance of the filling weights.

With $n-1 = 19$ degrees of freedom, we have $\chi^2_{.025} = 32.8523$ and $\chi^2_{.975} = 8.90655$. Using (11.6), the 95% confidence interval becomes

$$\frac{(20-1)(.0025)}{32.8523} \le \sigma^2 \le \frac{(20-1)(.0025)}{8.90655}$$

or

$$.001446 \le \sigma^2 \le .00533$$

By taking the square root of these terms, we can also find the 95% confidence interval for the population standard deviation σ:

$$.038 \le \sigma \le .073$$

Thus, we are 95% confident the population variance is between .001446 and .00533 and the population standard deviation is between .038 and .073.

■

EXAMPLE 11.5

Twenty-eight children were given a language test with the scores recorded on a scale of 0 to 100. The variance in the test scores serves as a measure of the homogeneity in language skills for the children. The sample standard deviation of the test scores was found to be $s = 12$. Provide a 90% confidence interval estimate of the variance and standard deviation of the test scores for the population of children.

With $n-1 = 27$ degrees of freedom, Table 11.1 shows that $\chi^2_{.95} = 16.1513$ and $\chi^2_{.05} = 40.1133$. Using the sample variance of $s^2 = (12)^2 = 144$ in (11.6), we have

$$\frac{(28-1)(144)}{40.1133} \le \sigma^2 \le \frac{(28-1)(144)}{16.1513}$$

or

$$96.93 \le \sigma^2 \le 240.72$$

This shows that a 90% confidence interval estimate of the population variance is 96.93 to 240.72. Taking the square root of the above values provides a 90% confidence interval estimate of the population standard deviation (9.85 to 15.52).

■

Hypothesis Tests About σ^2

Hypothesis tests about the value of a population variance are based on the value of $(n - 1)s^2/\sigma^2$ and the chi-square distribution with $n - 1$ degrees of freedom. The value of σ^2 specified in the null hypothesis is used in the denominator of $(n - 1)s^2/\sigma^2$. The rejection rule is based on the value of χ^2 in a manner similar to the use of z and t values in previous hypothesis-testing applications.

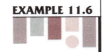

EXAMPLE 11.6

The St. Louis Metro Bus Company has recently made a concerted effort to improve reliability by encouraging its drivers to maintain consistent schedules. As a standard policy, the company expects arrival times at a bus stop to have low variability. Specifically, the company desires an arrival time standard deviation of 2 minutes or less, which indicates a variance of 4 or less. A sample of 10 arrival times shows a sample variance of 5. Using a .05 level of significance, should the company reject the hypothesis that the arrival time variance for the population is less than or equal to the allowable variance of 4?

The hypotheses for the study are

$$H_0: \sigma^2 \leq 4$$
$$H_a: \sigma^2 > 4$$

Rejection of H_0 will imply the company is not meeting the variance guideline.

With $\alpha = .05$, the one-tailed test decision rule is based on the upper-tail chi-square value of $\chi^2_{.05}$. With $n - 1 = 9$, Table 11.1 shows $\chi^2_{.05} = 16.919$. Therefore, if $(n - 1)s^2/\sigma^2$ is greater than 16.919, we will reject H_0. The chi-square distribution with the appropriate rejection region is shown in Figure 11.4.

The value of χ^2 is computed as follows:

$$\chi^2 = \frac{(n - 1)s^2}{\sigma^2} = \frac{(10 - 1)(5)}{4} = 11.25$$

FIGURE 11.4 Rejection Region for the St. Louis Metro Bus Example

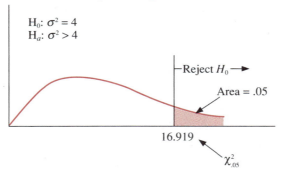

Comparing 11.25 to the $\chi^2_{.05} = 16.919$ value, we see that the null hypothesis cannot be rejected. Thus the sample of 10 bus arrivals provides insufficient evidence to conclude that the bus arrival time variance is not meeting the company standard.

■

EXAMPLE 11.7

Historically, the variance in test scores for individuals applying for drivers' licenses has been $\sigma^2 = 100$. A new examination has been designed. However, motor vehicle administrators believe that it is desirable for the variance in the test scores to remain at the historical level. A sample of 30 individuals are given the new version of the driver's examination; the sample variance is 64. Is there reason to believe the variance of the test scores has changed? Use $\alpha = .05$.

The hypothesis test is

$$H_0: \sigma^2 = 100$$

$$H_a: \sigma^2 \neq 100$$

The two-tailed test with $\alpha = .05$ requires the use of $\chi^2_{.975}$ and $\chi^2_{.025}$. With $n - 1 = 29$ degrees of freedom, Table 11.1 shows $\chi^2_{.975} = 16.0471$ and $\chi^2_{.025} = 45.7222$. Thus H_0 can be rejected if the value of χ^2 is less than 16.0471 or greater than 45.7222.

with $\sigma^2 = 100$, the value of χ^2 is

$$\chi^2 = \frac{(n - 1)s^2}{\sigma^2} = \frac{29(64)}{100} = 18.56$$

With this value of χ^2, we cannot reject H_0.

■

Notes and Comments

1. Although it can be shown that the sample variance s^2 is an unbiased estimator of σ^2, the standard deviation s is not an unbiased estimator of σ. However, for large samples the bias is small, and hence we use s to estimate σ.

2. If the population is not normal but the sample size is large ($n \geq 30$), the sampling distribution of s can be approximated by a normal distribution with mean σ and standard deviation $\sigma/\sqrt{2n}$. In such cases, the $(1 - \alpha)\%$ confidence interval for σ can be applied:

$$\frac{s}{1 + \dfrac{z_{\alpha/2}}{\sqrt{2n}}} \leq \sigma \leq \frac{s}{1 - \dfrac{z_{\alpha/2}}{\sqrt{2n}}} \tag{11.7}$$

1. The scores for a biology examination are normally distributed with a population standard deviation of $\sigma = 10$. A sample of 12 examinations will be selected and the sample standard deviation computed. With different samples possible, it is expected that the sample standard deviation will vary based on the sample selected.

    a. What is the sampling distribution of the quantity $(n - 1)s^2/\sigma^2$? Show a graph of this sampling distribution

    b. Using Table 11.1, what is the probability that the sample results will result in a value for $(n - 1)s^2/\sigma^2$ that is greater than or equal to 4.57481? Greater than or equal to 19.6751?

    c. Use the results from (b) to show that there is a .90 probability that the sample standard deviation s will be between 6.45 and 13.37.

2. For 20 randomly selected days, the standard deviation of the number of inmates in a county jail was 5.2. Place a 95% confidence interval on the population standard deviation of number of inmates in the jail.

3. The variance in unit weights is very critical in the pharmaceutical industry. For a specific drug, with unit weights measured in grams, a sample of 18 units provided a sample variance of $s^2 = .36$.

    a. Construct a 90% confidence interval estimate for the population variance for the weights of this drug.

    b. Construct a 90% confidence interval estimate for the population standard deviation.

4. As part of an 18-year follow-up study of the academic and psychosocial characteristics of low-birthweight adolescents, the reading proficiency of 24 children with low birthweights was measured. (*Social Biology*, 1988). The mean score for this group was 29.94, with a sample standard deviation of 9.94. Assume that reading proficiency scores follow a normal probability distribution.

    a. Develop a 95% confidence interval estimate of σ^2, the population variance of reading proficiency scores.

    b. Develop a 95% confidence interval estimate of σ.

5. A sample of cans of soups produced by Carle Foods resulted in the following weights, measured in ounces.

12.2	11.9	12.0	12.2
11.7	11.6	11.9	12.0
12.1	12.3	11.8	11.9

    Provide 95% confidence interval estimates for both the variance and the standard deviation of the population.

6. A certain part must be machined to very close tolerances. Production specifications call for a maximum standard deviation in the length of the part of .02 inches. The variance for a random sample of 30 parts was $s^2 = .0005$. Using $\alpha = .05$, test to see if the production specifications are being violated.

7. City Trucking, Inc. claims consistent delivery times for its routine customer deliveries. A random sample of 22 truck deliveries resulted in a sample variance of 1.5. Test to determine if the company is justified in claiming that the standard deviation in its delivery times is 1 hour or less. Use $\alpha = .10$.

8. The variance in the filling amounts for cups of soft drink from an automatic drink machine is an important consideration to the owner of the soft-drink service. If the variance is too large, overfilling and underfilling of cups will cause customer dissatisfaction with the service. An acceptable variance in filling amounts is $\sigma^2 = .25$ when filling amounts are measured in ounces. In a test of filling amounts for a particular machine, a sample of 18 cups resulted in a sample variance of .40.

a. Use a .05 level of significance. Do the sample results indicate that the filling mechanism should be replaced due to a large variance in filling amounts?

b. Provide a 90% confidence interval for the variance in the filling amounts for this machine.

9. Based upon a sample of 9 days selected at random over the past 6 months, a dentist has seen the following number of patients; 22, 25, 20, 18, 15, 22, 24, 19, and 26. Assume that the number of patients seen per day is normally distributed. Would analysis of this sample data support the claim that the variance in the number of patients per day is 10? Use a .10 level of significance. What is your conclusion?

11.2   Inferences About the Variances of Two Populations

In some statistical applications it is desirable to compare the variances of two populations. For instance, we might want to compare the variability in product quality resulting from two different production processes, the variability in assembly times for two assembly methods, or the variability in temperatures for two heating devices. In addition, recall that in Chapter 10 we developed a pooled variance estimate based on the assumption that the two populations had equal variances. Thus we might want to compare the variances of two populations to determine if the equal variance assumption, and thus pooling, can be justified.

The Sampling Distribution of s_1^2/s_2^2

In making comparisons about the variances of two normally distributed populations, we use data collected from two independent random samples, one from population 1 and another from population 2. The sample variances, s_1^2 and s_2^2 serve as the estimates of the corresponding population variances, σ_1^2 and σ_2^2. The statistic of interest is the ratio of the two sample variances, s_1^2/s_2^2. If the two populations involved both have normal probability distributions and have equal variances, the ratio s_1^2/s_2^2 has a special probability distribution known as an *F distribution*. Tables of areas or probabilities are available for this distribution.

The sampling distribution of s_1^2/s_2^2 is described as follows:

Sampling Distribution of s_1^2/s_2^2

Whenever independent simple random samples of sizes n_1 and n_2 are selected from normally distributed populations with equal variances, the ratio

$$F = \frac{s_1^2}{s_2^2} \tag{11.8}$$

has an F distribution with $n_1 - 1$ degrees of freedom for the numerator and $n_2 - 1$ degrees of freedom for the denominator, where s_1^2 is the sample variance for the random sample of n_1 items from population 1 and s_2^2 is the sample variance for the random sample of n_2 items from population 2.

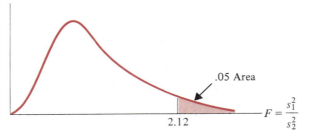

FIGURE 11.5 F Distribution with 20 Degrees of Freedom for the Numerator and 20 Degrees of Freedom for the Denominator

A graph of the sampling distribution of s_1^2/s_2^2 with 20 degrees of freedom associated with s_1^2 and 20 degrees of freedom associated with s_2^2 is shown in Figure 11.5. This distribution, which is an F distribution, shows what happens to the ratio s_1^2/s_2^2 as random samples of size $n_1 = 21$ and $n_2 = 21$ are taken from two normal populations with equal variances.

The F Distribution

Each specific F distribution depends upon the number of degrees of freedom associated with the numerator and the number of degrees of freedom associated with the denominator. We use F_α to denote the F value that provides an area of α in the upper tail of the F distribution. For instance, $F_{.05}$ provides an area in the upper tail of .05. Similarly, $F_{(1-\alpha)}$ provides an area of α in the lower tail of the F distribution. Thus $F_{.95}$ results in an area of .05 in the lower tail of the F distribution. Table 11.2 contains a table of $F_{.05}$ values for various numerator and denominator degrees of freedom. A more complete table for the F distribution is provided in Table 4 of Appendix B.

EXAMPLE 11.8

Suppose that a random sample of size 21 is selected from population 1 and a random sample of size 21 is selected from population 2. Assume both populations have normal distributions. Find $F_{.05}$ for the F distribution.

Referring to Table 11.2, we see that with 20 degrees of freedom in the numerator and 20 degrees of freedom for the denominator, $F_{.05} = 2.12$. This value is shown in Figure 11.5.

■

The tables for the F distribution in Appendix B provide F values with areas of .05, .025, and .01 in the upper tail of the distribution. However, F values that provide these same areas in the lower tail of the F distribution can be easily computed from the inverse relationship shown at the top of the next page.

EXAMPLE 11.9

Consider a case in which a random sample of size 25 has been taken from population 1 and a random sample of size 10 has been taken from population 2. Assume both populations are normally distributed. Find $F_{.05}$ and $F_{.95}$.

Refer to Table 11.2 in the column corresponding to 24 degrees of freedom and the row corresponding to 9 degrees of freedom. We find $F_{.05} = 2.90$.

$$F_{(1-\alpha)} = \frac{1}{F_\alpha} \qquad\qquad (11.9)$$

where $F_{(1-\alpha)}$ is from an F distribution with v_1 degrees of freedom in the numerator and v_2 degrees of freedom in the denominator and F_α is from an F distribution with v_2 degrees of freedom in the *numerator* and v_1 degrees of freedom in the *denominator*.

To find $F_{.95}$ we must employ the inverse relationship. From (11.9) we have

$$F_{.95} = \frac{1}{F_{.05}}$$

where the numerator and denominator degrees of freedom for $F_{.05}$ are reversed. Using Table 11.2, we find that for 9 numerator degrees of freedom and 24 denominator degrees of fredom, $F_{.05} = 2.30$. Therefore

$$F_{.95} = \frac{1}{F_{.05}} = \frac{1}{2.30} = .435$$

Here $F_{.95}$ has 24 numerator and 9 denominator degrees of freedom, respectively, since a sample of size 25 was taken from population 1 and a sample of size 10 was taken from population 2.

■

Hypothesis Tests About σ_1^2 and σ_2^2

Let us now see how the F distribution can be used for hypothesis tests concerning the variances of two normally distributed populations. Hypothesis tests about the variances of two populations are based on the value of s_1^2/s_2^2. The rejection rule is based on the F value in a manner similar to how z, t, and χ^2 values have been used in previous hypothesis-testing applications.

EXAMPLE 11.10

Dullus County Schools is renewing its school bus service contract for the coming year and must select one of the two bus companies, the Milbank Company or the Gulf Park Company. We will use the variance of pickup and delivery time as a primary measure of the quality of the bus service. Low variance values will indicate the more consistent and higher-quality service. If the population variances for the two services are the same, Dullus School administrators will select the company offering the better financial terms. However, if sample data on bus pickup and delivery times for the two companies indicate that a significant difference exists between the variances, the administrators may want to give special consideration to the company with the better or lower-variance service. The appropriate hypotheses and their associated conclusions and actions are as follows:

TABLE 11.2 Values of $F_{.05}$ for the F Distribution

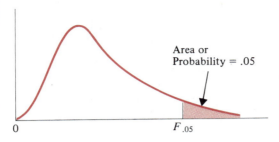

Area or
Probability = .05

0   $F_{.05}$

DENOMINATOR DEGREES OF FREEDOM	NUMERATOR DEGREES OF FREEDOM							
	1	2	3	4	5	6	7	8
1	161.4	199.5	215.7	224.6	230.2	234.0	236.8	238.9
2	18.51	19.00	19.16	19.25	19.30	19.33	19.35	19.37
3	10.13	9.55	9.28	9.12	9.01	8.94	8.89	8.85
4	7.71	6.94	6.59	6.39	6.26	6.16	6.09	6.04
5	6.61	5.79	5.41	5.19	5.05	4.95	4.88	4.82
6	5.99	5.14	4.76	4.53	4.39	4.28	4.21	4.15
7	5.59	4.74	4.35	4.12	3.97	3.87	3.79	3.73
8	5.32	4.46	4.07	3.84	3.69	3.58	3.50	3.44
9	5.12	4.26	3.86	3.63	3.48	3.37	3.29	3.23
10	4.96	4.10	3.71	3.48	3.33	3.22	3.14	3.07
11	4.84	3.98	3.59	3.36	3.20	3.09	3.01	2.95
12	4.75	3.89	3.49	3.26	3.11	3.00	2.91	2.85
13	4.67	3.81	3.41	3.18	3.03	2.92	2.83	2.77
14	4.60	3.74	3.34	3.11	2.96	2.85	2.76	2.70
15	4.54	3.68	3.29	3.06	2.90	2.79	2.71	2.64
16	4.49	3.63	3.24	3.01	2.85	2.74	2.66	2.59
17	4.45	3.59	3.20	2.96	2.81	2.70	2.61	2.55
18	4.41	3.55	3.16	2.93	2.77	2.66	2.58	2.51
19	4.38	3.52	3.13	2.90	2.74	2.63	2.54	2.48
20	4.35	3.49	3.10	2.87	2.71	2.60	2.51	2.45
21	4.32	3.47	3.07	2.84	2.68	2.57	2.49	2.42
22	4.30	3.44	3.05	2.82	2.66	2.55	2.46	2.40
23	4.28	3.42	3.03	2.80	2.64	2.53	2.44	2.37
24	4.26	3.40	3.01	2.78	2.62	2.51	2.42	2.36
25	4.24	3.39	2.99	2.76	2.60	2.49	2.40	2.34
26	4.23	3.37	2.98	2.74	2.59	2.47	2.39	2.32
27	4.21	3.35	2.96	2.73	2.57	2.46	2.37	2.31
28	4.20	3.34	2.95	2.71	2.56	2.45	2.36	2.29
29	4.18	3.33	2.93	2.70	2.55	2.43	2.35	2.28
30	4.17	3.32	2.92	2.69	2.53	2.42	2.33	2.27
40	4.08	3.23	2.84	2.61	2.45	2.34	2.25	2.18
60	4.00	3.15	2.76	2.53	2.37	2.25	2.17	2.10
120	3.92	3.07	2.68	2.45	2.29	2.17	2.09	2.02
∞	3.84	3.00	2.60	2.37	2.21	2.10	2.01	1.94

Entries in the table give $F_{.05}$ values, where .05 is the area or probability in the upper tail of the F distribution. For example, with 12 numerator degrees of freedom and 15 denominator degrees of freedom, $F_{.05} = 2.48$. Additional values for the F distribution can be found in Table 4 of Appendix B.

TABLE 11.2 (continued)

DENOMINATOR DEGREES OF FREEDOM	NUMERATOR DEGREES OF FREEDOM							
	1	2	3	4	5	6	7	8
1	240.5	241.9	243.9	245.9	248.0	249.1	250.1	251.1
2	19.38	19.40	19.41	19.43	19.45	19.45	19.46	19.47
3	8.81	8.79	8.74	8.70	8.66	8.64	8.62	8.59
4	6.00	5.96	5.91	5.86	5.80	5.77	5.75	5.72
5	4.77	4.74	4.68	4.62	4.56	4.53	4.50	4.46
6	4.10	4.06	4.00	3.94	3.87	3.84	3.81	3.77
7	3.68	3.64	3.57	3.51	3.44	3.41	3.38	3.34
8	3.39	3.35	3.28	3.22	3.15	3.12	3.08	3.04
9	3.18	3.14	3.07	3.01	2.94	2.90	2.86	2.83
10	3.02	2.98	2.91	2.85	2.77	2.74	2.70	2.66
11	2.90	2.85	2.79	2.72	2.65	2.61	2.57	2.53
12	2.80	2.75	2.69	2.62	2.54	2.51	2.47	2.43
13	2.71	2.67	2.60	2.53	2.46	2.42	2.38	2.34
14	2.65	2.60	2.53	2.46	2.39	2.35	2.31	2.27
15	2.59	2.54	2.48	2.40	2.33	2.29	2.25	2.20
16	2.54	2.49	2.42	2.35	2.28	2.24	2.19	2.15
17	2.49	2.45	2.38	2.31	2.23	2.19	2.15	2.10
18	2.46	2.41	2.34	2.27	2.19	2.15	2.11	2.06
19	2.42	2.38	2.31	2.23	2.16	2.11	2.07	2.03
20	2.39	2.35	2.28	2.20	2.12	2.08	2.04	1.99
21	2.37	2.32	2.25	2.18	2.10	2.05	2.01	1.96
22	2.34	2.30	2.23	2.15	2.07	2.03	1.98	1.94
23	2.32	2.27	2.20	2.13	2.05	2.01	1.96	1.91
24	2.30	2.25	2.18	2.11	2.03	1.98	1.94	1.89
25	2.28	2.24	2.16	2.09	2.01	1.96	1.92	1.87
26	2.27	2.22	2.15	2.07	1.99	1.95	1.90	1.85
27	2.25	2.20	2.13	2.06	1.97	1.93	1.88	1.84
28	2.24	2.19	2.12	2.04	1.96	1.91	1.87	1.82
29	2.22	2.18	2.10	2.03	1.94	1.90	1.85	1.81
30	2.21	2.16	2.09	2.01	1.93	1.89	1.84	1.79
40	2.12	2.08	2.00	1.92	1.84	1.79	1.74	1.69
60	2.04	1.99	1.92	1.84	1.75	1.70	1.65	1.59
120	1.96	1.91	1.83	1.75	1.66	1.61	1.55	1.50
∞	1.88	1.83	1.75	1.67	1.57	1.52	1.46	1.39

HYPOTHESIS	CONCLUSION AND ACTION
H_0: $\sigma_1^2 = \sigma_2^2$	No significant difference in quality of service; base service selection decision on financial terms
H_a: $\sigma_1^2 \neq \sigma_2^2$	Unequal quality of service; give special consideration to the low-variance service

A sample of 25 arrival times is available for the Milbank service (population 1) and a sample of 16 arrival times is available for the Gulf Park service (population 2). The F distribution with $n_1 - 1 = 24$ degrees of freedom and $n_2 - 1 = 15$ degrees of freedom is shown in Figure 11.6. Note that for $\alpha = .10$, the two-tailed rejection regions are indicated by the critical values at $F_{.95}$ and $F_{.05}$.

Using Table 11.2, we find that with 24 degrees of freedom in the numerator and 15 degrees of freedom in the denominator, $F_{.05} = 2.29$. While the $F_{.95}$ value is not available in this table, it can be found from the inverse relationship of 11.9. With 15 and 24 numerator and denominator degrees of freedom, respectively, Table 11.2 shows that $F_{.05} = 2.11$. Substituting into (11.9), we find that $F_{.95}$ for 24 and 15 numerator and denominator degrees of freedom, is

$$F_{.95} = \frac{1}{2.11} = .474$$

Thus the rejection rule for the hypothesis test is

Reject H_0 if $F < .474$ or if $F > 2.29$

Suppose that the samples of pickup and delivery times show a sample variance of $s_1^2 = 48$ for Milbank and a sample variance of $s_2^2 = 20$ for Gulf Park. The corresponding F value is

$$F = \frac{s_1^2}{s_2^2} = \frac{48}{20} = 2.4$$

Since $F = 2.40 > 2.29$, we reject H_0. The bus services appear to differ in terms of pickup and delivery-time variances. The recommendation is that the Dullus school ad-

FIGURE 11.6 F Distribution with 24 Degrees of Freedom for the Numerator and 15 Degrees of Freedom for the Denominator

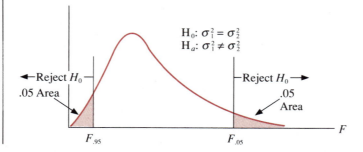

ministrators give special consideration to the better, or lower-variance, service offered by the Gulf Park Company.

■

One-tailed tests involving two population variances are also possible. Again the F distribution is used, with the one-tailed rejection region enabling us to concluded whether or not one population variance is significantly greater or significantly less than the other. Only upper-tail F values are needed. For any one-tailed test we set up the null hypothesis so that the rejection region is in the upper tail. This can be accomplished by a judicious choice of which population is labeled population 1. That is, the sample with the largest variance is treated as if it was selected from population 1. For example, one-tailed hypothesis tests about two population variances should be stated in the following form

$$H_0: \sigma_1^2 \leq \sigma_2^2$$
$$H_a: \sigma_1^2 > \sigma_2^2$$

In this form, rejection of H_0 will only occur when the value of s_1^2/s_2^2 is in the upper tail of the F distribution. Thus the population with the variance that involves "greater than" in the alternative hypothesis should be labeled population 1.

EXAMPLE 11.11

Samples of men and women were selected to participate in a study designed to measure attitudes about current political issues. The attitude scores of 31 men showed a sample variance of 80. The attitude scores of 41 women showed a sample variance of 120. From this evidence, can it be concluded that women demonstrate a greater variation in attitude about current political issues? Use $\alpha = .05$.

In formulating the hypotheses to be tested, we note that for women to demonstrate a significantly greater variance, women must show the larger variance in the alternative hypothesis. Thus the hypothesis test should be stated as

$$H_0: \sigma_{\text{women}}^2 \leq \sigma_{\text{men}}^2$$
$$H_a: \sigma_{\text{women}}^2 > \sigma_{\text{men}}^2$$

In conducting the F test, women will be population 1 and men will be population 2. Thus the corresponding F distribution has 40 degrees of freedom in the numerator and 30 degrees of freedom in the denominator. Table 11.2 shows that with these degrees of freedom, $F_{.05} = 1.79$. Computing the F value from the sample results, we have

$$F = \frac{s_1^2}{s_2^2} = \frac{120}{80} = 1.5$$

Since 1.5 is less than $F_{.05} = 1.79$, H_0 cannot be rejected. Thus the sample results do not support the position that women have significantly greater variation in attitudes about current political issues.

■

10. Assume that we have two automobile mechanics that are equal in ability and that the variances of their completion times for automobile repair jobs are the same. Suppose that we have a sample of six jobs performed by the first mechanic and eight jobs performed by the second mechanic. Furthermore, assume that we will compute the sample variances s_1^2 and s_2^2 for the completion times of the two mechanics.

    a. What is the sampling distribution of s_1^2/s_2^2? Show a graph of this sampling distribution.

    b. Use Table 11.2 to determine the probability that the ratio of the two sample variances is less than or equal to 3.97.

11. Two secretaries are each given eight typing assignments of equal difficulty. The sample standard deviations of the completion times were 3.8 minutes and 5.2 minutes, respectively. Do the data suggest that there is a difference in the variability of completion times for the two secretaries? Test the hypothesis at a .10 level of significance.

12. It is assumed that the variance of stopping distances of automobiles on wet pavement is substantially greater than the variance of stopping distances of automobiles on dry pavement. In testing this assumption, 16 automobiles traveling at the same speeds are tested with respect to stopping distances on wet pavement and stopping distances on dry pavement. On wet pavement, the standard deviation of stopping distances was 32 feet. On dry pavement, the standard deviation was 16 feet.

    a. At a .05 level of significance, do the sample data provide a clear indication that the variability in stopping distances on wet pavement is greater than the variability in stopping distances on dry pavement?

    b. What are the implications of your statistical conclusions in terms of driving safety recommendations?

13. Assume that ball bearings are produced on two different shifts. Sample results concerning the standard deviation of bearing sizes (in inches) from the two shifts are as follows:

FIRST SHIFT	SECOND SHIFT
$n_1 = 22$	$n_2 = 25$
$s_1 = .12$	$s_2 = .09$

    Are the sample results sufficient to conclude that the variances for the ball bearings differ for the two shifts? Test with $\alpha = .10$.

14. Consider the problem of estimating the difference between the mean checking account balances at two branches of the Clearview National Bank. The data collected from two independent random samples are as follows.

BRANCH BANK	NUMBER OF CHECKING ACCOUNTS	SAMPLE MEAN BALANCE	SAMPLE STANDARD DEVIATION
Cherry Grove	12	$\bar{x}_1 = \$1000$	$s_1 = \$150$
Beechmont	10	$\bar{x}_2 = \$920$	$s_2 = \$120$

    Recall that in order to use the t distribution to estimate the difference between means we assumed that the variances of the two populations are equal. This assumption is the basis for developing a pooled-variance estimate. Using $\alpha = .10$, test for the equality of the population variances. Do your results justify the use of a pooled-variance estimate?

15. Two new assembly methods showed the following variances in assembly times.

METHOD	SAMPLE SIZE	SAMPLE VARIANCE
A	31	$s_1^2 = 25$
B	25	$s_2^2 = 12$

Using $\alpha = .10$, are the two population variances equal.

16. As part of an 18-year follow-up study of the academic and psychosocial characteristics of low-birthweight adolescents, the math proficiency of 41 children with low birthweights and 41 children with full birthweights was measured. (*Social Biology,* 1988). The mean score for the low-birthweight group was 28.75, with a sample standard deviation of 11.21; the mean score for the full-birthweight group was 29.27, with a sample standard deviation of 9.6. Assume that reading proficiency scores follow a normal probability distribution. At the 10% level of significance, is there any difference between the population variances for the two groups?

17. Independent random samples of parts manufactured by two suppliers show the following results.

SUPPLIER	SAMPLE SIZE	SAMPLE VARIANCE OF PART SIZES
Durham Electric	41	$s_1^2 = 3.8$
Raleigh Electronics	31	$s_2^2 = 2.0$

A firm currently using the Durham supplier will continue to do so unless it can be shown that the Raleigh supplier provides a significantly lower variance in part sizes. Using $\alpha = .05$, provide a recommendation regarding whether or not the firm should change suppliers.

SUMMARY

In this chapter we have presented statistical procedures that can be used to make inferences about population variances. In the process we introduced two new probability distributions: the chi-square distribution and the F distribution. The chi-square distribution can be used as the basis for interval estimation and hypothesis tests concerning the variance of a normally distributed population. In particular, we showed that with random samples of size n selected from a normally distributed population, the quantity $(n - 1) s^2/\sigma^2$ has a chi-square distribution with $n - 1$ degrees of freedom.

We illustrated the use of the F distribution in making hypothesis tests concerning the variances of two normally distributed populations. In particular, we showed that with independent random samples of sizes n_1 and n_2 selected from two normal populations with equal variances, $\sigma_1^2 = \sigma_2^2$, the sampling distribution of the ratio of the two sample variances, s_1^2/s_2^2, has an F distribution with $n_1 - 1$ degrees of freedom for the numerator and $n_2 - 1$ degrees of freedom for the denominator.

INTERVAL ESTIMATION FOR A POPULATION VARIANCE

$$\frac{(n-1)s^2}{\chi^2_{\alpha/2}} \leq \sigma^2 \leq \frac{(n-1)s^2}{\chi^2_{(1-\alpha/2)}} \tag{11.6}$$

SAMPLING DISTRIBUTION OF s_1^2/s_2^2 when $\sigma_1^2 = \sigma_2^2$

$$F = \frac{s_1^2}{s_2^2} \tag{11.8}$$

REVIEW QUIZ

TRUE/FALSE

1. The sample variance should not be used as a point estimator of the population variance.

2. The sampling distribution of $(n-1)s^2/\sigma^2$ is a normal probability distribution.

3. The random variable for the chi-square probability distribution may not assume negative values.

4. The chi-square probability distribution is symmetric.

5. In order to determine the appropriate number of degrees of freedom for a chi-square distribution, one must know the sample size.

6. When applying the chi-square distribution, we must be able to assume the population being sampled follows a normal probability distribution.

7. For the F distribution the number of degrees of freedom in the numerator must be greater than or equal to the number of degrees of freedom in the denominator.

8. An F test can be used for a hypothesis test concerning the equality of two population variances.

MULTIPLE CHOICE

9. The sampling distribution of the quantity $(n-1)s^2/\sigma^2$ is the
 a. chi-square distribution
 b. normal distribution
 c. F distribution
 d. t distribution

10. The sampling distribution of the ratio of independent sample variances from two normally distributed populations with equal variances is the
 a. chi-square distribution
 b. normal distribution
 c. F distribution
 d. t distribution

11. When a sample variance of 25 is obtained from a sample of 10 items, the 90% confidence interval for a population variance is
   a. 12.3 to 57.1
   b. 13.3 to 67.7
   c. 14.1 to 46.25
   d. 15.3 to 53.98

12. Compared to a 95% confidence interval, a 99% confidence interval would be
   a. a narrower interval
   b. a wider interval
   c. the same width because you use the same data
   d. either narrower or wider

Use the following information for questions 13-15. These sample results were obtained for independent random samples from two normally distributed populations with equal variances.

	SAMPLE 1	SAMPLE 2
Sample size	10	16
Sample variance	25	20

13. Using a .05 level of significance, one critical value needed to use the F test for the equality of the population variances is:
   a. 2.59
   b. 3.01
   c. 3.12
   d. 3.89

14. The value of test statistic F is
   a. .13
   b. .5
   c. .8
   d. 1.25

15. Which conclusion would be reached for these data?
   a. There is a statistically significant difference between the variances of the two populations.
   b. There is no statistically significant difference between the variances of the two populations.
   c. Insufficient data; can't tell in this case.

SUPPLEMENTARY EXERCISES

18. Because of staffing decisions, management of the Gibson-Marimont Hotel is interested in the variability for the number of rooms occupied per day during a particular season of the year. A sample of 20 days of operation showed a sample mean of 290 rooms occupied per day and a sample standard deviation of 30 rooms.
   a. What is the point estimate of the population variance?
   b. Provide a 90% confidence interval estimate of the population variance.
   c. Provide a 90% confidence interval estimate of the population standard deviation.

19. A random sample of 30 days of sales for United Mufflers, Inc., resulted in a sample mean of 22.5 mufflers sold per day, with a sample standard deviation of 6. Provide 95% confidence interval estimates of the population variance and the population standard deviation.

20. Historical delivery times for Buffalo Trucking, Inc., have had a mean of 3 hours and a standard deviation of .5 hours. A sample of 22 deliveries over the past month resulted in a sample mean of 3.1 hours and a sample standard deviation of .75 hours.

    a. Do the sample results support the historical delivery variability of $\sigma^2 = (.5)^2 = .25$. Use $\alpha = .05$.

    b. Provide a 95% confidence interval estimate of both the population variance and the population standard deviation.

21. Part variability is very critical in manufacturing ball bearings, since large variances in the size of the ball bearings cause bearing failure and rapid wearout. Production standards call for a maximum variance of .0001 when the bearing sizes are measured in inches. A sample of 15 bearings resulted in a sample standard deviation of .014 inches.

    a. Using $\alpha = .10$, determine if the sample bearings were taken from a population having a variance of .0001 or less.

    b. Provide a 90% confidence interval estimate of the variance of the ball bearings in the population.

22. The filling variance for boxes of cereal is designed to be .02 or less. A sample of 41 boxes of cereal shows a sample standard deviation of .16 ounces. Using $\alpha = .05$, determine if the variance in the cereal box fillings is exceeding the standard.

23. A sample standard deviation for the number of passengers taking a particular airline flight is 8. A 95% confidence interval estimate of the standard deviation is 5.86 to 12.62.

    a. Was a sample size of 10 or 15 used in the above statistical analysis?

    b. If the sample standard deviation of $s = 8$ had been based on a sample of 25 flights, what change would you expect in the confidence interval for the population standard deviation? Compute a 95% confidence interval estimate of σ is a sample size 25 had been used.

24. The following sample data have been collected from two independent random samples:

POPULATION	SAMPLE SIZE	SAMPLE MEAN	SAMPLE VARIANCE
A	$n_A = 25$	$\bar{x}_A = 40$	$s_A^2 = 5$
B	$n_B = 21$	$\bar{x}_B = 50$	$s_B^2 = 11$

In a test for the difference between the two population means, the statistical analyst is considering using a pooled estimate of the population variance based on the assumption that the variances of the two populations are equal. Is pooling appropriate in this case? Use $\alpha = .10$.

25. A firm gives a mechanical aptitude test to all job applicants. A sample of 20 male applicants resulted in a sample variance of 80 for the test scores, and a sample of 16 female applicants resulted in a sample variance of 220. Using $\alpha = .05$, determine if the test score variances differ for male and female job applicants. If a difference in variances exist, which group has the higher variance in mechanical aptitude?

26. The accounting department analyzes the variances of the weekly unit costs reported by two production departments. A sample of 16 cost reports for each of the two departments resulted in cost variances of 2.3 and 5.4, respectively. Is this sample sufficient to conclude that the two production departments differ in terms of unit cost variances? Use $\alpha = .10$.

27. In using the *t* distribution to estimate the difference between two population means an analyst is interested in computing a pooled estimate of the variance of the populations. Pooling is justified only if it appears reasonable to assume that the two populations have equal variances. Use $\alpha = .05$ and the following data to determine if pooling is appropriate for this situation.

Sample 1   80, 72, 75, 90, 78, 75, 72, 85.

Sample 2   50, 48, 45, 60, 65, 66, 70, 54.

If pooling is appropriate, what is the pooled estimate of the population variance?

28. The respiratory rate in number of breaths per minute was recorded for seven patients who had undergone heart-lung transplants and for seven patients who had undergone only heart transplants. (*New England Journal of Medicine*, 1988). The data obtained are as follows:

| | **BREATHS PER MINUTE** | |
	Heart-Lung Transplant Recipients	Heart-Transplant Recipients
	Note: Mean $\pm$ Standard Deviation	
At rest	16 ± 5	14 ± 9
Maximal exercise	30 ± 7	40 ± 11

Assume that the distribution of respiratory rates can be approximated by a normal probability distribution.

a. Consider the at rest condition. At the 5% level of significance, is there any difference between the variances for the heart-lung and heart transplant populations?

b. Consider the at-rest condition. At the 5% level of significance, does there appear to be any difference between the mean respiratory rates for the two populations?

c. Consider the maximal exercise condition. At the 5% level of significance, is there any difference between the variances for the two populations?

d. Consider the maximal exercise condition. At the 5% level of significance, does there appear to be any difference between the mean respiratory rates for the two populations?

e. Overall, what conclusions can you make regarding these two populations?

COMPUTER EXERCISE

An Air Force introductory course in electronics is currently taught using computer-assisted instruction, with each student in the course working individually at a computer terminal. It has been proposed that a better approach to teaching the course would be to have a pair of students work together at each computer terminal. In addition to the fact that a greater number of students could be taught at the same time, the proposed method may have the positive effect of reducing overall training time due to the fact that the students can help each other learn. In order to test the proposed method, each of the students in an entering class of 120 students was randomly assigned to one of two groups. One group of 60 was taught using the current method; the second group of 60 was taught using the new method. The time in hours was recorded for each student in the sample. The data are shown.

PROPOSED METHOD	CURRENT METHOD	PROPOSED METHOD	CURRENT METHOD
75	73	75	79
76	77	71	79
75	72	75	74
73	76	72	79
74	65	77	72
73	72	78	75
75	71	74	75
73	73	76	71
74	67	75	74
76	82	77	80
77	68	76	82
77	72	71	76
76	69	77	76
72	76	74	77
77	79	82	81
75	76	75	74
76	80	73	82
77	73	72	75
72	79	76	72
75	77	73	76
74	75	77	63
75	77	75	79
70	70	73	75
74	73	78	74
78	78	75	76
80	73	74	77
72	76	72	74
76	72	77	70
76	61	76	72
73	76	78	78

QUESTIONS

1. Develop tabular and numerical measures to summarize the data.

2. Does there appear to be any difference between the mean training times for the two methods? Explain.

3. Does the variance in the training times appear to be a significant factor in terms of the difference between the two methods?

4. What conclusion can you reach about any differences in the two methods? What is your recommendation? Explain.

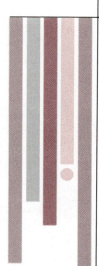

12

Experimental Design and the Analysis of Variance

What You Will Learn in This Chapter:

- How to use analysis of variance to test for equality of the means of three or more populations

- How the F distribution is used in analysis of variance

- What an analysis of variance table is and how it is constructed

- The assumptions of analysis of variance

- The completely randomized, randomized block, and factorial experimental designs

- How to make computations and draw conclusions for completely randomized, randomized block, and factorial experimental designs

Contents

High Self-Esteem: A Key to Academic Achievement

Self-esteem has been the subject of a great amount of research that reveals its importance in the school setting. Students with high self-esteem are more receptive to the educational process and respond in more positive directions toward the teacher, assignments, and school in general.

Robert H. Phillips, a researcher at Fordham University, reports on a study designed to investigate ways in which teacher interaction with elementary-school students can improve student self-esteem. A sample of 30 elementary-school students was selected from 10 low-income areas of New York City. Each student was assigned to one of three groups. The first group, the experimental group, was provided an environment in which the students were given positive reinforcement by the teacher for any positive statements the students made about themselves. For example, if a student offered a legitimate positive self-statement, the teacher was instructed to respond with statements such as, "I'm proud of you too," "Yes, you did do well," "You are doing beautifully," and so on. The second group, the control group, was provided the same physical environment but the teacher made no comment if students offered positive statements about themselves. The third group, referred to as the inventory group, received no special instructions and operated under normal conditions.

All students in the research project were given self-esteem tests, at the beginning and at the end of the 7-week study. A measure of how each student's self-esteem score changed during the study was recorded. The statistical technique known as analysis of variance was used to see if the three groups differed in terms of their change in self-esteem scores. The statistical conclusion was that there was a significant difference among the self-esteem

A teacher's individual attention and positive statements help students improve self-esteem and academic achievement.

scores for the three groups. The experimental group with its reinforcing teacher statements showed a significantly greater increase in self-esteem scores.

Because self-esteem is important in children's academic experiences, this research suggests that teacher responses designed to enhance student self-esteem should be part of regular educational programs. Future research investigations are planned to gain additional evidence concerning the degree to which improvements in self-esteem scores lead to corresponding improvements in academic achievement.

Based on "Increasing Positive Self-Referent Statements to Improve Self-Esteem in Low-Income Elementary School Children" by Robert H. Phillips, *The Journal of School Psychology* (Summer 1984).

Experiments are undertaken in order to discover something not yet known or examine the validity of a hypothesis. In an experimental study, the variables of interest are first identified. Then, one or more factors in the study are controlled so that data may be obtained about how the factors influence the variables. These data are then analyzed in order to reach some conclusion. The process of planning an experiment so that appropriate data can be collected and analyzed using statistical methods is referred to as the *statistical design of experiments*.

For example, a pharmaceutical firm might be interested in conducting an experiment designed to determine the effect of three different drugs, referred to as A, B, and C, on the cholesterol level of individuals. Note that this experiment involves just one variable, the cholesterol level of individuals, and one factor, the type of drug adminstered. Suppose that the objective of the experimenter is to determine if the mean change in cholesterol level for individuals is the same for all three drugs. To obtain data on the effect of the three drugs, a sample of individuals would be selected. In the simplest type of experimental design, one-third of the individuals would be given drug A, one-third would be given drug B, and one-third would be given drug C. Data on cholesterol level would then be collected for each group. Statistical analysis of the data would help determine if the mean change in cholesterol level is the same for all three drugs.

Every statistical problem involving an experiment such as the one just described consists of three parts: (1) designing the experiment; (2) collecting the data according to the principles of the experimental design that was selected; (3) analyzing the data using appropriate statistical methodology. These three parts are closely related, since how the data are collected and analyzed depends upon the type of design selected. In this chapter we discuss three different experimental designs: completely randomized designs, randomized block designs, and factorial designs.

In Chapter 11 we discussed how to test whether or not the means of two populations are equal. Recall that the test required the selection of an independent random sample from each of the two populations. The statistical approach that we introduce in this chapter, referred to as the *analysis of variance (ANOVA) procedure,* can be thought of as an extension of the Chapter 11 methods to the case of more than two populations. Using the ANOVA procedure, the variability in the data is partitioned into two or more components. A test statistic based upon this partitioning is then used to determine whether or not the the variability in the data is due to differences in the populations or random variability. For example, in "Statistics in the News," the ANOVA procedure was used to test whether or not the mean self-esteem scores for three groups of students were equal.

We begin the study of experimental design and the analysis of variance procedure with a very simple problem involving a completely randomized experimental design. Using this problem we introduce the experimental design terminology and illustrate how the basic calculations in the ANOVA procedure are performed. These ideas are then extended to randomized block and factorial designs.

12.1   Experimental Design: Concepts and Data Collection

In order to introduce the concepts of experimental design, let us consider the experiment described in the following example.

EXAMPLE 12.1

The Penn Yan School District has developed an 80-question reading-competency examination in order to identify eighth-grade students with reading deficiencies. For students who answer less than 55 questions correctly on this examination, three remedial programs have been proposed:

Program 1: A review class that meets 1 hour each week for 10 weeks.

Program 2: A personalized system of instruction using microcomputers.

Program 3: A formal 10-week class that meets for 1 hour each day.

Before making a final decision about which remedial program to adopt, the board of education has requested that further study be conducted to determine how the proposed programs affect reading-competency examination scores.

■

In Example 12.1, the reading-competency remedial program is referred to as a *factor.* Since there are three remedial programs, called *levels,* corresponding to this factor, we say that there are three *treatments* associated with the experiment: one treatment corresponds to Program 1, another to Program 2, and the third to Program 3. In general, a treatment corresponds to a level of a factor.*

The random variable of interest in this experiment is the reading-competency examination scores received by the students when they are retested. The primary statistical objective is to determine whether or not the mean reading-competency examination scores are the same for all three programs. In experimental design terminology, the random variable of interest is referred to as the *dependent variable,* the *response variable,* or the *response.*

Let us now turn our attention to how the experiment is actually conducted. To begin with, let us assume that a random sample of three students has been selected from the population of students who answered less than 55 questions correctly on the original reading competency examination. In experimental design terminology, the students are referred to as *experimental units.*

The experimental design that we will use for this experiment is referred to as a *completely randomized design.* This type of design requires that each of the three remedial programs, or treatments, be randomly assigned to one of the experimental units or students. For example, Program 1 might be randomly assigned to the second student, Program 2 to the first student, and Program 3 to the third student. The concept of randomization as illustrated in this example is an important principle of all experimental designs.

Note that the experiment as described would result in only one measurement of the reading-competency examination score for each treatment. In other words, we have a sample size of 1 corresponding to each treatment. In order to obtain additional data for each preparation program, we must repeat, or *replicate,* the basic experimental process. For example, suppose that instead of selecting just three students at random, we had selected 15 students. If *each* of the three remedial programs is randomly assigned to five students, we say that five replicates have been obtained. The concept of *replication* is another important principle of experimental design. Figure 12.1 shows the completely randomized design for the reading-competency experiment.

Data Collection

Once we are satisfied with the experimental design, we proceed by carrying out the experiment and collecting the data. In this case the students would receive additional reading instruction under their assigned remedial program. The students would then retake the reading-competency examination, and their scores would be recorded. Let us assume that this has been done and that the scores for the 15 students in the study are as shown

*The term *treatment* was originally used in experimental design because many of the applications were in agriculture, where the treatments often corresponded to different types of fertilizers applied to selected agricultural plots. Today the term is used in a more general context.

in Table 12.1. Using these data we calculate the sample mean score for each of the three remedial programs, as shown.

REMEDIAL PROGRAM	MEAN EXAM SCORE
Program 1	55
Program 2	68
Program 3	57

From these data it appears that Program 2 may result in higher retest scores than either of the other methods. However, before we make any final recommendations we must remember that each of the given means is based upon the test results of just five students. Thus we are looking at three samples drawn from the three populations representing all students who might participate in each program. The real issue, then, is

FIGURE 12.1   Completely Randomized Design for Evaluating the Remedial Programs

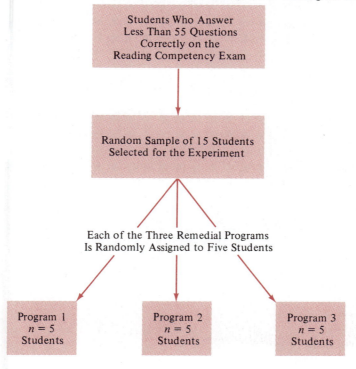

TABLE 12.1

Reading-Competency Examination Scores for the 15 Students Who Were Retested

		OBSERVATIONS				
	Program 1	48	54	57	54	62
Treatment	Program 2	73	63	66	64	74
	Program 3	51	63	61	54	56

whether or not the three sample means observed are different enough for us to conclude that the means of the populations corresponding to the three remedial programs are different. Let

μ_1 = mean retest score for the population of all students assigned to Program 1

μ_2 = mean retest score for the population of all students assigned to Program 2

μ_3 = mean retest score for the population of all students assigned to Program 3

We use $\bar{x}_1$, $\bar{x}_2$, and $\bar{x}_3$ to denote the corresponding sample means.

The experimental results for the three remedial programs yielded $\bar{x}_1 = 55$, $\bar{x}_2 = 68$, and $\bar{x}_3 = 57$. Although we will never know the actual values of μ_1, μ_2, and μ_3, we want to use the sample results to test the following hypotheses:

$$H_0: \quad \mu_1 = \mu_2 = \mu_3$$

$$H_a: \quad \text{Not all means are equal}$$

In the next section we show how the ANOVA procedure can be used to determine if there is a significant difference between the means of the three populations.

12.2   The Analysis of Variance Procedure for Completely Randomized Designs

The analysis of variance (ANOVA) procedure that we describe is designed to test the following hypotheses.

Hypotheses for Analysis of Variance

$$H_0: \quad \mu_1 = \mu_2 = \cdots = \mu_k$$

$$H_a: \quad \text{Not all } \mu_i \text{ are equal}$$

where

$$k = \text{number of populations or treatments}$$

We assume that a simple random sample of size n has been selected from each of the k populations and that the sample means $\bar{x}_1$, $\bar{x}_2$, . . . , $\bar{x}_k$ have been computed. In general, we refer to the sample mean corresponding to the ith population as $\bar{x}_i$ and the overall sample mean as $\bar{\bar{x}}$. That is,

$$\bar{x}_i = \frac{\sum\limits_{j} x_{ij}}{n} \tag{12.1}$$

$$\bar{\bar{x}} = \frac{\sum_i \sum_j x_{ij}}{n_T} \tag{12.2}$$

where

x_{ij} = jth observation corresponding to the ith treatment

n_T = total sample size for the experiment

Note that since the sample size is n for each of the k treatments, $n_T = kn$ and $\bar{\bar{x}} = \sum \bar{x}_i / k$.

EXAMPLE 12.1 (continued)

Since there are three remedial programs, or treatments, $k = 3$. Also, since each of the three programs is randomly assigned to five students, $n = 5$. Thus $n_T = (3)(5) = 15$.

■

Assumptions for Analysis of Variance

The ANOVA procedure is based upon the following two assumptions:

1. The response variable (the variable of interest for each population) has a normal probability distribution.
2. The variance of the response variable is the same for each population.

EXAMPLE 12.1 (continued)

Assumption 1 requires that the variable of interest, the retest score, be normally distributed for each of the three programs under study. Assumption 2 requires that the variance of retest scores be the same for each program.

■

We denote the common population variance (Assumption 2) of the response variable as σ^2. The basis for the methodology of the ANOVA procedure is the development of two independent estimates of σ^2. One estimate of σ^2 is based upon the differences *between* the treatment means and the overall sample mean. The other estimate is based upon the differences of observations *within* each treatment from the corresponding treatment mean. By comparing these two estimates of σ^2, we will be able to determine whether or not we can reject the null hypothesis that the population means are equal.

Between-Treatments Estimate of Population Variance

The procedure for developing an estimate of σ^2 based upon the differences between the treatment means and the overall sample mean depends on the assumption that the null hypothesis is true and that the two assumptions given earlier are valid. In this case, if we let μ denote the common population mean under the null hypothesis (that is $\mu_1 = \mu_2 = \cdots = \mu_k = \mu$), then all the sample observations would represent data values drawn from the same normal probability distribution with mean μ and variance σ^2. If we let $\bar{x}$ denote the mean of a simple random sample of size n selected from this probability distribution, then the sampling distribution of $\bar{x}$ is normally distributed with mean μ and variance $\sigma_{\bar{x}}^2 = \sigma^2 / n$. Figure 12.2 illustrates such a sampling distribution. Thus, under the null hypothesis, we can think of each $\bar{x}$ as a value drawn at random from this sampling

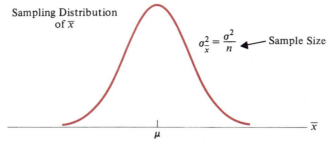

Sampling Distribution
of $\bar{x}$

$\sigma_{\bar{x}}^2 = \dfrac{\sigma^2}{n}$ ← Sample Size

μ

$\bar{x}$

Sample Means Come from
One Sampling Distribution
When H_0 Is True

FIGURE 12.2   Sampling Distribution of $\bar{x}$ Given the Null Hypothesis $\mu_1 = \mu_2 \cdots = \mu_k = \mu$ and a Sample of Size n

distribution. An estimate of the mean μ of this sampling distribution can be obtained by computing $\bar{x}$, the overall sample mean.

EXAMPLE 12.1
(continued)

For the retest scores in Table 12.1, we use all the data to find

$$\sum_i \sum_j x_{ij} = (48 + 54 + 57 + \cdots + 54 + 56) = 900$$

Then using 12.2 with $n_T = 15$, we compute the overall sample mean,

$$\bar{\bar{x}} = \frac{\sum_i \sum_j x_{ij}}{n_T} = \frac{900}{15} = 60$$

as an estimate of μ.

■

To estimate the variance of the sampling distribution of $\bar{x}$ (that is, $\sigma_{\bar{x}}^2$), we can use the variance of the individual sample means about the overall sample mean, $\bar{\bar{x}}$. For the case where each sample is the same size, the estimated variance, denoted by $s_{\bar{x}}^2$, is

$$s_{\bar{x}}^2 = \frac{\sum_i (\bar{x}_i = \bar{\bar{x}})^2}{k - 1} \qquad (12.3)$$

where $\bar{x}_i$ = mean of the ith sample.

EXAMPLE 12.1
(continued)

Since the previous computations of the three sample means showed that $\bar{x}_1 = 55$, $\bar{x}_2 = 68$, and $\bar{x}_3 = 57$, we have

$$\sum_i (\bar{x}_i - \bar{\bar{x}})^2 = (55 - 60)^2 + (68 - 60)^2 + (57 - 60)^2$$

$$= 25 + 64 + 9 = 98$$

With $k = 3$ populations, or treatments, we have $k - 1 = 2$. Thus with (12.3), the estimate of the variance of the sampling distribution becomes

$$s_{\bar{x}}^2 = \frac{\sum_i (\bar{x}_i - \bar{\bar{x}})^2}{k - 1} = \frac{98}{2} = 49$$

■

Since we know that $\sigma_{\bar{x}}^2 = \sigma^2/n$, solving for σ^2 gives

$$\sigma^2 = n\sigma_{\bar{x}}^2 \tag{12.4}$$

where n is the sample size involved in computing $\bar{x}$. Hence an estimate of the population variance, σ^2, can be obtained from (12.4) by multiplying n by the estimate of $\sigma_{\bar{x}}^2$.

EXAMPLE 12.1
(continued)

$$\text{Estimate of } \sigma^2 = n(\text{Estimate of } \sigma_{\bar{x}}^2) = ns_{\bar{x}}^2 \tag{12.5}$$

With all samples of size $n = 5$ and with $s_{\bar{x}}^2 = 49$, for the reading-competency experiment we have

$$\text{Estimate of } \sigma^2 = 5(49) = 245$$

■

This estimate of σ^2 is given the name *mean square between treatments* and is denoted by MSTR.

From (12.3) and (12.5) we see that MSTR can be written

$$\text{MSTR} = \frac{n\sum_i (\bar{x}_i - \bar{\bar{x}})^2}{k - 1} \tag{12.6}$$

The numerator of (12.6) is called the *sum of squares between treatments* and is denoted by SSTR. The denominator, $k - 1$, represents the *degrees of freedom* corresponding to SSTR.

Between-Treatments Estimate of Population Variance

$$\text{MSTR} = \frac{\text{SSTR}}{k - 1} \tag{12.7}$$

where

$$\text{SSTR} = n\sum_i (\bar{x}_i - \bar{\bar{x}})^2$$

**EXAMPLE 12.1
(continued)**

Using the previous calculations, we have the following values for SSTR, degrees of freedom, and MSTR:

$$\text{SSTR} = n\sum_i (\bar{x}_i - \bar{\bar{x}})^2 = 5(98) = 490, \qquad k - 1 = 3 - 1 = 2$$

and thus

$$\text{MSTR} = \frac{\text{SSTR}}{k - 1} = \frac{490}{2} = 245$$

Within-Treatments Estimate of Population Variance

We now develop a second estimate of σ^2 that is not based on the null hypothesis assumption that the population means are equal. Instead, it is based upon the variation of the sample observations "within" each treatment. This estimate is called the *mean square due to error or the mean square within treatments,* and is denoted by MSE.

If each of the k samples is a simple random sample, then each of the k sample variances provides an estimate of σ^2. For each sample, the sample variance is computed in the usual fashion.

$$\text{Variance of Sample } i = s_i^2 = \frac{\sum_j (x_{ij} - \bar{x}_i)^2}{n - 1} \tag{12.8}$$

Since each sample variance provides an individual estimate of σ^2, it would seem reasonable that the overall best estimate of σ^2 could be obtained by pooling the individual estimates. When the sample size corresponding to each treatment is the same, this can be done by simply averaging the individual estimates of σ^2. Thus

$$\text{MSE} = \frac{s_1^2 + s_2^2 + \cdots + s_k^2}{k}$$

Using (12.8) to substitute for each of the within-treatment variances, s_i^2, and simplifying algebraically yields the following result.

$$\begin{aligned}
\text{MSE} &= \frac{\sum_j (x_{1j} - \bar{x}_1)^2 + \sum_j (x_{2j} - \bar{x}_2)^2 + \cdots + \sum_j (x_{kj} - \bar{x}_k)^2}{k(n - 1)} \\
&= \frac{\sum_i \sum_j (x_{ij} - \bar{x}_i)^2}{kn - k} \\
&= \frac{\sum_i \sum_j (x_{ij} - \bar{x}_i)^2}{n_T - k}
\end{aligned} \tag{12.9}$$

The numerator in (12.9) is given the name *sum of squares within,* or *sum of squares due to error,* and is denoted by SSE. The denominator of MSE is the number of *degrees of freedom* associated with the within-treatment variance estimate. In general, then, we can compute MSE as follows.

Within-Treatments Estimate of Population Variance

$$\text{MSE} = \frac{\text{SSE}}{n_T - k} \tag{12.10}$$

where

$$\text{SSE} = \sum_i \sum_j (x_{ij} - \bar{x}_i)^2 \tag{12.11}$$

EXAMPLE 12.1 (continued)

In considering the retest scores for the five students assigned to Program 1, we obtain the following result.

$$\sum_j (x_{1j} - \bar{x}_1)^2 = (48 - 55)^2 + (54 - 55)^2 + (57 - 55)^2 + (54 - 55)^2 + (62 - 55)^2$$

$$= 49 + 1 + 4 + 1 + 49 = 104$$

Similarly, for Programs 2 and 3 we obtain

$$\sum_j (x_{2j} - \bar{x}_2)^2 = 106 \qquad \sum_j (x_{3j} - \bar{x}_3)^2 = 98$$

Hence

$$\text{SSE} = \sum_i \sum_j (x_{ij} - \bar{x}_i)^2$$
$$= \sum_j (x_{1j} - \bar{x}_1)^2 + \sum_j (x_{2j} - \bar{x}_2)^2 + \sum_j (x_{3j} - \bar{x}_3)^2$$
$$= 104 + 106 + 98 = 308$$

Thus, since $n_T - k = (3)(5) - 3 = 12$, using (12.10) we obtain

$$\text{MSE} = \frac{\text{SSE}}{n_T - k} = \frac{308}{12} = 25.667$$

■

Comparing the Variance Estimates: The *F* Test

We have now developed two estimates of σ^2. The first estimate (MSTR) is based upon the variation between treatment means, and the second estimate (MSE) is based upon the variation within each treatment. Recall that in order to compute MSTR we had to assume that the null hypothesis was true (that is, $\mu_1 = \mu_2 = \ldots = \mu_k$). However, in computing MSE this assumption was not required. In fact, regardless of whether or not the means of the *k* populations are equal, MSE will always provide an unbiased estimate of σ^2.

It can be shown that MSTR provides an unbiased estimate of σ^2 when H_0 is true. However, if the means of the k populations are not equal (that is, H_a is true), MSTR is not an unbiased estimate of σ^2. In fact, in this case MSTR overestimates σ^2. This is the key to the analysis of variance procedure: that is, we test the hypotheses

$$H_0: \quad \mu_1 = \mu_2 = \cdots = \mu_k$$

$$H_a: \quad \text{Not all } \mu_i \text{ are equal}$$

by comparing the two estimates of the population variance, MSTR and MSE. If MSTR and MSE are approximately equal, such that the ratio MSTR/MSE is near 1, we cannot reject H_0; in this case the statistical evidence does not lead us to conclude that the means are not all equal. However, if MSTR is much larger than MSE, such that the ratio MSTR/MSE is much larger than 1, we reject H_0 and conclude that the means are not all equal.

To obtain a better intuitive feel for why larger values of MSTR are obtained when the null hypothesis is false, recall the formula for computing MSTR.

$$\text{MSTR} = \frac{\text{SSTR}}{k-1} = \frac{n\sum_i (\bar{x}_i - \bar{\bar{x}})^2}{k-1} \tag{12.12}$$

Note that the numerator (SSTR) is based on the dispersion of the k sample means, $\bar{x}_i$'s, around the overall sample mean $\bar{\bar{x}}$. Now consider the two situations shown in Figure 12.3. In (a) we have the sampling distribution of $\bar{x}$ under the assumption that H_0 is true with $\mu_1 = \mu_2 = \mu_3 = \mu$. In this case, since each sample comes from the same population and since there is only one sampling distribution, MSTR as computed in (12.12) will provide an unbiased estimate of σ^2.

The second diagram, part (b) of Figure 12.3, shows a situation where H_0 is false and the three means μ_1, μ_2, and μ_3 are not the same. What happens to the value of SSTR in this case? Note that since the sample means $\bar{x}_1$, $\bar{x}_2$, and $\bar{x}_3$ are coming from distributions with different μ's, they have different sampling distributions and show a much larger dispersion than when H_0 is true. In this case the value of SSTR in (12.12) will be larger, causing the value of MSTR to overestimate the population variance σ^2.

Let us assume for the moment that the null hypothesis is true and that $\mu_1 = \mu_2 = \ldots = \mu_k$. In this case it can be shown that MSTR and MSE provide two independent unbiased estimates of σ^2. Recall from Chapter 11 that the sampling distribution of the ratio of two independent estimates of σ^2 for a normal population follows an F probability distribution. Thus if the null hypothesis is true and the ANOVA assumptions are valid, the sampling distribution of MSTR/MSE is an F distribution with numerator degrees of freedom equal to $k-1$ and denominator degrees of freedom equal to $n_T - k$.

On the other hand, if the means of the k populations are not all equal, the value of MSTR/MSE will be inflated because MSTR overestimates σ^2. Hence we will reject H_0 if the resulting value of MSTR/MSE appears to be too large to have been selected at random from an F distribution with degrees of freedom $k-1$ in the numerator and $n_T - k$ in the denominator. The value of F that will cause us to reject H_0 depends upon α, the level of significance. Once α is selected a critical value of F can be determined. Figure 12.4 shows the sampling distribution of MSTR/MSE and the rejection region associated with a level of significance equal to α. Note that F_α denotes the critical value.

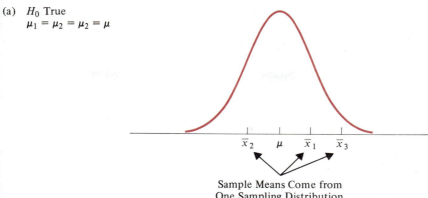

(a) H_0 True
$\mu_1 = \mu_2 = \mu_2 = \mu$

$\bar{x}_2$ μ $\bar{x}_1$ $\bar{x}_3$

Sample Means Come from
One Sampling Distribution
When H_0 Is True

(b) H_0 False
μ_i's Not All Equal

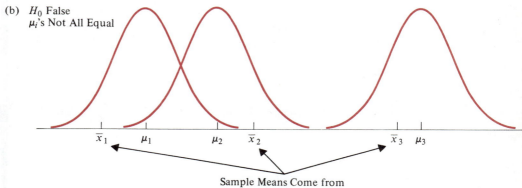

$\bar{x}_1$ μ_1 μ_2 $\bar{x}_2$ $\bar{x}_3$ μ_3

Sample Means Come from
Different Sampling Distributions
When H_0 Is Not True

FIGURE 12.3   Examples of the Sampling Distribution of $\bar{x}$ for the Cases of H_0 True and H_0 False

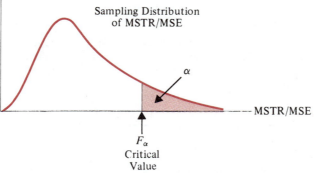

Sampling Distribution
of MSTR/MSE

α

MSTR/MSE

F_α
Critical
Value

FIGURE 12.4   Sampling Distribution of MSTR/MSE; the Critical Value for Rejecting the Null Hypothesis of Equality of Means is F_α

EXAMPLE 12.1
(continued)

Let us finish the ANOVA procedure for the reading-competency experiment. Assume that the board of education was willing to accept a type I error probability of $\alpha = .05$. From Table 4 of Appendix B we can determine the critical value by locating the value corresponding to numerator degrees of freedom equal to $k - 1 = 3 - 1 = 2$ and denominator degrees of freedom equal to $n_T - k = 15 - 3 = 12$. Thus we obtain the value $F_{.05} = 3.89$. Hence the appropriate rejection rule for the problem is written

$$\text{Reject } H_0 \text{ if MSTR/MSE} > 3.89$$

where

$$H_0: \quad \mu_1 = \mu_2 = \mu_3$$

$$H_a: \quad \text{Not all } \mu_i \text{ are equal}$$

Since MSTR/MSE $= 245/25.667 = 9.55$ is greater than the critical value $F_{.05} = 3.89$, there is sufficient statistical evidence to reject the null hypothesis that the means of the three programs are the same.

■

EXERCISES

1. The Jacobs Chemical Company wants to estimate the mean time (minutes) required to mix a batch of material on machines produced by three different manufacturers. In order to limit the cost of testing, four batches of material were mixed on machines produced by each of the three manufacturers. The time needed to mix the material was recorded. The times in minutes are shown.

MANUFACTURER 1	MANUFACTURER 2	MANUFACTURER 3
20	28	20
26	26	19
24	31	23
22	27	22

Use these data and test to see if the mean time needed to mix a batch of material is the same for each manufacturer. Use $\alpha = .05$.

2. Four different paints are advertised as having the same drying time. In order to check the manufacturers' claims, five paint samples were tested for each make of paint. The time in minutes until the paint was dry enough for a second coat to be applied was recorded. The following data were obtained.

PAINT 1	PAINT 2	PAINT 3	PAINT 4
128	144	133	150
137	133	143	142
135	142	137	135
124	146	136	140
141	130	131	153

At the $\alpha = .05$ level of significance, test to see if the mean drying time is the same for each type of paint.

3. Three different brands of radial tires were compared for wear characteristics. For each brand of tire, 10 tires were randomly selected and subjected to standard wear-testing procedures. The average mileage obtained for each brand of tire and the sample standard deviations are shown.

	BRAND A	BRAND B	BRAND C
Average mileage	36,400	38,200	33,100
Sample standard deviation	1,650	1,800	1,500

Using $\alpha = .05$, test to see if the mean mileage for all three brands of tires is the same.

4. Three top-of-the-line intermediate-sized automobiles manufactured in the United States have been test driven and compared on a variety of criteria by a well-known automotive magazine. In the area of gasoline mileage performance, five automobiles of each brand were each test-driven 500 miles; the miles per gallon data obtained are shown.

MILES PER GALLON DATA					
Automobile A	19	21	20	19	21
Automobile B	19	20	22	21	23
Automobile C	24	26	23	25	27

Use the analysis of variance procedure with $\alpha = .05$ to determine if there is a significant difference in the mean miles per gallon for the three types of automobiles.

5. Three brands of paper towels were tested for their abilities to absorb water. Equal-sized towels were used, with four sections of towels tested per brand. The absorbency rating data are below. Using a .05 level of significance, does there appear to be a difference in the ability of the brands to absorb water?

BRAND	ABSORBENCY RATING			
X	91	100	88	89
Y	99	96	94	99
Z	83	88	89	76

6. In order to study the effect of temperature upon yield in a chemical process, five batches were produced under each of three temperature levels. The results are given. Using a .05 level of significance, test to see if the temperature level appears to have an effect upon the mean yield of the process.

TEMPERATURE (°C)	YIELD				
50	34	24	36	39	32
60	30	31	34	23	27
70	23	28	28	30	31

12.3 The ANOVA Table and Other Considerations

A convenient way to summarize the computations and results of the analysis of variance procedure is to develop an *analysis of variance (ANOVA) table*. In order to present the ANOVA table for the reading-competency experiment however, we need to perform some additional calculations with the data collected (see Table 12.1). Treating the entire data set as one sample of 15 observations, we have already found that the overall sample mean is $\bar{\bar{x}} = 60$. Using the entire data set as one sample, let us now compute an overall sample variance.

$$s^2 = \frac{\sum_i \sum_j (x_{ij} - \bar{\bar{x}})^2}{n_T - 1} \qquad (12.13)$$

The numerator of (12.13) is called the *total sum of squares about the mean* (SST), and the denominator represents the degrees of freedom associated with this total sum of squares.

Total Sum of Squares About the Mean

$$\text{SST} = \sum_i \sum_j (x_{ij} - \bar{\bar{x}})^2 \qquad (12.14)$$

where

$$\text{Degrees of Freedom for Total Sum of Squares} = n_T - 1$$

EXAMPLE 12.1 (continued)

Check for yourself that the 15 values in Table 12.1 result in the following values for SST and $n_T - 1$.

$$\text{SST} = \sum_i \sum_j (x_{ij} - \bar{\bar{x}})^2$$

$$= (48 - 60)^2 + (54 - 60)^2 + \cdots + (54 - 60)^2 + (56 - 60)^2$$

$$= 798$$

$$n_T - 1 = 15 - 1 = 14$$

We can now develop the ANOVA table for a completely randomized design. Table 12.2 shows a general ANOVA table, and Table 12.3 shows the ANOVA table for the reading-competency experiment.

Note that the rows of the table contain information concerning the two sources of variation: between treatments, which refers to the between-group variation, and error, which refers to the within-group variation. The "Sum of Squares" column and the "Degrees of Freedom" column provide the corresponding values as defined in the previous section. The column labeled "Mean Square" is simply the sum of the squares divided by the corresponding degrees of freedom. In Section 12.2 we showed that when we divided the sums of squares by the corresponding degrees of freedom, we obtained two estimates of the population variance (MSTR and MSE). The ANOVA table simply summarizes these values in the "Mean Square" column.

Finally, the last column in the table contains the F value corresponding to MSTR/MSE. Since the variance estimates are found in the "Mean Square" column, the F value is computed by dividing the mean square treatments (MSTR) by the mean square error (MSE). That is, $F = $ MSTR/MSE.

What observation can you make from the "Sum of Squares" columns in Tables 12.2 and 12.3? Note in particular that the following condition holds.

$$SST = SSTR + SSE \qquad (12.15)$$

$$\sum_i \sum_j (x_{ij} - \bar{\bar{x}})^2 = n \sum_i (x_i - \bar{\bar{x}})^2 + \sum_i \sum_j (x_{ij} - \bar{x}_i)^2 \qquad (12.16)$$

Thus we see that SST can be partitioned into two sums—one called the sum of squares between treatments and one called the sum of squares due to error. This is known as *partitioning* the sum of squares.

Note also that the degrees of freedom corresponding to SST ($n_T - 1 = 14$) can be partitioned into the degrees of freedom corresponding to SSTR ($k - 1 = 2$) and the degrees of freedom corresponding to SSE ($n_T - k = 12$). The analysis of variance procedure can be viewed as the process of partitioning the total sum of squares and degrees of freedom into their corresponding sources: between treatments and error. Dividing the sum of squares by the appropriate degrees of freedom provides the variance estimates and the F value used to test the hypothesis of equal population or treatment means.

TABLE 12.2

ANOVA Table for a Completely Randomized Design

SOURCE OF VARIATION	SUM OF SQUARES	DEGREES OF FREEDOM	MEAN SQUARE	F
Treatments	SSTR	$k - 1$	$MSTR = \dfrac{SSTR}{k - 1}$	$\dfrac{MSTR}{MSE}$
Error	SSE	$n_T - k$	$MSE = \dfrac{SSE}{n_T - k}$	
Total	SST	$n_T - 1$		

TABLE 12.3

ANOVA Table for Reading-Comprehension Experiment

SOURCE OF VARIATION	SUM OF SQUARES	DEGREES OF FREEDOM	MEAN SQUARE	F
Treatments	490	2	245	9.55
Error	308	12	25.667	
Total	798	14		

Unbalanced Designs

The formula presented previously for computing SSTR applies only to *balanced designs*. Balanced designs are ones in which the sample size is the same for each treatment. Any experimental design for which the sample size is not the same for each treatment is said to be *unbalanced*. Although we would prefer a balanced design, in some cases we must work with unequal sample sizes.* For unbalanced designs the analysis of variance procedure described above may be used with the following modification of the formula for SSTR.

$$SSTR = \sum_i n_i(\bar{x}_i - \bar{\bar{x}})^2 \qquad (12.17)$$

where

$$n_i = \text{sample size for the } i\text{th treatment}$$

In such cases

$$n_T = n_1 + n_2 + \cdots + n_k$$
$$= \sum_i n_i$$

Note that (12.17) yields the same formula we used earlier when the sample size is the same for each treatment; that is, when $n_i = n$.

A Summary of the ANOVA Table Computations

Equation (12.14) provides a formula for computing the total sum of squares (SST) and (12.17) provides a formula for computing the sum of squares between treatments (SSTR). Since SST = SSTR + SSE, we can obtain SSE by subtraction if we know the values of SST and SSTR. That is,

$$SSE = SST - SSTR \qquad (12.18)$$

This result suggests the following three-step procedure for computing the sums of squares for the ANOVA table.

Step 1 Compute the total sum of squares (SST) using (12.14):

$$SST = \sum_i \sum_j (x_{ij} - \bar{\bar{x}})^2$$

Step 2 Compute the sum of squares between treatments (SSTR) using (12.17):

$$SSTR = \sum_i n_i(\bar{x}_i - \bar{\bar{x}})^2$$

Step 3 Compute the sum of squares due to error (SSE) using (12.18):

$$SSE = SST - SSTR$$

---

*With a balanced design the F statistic is less sensitive to small departures from the assumption than σ^2 is the same for each population. In addition choosing equal sample sizes maximizes the power of the test.

```
ANALYSIS OF VARIANCE
SOURCE    DF      SS       MS       F      P
FACTOR    2     490.0    245.0    9.55   0.003
ERROR    12     308.0     25.7
TOTAL    14     798.0

LEVEL     N     MEAN     STDEV
Prog. 1   5    55.000    5.099
Prog. 2   5    68.000    5.148
Prog. 3   5    57.000    4.950
```

FIGURE 12.5 Computer Output for the Reading-Competency Experiment

The advantage of this procedure is that SSE can be computed without having to perform all the computations required when using equation (12.11).

Computer Results for Analysis of Variance

The computational aspect of the analysis of variance procedure is devoted primarily to computing the appropriate sums of squares. When a hand calculator is used to compute the sum of squares, some computational help can be obtained by using alternate forms of the sums-of-squares formulas. In Appendix 12.1 we provide a step-by-step procedure that uses these revised formulas to compute the sums of squares for a completely randomized design.

Because of the widespread availability of statistical computer packages, problems involving large sample sizes and/or large numbers of treatments can be routinely solved. In Figure 12.5 we show part of the output obtained from the Minitab computer package when it was used to perform the ANOVA calculations for the reading-competency experiment.

We see that the computer output contains the familiar ANOVA table format. It should prove easy for you to interpret. Comparing Figure 12.5 with Table 12.3, we see that the same information is available, although some of the headings are a little different. "SOURCE" is the heading used for the source of variation column, "FACTOR" identifies the between-treatments row, and the sum of squares and degrees of freedom columns are interchanged. Note also that below the ANOVA table the computer output contains all of the sample means, their standard deviations, and respective sample sizes. The "F" column contains the F statistic used for the hypothesis test. Next to the F statistic is the p-value showing the level of significance of the sample results. Since the p-value provides the probability of obtaining a sample result more unlikely than what is observed, the p-value of .003 indicates that for a level of significance of $\alpha = .05$, the null hypothesis that $\mu_1 = \mu_2 = \mu_3$ should be rejected.

EXERCISES

7. In a completely randomized experimental design, seven experimental units were used for each of the five levels of the factor. Complete the ANOVA table shown.

SOURCE OF VARIATION	SUM OF SQUARES	DEGREES OF FREEDOM	MEAN SQUARES	F
Between treatments	300			
Error				
Total	480			

8. In an experiment designed to test the output levels of three different machines, the following results were obtained: SST $= 400$, SSTR $= 150$, $n_T = 18$. Set up the ANOVA table and test for any significant difference between the mean output levels of the three machines. Use $\alpha = .05$.

9. Managers at all levels of an organization need to have the information necessary to perform their respective tasks. A recent study investigated the effect the source has on the dissemination of the information (*Journal of Management Information Systems,* Fall 1988). In the particular study the sources of information were a superior, a peer, and a subordinate. In each case a measure of dissemination was obtained with higher values indicating greater dissemination of the information. Using $\alpha = .05$, and the data shown below, test whether or not the source of information significantly affects dissemination. What is your conclusion and what does this suggest about the use and dissemination of information?

SUPERIOR	PEER	SUBORDINATE
8	6	6
5	6	5
4	7	7
6	5	4
6	3	3
7	4	5
5	7	7
5	6	5

10. Three different methods for assembling a product were proposed by an industrial engineer. To investigate the number of units assembled correctly using each method, 30 employees were randomly selected and randomly assigned to the three proposed methods such that 10 workers were associated with each method. The number of units assembled correctly was recorded, and the analysis of variance procedure was applied to the resulting data set. The following results were obtained: SST $= 10,800$, SSTR $= 4560$.
   a. Set up the ANOVA table for this problem.
   b. Using $\alpha = .05$, test for any significant difference in the means for the three assembly methods.

11. In an experiment designed to test the breaking strength of four types of cables, the following results were obtained: SST $= 85.05$, SSTR $= 61.64$, $n_T = 24$. Set up the ANOVA table and test for any significant difference in the mean breaking strength of the four cables. Use $\alpha = .05$.

12. To test for any significant difference in the time between breakdowns for four machines, the following data were obtained:

	TIME (HOURS)					
Machine 1	6.4	7.8	5.3	7.4	8.4	7.3
Machine 2	8.7	7.4	9.4	10.1	9.2	9.8
Machine 3	11.1	10.3	9.7	10.3	9.2	8.8
Machine 4	9.9	12.8	12.1	10.8	11.3	11.5

At the $\alpha = .05$ level of significance, is there any difference in the mean time between breakdowns among the four machines?

13. In order to study the effect of temperature upon yield in a chemical process, five batches were produced under each of three temperature levels. The results are given. Construct an analysis of variance table. Using a .05 level of significance, test to see if the temperature level appears to have an effect upon the mean yield of the process.

TEMPERATURE (°C)	YIELD				
50	34	24	36	39	32
60	30	31	34	23	27
70	23	28	28	30	31

14. In an unbalanced experimental design, 12 experimental units were used for the first treatment, 15 experimental units were used for the second treatment, 20 experimental units were used for the third treatment, and 10 experimental units were used for the fourth treatment. Complete the analysis of variance table shown. Using a .05 level of significance, is there a significant difference between the treatments?

SOURCE	SUM OF SQUARES	DEGREES OF FREEDOM	MEAN SQUARE	F
Between treatments	1200			
Error				
Total	1800			

15. Develop the analysis of variance computations for the following unbalanced experimental design. Using $\alpha = .05$, is there a significant difference between the treatment means?

TREATMENT	VALUE OF OBSERVATION									
A	136	120	113	107	131	114	129	102		
B	107	114	120	104	107	109	97	114	104	94
C	92	82	85	101	89	117	110	120	98	106

12.4   Inferences About Individual Treatment Means

For the reading-competency experiment (Example 12.1) we rejected the null hypothesis that the means of the three remedial programs are the same. Sometimes we may be satisfied with this conclusion, but in other cases we may want to go a step further and determine where the differences occur. The purpose of this section is to illustrate a procedure that can be used to test for the equality of two treatment means; this procedure should be used only in cases where we have already concluded that the treatment means are not all the same.

A simple test for a difference between two treatment means is based on the use of the t distribution (as presented for the two population case in Chapter 10). The mean square error (MSE) is used as an unbiased estimate of the population variance σ^2. A confidence interval estimate of the difference between the two treatment means is given by the following expression:

Interval Estimate of the Difference Between Two Treatment Means

$$\bar{x}_1 - \bar{x}_2 \pm t_{\alpha/2} \sqrt{\text{MSE}\left(\frac{1}{n_1} + \frac{1}{n_2}\right)} \qquad (12.19)$$

where

$\bar{x}_1$ = sample mean for the first treatment

$\bar{x}_2$ = sample mean for the second treatment

$t_{\alpha/2}$ = t value based upon the degrees of freedom for error (see ANOVA table "Degrees of Freedom" column)

MSE = mean square error (see ANOVA table)

n_1 = sample size for the first treatment

n_2 = sample size for the second treatment

If the confidence interval in (12.19) includes the value 0, we have to conclude that there is no significant difference between the treatment means. In this case we cannot reject the hypothesis that no difference exists between the treatment means. However, if the confidence interval does not include the value 0, we conclude that there is a difference between the treatment means.

EXAMPLE 12.1 (continued)

Recall that the ANOVA procedure enabled us to conclude that the treatment means are not the same. Let us now use the preceding procedure to test to see if there is a significant difference between the means of Programs 1 and 3. Recall that the analysis of the data showed that the sample mean retest score for Program 1 was 55 and the sample mean retest score for Program 3 was 57. But is this sample evidence sufficient to justify the conclusion *that a significant statistical difference* exists between the mean retest scores for the two programs? To use (12.17) to provide a confidence interval that can answer this question, the following data are needed:

$$\bar{x}_1 = 55 \text{ (Program 1)} \quad \bar{x}_3 = 57 \text{ (Program 3)}$$

With $\alpha = .05$, $t_{.025} = 2.179$ (*Note:* 12 degrees of freedom for error)

$$\text{MSE} = 25.667$$

$$n_1 = n_2 = 5$$

Using these data and (12.17) we have

$$55 - 57 \pm 2.179 \sqrt{25.667\left(\frac{1}{5} + \frac{1}{5}\right)}$$

$$-2 \pm 2.179(3.204)$$

$$-2 \pm 6.98$$

Thus a 95% confidence interval for the difference between Program 1 and Program 3 is given by -8.98 to 4.98. Since this interval includes 0, we are unable to reject the hypothesis that the two programs provide the same mean score.

Using the analysis of variance procedure we concluded that the mean retest scores for the three remedial programs are not all the same. However, the preceding result provides support for concluding that the difference in the three means is not due to a difference between the means for Programs 1 and 3. With $\bar{x}_2 = 68$, Program 2 appears to have the highest mean retest score. Thus we should feel comfortable in concluding that the difference in population means is due to Program 2.

A word of caution is needed at this point: This test of an individual difference should be applied only if we reject the null hypothesis of equal population means. That is, while it may appear natural to use (12.19) to compare all possible pairs of treatment means (Program 1 versus Program 2, Program 1 versus Program 3, Program 2 versus Program 3), statistical problems can occur with this sequential approach. If a null hypothesis is true (that is, the two population means are equal) and a test is conducted using (12.19) with a type I error probability of $\alpha = .10$, there is a probability of .10 of rejecting the null hypothesis when it is really true; hence the probability of making a correct decision is .90. If two tests are conducted in this manner, the probability that a correct decision is made on both tests is $(.90)(.90) = .81$. Thus the probability that *at least* one of the tests would result in rejecting a true null hypothesis is $1 - .81 = .19$.* Thus we see that the actual type I error probability using a sequential testing procedure to test two hypotheses at the .10 significance level is .19 and not .10. For three tests at the .10 level of significance, the type I error probability is actually $1 - (.90)(.90)(.90) = 1 - .729 = .271$. Note that the probability of making a type I error increases rapidly as the number of multiple tests increases.

A simple procedure to adjust (approximately) for this increasing probability of making the type I error is to reduce the α level for each separate test. For instance, with an α level of significance desired for all tests taken together and m tests to be made, we would use α/m as the probability of making a type I error on any one test. Then the overall probability of making a type I error will approximately equal α.

Because of the difficulty of the type I error increasing when making multiple tests of difference between individual means, a variety of specialized tests have been developed. Often these tests carry the name of the developer. The better known tests for multiple

*This assumes that the two tests are independent and hence the joint probability of the two events can be obtained simply by multiplying the individual probabilities. In fact the two tests are not independent (MSE is used in each test), and hence the error involved is even greater than that shown.

comparisons include the Duncan multiple range test, the Newman-Kuels test, Tukey's test, and Scheffe's method. References in the bibliography provide details for these methods. It is recommended that these tests be considered whenever multiple comparisons among treatment means are expected to be a major concern in the study.

EXERCISES

16. Refer to Exercise 1. Use the procedure described in this section to test for the equality of the mean for manufacturers 1 and 3. Use $\alpha = .05$. What conclusion can you make after carrying out this test?

17. Refer to Exercise 4. Use the procedure described in this section to test for the equality of the mean mileage for automobiles A and B. What general conclusion can you make after carrying out this test? Use $\alpha = .05$.

18. Refer to Exercise 5. Use the procedure described in this section to test for the equality of brands Y and Z. What general conclusions can you make after carrying out this test? Use $\alpha = .05$.

12.5 Randomized Block Design

Thus far we have considered the completely randomized experimental design. Recall that in order to test for a difference in means, we compute an F value using the ratio

$$F = \frac{\text{MSTR}}{\text{MSE}} \tag{12.20}$$

A problem can arise whenever differences due to extraneous factors (ones not considered) cause the MSE term in this ratio to become large. In such cases the F value can become small, signaling no difference between treatment means, when in fact such a difference exists.

In this section we present an experimental design referred to as a *randomized block design*. The purpose of this design is to remove some of the extraneous sources of variation from the MSE term. This design tends to provide a more powerful hypothesis test in terms of the ability to detect differences between treatment means. To illustrate, let us consider a stress study for air traffic controllers.

EXAMPLE 12.2

A study directed at measuring the fatigue and stress on air traffic controllers has resulted in proposals for modification and redesign of the controller's work station. After consideration of several designs for the work station, three specific alternatives have been selected as having the best potential for reducing controller stress. The key question is: To what extent do the three work station alternatives differ in terms of their effect on controller stress? To answer this question we need to design an experiment that will provide measurements of air traffic controller stress under each alternative.

In a completely randomized design a random sample of controllers would be assigned to each work station alternative. However, it is believed that controllers differ substantially in terms of their ability to handle stressful situations. What is high stress to one controller might be only moderate or even low stress to another. Thus when considering the within-group source of variation (MSE) we must realize that this variation includes both random error and error due to individual controller differences. In fact, for this study management expected controller variability to be a major contributor to the MSE term.

One way to separate the effect of the individual differences is to use the randomized block design. This design will identify the variability stemming from individual controller differences and remove it from the MSE term. The randomized block design calls for a single sample of controllers. Each controller in the sample is tested using each of the three work station alternatives. In experimental design terminology, the work station is the *factor of interest,* and the controllers are referred to as the *blocks*. The three treatments or populations associated with the work station factor correspond to the three work station alternatives. For simplicity, we will refer to these alternatives as System A, System B, and System C.

In general, a block refers to a relatively homogeneous experimental unit. The term *blocking* refers to the process of using blocks in order to ensure that comparisons among the treatments are made within relatively homogeneous experimental units.

The *randomized* aspect of the randomized block design refers to the fact that the order in which the treatments (systems) are assigned to the blocks (controllers) is chosen randomly. If every controller were to test the three systems in the same order, any observed difference in systems might be due to the order of the test rather than to true differences in the systems.

To provide the necessary data, the three types of work stations were installed at the Cleveland Control Center in Oberlin, Ohio. Six controllers were selected at random and assigned to operate each of the systems. A follow-up interview and a medical examination of each controller participating in the study provided a measure of the stress for each controller on each system. The data are shown in Table 12.4.

A summary of the stress data collected is shown in Table 12.5. In this table we have included column totals (blocks) and row totals (treatments) as well as some sample means that will be helpful in making the sum of squares computations for the ANOVA procedure. Since lower stress values are viewed as better, the sample data available would seem to favor System B with its mean stress rating of 13. However, the usual question remains: Do the sample results justify the conclusion that the mean stress levels for the three systems differ? That is, are the differences statistically significant? An analysis of variance computation similar to the one performed for the completely randomized design can be used to answer this statistical question.

The ANOVA Procedure for a Randomized Block Design

The ANOVA procedure for the randomized block design requires us to partition the sum of squares total (SST) into three groups: sum of squares between treatments, sum of squares due to blocks, and sum of squares due to error. The formula for this partitioning is as follows:

$$SST = SSTR + SSB + SSE \qquad (12.21)$$

TABLE 12.4 A Randomized Block Design for the Air Traffic Controllers Stress Test

		Controller 1	Controller 2	Controller 3	BLOCKS Controller 4	Controller 5	Controller 6
Treatments	System A	15	14	10	13	16	13
	System B	15	14	11	12	13	13
	System C	18	14	15	17	16	13

Stress Value

TABLE 12.5 Summary of Stress Data for the Air Traffic Controllers Stress Test

		Controller 1	Controller 2	BLOCKS Controller 3	Controller 4	Controller 5	Controller 6	ROW OR TREATMENT TOTALS	TREATMENT MEANS
Treatments	System A	15	14	10	13	16	13	81	$\bar{x}_{1.} = 81/6 = 13.5$
	System B	15	14	11	12	13	13	78	$\bar{x}_{2.} = 78/6 = 13.0$
	System C	18	14	15	17	16	13	93	$\bar{x}_{3.} = 93/6 = 15.5$
Column or Block Totals		48	42	36	42	45	39	252	$\bar{\bar{x}} = \dfrac{252}{18} = 14.0$
Block Means		$\bar{x}_{.1} = \dfrac{48}{3}$ $= 16.0$	$\bar{x}_{.2} = \dfrac{42}{3}$ $= 14.0$	$\bar{x}_{.3} = \dfrac{36}{3}$ $= 12.0$	$\bar{x}_{.4} = \dfrac{42}{3}$ $= 14.0$	$\bar{x}_{.5} = \dfrac{45}{3}$ $= 15.0$	$\bar{x}_{.6} = \dfrac{39}{3}$ $= 13.0$		

This sum of squares partition is summarized in the ANOVA table for the randomized block design as shown in Table 12.6. The notation used in this table is as follows.

$$k = \text{the number of treatments}$$

$$b = \text{the number of blocks}$$

$$n_T = \text{the total sample size } (n_T = kb)$$

Note that the ANOVA table in Table 12.6 also shows that the $n_T - 1$ total degrees of freedom are partitioned such that $k - 1$ go to treatments, $b - 1$ go to blocks, and $(k - 1)(b - 1)$ go to the error term. The mean square column shows the sum of squares divided by the degrees of freedom, and $F = \text{MSTR/MSE}$ is the F ratio used to test for a significant difference among the treatment means. The primary contribution of the randomized block design is that by including blocks we have removed the individual controller differences from the MSE term and obtained a more powerful test for the stress differences in the three work station alternatives.

Computations and Conclusions

To compute the F statistic needed to test for a difference among treatment means using a randomized block design, we need to compute MSTR and MSE. To calculate these two mean squares, we must first compute SSTR and SSE; in doing so, we will also compute SSB and SST. To simplify the presentation, we will perform the calculations using four steps. In addition to k, b, and n_T as previously defined, the following notation is used:

$$x_{ij} = \text{value of the observation under treatment } i \text{ in block } j$$

$$\bar{x}_{i.} = \text{sample mean of the } i\text{th treatment}$$

$$\bar{x}_{.j} = \text{sample mean for the } j\text{th block}$$

$$\bar{\bar{x}} = \text{overall sample mean}$$

TABLE 12.6

ANOVA Table for the Randomized Block Design with k Treatments and b Blocks

SOURCE OF VARIATION	SUM OF SQUARES	DEGREES OF FREEDOM	MEAN SQUARE	F
Treatments	SSTR	$k - 1$	$\text{MSTR} = \dfrac{\text{SSTR}}{k-1}$	$\dfrac{\text{MSTR}}{\text{MSE}}$
Blocks	SSB	$b - 1$	$\text{MSB} = \dfrac{\text{SSB}}{b-1}$	
Error	SSE	$(k - 1)(b - 1)$	$\text{MSE} = \dfrac{\text{SSE}}{(k-1)(b-1)}$	
Total	SST	$n_T - 1$		

STEP 1 Compute the total sum of squares (SST):

$$\text{SST} = \sum_i \sum_j (x_{ij} - \bar{\bar{x}})^2 \tag{12.22}$$

STEP 2 Compute the sum of squares between treatments (SSTR):

$$\text{SSTR} = b \sum_i (\bar{x}_{i.} - \bar{\bar{x}})^2 \tag{12.23}$$

STEP 3 Compute the sum of squares due to blocks (SSB):

$$\text{SSB} = k \sum_j (\bar{x}_{.j} - \bar{\bar{x}})^2 \tag{12.24}$$

STEP 4 Compute the sum of squares due to error (SSE):

$$\text{SSE} = \text{SST} - \text{SSTR} - \text{SSB} \tag{12.25}$$

EXAMPLE 12.2 (continued)

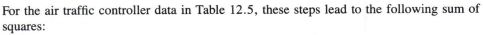

For the air traffic controller data in Table 12.5, these steps lead to the following sum of squares:

STEP 1 $\text{SST} = (15 - 14)^2 + (14 - 14)^2 + (10 - 14)^2 + \ldots + (13 - 14)^2 = 70$

STEP 2 $\text{SSTR} = 6[(13.5 - 14)^2 + (13.0 - 14)^2 + (15.5 - 14)^2] = 21$

STEP 3 $\text{SSB} = 3[(16 - 14)^2 + (14 - 14)^2 + (12 - 14)^2 + (14 - 14)^2 +$
$(15 - 14)^2 + (13 - 14)^2] = 30$

STEP 4 $\text{SSE} = 70 - 21 - 30 = 19$

These sums of squares divided by their degrees of freedom provide the corresponding mean square values shown in Table 12.7. The F ratio used to test for differences between treatment means is $\text{MSTR}/\text{MSE} = 10.5/1.9 = 5.53$. Checking the F values in Table 4 of Appendix B, we find that the critical F value at $\alpha = .05$ (2 numerator degrees of freedom and 10 denominator degrees of freedom) is 4.10. With $F = 5.53$, we reject the null hypothesis $H_0: \mu_1 = \mu_2 = \mu_3$ and conclude that the work station designs differ in terms of their mean stress effects on air traffic controllers.

■

TABLE 12.7

ANOVA Table for the Air Traffic Controller Stress Test

SOURCE OF VARIATION	SUM OF SQUARES	DEGREES OF FREEDOM	MEAN SQUARE	F
Treatments	21	2	10.5	$\dfrac{10.5}{1.9} = 5.53$
Blocks	30	5	6.0	
Error	19	10	1.9	
Total	70	17		

Before leaving this section let us make some general comments about the randomized block design. The blocking as described in this section is referred to as a *complete* block design; the word complete indicates that each block is subjected to all k treatments. That is, all controllers (blocks) were tested using all three systems (treatments). Experimental designs employing blocking where some but not all treatments are applied to each block are referred to as *incomplete* block designs. A discussion of incomplete block designs is outside the scope of this text.

In addition, note that in the air traffic controller stress test each controller in the study was required to use all three systems. While this guarantees a complete block design, in some cases blocking is carried out with "similar" experimental units in each block. For example, assume that in a pretest of air traffic controllers the population of controllers was divided into groups ranging from extremely high stress individuals to extremely low stress individuals. The blocking could still have been accomplished by having three controllers from each of the stress classifications participate in the study. Each block would then be formed from three controllers in the same stress class. The randomized aspect of the block design would be conducted by randomly assigning the three controllers in each block to the three systems.

Finally, note that the ANOVA table shown in Table 12.7 provides an F value to test for treatment effects but *not* for blocks. The reason is that the experiment was designed to test a single factor—work station design. The blocking based on individual stress differences was conducted in order to remove this variation from the MSE term. However, the study was not designed to test specifically for individual differences in stress.

Some analysts compute $F = $ MSB/MSE and use this statistic to test for significance of the blocks. Then they use the result as a guide to whether or not this type of blocking would be desired in future experiments. However, if individual stress difference is to be a factor in the study, a different experimental design should be used. A test of significance on blocks should not be performed to attempt to draw such a conclusion about a second factor.

In closing this section, we also note that alternate formulas for SST, SSTR, and SSB can be developed that can ease the computational burden when using hand calculation. In Appendix 12.2 we have included a step-by-step procedure that illustrates the use of these alternate formulas.

EXERCISES

19. An automobile dealer conducted a test to determine if the time needed to complete a minor engine tune-up depends upon whether a computerized engine analyzer or an electronic analyzer is used. Because tuneup time varies among compact, intermediate, and full-sized cars, the three types of cars were used as blocks in the experiment. The data obtained are shown.

		CAR TYPE		
		Compact	Intermediate	Full-size
Analyzer	Computerized	50	55	63
	Electronic	42	44	46

Time (minutes)

Using $\alpha = .05$, test for any significant differences.

20. An important factor in selecting software for word processing and data base management systems is the time required to learn how to use a particular system. In order to evaluate three file management systems, a firm designed a test involving five different word processing operators. Since operator variability was believed to be a significant factor, each of the five operators was trained on each of the three file management systems. The data obtained are shown.

	OPERATOR				
	1	2	3	4	5
System A	16	19	14	13	18
System B	16	17	13	12	17
System C	24	22	19	18	22

Time (hours)

Using $\alpha = .05$, test to see if there is any difference in training time for the three systems.

21. The following data were obtained for a randomized block design involving five treatments and three blocks: SST = 430, SSTR = 310, SSB = 85. Set up the ANOVA table and test for any significant differences. Use $\alpha = .05$.

22. Five different auditing procedures were compared with respect to total audit time. To control for possible variation due to the person conducting the audit, four accountants were selected randomly and treated as blocks in the experiment. The following values were obtained using the ANOVA procedure: SST = 100, SSTR = 45, SSB = 36. Using $\alpha = .05$, test to see if there is any significant difference in total audit time stemming from the auditing procedure used.

23. An experiment has been conducted for four treatments using eight blocks. Complete the analysis of variance table shown below.

SOURCE	SUM OF SQUARES	DEGREES OF FREEDOM	MEAN SQUARE	F
Between treatments	900			
Blocks	400			
Error				
Total	1800			

Using $\alpha = .05$, test for any significant differences.

24. Consider the following experimental results of a randomized block design. Make the calculations necessary to set up the analysis of variance table, and using $\alpha = .05$, test for any significant differences.

		BLOCKS				
		1	2	3	4	5
	A	10	12	18	20	8
Treatment	B	9	6	15	18	7
	C	8	5	14	18	8

Factorial Experiments

The experimental designs we have considered thus far enable statistical conclusions to be drawn about one factor. However, in some experiments we want to draw conclusions about more than one variable or factor. *Factorial experiments* and their corresponding ANOVA computations are valuable designs when simultaneous conclusions are required about two or more factors. The term *factorial* is used because the experimental conditions include all possible combinations of the factors involved. For example, if there are *a* levels of factor A and *b* levels of factors B, the experiment will involve collecting data on *ab* treatment combinations. In this section we will show the analysis of a 2-factor factorial experiment. This basic approach can be extended to experiments involving more than two factors.

EXAMPLE 12.3

To illustrate a 2-factor factorial experiment, we will consider an experiment involving the Graduate Management Admissions Test (GMAT). The GMAT is a standardized test used by graduate schools of business to evaluate an applicant's ability to pursue successfully a graduate program in business. Scores on the GMAT range from 200 to 800, with higher scores implying a higher aptitude.

In an attempt to improve the performance of undergraduate students on the exam, a university is considering the use of a GMAT preparation program. The following three preparation programs have been proposed:

1. A 3-hour review session covering the types of questions generally asked on the GMAT.
2. A 1-day program covering relevant exam material, along with the taking and grading of a sample exam.
3. An intensive 10-week course involving the identification of each student's weaknesses and the setting up of individualized programs for improvement.

Thus, one factor in this study is the GMAT preparation program, which has three treatments: 3-hour review, 1-day program, and 10-week course.

In discussing the use of these programs with the provost of the university (the senior academic administrator), the deans of the College of Business, the College of Engineering, and the College of Arts and Sciences suggested that another factor that should be considered in the experiment is the undergraduate college of the student. Their reason for including this additional factor is that they suspect that the student's undergraduate college may affect the GMAT score as much as—or even more than—the type of preparation program. Since the university has just these three colleges, this factor also has three treatments: business, engineering, and arts and sciences.

Before making a final decision as to the preparation program to adopt, the provost has requested that further study be conducted in order to determine how the proposed programs (Factor A) and the student's undergraduate college (Factor B) affect GMAT scores. Because Factor A has three treatments and Factor B has three treatments, the factorial design for this experiment will have a total of $3 \times 3 = 9$ treatment combinations. These treatment combinations are summarized in Table 12.8.

Assume that a sample of 2 students will be selected corresponding to each of the 9 treatment combinations shown in Table 12.8: two business students will take the 3-hour review, 2 will take the 1-day program, and 2 will take the 10-week course. In addition, 2 engineering students and 2 arts and sciences students will take each of the three preparation programs. In experimental design terminology the sample size of 2 for each treatment combination indicates that we have 2 replications. Additional replications and an increased sample size could easily be made, but we elected not to do so in order to minimize the computational aspects for this illustration.

This experimental design requires that 6 students who plan to take the GMAT be randomly selected from *each* of the 3 undergraduate colleges. Then 2 students from each college should be assigned randomly to each preparation program, resulting in a total of 18 students being used in the study.

Let us assume that the students have been randomly selected, have participated in the preparation program, and have taken the GMAT. The scores obtained are shown in Table 12.9.

The analysis of variance computations using the data in Table 12.9 will provide answers to the following questions:

Main effect (factor A): Do the preparation programs differ in terms of effect on GMAT scores?

Main effect (factor B): Do the undergraduate colleges differ in terms of student ability to perform on the GMAT?

Interaction effect (factor A and B): Do students in some colleges do better on one type of preparation program while others do better on a different type of preparation program?

The term *interaction* refers to a new effect that we can now study because we have used a factorial experiment. If the interaction effect has a significant impact on the GMAT scores, it will mean that the effect of the type of preparation program depends on the undergraduate college.

TABLE 12.8

Nine Treatment Combinations for the 2-Factor GMAT Experiment

		Factor B: College		
	Treatments	Business	Engineering	Arts and Sciences
Factor A:	3-hour review	1	2	3
Preparation	1-day program	4	5	6
Program	10-week course	7	8	9

TABLE 12.9

GMAT Scoures for the 2-Factor Experiment

		Factor B: College		
		Business	Engineering	Arts and Sciences
	3-hour review	500	540	480
		580	460	400
Factor A:	1-day program	460	560	420
Preparation		540	620	480
Program	10-week course	560	600	480
		600	580	410

The ANOVA Procedure

The ANOVA procedure for the 2-factor factorial experiment is similar to the completely randomized experiment and the randomized block experiment in that we once again partition the sum of squares and the degrees of freedom into their respective sources. The formula for partitioning the sum of squares for the 2-factor factorial experiments is as follows:

$$SST = SSA + SSB + SSAB + SSE \tag{12.26}$$

The partitioning of the sum of squares and degrees of freedom is summarized in Table 12.10. The following notation is used:

a = number of levels of factor A

b = number of levels of factor B

r = number of replications

n_T = total number of observations taken in the experiment; $n_T = abr$

Computations and Conclusions

To compute the F statistics needed to test for the significance of factor A, factor B, and interaction, we need to compute MSA, MSB, MSAB, and MSE. To calculate these four mean squares, we must first compute SSA, SSB, SSAB, and SSE; in doing so we will also compute SST. To simplify the presentation, we will perform the calculations using five steps. In addition to a, b, r, and n_T as previously defined, the following notation is used:

x_{ijk} = observation corresponding to the kth replicate taken from treatment i of factor A and treatment j of factor B

$\bar{x}_{i\cdot}$ = sample mean for the observations corresponding to treatment i (factor A)

TABLE 12.10

ANOVA Table for the 2-Factor Factorial Experiment with r Replications

SOURCE OF VARIATION	SUM OF SQUARES	DEGREES OF FREEDOM	MEAN SQUARE	F
Factor A	SSA	$a - 1$	$MSA = \dfrac{SSA}{a - 1}$	$\dfrac{MSA}{MSE}$
Factor B	SSB	$b - 1$	$MSB = \dfrac{SSB}{b - 1}$	$\dfrac{MSB}{MSE}$
Interaction	SSAB	$(a - 1)(b - 1)$	$MSAB = \dfrac{SSAB}{(a - 1)(b - 1)}$	$\dfrac{MSAB}{MSE}$
Error	SSE	$ab(r - 1)$	$MSE = \dfrac{SSE}{ab(r - 1)}$	
Total	SST	$n_T - 1$		

$\bar{x}_{.j}$ = sample mean for the observations corresponding to treatment j (factor B)

$\bar{x}_{ij}$ = sample mean for the observations corresponding to the combination of treatment i (factor A) and treatment j (factor B)

$\bar{\bar{x}}$ = sample mean for all n_T observations

STEP 1 Compute the total sum of squares:

$$SST = \sum_i \sum_j \sum_k (x_{ijk} - \bar{\bar{x}})^2 \qquad (12.27)$$

STEP 2 Compute the sum of squares for factor A:

$$SSA = br \sum_i (\bar{x}_{i.} - \bar{\bar{x}})^2 \qquad (12.28)$$

STEP 3 Compute the sum of squares for factor B:

$$SSB = ar \sum_j (\bar{x}_{.j} - \bar{\bar{x}})^2 \qquad (12.29)$$

STEP 4 Compute the sum of squares for interaction:

$$SSAB = r \sum_i \sum_j (\bar{x}_{ij} - \bar{x}_{i.} - \bar{x}_{.j} + \bar{\bar{x}})^2 \qquad (12.30)$$

STEP 5 Compute the sum of squares due to error:

$$SSE = SST - SSA - SSB - SSAB \qquad (12.31)$$

EXAMPLE 12.3 (continued)

Table 12.11 shows the data collected in the experiment, along with the various sums that will help us with the sum of squares computations. Using (12.26) to (12.29) we have the following sum of squares for the GMAT 2-factor factorial experiment:

STEP 1 SST = $(500 - 515)^2 + (580 - 515)^2 + (540 - 515)^2 + \cdots$
$+ (410 - 515)^2 = 82,450$

STEP 2 SSA = $(3)(2)[(493.33 - 515)^2 + (513.33 - 515)^2 + (538.33 - 515)^2]$
$= 6100$

STEP 3 SSB = $(3)(2)[(540 - 515)^2 + (560 - 515)^2 + (445 - 515)^2] = 45,300$

STEP 4 SSAB = $2[(540 - 493.33 - 540 + 515)^2 + (500 - 493.33 - 560 + 515)^2$
$+ \cdots + (445 - 538.33 - 445 + 515)^2] = 11,200$

STEP 5 SSE = $82,450 - 6100 - 45,300 - 11,200 = 19,850$

These sums of squares divided by their corresponding degrees of freedom, as shown in Table 12.12, provide the appropriate mean square values for testing the two main effects (preparation program and undergraduate college) and the interaction effect. The F ratio used to test for differences among preparation programs is 1.38. The critical F value at $\alpha = .05$ (with 2 numerator degrees of freedom and 9 denominator degrees of freedom) is 4.26. With $F = 1.38$, we cannot reject the null hypothesis and must conclude that there is not a significant difference in the mean GMAT scores for the 3 preparation programs. However, for the undergraduate college effect, $F = 10.27$ exceeds the critical

TABLE 12.11   GMAT Summary Data for the 2-Factor Experiment

Treatment Combination Totals	Factor B: College			Row Totals	Factor A Means
	Business	Engineering	Arts and Science		
3-hour review	$\dfrac{\begin{array}{r}500\\580\end{array}}{1080}$ $\bar{x}_{11} = \dfrac{1080}{2} = 540$	$\dfrac{\begin{array}{r}540\\460\end{array}}{1000}$ $\bar{x}_{12} = \dfrac{1000}{2} = 500$	$\dfrac{\begin{array}{r}480\\400\end{array}}{880}$ $\bar{x}_{13} = \dfrac{880}{2} = 440$	2960	$\bar{x}_{1\cdot} = \dfrac{2960}{6} = 493.33$
Factor A: Preparation Program **1-day program**	$\dfrac{\begin{array}{r}460\\540\end{array}}{1000}$ $\bar{x}_{21} = \dfrac{1000}{2} = 500$	$\dfrac{\begin{array}{r}560\\620\end{array}}{1180}$ $\bar{x}_{22} = \dfrac{1180}{2} = 590$	$\dfrac{\begin{array}{r}420\\480\end{array}}{900}$ $\bar{x}_{23} = \dfrac{900}{2} = 450$	3080	$\bar{x}_{2\cdot} = \dfrac{3080}{6} = 513.33$
10-week course	$\dfrac{\begin{array}{r}560\\600\end{array}}{1160}$ $\bar{x}_{31} = \dfrac{1160}{2} = 580$	$\dfrac{\begin{array}{r}600\\580\end{array}}{1180}$ $\bar{x}_{32} = \dfrac{1180}{2} = 590$	$\dfrac{\begin{array}{r}480\\410\end{array}}{890}$ $\bar{x}_{33} = \dfrac{890}{2} = 445$	3230	$\bar{x}_{3\cdot} = \dfrac{3230}{6} = 538.33$
Column totals	3240	3360	2670	9270 ← Overall Total	$\bar{\bar{x}} = \dfrac{9270}{18} = 515$
Factor B Means	$\bar{x}_{\cdot 1} = \dfrac{3240}{6}$ $= 540$	$\bar{x}_{\cdot 2} = \dfrac{3360}{6}$ $= 560$	$\bar{x}_{\cdot 3} = \dfrac{2670}{6}$ $= 445$		

F value of 4.26. Thus the analysis of variance results allow us to conclude that there is a difference in the mean GMAT test scores among the three undergraduate colleges; that is, the three undergraduate colleges do not provide the same preparation for performance on the GMAT. Finally, the interaction F value of $F = 1.27$ (critical F value = 3.63 at $\alpha = .05$) means that we cannot identify a significant interaction effect. Thus there is no reason to believe that the 3 preparation programs differ in their ability to prepare students from the different colleges for the GMAT.

Undergraduate college was found to be a significant factor. Checking the calculations in Table 12.11 we see that the sample means are as follows: business students $\bar{x}_{.1} = 540$, engineering students $\bar{x}_{.2} = 560$, and arts and sciences students $\bar{x}_{.3} = 445$. Tests on individual treatment means can be conducted; yet after reviewing the three sample means we would anticipate no difference in preparation for business and engineering graduates. However, the arts and sciences students appear to be significantly less prepared for the GMAT than students in the other colleges. Perhaps this observation will lead the provost to consider other options for assisting these students in preparing for the Graduate Management Admission Test.

■

Because of the computational effort involved in any modest to large-size factorial experiment, the computer usually plays an important role in making and summarizing the analysis of variance computations. The computer printout for the analysis of variance of the GMAT 2-factor factorial experiment is shown in Figure 12.6.

FIGURE 12.6 Computer Output for 2-Factor Design

```
ANALYSIS OF VARIANCE   GMAT

SOURCE          DF        SS        MS
FACTOR A         2      6100      3050
FACTOR B         2     45300     22650
INTERACTION      4     11200      2800
ERROR            9     19850      2206
TOTAL           17     82450
```

TABLE 12.12

ANOVA Table for the 2-Factor GMAT Experiment

SOURCE OF VARIATION	SUM OF SQUARES	DEGREES OF FREEDOM	MEAN SQUARE	F
Factor A	6,100	2	3,050	3050/2206 = 1.38
Factor B	45,300	2	22,650	22,650/2206 = 10.27
Interaction	11,200	4	2,800	2800/2206 = 1.27
Error	19,850	9	2,206	
Total	82,450	17		

25. A mail-order catalog firm designed a factorial experiment to test the effect of the size of a magazine advertisment and the advertisement design on the number of catalog requests received (1000s). Three advertising designs and two different-size advertisements were considered. The data obtained are shown:

		SIZE OF ADVERTISEMENT	
		Small	Large
Design	A	8 12	12 8
	B	22 14	26 30
	C	10 18	18 14

Use the ANOVA procedure for factorial designs to test for any significant effects due to type of design, size of advertisement, or interaction. Use $\alpha = .05$.

26. An amusement park has been studying methods for decreasing the waiting time on rides by loading and unloading rides more efficiently. Two alternative loading/unloading methods have been proposed. To account for potential differences due to the type of ride and the possible interaction between the method of loading and unloading and the type of ride, a factorial experiment was designed. Using the data shown below, test for any significant effect due to the loading and unloading method, the type of ride, and interaction. Use $\alpha = .05$.

	TYPE OF RIDE		
	Roller Coaster	Screaming Demon	Log Flume
Method 1	41 43	52 44	50 46
Method 2	49 51	50 46	48 44

Waiting Time (minutes)

27. The calculations for a factorial experiment involving 4 levels of factor A, 3 levels of factor B, and 3 replications resulted in the following data: SST = 280, SSA = 26, SSB = 23, SSAB = 175. Set up the ANOVA table and test for any significant main effect and any interaction effect. Use $\alpha = .05$.

SUMMARY

In this chapter we discussed several types of experimental designs that can be used to test for differences among means of several populations or treatments. Specifically, we introduced the single-factor completely randomized, the randomized block, and the 2-factor factorial experimental designs. The completely randomized design and the randomized block designs were used to draw conclusions about differences in the means of a single factor. The primary purpose of the blocking in the randomized block design was to remove extraneous sources of variation from the error term. This blocking provided a better estimate of the error variance and a better test to determine whether or not the population or treatment means of the single factor differed significantly.

Although the various experimental designs required different formulas and computations, we showed that the basis for the statistical tests is the development of independent estimates of the population variance σ^2. In the single factor case one estimator, MSTR, is based upon the variation between the treatments. This value provides and unbiased estimate of σ^2 only if the means μ_1, μ_2, . . . μ_k are all equal. A second estimator, MSE, is based upon the variation of the observations within each sample; this estimator will always provide an unbiased estimate of σ^2. By computing F = MSTR/MSE and using the F distribution, we developed a rejection rule for determining whether or not to reject the null hypothesis that the treatment means are equal. In all the experimental designs considered, the partitioning of the sum of squares and degrees of freedom into their various sources enabled us to compute the appropriate values for making the analysis of variance calculations and tests.

Whenever an analysis of variance conclusion results in the rejection of the equal-means hypothesis (H_0), we may want to consider testing for a difference between the individual treatment means. We showed how the t distribution test could be used to compare two treatment means. However, we warned against indiscriminate use of this testing procedure because of the increasing probability of making a type I error. By simultaneously making several tests for individual differences, the probability of erroneously claiming that a difference exists increases. Several specialized tests, discussed in more advanced tests, are available if multiple comparisons of the treatment means are to be considered.

GLOSSARY

Statistical design of experiments The process of planning an experiment so that appropriate data can be collected and analyzed using statistical methods.

Analysis of variance (ANOVA) procedure A statistical approach for determining whether or not the means of several different populations are equal.

Factor Another word for the variable of interest in an ANOVA procedure.

Treatments Different levels of a factor.

Experimental units The objects involved in the experiment.

Completely randomized design An experimental design where the treatments are randomly assigned to the experimental units.

Replication The number of times each experimental condition is repeated in an experiment.

Mean square The sum of squares divided by its corresponding degrees of freedom.

ANOVA table A table used to summarize the analysis of variance computations and results. It contains columns showing the source of variation, the degrees of freedom, the sum of squares, the mean squares, and the F values.

Partitioning The process of allocating the total sum of squares and degrees of freedom into the various components.

Block A relatively homogeneous experimental unit.

Blocking The process of using blocks in order to ensure that comparisons among the treatments are made within relatively homogeneous experimental units.

Randomized block design An experimental design employing blocking. The experimental unit(s) within a block are randomly ordered for the treatments.

Factorial experiments An experimental design that permits statistical conclusions about two or more factors. All levels of each factor are considered with all levels of the other factors in order to specify the experimental conditions for the experiment.

Interaction   The response produced when the levels of one factor interact with the levels of another factor in influencing the response variable.

Completely Randomized Designs

SUM OF SQUARES ABOUT THE MEAN

$$\text{SST} = \sum_i \sum_j (x_{ij} - \bar{\bar{x}})^2 \tag{12.14}$$

PARTITIONING THE SUM OF SQUARES

$$\text{SST} = \text{SSTR} + \text{SSE} \tag{12.15}$$

SUM OF SQUARES BETWEEN TREATMENTS

$$\text{SSTR} = \sum_i n_i(\bar{x}_i - \bar{\bar{x}})^2 \tag{12.17}$$

SUM OF SQUARES DUE TO ERROR

$$\text{SSE} = \text{SST} - \text{SSTR} \tag{12.18}$$

THE *F* VALUE

$$F = \frac{\text{MSTR}}{\text{MSE}} \tag{12.20}$$

Randomized Block Designs

TOTAL SUM OF SQUARES

$$\text{SST} = \sum_i \sum_j (x_{ij} - \bar{\bar{x}})^2 \tag{12.22}$$

SUM OF SQUARES BETWEEN TREATMENTS

$$\text{SSTR} = b \sum_i (\bar{x}_{i\cdot} - \bar{\bar{x}})^2 \tag{12.23}$$

SUM OF SQUARES DUE TO BLOCKS

$$\text{SSB} = k \sum_j (\bar{x}_{\cdot j} - \bar{\bar{x}})^2 \tag{12.24}$$

SUM OF SQUARES DUE TO ERROR

$$\text{SSE} = \text{SST} - \text{SSTR} - \text{SSB} \tag{12.25}$$

Factorial Experiments

TOTAL SUM OF SQUARES

$$\text{SST} = \sum_i \sum_j \sum_k (x_{ijk} - \bar{\bar{x}})^2 \tag{12.27}$$

SUM OF SQUARES FOR FACTOR A

$$\text{SSA} = br \sum_i (\bar{x}_{i\cdot} - \bar{\bar{x}})^2 \tag{12.28}$$

SUM OF SQUARES FOR FACTOR B

$$\text{SSB} = ar \sum_j (\bar{x}_{\cdot j} - \bar{\bar{x}})^2 \tag{12.29}$$

SUM OF SQUARE FOR INTERACTION

$$SSAB = r \sum_i \sum_j (\bar{x}_{ij} - \bar{x}_{i.} - \bar{x}_{.j} + \bar{\bar{x}})^2 \qquad (12.30)$$

SUM OF SQUARES DUE TO ERROR

$$SSE = SST - SSA - SSB - SSAB \qquad (12.31)$$

REVIEW QUIZ

TRUE/FALSE

1. In a completely randomized design, the treatments must be randomly assigned to the experimental units.

2. In an experimental design involving three treatments, adding a replication requires adding three experimental units.

3. The analysis of variance procedure can be used to test the null hypothesis that k population means are equal.

4. The only assumption required for the ANOVA test on the difference of k population means is that the response variable is normally distributed.

5. In computing the between-treatments estimate of the population variance, we cannot assume that the null hypothesis is true.

6. The degrees of freedom associated with the sum of squares between treatments is the same as the number of treatments.

7. The within-treatments estimate of the population variance is called the mean square due to error.

8. The sum of squares due to error has a number of degrees of freedom equal to the total sample size minus the number of treatments minus 1.

9. In a completely randomized design the degrees of freedom associated with the total sum of squares is the sum of the degrees of freedom for SSTR and the degrees of freedom for SSE.

10. An unbalanced experimental design is one in which the number of experimental units exceeds the number of treatments.

11. Whenever a randomized block design is used, the F test should not be used.

12. If the null hypothesis that k population means are equal is rejected, we can then use a chi-square test to determine which individual means differ.

13. In a 2-factor factorial experiment, the total sum of squares is partitioned into two components: main effect factor A and main effect factor B.

14. In a factorial experiment, the number of levels of factor A must be equal to the number of levels of factor B.

MULTIPLE CHOICE

Use the following information to answer questions 15–19. A statistics professor wishes to know the effect of class format on student learning, as measured by improvement on examination scores from the beginning to the end of the semester. The five class formats to be studied reflect different emphases on homework problems and computer exercises. Sixty students are randomly selected for this study; 12 students are randomly assigned to each class format.

15. In this example, the term *factor* is illustrated by
   a. the change in exam scores
   b. the class formats
   c. the different emphases on homework problems and computer exercises.
   d. the 12 students in the sample from each class format
   e. the 60 students in the random sample

16. The term *treatment* is illustrated by
   a. the change in exam scores
   b. the class formats
   c. the different amounts of homework and computer work in the different class formats
   d. the 12 students assigned to each class format
   e. the 60 students in the random sample

17. The term *replication* is illustrated by
   a. the change in exam scores
   b. the class formats
   c. the different amounts of homework and computer work in the different class formats
   d. the 12 students assigned to each class format
   e. the 60 students in the random sample

18. The term *response* is illustrated by
   a. the change in exam scores
   b. the class formats
   c. the different amounts of homework and computer work in the different class formats
   d. the 12 students in the sample from each class format
   e. the 60 students in the random sample

19. This example best reflects which experimental design?
   a. randomized block design
   b. completely randomized design
   c. factorial experiments
   d. individual treatment means

20. What is the sampling distribution used in the test for the equality of more than two population means?
   a. normal distribution
   b. t distribution
   c. chi-square distribution
   d. F distribution

21. Which of the following is a required condition for using an ANOVA procedure on data from several populations?
   a. The data is obtained from independently selected samples.
   b. The populations are all normally distributed.

c. The populations have the same variance.

d. All of the above are necessary conditions.

22. An ANOVA procedure is used for data that was obtained from three sample groups, each comprised of four observations. The degrees of freedom for the critical value of F is

   a. 11

   b. 2, 9

   c. 2, 11

   d. 3, 11

23. A factorial experiment that involved four levels of factor A, five levels of factor B, and three replications would have error degrees of freedom equal to

   a. 60

   b. 24

   c. 40

   d. 59

24. To compute the F statistics needed to test the significance of factor A, factor B, and interaction in a 2-factorial experiment, we need to compute

   a. MSA and MSB

   b. MSA, MSB, and MSAB

   c. MSA, MSB, and MSE

   d. MSA, MSB, MSAB, and MSE

   e. MSAB and MSE

SUPPLEMENTARY EXERCISES

28. In your own words explain what the ANOVA procedure is used for.

29. What has to be true in order for MSTR to provide a good estimate of σ^2? Explain.

30. Why do we assume that the populations sampled all have the same variance when we apply the ANOVA procedure?

31. Explain why MSTR and MSE provide two independent estimates of σ^2.

32. Explain why MSTR provides an inflated estimate of σ^2 when the population means are not the same.

33. A simple random sample of the asking price ($1000s) of four houses currently for sale in each of two residential areas resulted in the following data.

AREA 1	AREA 2
92	90
89	102
98	96
105	88

   a. Use the procedure developed in Chapter 10 and test if the mean asking price is the same in both areas. Use $\alpha = .05$.

   b. Use the ANOVA procedure to test if the mean asking price is the same. Compare your analysis with (a). Use $\alpha = .05$.

34. Suppose that in Exercise 33 data were collected for another residential area. The asking prices for the simple random sample from the third area were as follows: $81,000, $86,000, $75,000, and $90,000. Are the mean asking prices for all three areas the same? Use $\alpha = .05$.

35. An analysis of the number of units sold by ten salespersons in each of four sales territories resulted in the following data.

	SALES TERRITORY			
	1	2	3	4
Number of salespersons	10	10	10	10
Average number sold ($\bar{x}$)	130	120	132	114
Sample variance (s^2)	72	64	69	67

Test at the $\alpha = .05$ level if there is any significant difference in the mean number of units sold in the four sales territories.

36. Suppose that in Exercise 35 the number of salespersons in each territory was as follows: $n_1 = 10$, $n_2 = 12$, $n_3 = 10$, and $n_4 = 15$. Using the same data for $\bar{x}$ and s^2 as given in Exercise 35, test at the $\alpha = .05$ level if there is any significant difference in the mean number of units sold in the four sales territories.

37. Consider an analysis of variance for the three-group data shown.

GROUP 1	GROUP 2	GROUP 3
54	63	73
60	60	71
57	59	77
55	59	64
69	69	70

a. What are the hypotheses that will be tested using the analysis of variance procedure?
b. What assumptions are made?
c. Calculate the sum of squares.
d. Show the analysis of variance table.

38. Three different assembly methods have been proposed for a new product. In order to determine which assembly method results in the greatest number of parts produced per hour, 30 workers were randomly selected and assigned to use one of the proposed methods. The number of units produced by each worker is given.

Method A	97	73	93	100	73	91	100	86	92	95
Method B	93	100	93	55	77	91	85	73	90	83
Method C	99	94	87	66	59	75	84	72	88	86

Using $\alpha = .05$, test to see if the mean number of parts produced with each method is the same.

39. In order to test to see if there is any significant difference in the mean number of units produced per week by each of three production methods, the following data were collected.

METHOD 1	METHOD 2	METHOD 3
58	52	48
64	63	57

METHOD 1	METHOD 2	METHOD 3
55	65	59
66	58	47
67	62	49

At the $\alpha = .05$ level of significance is there any difference in the means for the three methods?

40. Pappashales Restaurant is considering introducing a new specialty sandwich. For a determination of the effect of sandwich price on sales, the new sandwich was test-marketed at three prices in selected company restaurants. The following data, in terms of the number of sandwiches sold per day, were obtained.

$1.49	$1.79	$1.99
925	910	860
850	845	935
930	905	820
955	860	845

At the $\alpha = .05$ level of significance is there any difference in the mean number of sandwiches sold per day for the three prices? What should management of Pappashales do?

41. Hargreaves Automotive Parts, Inc. would like to compare the mileage for four different types of brake linings. Thirty linings of each type were produced and placed on a fleet of rental cars. The number of miles that each brake lining lasted until it no longer met the required federal safety standard was recorded, and an average value was computed for each type of lining. The following data were obtained.

	SAMPLE SIZE	SAMPLE MEAN	STANDARD DEVIATION
Type A	30	32,000	1,450
Type B	30	27,500	1,525
Type C	30	34,200	1,650
Type D	30	30,300	1,400

Using $\alpha = .05$, test to see if the corresponding population means are equal.

42. In a study of the educational skills of students in various school environments, samples of students were taken from inner-city schools, suburban schools, and rural schools. Test scores on reading skills of the students are shown below. Using a .05 level of significance, test for a significant difference between the mean reading scores of the three groups.

SCHOOL SYSTEM	MEAN READING SCORES									
Inner city	93	62	99	84	67	77	79	76	78	95
Suburban	72	79	75	65	62	69	93	80	84	81
Rural	94	92	95	78	67	87	95	90	95	97

43. A manufacturer of batteries for electronic toys and calculators is considering three new battery designs. An attempt was made to determine if the mean lifetime in hours is the same for each of the three designs. The following battery lifetime data were collected.

DESIGN A	DESIGN B	DESIGN C
78	112	115
98	99	101
88	101	100
96	116	120

Test to see if the population means are equal. Use $\alpha = .05$.

44. At the end of each quarter, college students submit course evaluations to university administrators. An overall evaluation of each course an instructor teaches is then computed. Currently, four faculty are being considered for a teacher-of-the-year award. The overall course evaluation rating for each course taught by each instructor is shown below.

INSTRUCTOR	COURSE EVALUATION RATING							
Black	88	80	79	68	96	69			
Jennings	87	97	82	85	99	99	85	94	
Swanson	88	76	68	82	85	82	84	83	81
Wilson	80	85	56	71	89	87			

a. Do the data support the conclusion that student evaluations indicate a difference in teaching abilities among the four candidates? Use a .05 level of significance.

b. If a significant difference exists, use a test on individual treatment means to select the higher-rated instructors.

45. Refer to Exercise 43. Use the procedure described in Section 12.4 to test for the equality of the mean for Design A and Design B. What conclusion can you make after carrying out the above test? Use $\alpha = .05$.

46. In Exercise 40, would it make sense to do a test on individual treatment means as described in Section 12.4? Explain.

47. A research firm tests the miles per gallon obtained with three brands of gasoline. Because of different gasoline performance characteristics in different brands of automobiles, five brands of automobiles are selected and treated as blocks in the experiment. That is, each brand of automobile is tested with each type of gasoline. The results of the experiment are shown below:

Gasoline Brands	BLOCKS: AUTOMOBILES				
	A	B	C	D	E
I	18	24	30	22	20
II	21	26	29	25	23
III	20	27	34	24	24

Miles per gallon

With $\alpha = .05$, is there a significant difference in the mean miles per gallon characteristics of the three brands of gasoline?

48. Analyze the experimental data provided in Exercise 47 using the ANOVA procedure for completely randomized designs. Compare your findings with those obtained in Exercise 47. What is the advantage of attempting to remove the block effect?

49. The following data were obtained for a randomized block design involving three treatments and four blocks: SST = 148, SSTR = 84, SSB = 50. Set up the ANOVA table and test for any significant differences. Use $\alpha = .05$.

50. Three different road-repair compounds were tested at four different highway locations. At each location, three sections of the road were repaired, with each section using one of the three compounds. Data were then collected on the number of days of traffic usage before additional repair was required. These data are as follows.

Compound	LOCATION			
	1	2	3	4
A	99	73	85	103
B	82	72	85	97
C	81	79	82	86

Using $\alpha = .01$, test to see if there is a significant difference in the compounds.

51. Four types of fertilizers were studied in an attempt to determine the effect of the fertilizers on the yield of corn crops. Five locations (blocks) were selected for test purposes. Each location represented a different type of soil condition that might be present for the corn. The four fertilizers were tested at each location and the yield data collected. Using the data shown and a .01 level of significance, determine if there is evidence that different fertilizers will provide a different corn crop yield.

Fertilizers	LOCATION				
	1	2	3	4	5
A	21	24	26	22	27
B	20	22	23	21	24
C	16	23	22	21	23
D	15	20	22	21	22

52. A factorial experiment was designed to test for any significant differences in the time needed to perform English to foreign language translations using two computerized language translators. Since the type of language translated was also considered a significant factor, translations were made using both systems for three different languages: Spanish, French, and German. Use the data shown:

	LANGUAGE		
	Spanish	French	German
System 1	8	10	12
	12	14	16
System 2	6	14	16
	10	16	22

Translation Time (hours)

Test for any significant differences due to language translator, type of language, and interaction. Use $\alpha = .05$.

53. A manufacturing company designed a factorial experiment in order to determine if the number of defective parts produced by two machines differed. It was believed that the number of defective parts produced also depends upon whether or not raw material needed by the machine was loaded manually or using an automatic feed system. Using $\alpha = .05$ and the following data, to test for any significant effect due to machine, loading system, and interaction.

	LOADING SYSTEM	
	Manual	Automatic
Machine 1	30	30
	34	26
Machine 2	20	24
	22	28

Number of Defective Parts

COMPUTER EXERCISE

As part of a long-term study of individuals 65 years of age or older, sociologists and physicians at Upstate Medical Center recently conducted an experiment to study the relationship between geographic location and depression levels. A sample of 60 individuals was selected; 20 of the individuals were lifetime residents of Florida, 20 were lifetime residents of New York, and 20 were lifetime residents of North Carolina. To account for any possible effects that might be due to the health status of the individual, 50% of the subjects that were sampled from each state had a chronic health condition (arthritis, hypertension, hearing loss, or heart ailment), and 50% did not have a chronic health condition.

Each individual who participated in the study was given a standardized test in order to measure depression. Higher test scores indicate higher levels of depression. The data are shown. A code of 1 is used to indicate that the subject had a chronic health condition and a code of 2 is used to indicate that the subject did not have a chronic health condition.

FLORIDA		NEW YORK		NORTH CAROLINA	
Score	Condition	Score	Condition	Score	Condition
7	2	9	2	14	1
10	1	19	1	14	1
12	1	10	1	20	1
19	1	11	2	15	1
11	1	8	2	6	2
2	2	17	1	5	2
4	2	15	1	12	2
13	1	11	1	8	2
14	1	9	2	12	1
17	1	6	2	9	2
8	2	7	2	16	1
5	2	12	1	12	1
11	1	20	1	7	2
18	1	10	2	13	1
7	2	13	1	6	2
4	2	10	2	15	1
3	2	12	1	4	2
4	2	15	1	6	2
1	2	4	2	6	2
9	1	10	2	10	1

QUESTIONS

1. Use numerical and graphical measures to summarize the data.

2. Ignoring the data on health condition, does there appear to be a significant difference in depression levels for residents of the three states?

3. Considering health condition as a block, does there appear to be a significant difference in depression levels for residents of the three states? Explain.

APPENDIX 12.1
Computational Procedure for a Completely Randomized Design

The following step-by-step procedure is designed to ease the burden in computing the appropriate sums of squares for completely randomized designs. The formulas shown can be applied to both balanced and unbalanced designs.

Notation

x_{ij} = value of the jth observation under treatment i

T_i = sum of all observations for treatment i

T = sum of all observations

n_i = sample size for the ith treatment

n_T = total sample size for the experiment

Procedure

STEP 1 Compute the sum of squares about the mean (SST):

$$\text{SST} = \sum_i \sum_j x_{ij}^2 - \frac{T^2}{n_T}$$

STEP 2 Compute the sum of squares due to treatments (SSTR):

$$\text{SSTR} = \sum_i \frac{T_i^2}{n_i} - \frac{T^2}{n_T}$$

STEP 3 Compute the sum of squares due to error (SSE):

$$\text{SSE} = \text{SST} - \text{SSTR}$$

EXAMPLE

Using this computational procedure for the reading-competency data in Table 12.1, we obtain the following results:

STEP 1 $\text{SST} = 54,798 - \dfrac{(900)^2}{15} = 798$

STEP 2 $\text{SSTR} = \dfrac{(275)^2}{5} + \dfrac{(340)^2}{5} + \dfrac{(285)^2}{5} - \dfrac{(900)^2}{15} = 490$

STEP 3 $\text{SSE} = 798 - 490 = 308$

APPENDIX 12.2
Computational Procedure for a Randomized Block Design

The following step-by-step procedure is designed to ease the burden in computing the appropriate sums of squares for randomized block designs.

Notation

$$x_{ij} = \text{value of the observation under treatment } i \text{ in block } j$$

$$T_{i\cdot} = \text{the total of all observations in treatment } i$$

$$T_{\cdot j} = \text{the total of all observations in block } j$$

$$T = \text{the total of all observations}$$

$$k = \text{the number of treatments}$$

$$b = \text{the number of blocks}$$

$$n_T = \text{the total sample size } (n_T = kb)$$

Procedure

STEP 1 Compute the total sum of squares (SST):

$$\text{SST} = \sum_i \sum_j x_{ij}^2 - \frac{T^2}{n_T}$$

STEP 2 Compute the sum of squares between treatments (SSTR):

$$\text{SSTR} = \frac{\sum_i T_{i\cdot}^2}{b} - \frac{T^2}{n_T}$$

STEP 3 Compute the sum of squares due to blocks (SSB):

$$\text{SSB} = \frac{\sum_j T_{\cdot j}^2}{k} - \frac{T^2}{n_T}$$

STEP 4 Compute the sum of squares due to error (SSE):

$$SSE = SST - SSTR - SSB$$

For the air traffic controller data in Table 12.5, these steps lead to the sum of squares:

STEP 1 $SST = 3598 - \dfrac{(252)^2}{18} = 70$

STEP 2 $SSTR = \dfrac{(81)^2 + (78)^2 + (93)^2}{6} - \dfrac{(252)^2}{18} = 21$

STEP 3 $SSB = \dfrac{(48)^2 + (42)^2 + \cdots + (39)^2}{3} - \dfrac{(252)^2}{18} = 30$

STEP 4 $SSE = 70 - 21 - 30 = 19$

APPENDIX 12.3
Computational Procedure for 2-Factor Factorial Design

Notation

x_{ijk} = observation corresponding to the kth replicate taken from treatment i of factor A and treatment j of factor B

$T_{i\cdot}$ = total of all observations in treatment i (factor A)

$T_{\cdot j}$ = total of all observations in treatment j (factor B)

T_{ij} = total of all observations in the combination of treatment i (factor A) and treatment j (factor B)

T = total of all observations

a = number of levels of factor A

b = number of levels of factor B

r = number of replications

n_T = total number of observations; $n_T = abr$

Procedure

STEP 1 Compute the total sum of squares:

$$\text{SST} = \sum_i \sum_j \sum_k x_{ijk}^2 - \frac{T^2}{n_T}$$

STEP 2 Compute the sum of squares for factor A:

$$\text{SSA} = \frac{\sum_i T_{i\cdot}^2}{br} - \frac{T^2}{n_T}$$

STEP 3 Compute the sum of squares for factor B:

$$\text{SSB} = \frac{\sum_j T_{\cdot j}^2}{ar} - \frac{T^2}{n_T}$$

STEP 4 Compute the sum of squares for the interaction:

$$SSAB = \frac{\displaystyle\sum_i \sum_j T_{ij}^2}{r} - \frac{T^2}{n_T} - SSA - SSB$$

STEP 5 Compute the sum of squares due to error:

$$SSE = SST - SSA - SSB - SSAB$$

For the GMAT data in Table 12.11, these steps lead to the following sum of squares:

STEP 1 $SST = 4{,}856{,}500 - \dfrac{(9270)^2}{18} = 82{,}450$

STEP 2 $SSA = \dfrac{(2960)^2 + (3080)^2 + (3230)^2}{6} - \dfrac{(9270)^2}{18} = 6100$

STEP 3 $SSB = \dfrac{(3240)^2 + (3360)^2 + (2670)^2}{6} - \dfrac{(9270)^2}{18} = 45{,}300$

STEP 4 $SSAB = \dfrac{(1080)^2 + (1000)^2 + \cdots + (890)^2}{2} - \dfrac{(9270)^2}{18}$
$- 6100 - 45{,}300 = 11{,}200$

STEP 5 $SSE = 82{,}450 - 6100 - 45{,}300 - 11{,}200 = 19{,}850$

13

Linear Regression and Correlation

What You Will Learn in This Chapter:

- How to use the least squares method to develop an equation that estimates the relationship between two variables

- How to compute and interpret the coefficient of determination

- How to use the t and F distributions to test for significant relationships between variables

- How to use a regression equation for estimation and prediction

- How to compute and interpret the sample correlation coefficient

Contents

College Basketball Computer Ratings

Vast amounts of data are compiled for college sports teams. For example, data for football teams include statistics on running plays, passing plays, running yardage, passing yardage, number of interceptions, number of fumbles, and so on. For basketball teams data are recorded on points scored, field-goal percentages, turnovers, rebounds, and so on. These and other data are used by individuals and sports publications to develop rankings and strength indexes for teams.

USA Today publishes college basketball computer ratings for the 294 Division I basketball teams. The rating system was developed by Jeff Sagarin, a 1970 Massachusetts Institute of Technology mathematics graduate. Using his system, a strength rating is computed for each team. Then, the 294 teams are ranked according to their strength ratings.

The strength ratings can also be used to predict the victory margin for games. For the visiting team, the predicted score is its strength rating. For the home team, the predicted score is the strength rating plus 4½. For each game, the actual victory margin is the winning team's score minus the losing team's score. The predicted victory margin is the predicted score for the winning team minus the predicted score for the losing team.

A sample of 10 college basketball games was taken to investigate the accuracy of the predicted victory margin. The following table shows the actual score and the strength rating for each team. A regression model was developed with the predicted victory margin as the independent variable and the actual victory margin as the dependent variable.

Indiana's Dean Garrett goes up and blocks a shot by Nevada–Las Vegas' Jarvis Basnight in an NCAA tournament game.

Results showed a significant correlation between the predicted and actual victory margins.

Based on "College Basketball Computer Ratings," *USA Today* (December 13, 1988).

Results of Ten Basketball Games December 13–15, 1988
(Strength Ratings in Parentheses)

VISITING TEAM	SCORE	HOME TEAM	SCORE
Eastern Michigan (77.06)	57	Michigan (101.07)	80
Jackson State (65.75)	71	Iowa (96.63)	86
Georgia Southern (80.01)	80	Eastern Kentucky (61.24)	69
Seton Hall (92.66)	96	Rutgers (72.22)	70
Niagara (70.32)	78	St. Bonaventure (70.57)	81
Fairfield (63.01)	48	Connecticutt (85.15)	71
S. Carolina St. (73.29)	70	Clemson (80.13)	93
Monmouth (57.43)	70	Maryland (80.53)	74
Illinois-Chicago (73.95)	74	Michigan State (83.45)	96
Oral Roberts (71.37)	75	Georgetown (87.48)	91

Regression and correlation analysis are statistical methods developed to study relationships between and among variables. An objective of regression analysis is to use one or more independent variables to predict the value of a dependent variable. A mathematical equation, developed by the method of least squares, is used for this purpose. In correlation analysis we are not concerned with identifying a mathematical equation relating an independent and dependent variable; we are concerned only with determining the extent to which two variables are linearly related. Correlation analysis is a procedure for making this determination and, if such a relationship exists, providing a measure of the relative strength of the relationship.

In this chapter, we consider regression analysis situations involving one independent and one dependent variable for which the relationship between the variables is approximated by a straight line. This is called *simple linear regression* analysis. Regression analysis involving two or more independent variables is called *multiple regression* analysis; multiple regression is covered in Chapters 14 and 15.

13.1   The Least Squares Method

In this section, we show how the least squares method can be used to develop a linear equation relating two variables. The variable that is being predicted by the equation is called the *dependent* variable. The variable being used to predict the value of the dependent variable is called the *independent* variable. Common notation is to use y to denote the dependent variable and x to denote the independent variable.

EXAMPLE 13.1

New Hampshire Gas and Electric would like to develop an equation that can be used to predict the daily demand for electricity using the temperature forecast. The utility's desire to predict demand for electricity would suggest making daily demand for electricity the dependent variable for the analysis and the temperature forecast the independent variable. Thus, y = the daily demand for electricity and x = the temperature forecast.

■

EXAMPLE 13.2

A medical laboratory at Duke University estimates the amount of protein in liver samples through the use of regression analysis. A spectrometer emitting light shines through a subtance containing the sample, and the amount of light absorbed is used to estimate the amount of protein in the sample. The amount of light absorbed is the independent variable (x); the amount of protein is the dependent variable (y).

■

A *scatter diagram* is a graph of the data involving two variables. Values for the independent variable are shown on the horizontal axis and the corresponding values for the dependent variable are shown on the vertical axis. The scatter diagram provides an overview of the data and enables us to draw preliminary conclusions about a possible relationship between the variables.

EXAMPLE 13.3

The instructor in a freshman computer science course is interested in the relationship between time spent using the computer system (x) and the final exam score (y). Data collected for a sample of 10 students who took the course last quarter are presented in Table 13.1

Figure 13.1 shows the scatter diagram for the data presented in Table 13.1. For each observation, the number of hours spent using the computer system is shown on the horizontal axis, and the corresponding final exam score is shown on the vertical axis; thus, the scatter diagram contains 10 points. What preliminary conclusions can we draw from Figure 13.1? It appears that fewer hours spent using the computer system are associated with lower final exam scores and that more hours spent using the computer system are associated with higher final exam scores. It also appears that the relationship between the two variables can be approximated by a straight line. In Figure 13.2 we have drawn a straight line through the data that appears to provide a good linear approximation of the relationship between the variables. The equation for this line is $y = 10 + .75x$.

FIGURE 13.1   Scatter Diagram for Example 13.3

Hours Spent Using The Computer System

TABLE 13.1

Data for Example 13.3

x = HOURS USING COMPUTER SYSTEM	y = FINAL EXAM SCORE
45	40
30	35
90	75
60	65
105	90
65	50
90	90
80	80
55	45
75	65

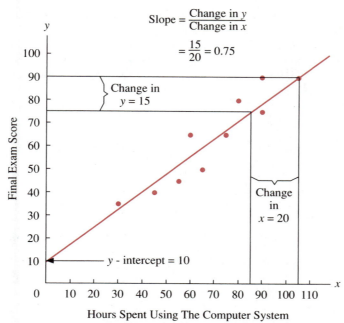

$$\text{Slope} = \frac{\text{Change in } y}{\text{Change in } x}$$

$$= \frac{15}{20} = 0.75$$

Change in $y = 15$

Change in $x = 20$

y - intercept = 10

Hours Spent Using The Computer System

FIGURE 13.2   Straight-Line Approximation for Example 13.3

The *y*-intercept, the point at which the line intersects the *y*-axis, is 10 and the slope, the amount of change in *y* per unit change in *x,* is 0.75.

■

Clearly, there are many other straight lines that we could have drawn in Figure 13.2 to represent the relationship between *x* and *y*. The question is, Which of the straight lines that could be drawn "best" represents the relationship?

The *least squares method* is a procedure that is used to find the straight line that provides the best approximation for the relationship between the independent and dependent variables. We refer to the equation of the line developed using the least squares method as the *estimated regression line,* or the *estimated regression equation*.

Estimated Regression Equation

$$\hat{y} = b_0 + b_1 x \tag{13.1}$$

where

b_0 = *y*-intercept of the line

b_1 = slope of the line

$\hat{y}$ = estimated value of the dependent variable

For any particular value of the independent variable, denoted x_i, the corresponding value on the estimated regression line is denoted by $\hat{y}_i = b_0 + b_1 x_i$.

Application of the least squares method provides the values of b_0 and b_1 that make the sum of the squared deviations between the observed values of the dependent variable (y_i) and the estimated values of the dependent variable $(\hat{y}_i)$ a minimum. The criterion for the least squares method is given by (13.2).

Least Squares Criterion

$$\min \Sigma (y_i - \hat{y}_i)^2 \tag{13.2}$$

where

y_i = observed value of the dependent variable for the ith observation

$\hat{y}_i$ = estimated value of the dependent variable for the ith observation.

Using differential calculus, it can be shown (see the appendix to this chapter) that the values of b_0 and b_1 that minimize (13.2) can be found using (13.3) and (13.4).

Slope and y-Intercept for the Estimated Regression Equation

$$b_1 = \frac{\Sigma(x_i - \bar{x})(y_i - \bar{y})}{\Sigma(x_i - \bar{x})^2} = \frac{\Sigma x_i y_i - (\Sigma x_i \Sigma y_i)/n}{\Sigma x_i^2 - (\Sigma x_i)^2/n} \tag{13.3}$$

$$b_0 = \bar{y} - b_1 \bar{x} \tag{13.4}$$

where

x_i = value of the independent variable for the ith observation

y_i = value of the dependent variable for the ith observation

$\bar{x}$ = mean value for the independent variable

$\bar{y}$ = mean value for the dependent variable

n = total number of observations

The last term in (13.3) is normally used for computing b_1 with a calculator because it avoids the tedious calculations involving the computation of each $(x_i - \bar{x})$ and $(y_i - \bar{y})$. However, to avoid rounding errors, it is best to carry as many significant digits as possible in the calculation.

**EXAMPLE 13.3
(continued)**

The calculations shown below are used to compute the values of b_1 and b_0 using (13.3) and (13.4).

x_i	y_i	$x_i y_i$	x_i^2
45	40	1,800	2,025
30	35	1,050	900
90	75	6,750	8,100
60	65	3,900	3,600
105	90	9,450	11,025
65	50	3,250	4,225
90	90	8,100	8,100
80	80	6,400	6,400
55	45	2,475	3,025
75	65	4,875	5,625
695	635	48,050	53,025

Hence, we obtain

$$b_1 = \frac{48,050 - (695)(635)/10}{53,025 - (695)^2/10}$$

$$= \frac{3917.5}{4722.5} = .8295$$

$$b_0 = 63.5 - .8295(69.5)$$

$$= 5.8498$$

Rounding the values for b_0 and b_1 to two decimal places, we have $\hat{y} = 5.85 + .83x$ as the estimated regression line. In Figure 13.3 we show the graph of this equation. ■

The slope of the estimated regression equation ($b_1 = 5.85$) is positive, implying that as the number of hours spent using the computer system increases, the score on the final exam increases. In fact, we can conclude that each additional hour spent using the computer system is associated with an increase of 5.85 points in the expected final exam score.

If we believe that the least squares estimated regression equation adequately describes the relationship between x and y, then it would seem reasonable to use the estimated regression equation to predict the value of y for a given value of x. For example, if we wanted to predict the final exam score for a student who spent 80 hours on the computer system, we would compute

$$\hat{y} = 5.85 + .83(80)$$

$$= 72.25$$

Hence, we would predict a final exam score of about 72. In the following sections we will discuss methods for assessing the appropriateness of using the estimated regression equation for estimation and prediction.

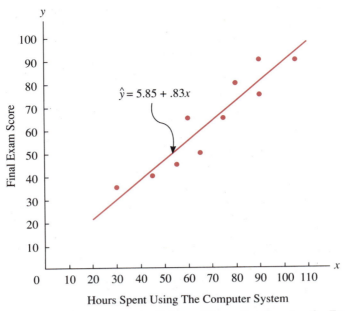

FIGURE 13.3 Graph of the Estimated Regression Equation for Example 13.3: $\hat{y} = 5.85 + .83x$.

Notes and Comments

The least squares method provides an estimated regression equation that minimizes the sum of squared deviations between the observed values of the dependent variable (y_i) and the estimated values of the dependent variable ($\hat{y}_i$). This is the least squares criterion for choosing the equation that provides the best fit. If some other criterion were used, such as minimizing the sum of the absolute deviations between y_i and $\hat{y}_i$, a different equation would be obtained. In practice, the least squares method is the most widely used approach.

EXERCISES

1. The following data have been collected to estimate the relationship between two variables.

OBSERVATION	x_i	y_i
1	22	31
2	52	38
3	38	35
4	64	51
5	26	37
6	12	22
7	42	45
8	25	25

a. Develop a scatter diagram for these data.
b. Use the method of least squares to develop an estimated regression line for the relationship.
c. Draw your estimated regression line on the same graph as the scatter diagram. Does it appear to provide a good fit?

2. Given are five observations taken for two variables, x and y.

OBSERVATION	x_i	y_i
1	2	25
2	3	25
3	5	20
4	1	30
5	8	16

a. Develop a scatter diagram for these data.
b. Use the method of least squares to compute an estimated regression line for these data.
c. Draw the estimated regression line so it passes through the points on the scatter diagram. Does it appear to provide a good fit?

3. The following data were collected regarding the height (inches) of women swimmers and their weight (pounds).

Height	68	64	62	65	66
Weight	132	108	102	115	128

a. Develop a scatter diagram for these data with height on the horizontal axis.
b. What does the scatter diagram developed in (a) indicate about the relationship between the two variables?
c. Try to approximate the relationship between height and weight by drawing a straight line through the data.
d. Develop the estimated regression line by computing the values of b_0 and b_1 using (13.3) and (13.4).
e. If a swimmer's height is 63 inches, what would you estimate her weight to be?

4. Eddie's Restaurants in Des Moines, Iowa, collected the following data on the relationship between advertising and sales at a sample of seven restaurants.

ADVERTISING EXPENDITURES ($1000s)	SALES ($1000s)
1.0	19.0
2.0	32.0
4.0	44.0
6.0	40.0
10.0	52.0
14.0	53.0
20.0	54.0

Develop a scatter diagram for these data with advertising expenditures on the horizontal axis and sales on the vertical axis. Does there appear to be a linear relationship?

5. Shown are some data that a sales manager has collected concerning annual sales and years of experience.

SALESPERSON	YEARS OF EXPERIENCE	ANNUAL SALES ($1000s)
1	1	80
2	3	97
3	4	92
4	4	102
5	6	103
6	8	111
7	10	119
8	10	123
9	11	117
10	13	136

a. Develop a scatter diagram for these data with years of experience on the horizontal axis.
b. Use the method of least squares to compute an estimated regression line for the relationship between years of experience and annual sales.

6. Tyler Realty collected the following data regarding the selling price of new homes and the size of the homes measured in terms of square footage of living space.

SQUARE FOOTAGE	SELLING PRICE
2500	$136,000
2400	$118,000
1800	$101,000
3000	$160,000
2300	$121,000

a. Develop a scatter diagram for these data with square footage on the horizontal axis.
b. Try to approximate the relationship between square footage and selling price by drawing a straight line through the data.
c. Does there appear to be a linear relationship?
d. Develop an estimated regression line using the least squares method.
e. Predict the selling price for a home with 2700 square feet.

7. The owner of a local grocery store varied the price of a product for six consecutive weeks. The following data show the price per unit and the number sold that week.

PRICE	NUMBER SOLD
$.60	220
$.62	200
$.58	280
$.60	250
$.64	190
$.62	240

a. Develop a scatter diagram for these data with price on the horizontal axis.
b. What does the scatter diagram developed in (a) indicate about the relationship between price and the number of units sold?
c. Develop an estimated regression line that can be used to predict the number of units sold, given the price.
d. Predict the number of units sold at a price of $.63.

e. Use the estimated regression line to predict the number of units sold at a price of \$.64. How closely does this predicted value compare to the number of units the grocer actually sold at a price of \$.64?

8. A university medical center has developed a test designed to measure a patient's stress level. The test is designed so that higher scores on the test correspond to higher levels of stress. As part of a research study, the blood pressure (low reading) of patients who took the test was recorded. The following results were obtained.

STRESS TEST SCORE	BLOOD PRESSURE
53	70
94	91
64	78
73	78
82	85
90	84

a. Develop a scatter diagram for these data with stress test score on the horizontal axis. Does a linear relationship between the two variables appear to be appropriate?
b. Develop the estimated least squares line for these data.
c. Estimate an individual's blood pressure if he or she scored 85 on the stress test.

13.2   The Coefficient of Determination

As shown, the least squares method can be used to generate an estimated regression line. We shall now develop a means of measuring the goodness of fit of the line to the data. Recall that the least squares method provides the values of b_0 and b_1 that minimize the sum of squared deviations between the observed values of the dependent variable (y_i) and the estimated values of the dependent variable $(\hat{y}_i)$. Note that the deviations between y_i and $\hat{y}_i$ for any data point, or observation, actually represent the errors in using $\hat{y}_i$ to estimate y_i. That is, for any observation the error is $y_i - \hat{y}_i$. This difference is also referred to as a *residual*. Thus the resulting sum of squared deviations is referred to as the *sum of squares due to error,* or the *residual sum of squares*. We use SSE to represent this quantity.

> **Sum of Squares Due to Error**
>
> $$SSE = \Sigma(y_i - \hat{y}_i)^2 \qquad (13.5)$$

EXAMPLE 13.3 (continued)

The calculations required to compute SSE for the data presented in Example 13.3 are shown.

x_i = HOURS USING THE COMPUTER SYSTEM	y_i = FINAL EXAM SCORE	$\hat{y}_i = 5.85 + .83x_i$	$y_i - \hat{y}_i$	$(y_i - \hat{y}_i)^2$
45	40	43.20	−3.20	10.24
30	35	30.75	4.25	18.06
90	75	80.55	−5.55	30.80
60	65	55.65	9.35	87.42
105	90	93.00	−3.00	9.00
65	50	59.80	−9.80	96.04
90	90	80.55	9.45	89.30
80	80	72.25	7.75	60.06
55	45	51.50	−6.50	42.25
75	65	68.10	−3.10	9.61
				452.78 ← SSE

Thus SSE = 452.78 is a measure of the error in using the estimated regression line $\hat{y} = 5.85 + .83x$ to predict y.

■

Now suppose we were asked to develop an estimate of the final exam score for Example 13.3 without using the number of hours spent on the computer system. We could not use the estimated regression line and would probably use the value of the sample mean, $\bar{y} = 635/10 = 63.5$, as the best estimate of the final exam score for an individual student. The error in using $\bar{y}$ to predict y_i is given by $y_i - \bar{y}$. The corresponding sum of squared deviations about the mean is referred to as the *total sum of squares*.

Total Sum of Squares

$$SST = \Sigma(y_i - \bar{y})^2 \tag{13.6}$$

EXAMPLE 13.3 (continued)

The calculations required to compute SST are shown.

y_i = FINAL EXAM SCORE	$y_i - \bar{y}$ $(y_i - 63.5)$	$(y_i - \bar{y}_i)^2$
40	−23.5	552.25
35	−28.5	812.25
75	11.5	132.25
65	1.5	2.25
90	26.5	702.25
50	−13.5	182.25
90	26.5	702.25
80	16.5	272.25
45	−18.5	342.25
65	1.5	2.25
		3702.50 ← SST

9. Refer again to the data in Exercise 1.
   a. Compute SSE, SST, and SSR using (13.5), (13.6), and (13.8).
   b. Recompute SSR and SST using (13.10) and (13.11). Do you get the same results as in (a)?
   c. Compute the coefficient of determination, r^2. Comment on the goodness of fit.

10. Refer again to the data in Exercise 2.
    a. Compute SSR and SST using (13.10) and (13.11).
    b. What percentage of the total sum of squares is accounted for by the estimated regression line?

11. Refer again to the data in Exercise 3.
    a. Compute SSE, SST, and SSR using (13.5), (13.6), and (13.8).
    b. Recompute SSR and SST using (13.10) and (13.11). Do you get the same results as in (a)?
    c. Compute the coefficient of determination, r^2. Comment on the goodness of fit.

12. Refer again to the data in Exercise 4.
    a. Compute SSR and SST using (13.10) and (13.11).
    b. What percentage of the total sum of squares is accounted for by the fitted regression line?

13. A medical laboratory at Duke University (see Example 13.2) estimates the amount of protein in liver samples through the use of a regression model. A spectrometer emitting light shines through a substance containing the sample, and the amount of light absorbed is used to estimate the amount of protein in the sample. A new regression formula is developed daily because of differing amounts of dye in the solution. On one day six samples resulted in the following data.

ABSORBANCE READING (x)	MILLIGRAMS OF PROTEIN (y)
.509	0
.756	20
1.020	40
1.400	80
1.570	100
1.790	127

   a. Use these data to develop an estimated regression line relating the light absorbance reading to milligrams of protein present in the sample.
   b. Compute r^2. Would you feel comfortable using this regression model to estimate the amount of protein in a sample?
   c. In a sample just received the light absorbance reading was 0.941. Estimate the amount of protein in the sample.

14. The data from Exercise 2 are repeated here.

OBSERVATION	x_i	y_i
1	2	25
2	3	25
3	5	20
4	1	30
5	8	16

a. Compute $\bar{x}$ and $\bar{y}$.
b. Substitute the values of $\bar{x}$ and $\bar{y}$ for x *and* $\hat{y}$ in the estimated regression equation. Do these values satisfy the equation?
c. Will the least squares line always pass through the point corresponding to $(\bar{x}, \bar{y})$? Why or why not?

13.3 The Regression Model and its Assumptions

An important concept that must be understood before we consider testing for significance in regression analysis involves the distinction between a *deterministic model* and a *probabilistic model*. In a deterministic model the relationship between the dependent variable y and the independent variable x is such that if we specify the value of the independent variable, the value of the dependent variable can be determined *exactly*.

EXAMPLE 13.4

To illustrate a deterministic relationship between two variables, suppose that a major oil company leases a service station under a contractual agreement of $500 per month plus 10% of the gross sales. The relationship between the dealer's monthly payment (y) and the gross sales value (x) can be expressed as

$$y = 500 + .10x$$

With this relationship a June gross sales of $6000 would provide a monthly payment of $y = 500 + .10(6000) = \$1100$, and a July gross sales of $7200 would provide a monthly payment of $y = 500 + .10\ (7200) = \$1220$. This type of relationship is deterministic: Once the gross sales value (x) is specified, the monthly payment (y) is determined exactly. Figure 13.5 shows graphically the relationship between gross sales and monthly payment. ■

EXAMPLE 13.3 (continued)

To illustrate a relationship between two variables that is probabilistic, recall the situation in Example 13.3; the data for this example are presented in Table 13.1. Note that two students spent 90 hours using the computer system; the final exam score for one of these students was 75 and the final exam score for the other was 90. Thus, we see that the relationship between y and x cannot be deterministic since different values of y are observed for the same value of x. Since the value of y cannot be determined exactly from the value of x, we say that the model relating x and y is probabilistic. ■

If the relationship between the two variables is probabilistic and the scatter diagram indicates that the relationship between x and y can be approximated by a straight line, we will make the assumption that the following *regression model* describes the actual relationship between the variables:

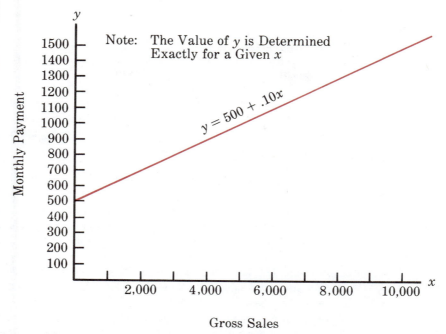

FIGURE 13.5   Illustration of a Deterministic Relationship

Regression Model

$$y = \beta_0 + \beta_1 x + \epsilon \tag{13.12}$$

where

β_0 = y-intercept of the line given by $\beta_0 + \beta_1 x$

β_1 = the slope of the line given by $\beta_0 + \beta_1 x$

ϵ = the error or deviation of the actual y value from the line given by $\beta_0 + \beta_1 x$

Using (13.12) as a model of the relationship between x *and* y, we are saying that we believe the two variables are related in such a fashion that the line given by $\beta_0 + \beta_1 x$ provides a good approximation of the y value at each x. However, to identify the exact value of y we must also consider the error term ϵ (the Greek letter epsilon) that corresponds to how far the actual y value is above or below the line $\beta_0 + \beta_1 x$. In the regression model, the independent variable x is treated as being known; the model is used to predict y given knowledge of x. We refer to β_0 (the y-intercept) and β_1 (the slope) as the *parameters* of the model.

The following assumptions are made about the error term ϵ in the regression model $y = \beta_0 + \beta_1 x + \epsilon$:

Assumptions About the Error Time ϵ in the Regression Model $y = \beta_0 + \beta_1 x + \epsilon$

---

1. The error term, ϵ, is a random variable with mean or expected value of 0; that is, $E(\epsilon) = 0$.
   Implication: Since β_0 and β_1 are constants, $E(\beta_0) = \beta_0$ and $E(\beta_1) = \beta_1$; thus for a given value of x, the expected value of y is

$$E(y) = \beta_0 + \beta_1 x \qquad (13.13)$$

   Equation (13.13) is referred to as the *regression equation*.
2. The variance of ϵ is denoted by σ^2 and is the same for all values of x.
   Implication: The variance of y equals σ^2 and is the same for all values of x.
3. The values of ϵ are independent.
   Implication: The value of ϵ for a particular value of x is not related to the value of ϵ for any other value of x; thus, the value of y for a particular value of x is not related to the value of y for any other value of x.
4. The error term ϵ is a normally distributed random variable.
   Implication: Since y is a linear function of ϵ, y is also a normally distributed random variable.

Figure 13.6 provides a graphical interpretation of the model assumptions and their implications. As shown in Figure 13.6 the value of $E(y)$ changes according to the specific value of x considered. However, note that regardless of the x value, the probability distribution of the errors *(ϵ)* and hence the probability distributions of y for a given value of x are normal distributions, each with the same shape and hence the same variance. The specific value of the error ϵ at any particular point depends upon whether the actual value of y is greater than or less than $E(y)$.

At this point we must keep in mind that we are also making an assumption or hypothesis about the form of the regression model and the associated regression equation. That is, we have assumed that a straight line represented by $\beta_0 + \beta_1 x$ is the basis for the relationship between the variables. We must not lose sight of the fact that some other model, for instance $\beta_0 + \beta_1 x^2$, may turn out to be a better model for the underlying relationship. After using the sample data to estimate the parameters of the regression model (β_0 and β_1), we will want to conduct further analysis to determine whether or not the specific model that we have assumed appears to be valid.

The Relationship Between the Regression Equation and the Estimated Regression Equation

Recall from Chapter 8 that when data were available for just one variable, the objective was to use a sample statistic (e.g., the sample mean) to make inferences about the corresponding population parameter (e.g., population mean). When we discussed the least squares method in Section 13.1, we presented formulas for computing the y-intercept (b_0) and the slope (b_1) of the estimated regression equation. The value of b_0 is a sample

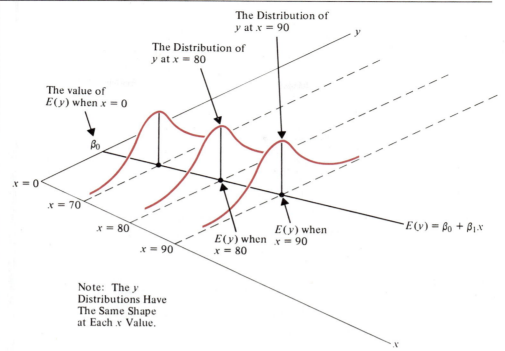

FIGURE 13.6   Illustration of Regression Assumptions

statistic that provides an estimate of the β_0 parameter in the regression model, and the value of b_1 is a sample statistic that provides an estimate of the β_1 parameter. Thus, since the regression equation is $E(y) = \beta_0 + \beta_1 x$, the best estimate of the regression equation is provided by the estimated regression equation $\hat{y} = b_0 + b_1 x$. Hence, $\hat{y}$ provides the estimate of $E(y)$. Figure 13.7 summarizes these concepts.

**EXAMPLE 13.3
(continued)**

Let us return to the example involving the relationship between time spent using the computer system (x) and the final exam score (y). Consider students who have spent 80 hours on the computer system $(x = 80)$; when $x = 80$, the value of $E(y)$ in (13.13) represents the mean final exam score for all such students. But, since we never know what the values of β_0 and β_1 are, in practice, it is not possible to compute $E(y)$. Thus, we use the estimated regression equation $\hat{y} = 5.85 + .83x$ to provide an estimate of $E(y)$ when $x = 80$. Doing so, we get $\hat{y} = 5.85 + .83(80) = 72.25$.

■

13.4

Testing For Significance

In Section 13.2 we saw how the coefficient of determination (r^2) could be used as a measure of the goodness of fit of the estimated regression line. Larger values of r^2 indicate

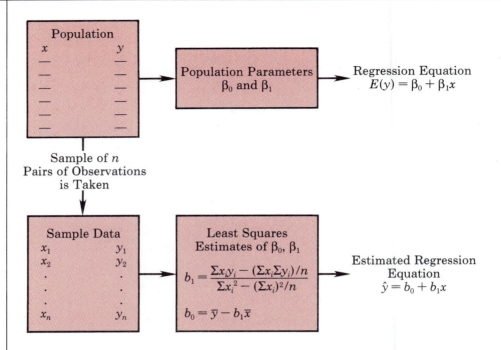

FIGURE 13.7   Estimating the Population Regression Equation Using Sample Data

a better fit. However, this measure does not allow us to conclude that the relationship is or is not statistically significant. In order to draw conclusions concerning statistical significance, we must take the sample size into consideration. In this section we show how to conduct significance tests that allow us to draw conclusions about the significance of a regression relationship.

An Estimate of σ^2

In Section 13.2 we used the sum of squares due to error, SSE, as a measure of the variability of the actual observations about the estimated regression line. This quantity is also used to develop an estimate of σ^2, the population variance of the y values about the regression line. Recall from our earlier definition of sample variance that we divided the sum of the squared deviations about the sample mean by $n - 1$ to obtain an unbiased estimate of the population variance. We used $n - 1$ instead of n because 1 degree of freedom was lost when the sample mean was used to compute the sum of the squared deviations about the mean. In other words, 1 degree of freedom was lost because one parameter used in computing the sum of squares, the population mean, had to be estimated from the sample data. In regression analysis we must estimate the parameters β_0 and β_1 to compute the sum of squares due to error. For this reason, 2 degrees of freedom are lost; hence we must divide SSE by $n - 2$ to obtain an unbiased estimate of σ^2. The estimate obtained, called the *mean square due to error*, is denoted by MSE. Since MSE is an estimate of σ^2, the notation s^2 is also used.

Mean Square Error (Estimate of σ^2)

$$s^2 = \text{MSE} = \frac{\text{SSE}}{n-2} \qquad (13.14)$$

EXAMPLE 13.3
(continued)

Recall that in Example 13.3, SSE = 452.78. Thus,

$$s^2 = \text{MSE} = \frac{\text{SSE}}{n-2} = \frac{452.78}{8} = 56.60$$

■

In the discussion that follows, we use this unbiased estimate of σ^2 in tests for the significance of the regression line.

t Test

Recall that the underlying regression equation is assumed to be $E(y) = \beta_0 + \beta_1 x$. If there really exists a relationship of this form between x and y, β_1 would have to differ from 0. Thus a conclusion regarding the significance of the regression relationship can be tested using the following hypotheses:

$$H_0: \quad \beta_1 = 0$$

$$H_a: \quad \beta_1 \neq 0$$

Before presenting the *t* test, we need to consider the properties of b_1, the least squares estimator of β_1. First, let us consider what would have happened if we had used a different random sample for the same regression study. For example, suppose that in Example 13.3 we had used the final exam records of 10 different students. A regression analysis of this new data, or sample, probably would result in an estimated regression line similar to the previous estimated regression line, $\hat{y} = 5.85 + .83x$. However, it is doubtful that we would obtain exactly the same values for b_0 and b_1. Thus b_0 and b_1 are themselves variables whose values depend upon the data items (the values of x_i and y_i) included in the sample. Recall the discussion of sampling distributions in Chapter 7; b_0 and b_1 must have their own sampling distributions. The properties of the sampling distribution for b_1 are defined as follows.

Sampling Distribution of b_1

Distribution Form: Normal

Mean: $E(b_1) = \beta_1$ (13.15)

Variance: $\sigma^2_{b_1} = \sigma^2 \left(\dfrac{1}{\sum x_i^2 - (\sum x_i)^2/n} \right)$

Since we do not know the value of σ^2, we develop an estimate of $\sigma_{b_1}^2$, denoted by $s_{b_1}^2$, by first estimating σ^2 with s^2. Thus we obtain the following.

Estimate of the Variance of b_1

$$s_{b_1}^2 = s^2 \left(\frac{1}{\Sigma x_i^2 - (\Sigma x_i)^2/n} \right)$$

(13.16)

EXAMPLE 13.3
(continued)

For Example 13.3, $s^2 = MSE = 56.60$. Thus,

$$s_{b_1}^2 = 56.60 \left(\frac{1}{53,025 - (695)^2/10} \right)$$

$$= 56.60 \left(\frac{1}{4722.5} \right)$$

$$= .0120$$

Hence

$$s_{b_1} = \sqrt{.0120} = .1095$$

The t test regarding β_1 is based on the fact that

$$\frac{b_1 - \beta_1}{s_{b_1}}$$

has a t distribution with $n - 2$ degrees of freedom. If the null hypothesis is true, then $\beta_1 = 0$ and b_1/s_{b_1} has a t distribution with $n - 2$ degrees of freedom. Using b_1/s_{b_1} as a test statistic, we use the following rejection rule to test H_0: $\beta_1 = 0$ versus H_a: $\beta_1 \neq 0$.

$$\text{Reject } H_0 \text{ if } \frac{b_1}{s_{b_1}} < - t_{\alpha/2}$$

$$\text{or } \frac{b_1}{s_{b_1}} > t_{\alpha/2}$$

where $t_{\alpha/2}$ has $n - 2$ degrees of freedom and where α is the level of significance for the test.

EXAMPLE 13.3
(continued)

Recall that $b_1 = .83$ and $s_{b_1} = .1095$. Thus we have $b_1/s_{b_1} = .83/.1095 = 7.58$. From Table 2 of Appendix B we find that the upper-tail t value corresponding to $\alpha = .01$ and $n - 2 = 8$ degrees of freedom is $t_{.005} = 3.355$. Since $b_1/s_{b_1} = 7.58 > 3.355$, we reject

H_0 at the .01 level of significance; β_1 is not equal to zero. Thus we conclude that there is a statistically significant relationship between time spent on the computer system and final exam score.

■

F Test

The t test has been used to test the null hypothesis H_0: $\beta_1 = 0$. An F test can also be used. In regression analysis with only one independent variable the t test and the F test yield the same results. But with more than one independent variable, only the F test can be used to test for a significant relationship between a dependent variable and a set of independent variables. Here we will introduce the F test and show that it leads to the same conclusion as the t test. In Chapter 14 we illustrate the use of the F test for multiple regression analysis.

The hypotheses we are testing are the same as before.

$$H_0: \quad \beta_1 = 0$$

$$H_a: \quad \beta_1 \neq 0$$

The logic behind the use of the F test for determining whether or not the relationship between x and y is statistically significant is based upon the use of two independent estimates of σ^2. We have just seen that s^2 provides an estimate of σ^2. If the null hypothesis H_0: $\beta_1 = 0$ is true, the mean square due to regression (denoted MSR) provides another *independent* estimate of σ^2.

To compute MSR, we first note that for any sum of squares the mean square is the sum of squares divided by its degrees of freedom. The number of degrees of freedom for the sum of squares due to regression, SSR, is always equal to the number of independent variables. Since in this chapter we are concerned only with models involving one independent variable, the number of regression degrees of freedom is 1. Using DF as an abbreviation for degrees of freedom, we can write

$$\text{MSR} = \frac{\text{SSR}}{\text{Regression DF}} = \frac{\text{SSR}}{\text{Number of Independent Variables}} \quad (13.17)$$

For Example 13.3, we find that MSR = SSR/1 = 3249.72.

If the null hypothesis (H_0: $\beta_1 = 0$) is true, MSR and MSE are two independent estimates of σ^2. In this case the sampling distribution of MSR/MSE follows an F distribution with numerator degrees of freedom equal to 1 and denominator degrees of freedom equal to $n - 2$. The test concerning the significance of the regression relationship is based on the following F statistic:

$$F = \frac{\text{MSR}}{\text{MSE}} \quad (13.18)$$

Given any sample size, the numerator of the F statistic will increase as more of the variability in y is explained by the regression model and decrease as less is explained. Similarly, the denominator will increase if there is more variability about the estimated regression line and decrease if there is less variability. Thus one would intuitively expect large values of $F = \text{MSR}/\text{MSE}$ to cast doubt on the null hypotheses and lead us to the conclusion that $\beta_1 \neq 0$ and there is a significant relationship between x and y. Indeed, this is correct; large values of F lead to rejection of H_0 and the conclusion that the relationship is statistically significant.

EXAMPLE 13.3
(continued)

Let us now conduct the F test for Example 13.3. Assume that the level of significance is $\alpha = .01$. From Table 4 of Appendix B we can determine the critical F value by locating the value corresponding to numerator degrees of freedom equal to 1 (the number of independent variables) and denominator degrees of freedom equal to $n - 2 = 10 - 2 = 8$. Thus we obtain $F = 11.26$. Hence the appropriate rejection rule is written

$$\text{Reject } H_0 \text{ if } \frac{\text{MSR}}{\text{MSE}} > 11.26$$

Since $\text{MSR}/\text{MSE} = 3249.72/56.60 = 57.42$ is greater than the critical value ($F_{.01} = 11.26$), we can reject H_0 and conclude that there is a statistically significant relationship between the final exam score and the number of hours spent on the computer system.

■

Cautionary Note Regarding Testing for Significance

We caution here that rejection of H_0 does not yet permit us to conclude that the relationship between x and y is *linear*. However, it is valid to conclude that x and y are related and that a linear approximation explains a significant amount of the variability in y over the range of x values observed in the sample. To illustrate this qualification we call your attention to Figure 13.8, where an F test (on $\beta_1 = 0$) yielded the conclusion that x and y were related. The figure shows that the actual relationship is nonlinear. In the graph·

FIGURE 13.8 Example of a Linear Approximation of a Nonlinear Relationship

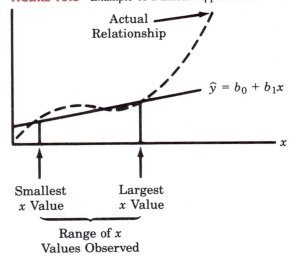

we see that the linear approximation is very good for the values of x used in developing the least squares line, but it is very bad for larger values of x.

Given a significant relationship, we should feel confident in using the regression equation for predictions corresponding to x values within the range of the x values for the sample. For Example 13.3 this corresponds to values of x between 30 and 105. But unless there are reasons to believe the model is valid beyond this range, predictions outside the range of the independent variable should be made with caution. For Example 13.3, since the regression relationship has been found significant at the .01 level, we should feel confident using it to predict the final exam score whenever the number of hours spent using the computer system is between 30 and 105.

Notes and Comments

1. The assumptions made about the error term (Section 13.3) are what permit the tests of statistical significance in this section. The properties of the sampling distribution of b_1 and the subsequent F and t tests directly follow from these assumptions.

2. Do not confuse statistical significance with practical significance. With very large sample sizes it is possible to obtain statistically significant results for small values of b_1; in such cases one must exercise care in concluding that the relationship has practical significance.

3. The reason that the F test and the t test yield the same result *for simple linear regression* is that $F = t^2$. The critical value for the F test is the square of the critical value for the t test, and the test statistic for the F test is the square of the test statistic for the t test.

15. The data from Exercise 2 are repeated here.

OBSERVATION	x_i	y_i
1	2	25
2	3	25
3	5	20
4	1	30
5	8	16

a. Compute MSE as an estimate of σ^2.
b. Compute an estimate of the variance of b_1.
c. Use the t test to test the hypotheses

$$H_0: \quad \beta_1 = 0$$
$$H_a: \quad \beta_1 \neq 0$$

at the .05 level of significance.
d. Use the F test to test the hypotheses in (c) at the $\alpha = .05$ level of significance.

16. Given are five observations collected in a regression study on two variables.

OBSERVATION	x_i	y_i
1	2	2
2	4	3
3	5	2
4	7	6
5	8	4

    a. Develop the estimated regression line for these data.
    b. Use the t test to test the hypotheses at the $\alpha = .05$ level of significance.

$$H_0: \quad \beta_1 = 0$$
$$H_a: \quad \beta_1 \neq 0$$

    c. Use the F test to test the hypotheses in (b) at the $\alpha = .05$ level of significance.

17. Refer to Exercise 1, where an estimated regression line was developed using 8 observations on 2 variables. SSR + SSE, and SST were computed in Exercise 9.
    a. Compute an estimate of σ^2.
    b. Compute $s_{b_1}^2$ and s_{b_1}.
    c. Use the t test to determine if x and y are related at the $\alpha = .05$ level of significance.
    d. Use the F test to determine if x and y are related at the $\alpha = .05$ level of significance.
    e. Compute t^2. Is it the same as F?

18. Refer to Exercise 6, where an estimated regression line relating square footage to selling prices of new homes was developed. Test whether or not selling price and square footage are related at the $\alpha = .01$ level of significance.

19. Refer to Exercise 7, where an estimated regression line relating the price of a product and the number of units sold was developed. Test whether or not price and the number of units sold are related at the $\alpha = .05$ level of significance.

20. Refer to Exercise 5, where an estimated regression line relating years of experience and annual sales was developed. At the $\alpha = .05$ level of significance, determine whether or not annual sales and years of experience are related.

21. Refer to Exercise 13, where an estimated regression line relating light absorbance readings and milligrams of protein present in a liver sample was developed. Test whether or not the absorbance readings and amount of protein present are related at the $\alpha = .01$ level of significance.

13.5   Estimation and Prediction

For Example 13.3 we concluded that the final exam grade (y) and the number of hours spent on the computer system (x) were related. Moreover, the estimated regression line $\hat{y} = 5.85 + .83x$ appears to describe adequately the relationship between x and y. Now we can begin to use the estimated regression line to develop estimates of y for a given value of x.

There are two types of estimates to consider. The first is an estimate of the mean value of y for a particular value of x. The second type of estimate is appropriate in situations where we want to predict an individual value of y corresponding to a particular value of x. In general, the point estimate of the mean value of y for a particular value of x is *the corresponding value of $\hat{y}$* given by the estimated regression equation. We denote the particular value of x by x_p, the mean value of y at x_p by $E(y_p)$, and the estimate of $E(y_p)$ by $\hat{y}_p = b_0 + b_1 x_p$. The point estimate for an individual value of y is also given by $\hat{y}_p = b_0 + b_1 x_p$.

EXAMPLE 13.3
(continued)

Suppose we want to develop a point estimate for the mean final exam score of *all* students who spend 85 hours on the computer system. The point estimate of $E(y)$ in this case is $\hat{y} = 5.85 + .83x = 5.85 + .83(85) = 76.40$.

Next, let us consider developing a prediction of the final exam score for David Edmunds, *one* particular student who has spent 85 hours on the computer system. The prediction of his final exam score is also given by $\hat{y} = 5.85 + .83x = 5.85 + .83(85) = 76.40$. Thus we see that the point estimate for the mean value of y for a particular value of x is the same as the point estimate of an individual value of y for a particular value of x.

■

Instead of computing a point estimate of the mean value of y for a particular value of x, suppose we wanted to develop an interval estimate of the mean value of y. We refer to this type of interval estimate as a *confidence interval estimate*. The second type of interval estimate, referred to as a *prediction interval estimate*, is applicable for the case where we want to predict an individual value of y.

Confidence Interval Estimate of the Mean Value of y

As we stated above, the point estimate of the mean value of y for a particular value of x is $\hat{y}_p = b_0 + b_1 x_p$. Since b_0 and b_1, are only estimates of β_0 and β_1, we cannot expect that the estimated value $\hat{y}_p$ will exactly equal $E(y_p)$. If we want to make an inference about how close $\hat{y}_p$ is to the true mean value $E(y_p)$, however, we will have to consider the variability that exists when we develop estimates based on the estimated regression equation. Statisticians have developed the following estimate of the variance of $\hat{y}_p$:

$$\text{Estimated variance of } \hat{y}_p = s_{\hat{y}_p}^2 = s^2 \left(\frac{1}{n} + \frac{(x_p - \bar{x})^2}{\Sigma x_i^2 - (\Sigma x_i)^2/n} \right) \qquad (13.19)$$

The confidence interval estimate of $E(y_p)$ is as follows:

Confidence Interval Estimate of $E(y_p)$

$$\hat{y}_p \pm t_{\alpha/2}\, s_{\hat{y}_p} \qquad (13.20)$$

where the confidence coefficient is $1 - \alpha$ and the t value has $n - 2$ degrees of freedom.

EXAMPLE 13.3
(continued)

Suppose that we wanted to develop a 95% confidence interval estimate of the mean value of y when $x = 85$. The estimated variance of $\hat{y}_p$ for students who spent 85 hours on the computer system is

$$s^2_{\hat{y}_p} = 56.6 \left[\frac{1}{10} + \frac{(85 - 69.5)^2}{53,025 - (695)^2/10} \right]$$

$$= 56.6 \, (.15087)$$

$$= 8.5392$$

Thus,

$$s_{\hat{y}_p} = \sqrt{s^2_{\hat{y}_p}} = \sqrt{8.5392} = 2.922$$

Thus, to develop a 95% confidence interval estimate of the expected final exam score for all students who spend 85 hours on the computer system, we need to find the t value from Table 2 of Appendix B corresponding to $n - 2 = 10 - 2 = 8$ degrees of freedom and $\alpha = .05$. Doing so, we find $t_{.025} = 2.306$. Since $\hat{y} = 5.85 + .83(85) = 76.4$, the resulting confidence interval is

$$76.4 \pm 2.306 \, s_{\hat{y}_p}$$

$$76.4 \pm 2.306(2.922)$$

$$76.4 \pm 6.74$$

Thus, we obtain 69.66 to 83.14 as the 95% confidence interval for the mean final exam score when $x = 85$.

■

Note that the estimate variance of $\hat{y}_p$ [see (13.19)] is smallest when the given value of $x_p = \bar{x}$. In this case (13.19) becomes

$$s^2_{\hat{y}_p} = s^2 \left(\frac{1}{n} + \frac{(\bar{x} - \bar{x})^2}{\Sigma x_i^2 - (\Sigma x_i)^2/n} \right)$$

$$= \frac{s^2}{n}$$

which implies that we can expect to make our best estimates at the mean of the independent variable.

Prediction Interval Estimate for an Individual Value of y

In the preceding discussion we showed how to develop a confidence interval estimate of the mean value of y for a particular value of x. Now we turn to the problem of developing an interval estimate for an individual value of y corresponding to a particular value of x.

In order to develop a prediction interval estimate, we must first determine the variance associated with using $\hat{y}_p$ as an estimate of a particular value of y when $x = x_p$. This variance is made up of the sum of the following two components:

1. The variance of individual y values about the mean $E(\hat{y}_p)$, an estimate of which is given by s^2.
2. The variance associated with using $\hat{y}_p$ to estimate $E(y_p)$, which we previously found to be $s_{\hat{y}_p}^2$.

Statisticians have shown that the variance of the estimate of an individual value of y_p, which we denote s_{ind}^2, is given by

$$
\begin{aligned}
s_{\text{ind}}^2 &= s^2 + s_{\hat{y}_p}^2 \\
&= s^2 + s^2 \left(\frac{1}{n} + \frac{(x_p - \bar{x})^2}{\Sigma x_i^2 - (\Sigma x_i)^2/n} \right) \qquad (13.21) \\
&= s^2 \left(1 + \frac{1}{n} + \frac{(x_p - \bar{x})^2}{\Sigma x_i^2 - (\Sigma x_i)^2/n} \right)
\end{aligned}
$$

The prediction interval estimate of y_p is given by (13.22):

Prediction Interval Estimate of y_p

$$\hat{y}_p \pm t_{\alpha/2}\, s_{\text{ind}} \qquad (13.22)$$

where the confidence coefficient is $1 - \alpha$ and the t value has $n - 2$ degrees of freedom.

EXAMPLE 13.3 (continued)

To obtain a 95% prediction interval estimate when $x = 85$, we must first compute the value of s_{ind}.

$$
s_{\text{ind}}^2 = 56.6 \left[1 + \frac{1}{10} + \frac{(85 - 69.5)^2}{53.025 - (695)^2/10} \right]
$$

$$
= 56.6\,(1.15087)
$$

$$
= 65.1392
$$

Thus,

$$
s_{\text{ind}} = \sqrt{s_{\text{ind}}^2} = \sqrt{65.1392} = 8.071
$$

Since $\hat{y}_p = 5.85 + .83(85) = 76.4$ and $t_{.025} = 2.306$, the 95% prediction interval estimate is

$$
76.4 \pm 2.306\, s_{\text{ind}}
$$

$$76.4 \pm 2.306(8.071)$$

$$76.4 \pm 18.61$$

Thus, 57.79 to 95.01 is the 95% prediction for the final exam score for an individual student who has spent 85 hours on the computer system. Note that the prediction interval is wider than the confidence interval for the mean final exam score. This difference simply reflects the fact that we are able to predict the mean final exam score with more precision than we can the final exam score for any one particular individual.

■

EXERCISES

22. As an extension of Exercise 17, develop a 95% confidence interval for estimating $E(y)$ when $x = 38$.

23. As an extension of Exercise 6, develop a 90% confidence interval for predicting the mean selling price for homes with 2,200 square feet of living space.

24. As an extension of Exercise 3, develop a 95% confidence interval for estimating the weight of Jane Heller, who is 5 feet 7 inches tall.

25. As an extension to Exercise 6, develop a 95% confidence interval for the selling price of a home on Highland Terrace with 2,800 square feet.

26. State in your own words why a smaller confidence interval is obtained when a mean value is predicted than when an individual value is predicted.

27. A study conducted by the Department of Transportation regarding driving speed and mileage for mid-size automobiles resulted in the following data.

Driving speed	30	50	40	55	30	25	60	25	50	55
Mileage	28	25	25	23	30	32	21	35	26	25

a. Determine the estimated regression line that relates mileage to the driving speed.
b. At the $\alpha = .05$ level of significance, determine whether or not mileage and driving speed are related.
c. Did the estimate regression line provide a good fit to the data?
d. Develop a 95% confidence interval for estimating the mean mileage for cars that are driven at 50 miles per hour.
e. If we were interested in one specific car that was driven at 50 miles per hour, how would the estimate of mileage change as compared to the estimate developed in (d)?

13.6   Computer Solution of Regression Problems

Performing all the computations associated with regression analysis can be quite time-consuming. In this section we discuss how the computational burden can be simplified by using a computer software package. The general procedure followed in using computer packages is for the user to input the data (x and y values for the sample) together with some instructions concerning the types of analyses that are required. The software package performs the analyses and prints the results in an output report. Before discussing the details of this approach, we discuss the use of the analysis of variance (ANOVA) table as a device for summarizing the calculations performed in regression analysis. The ANOVA table is an important component of the output report produced by most software packages.

The ANOVA Table

In Chapter 12 we saw how the ANOVA table could provide a convenient summary of the computational aspects of analysis of variance. In regression analysis a similar table can be developed. Table 13.2 shows the general form of the ANOVA table for 2-variable regression studies and Table 13.3 shows the ANOVA table for Example 13.3. It can be seen that the relationship that holds among the sums of squares (that is, SST = SSR + SSE) also holds for the degrees of freedom. That is,

$$\text{Total DF} = \text{Regression DF} + \text{Error DF}$$

Computer Output

In Figure 13.9 we show the Minitab computer output for Example 13.3. The dependent variable is labeled "SCORE" and the independent variable x is labeled "HOURS." The interpretation of the output is as follows:

1. Minitab prints the estimated regression equation as SCORE = 5.85 + 0.830 HOURS.
2. A table is printed that shows the values of the coefficients b_0 and b_1, the standard deviation of each coefficient, the t value obtained by dividing each coefficient value

TABLE 13.2

General Form of the ANOVA Table for 2-Variable Regression Analysis

SOURCE OF VARIATION	SUM OF SQUARES	DEGREES OF FREEDOM	MEAN SQUARE
Regression	SSR	1	$\text{MSR} = \dfrac{\text{SSR}}{1}$
Error	SSE	$n - 2$	$\text{MSE} = \dfrac{\text{SSE}}{n - 2}$
Total	SST	$n - 1$	

TABLE 13.3

ANOVA Table for Example 13.3

SOURCE OF VARIATION	SUM OF SQUARES	DEGREES OF FREEDOM	MEAN SQUARE
Regression	3249.72	1	$\dfrac{3249.72}{1} = 3249.72$
Error	452.78	8	$\dfrac{452.78}{8} = 56.6$
Total	3702.50	9	

```
The regression equation is
SCORE = 5.85 + 0.830 HOURS

Predictor        Coef       Stdev     t-ratio        p
Constant        5.847       7.972        0.73    0.484
HOURS          0.8295      0.1095        7.58    0.000

s = 7.523      R-sq = 87.8%     R-sq(adj) = 86.2%

Analysis of Variance

SOURCE         DF          SS          MS         F        p
Regression      1      3249.7      3249.7     57.42    0.000
Error           8       452.8        56.6
Total           9      3702.5
```

FIGURE 13.9 Minitab Output for Example 13.3

by its standard deviation, and the p-value associated with the t-value. Thus, to test H_0: $\beta_1 = 0$ versus H_a: $\beta_1 \neq 0$, we could compare 7.58 (located in the "t-ratio" column) to the appropriate critical value. This procedure was described in Section 13.4. Alternatively, we could use the p-value provided by Minitab to perform the same test. Recall from Chapter 8 that the p-value is the probability of obtaining a sample result more unlikely than what is observed. Since the p-value in this case is 0 (to three decimal places), the sample results indicate that the null hypothesis (H_0: $\beta_1 = 0$) should be rejected.

3. Minitab prints the estimate of σ, $s = 7.523$, as well as information regarding the goodness of fit. Note that "R-sq = 87.8%" is the coefficient of determination, which we denoted by r^2. The output "R-sq (adj) = 86.2%" is discussed in Chapter 14.

4. The ANOVA table is printed below the heading "Analysis of Variance." Note that MSR is given as 3249.7 and MSE as 56.6. The ratio of these two values provides the F value of 57.42; in Section 13.4 we showed how the F value can be used to determine if there is a significant statistical relationship between SCORE and HOURS. Minitab also prints the p-value associated with this F test. Since the p-value is 0 (to three decimal places), the relationship is judged statistically significant.

As you can see, the computer output from Minitab is fairly easy to interpret given our current background in regression analysis. Many other computer packages, most with rather cryptic names such as SPSS, SAS, BMDP, and so on, are available for solving regression problems. After a brief period of familiarity with the control language associated with each of these packages, computer output such as that shown in Figure 13.9 can be obtained easily. When large data sets are involved, computer packages provide the only practical means for solving regression problems.

EXERCISES

28. A medical laboratory at the University of Cincinnati Medical School uses an estimated regression line to perform creatinine assays. The amount of creatinine found in a sample is used to measure the rate at which plasma is filtered in the kidneys. A solution containing the sample is placed in a test tube. Light from a spectrometer is shined through the solution; the amount of light absorbed *(x)* is used to estimate the amount of creatinine in the solution *(y)*. Shown next are data collected by Dr. Robert Banks.

MICROGRAMS OF CREATININE	LIGHT ABSORBANCE	MICROGRAMS OF CREATININE	LIGHT ABSORBANCE
5	0.101	100	1.670
10	0.183	120	1.971
20	0.359	140	2.258
40	0.711	160	2.520
60	1.038		
80	1.376	180	2.746

a. Use a computer package to develop an estimated regression line relating the light absorbance reading to micrograms of creatinine present in the sample.
b. Test the significance of the relationship using α .05.
c. Did the model provide a good fit?
d. Develop a 95% confidence interval estimate of the mean amount of creatinine when x = 1.5.

29. The commercial division of a real estate firm is conducting a regression analysis of the relationship between x, annual gross rents ($1000s), and y, selling prices ($1000s) for apartment buildings. Data have been collected on a number of properties recently sold, and the accompanying output has been obtained in a computer run.

```
The regression equation is
Y = 20.0 + 7.21X

Predictor      Coef      Stdev     t-ratio
Constant     20.000     3.2213      6.21
X             7.210     1.3626      5.29

Analysis of Variance

SOURCE        DF        SS
Regression     1     41587.3
Error          7
Total          8     51984.1
```

a. How many apartment buildings were in the sample?
b. Write the estimated regression line.
c. What is the value of s_{b_1}?
d. Test the significance of the relationship at an α = .05 level of significance.
e. Estimate the selling price of an apartment building with gross annual rents of $50,000.

30. Shown is a portion of the computer output for a regression analysis relating maintenance expense (dollars per month) to usage (hours per week) of a particular brand of computer terminal.

```
The regression equation is
Y = 6.1092 + .8951X

Predictor      Coef      Stdev
Constant     6.1092     0.9361
X            0.8951     0.1490

Analysis of Variance

SOURCE        DF        SS         MS
Regression     1     1575.76    1575.76
Error          8      349.14      43.64
Total          9     1924.90
```

a. Write the estimated regression line.

b. Test to see if monthly maintenance expense is related to usage at the .01 level of significance.

c. Use the estimated regression line to predict monthly maintenance expense for any terminal that is used 25 hours per week.

31. A regression model relating x, number of salespersons at a branch office, toy , annual sales at the office ($1000's), has been developed. Shown is the computer output from a regression analysis of the data.

```
The regression equation is
Y = 80.0 + 50.00X

Predictor       Coef        Stdev       t-ratio
Constant        80.0        11.333        7.06
X               50.0         5.482        9.12

Analysis of Variance

SOURCE      DF          SS          MS
Regression   1        6828.6      6828.6
Error       28        2298.8       82.1
Total       29        9127.4
```

a. Write the estimated regression line.

b. How many branch offices were involved in the study?

c. Test the significance of the relationship at an $\alpha = .05$ level of significance.

d. Predict the annual sales at the Memphis branch office. This branch has 12 salespersons.

32. The following data show the dollar value of prescriptions (in thousands) for 13 pharmacies in Iowa and the population of the city served by the given pharmacy ("The Use of Categorical Variables in Data Envelopment Analysis," R. Banker and R. Morey, *Management Science*, December 1986).

VALUE ($1000s)	POPULATION	VALUE ($1000s)	POPULATION
61	1410	56	1016
92	1523	45	1070
93	1354	183	1694
45	822	156	1910
50	746	120	1745
29	1281	75	1353
122	1016		

a. Use a computer package to develop a scatter diagram for these data; plot population on the horizontal axis.

b. Does there appear to be any relationship between these two variables?

c. Use the computer package to develop the estimated regression line that could be used to predict the dollar value of prescriptions given the population of the city.

d. Test the significance of the relationship at an $\alpha = .05$ level of significance.

e. Predict the dollar value for a particular city with a population of 1500 people. Use $\alpha = .05$.

13.7 The Analysis of Residuals

For each observation in a regression analysis, there is a residual; it is the difference between the observed value of the dependent variable, y_i, and the value predicted by the regression equation, $\hat{y}_i$. The residual for observation i, $y_i - \hat{y}_i$, is an estimate of the error resulting from using the estimated regression equation to predict the value of y_i.

The analysis of residuals plays an important role in validating the assumptions made in regression analysis. In Section 13.4 we showed how hypothesis testing can be used to determine whether or not a regression relationship is statistically significant. Hypothesis tests concerning regression relationships are based on the assumptions made about the regression model. If the assumptions made are not satisfied, the hypothesis tests are not valid and the estimated regression equation should not be used. However, keep in mind that the regression model is being used only as an approximation of reality, so good judgment must be used to determine whether or not an assumption violation is severe enough to invalidate the model.

There are two key issues in verifying that the assumptions are satisfied in a regression model. Are the four assumptions concerning the error term, ϵ, satisfied and is the form we have assumed for the model appropriate? In this chapter, we restrict our discussion of residual analysis to these two issues concerning model validity. In Chapter 14 and 15, we show some further uses of residual analysis.

Regression analysis begins with an assumption concerning the appropriate form of the regression model. The simple linear regression model assumes the form

$$y = \beta_0 + \beta_1 x + \epsilon$$

With this form, y is a linear function of x. The assumptions regarding the error term (presented in Section 13.3) are as follows:

1. $E(\epsilon) = 0$.
2. The variance of ϵ, denoted by σ^2, is the same for all values of x.
3. The values of ϵ are independent.
4. ϵ is a normally distributed random variable.

Validating the assumptions concerning the error term, ϵ, means using the residuals to check to see if these assumptions seem reasonable.

Recall that the first assumption concerning ϵ implies that the regression equation is

$$E(y) = \beta_0 + \beta_1 x$$

This regression equation shows a linear relationship between x and the expected value of y, $E(y)$. Validating the assumption concerning model form means satisfying ourselves that the relationship between the independent and dependent variable is adequately represented by the regression equation. It is possible that the true relationship between x and y is curvilinear and/or that more independent variables should have been included (multiple regression). We will see how the statistician uses residual analysis to recognize when this assumption concerning model form might be violated.

The residuals, $y_i - \hat{y}_i$, are estimates of ϵ; with n observations in a regression analysis, we have n residuals. Residual plots are graphical presentations of the residuals that help

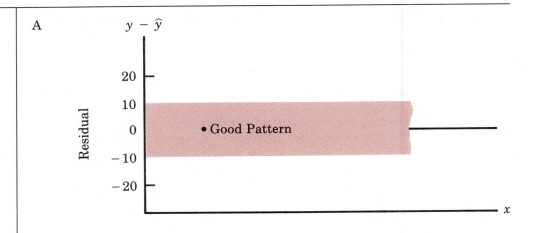

A

$y - \hat{y}$

Residual

20

10

0 • Good Pattern

−10

−20

x

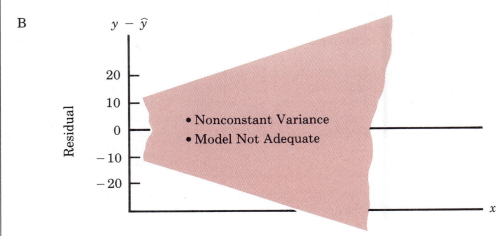

B

$y - \hat{y}$

Residual

20

10

0 • Nonconstant Variance

−10 • Model Not Adequate

−20

x

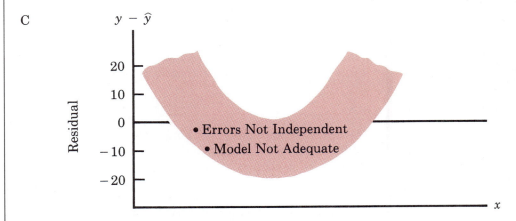

C

$y - \hat{y}$

Residual

20

10

0 • Errors Not Independent

−10 • Model Not Adequate

−20

x

FIGURE 13.10 Residual Plot from Three Regression Studies

reveal patterns and thus help determine whether or not the assumptions concerning ϵ are satisfied. Three of the most common residual plots are:

1. A plot of the residuals against the independent variable, x.
2. A plot of the residuals against the predicted value of the dependent variable, $\hat{y}$.
3. A standardized plot in which each residual is replaced by its z-score (i.e., the mean is subtracted and the result is divided by the standard error).

In this chapter we show how to develop a plot of the residuals against the independent variable, x. In chapters 14 and 15 we will discuss the two other types of residual plots.

Residual Plot Against x

A residual plot against the independent variable, x, is constructed by placing x on the horizontal axis and the residuals on the vertical axis. A residual is plotted for each observation; the first coordinate is x_i, the second is the residual, $y_i - \hat{y}_i$. Figure 13.10 shows some of the patterns statisticians look for when analyzing residuals. Panel A shows the type of plot to expect when the assumptions regarding ϵ are satisfied. The patterns shown in Panels B and C indicate a violation of one or more assumptions.

If the assumption that the variance of ϵ is the same for all values of x is valid, the residual plot should give an overall impression of a horizontal band of points. Panel A of Figure 13.10 shows the type of pattern to be expected in this case. On the other hand, if the variance of ϵ is not constant—for example, the variability about the regression line is greater for larger values of x—we would observe a pattern such as that of Panel B of Figure 13.10. Finally, if we observe a residual pattern such as that of Panel C of Figure 13.10, we would conclude that the error terms are not independent. The assumption of a linear relationship between x and y is not appropriate, or perhaps a multiple regression model is needed.

EXAMPLE 13.3 (continued)

A plot of the residuals against the independent variable x for Example 13.3 is shown in Figure 13.11 (the residuals are computed in Table 13.4). Looking at Figure 13.11, we see that the residuals appear to follow the pattern of Panel A of Figure 13.10. We thus conclude that the assumptions are satisfied and that the simple linear regression model for Example 13.3 is valid.

■

TABLE 13.4

Computation of the Residuals for Example 13.3

x_i = HOURS USING THE COMPUTER SYSTEM	y_i = FINAL EXAM SCORE	$\hat{y}_i = 5.85 + .83x$	$y_i - \hat{y}_i$
45	40	43.20	−3.20
30	35	30.75	4.25
90	75	80.55	−5.55
60	65	55.65	9.35
105	90	93.00	−3.00
65	50	59.80	−9.80
90	90	80.55	9.45
80	80	72.25	7.75
55	45	51.50	−6.50
75	65	68.10	−3.10

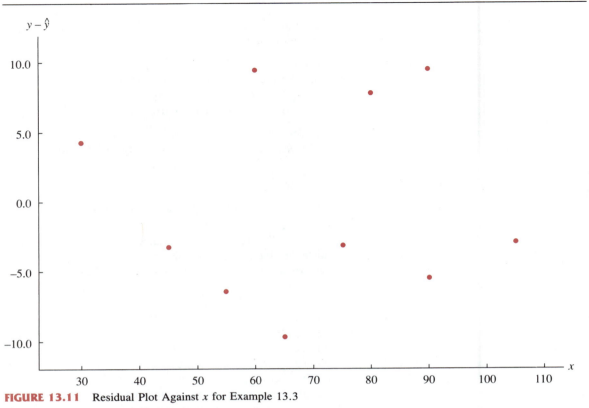

FIGURE 13.11 Residual Plot Against x for Example 13.3

33. In Exercise 4 data concerning advertising expenditures and sales at Eddie's Restaurants were given. These data are repeated here:

ADVERTISING EXPENDITURES ($1000s)	SALES ($1000s)
1.0	19.0
2.0	32.0
4.0	44.0
6.0	40.0
10.0	52.0
14.0	53.0
20.0	54.0

a. Let x equal advertising expenditures ($1000s) and y equal sales ($1000s). Use the method of least squares to develop a straight line approximation to the relationship between the two variables.

b. Test whether or not sales and advertising expenditures are related at the $\alpha = .05$ level of significance.

c. Prepare a plot of the residuals against x. Use the result of part (a) to obtain the values of $\hat{y}$.

d. What conclusions can you draw from residual analysis? Should this model be used, or should we look for a better one?

34. The following data were used in a regression study:

OBSERVATION	x_i	y_i
1	2	4
2	3	5
3	4	4
4	5	6
5	7	4
6	7	6
7	7	9
8	8	5
9	9	11

a. Develop an estimated regression equation for this data.

b. Construct a plot of the residuals. Do the assumptions concerning the error terms seem to be satisfied?

35. Refer to Exercise 5, where an estimated regression equation relating years of experience and annual sales was developed.

a. Compute the residuals and construct a residual plot for this problem.

b. Do the assumptions concerning the error terms seem reasonable in light of the residual plot?

36. The following data show the number of employees and the yearly revenues for the ten largest wholesale bakers *(Louis Rukeyser's Business Almanac)*.

COMPANY	EMPLOYEES	REVENUES ($1,000,000s)
Nabisco Brands USA	9,500	1,734
Continental Baking Co.	22,400	1,600
Campbell Taggart, Inc.	19,000	1,044
Keebler Company	8,943	988
Interstate Bakeries Corp.	11,200	704
Flowers Industries, Inc.	10,200	557
Sunshine Biscuits, Inc.	5,000	490
American Bakeries Co.	6,600	461
Entenmann's Inc.	3,734	450
Kitchens of Sara Lee	1,550	405

a. Use a computer package to develop an estimated regression equation relating revenues (y) to the number of employees (x).

b. Construct a plot of the residuals against the independent variable.

c. Do the assumptions concerning the error terms and model form seem reasonable in light of the residual plot?

13.8　Correlation Analysis

As we indicated in the introduction to this chapter, there are some situations in which the decision maker is not as concerned with the equation that relates two variables as in measuring the extent to which the two variables are related. In such cases a statistical technique referred to as correlation analysis can be used to determine the strength of the relationship between the two variables.* The output of a correlation study is a number referred to as the correlation coefficient. Because of the way in which it is defined, values of the correlation coefficient are always between -1 and $+1$. A value of $+1$ indicates that x and y are perfectly related in a positive linear sense. That is, all the points lie on a straight line that has a positive slope. A value of -1 indicates that x and y are perfectly related in a negative linear sense. That is, all the points lie on a straight line that has a negative slope. Values of the correlation coefficient close to zero indicate that x and y are not linearly related.

EXAMPLE 13.5

To provide an illustration of correlation analysis, we consider the situation of a stereo and sound-equipment store located in San Francisco. Management would like to investigate whether or not there is any relationship between the number of commercials (x) shown on Friday evening television and the resulting sales volume on Saturday (y), measured in hundreds of dollars. The sample data that were obtained are shown in Table 13.5.

In Figure 13.12 we show a scatter diagram of this data. The scatter diagram appears to indicate that there is a positive linear relationship between x and y. To measure the degree of linear association between these two variables, we first define a measure of linear association known as the *covariance*.

■

Covariance

Sample covariance is defined as follows.

Sample Covariance

$$s_{xy} = \frac{\Sigma(x_i - \bar{x})(y_i - \bar{y})}{n - 1} \qquad (13.23)$$

In this formula each x_i value is paired with a y_i value. We then sum the products obtained by multiplying the deviation of each x_i from its sample mean, $\bar{x}$, times the deviation of the corresponding y_i from its sample mean, $\bar{y}$; this sum is then divided by $n - 1$.

*In correlation analysis it is assumed that x and y are both random variables.

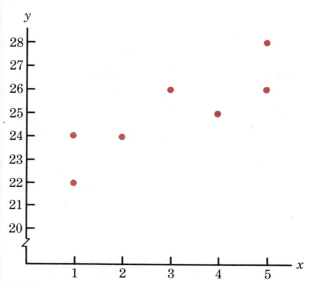

FIGURE 13.12 Scatter Diagram for the Stereo and Sound-Equipment Problem

EXAMPLE 13.5 (continued)

To measure the strength of the linear relationship between the number of commercials (x) and the sales volume (y), we can use (13.23) to compute the sample covariance. The calculations shown in Table 13.6 illustrate the computation of $\Sigma(x_i - \bar{x})(y_i - \bar{y})$. Note that $\bar{x} = 21/7 = 3$ and $\bar{y} = 175/7 = 25$. Using (13.23) we obtain

$$s_{xy} = \frac{\Sigma(x_i - \bar{x})(y_i - \bar{y})}{n-1} = \frac{17}{6} = 2.8333$$

The formula for computing the covariance of a population of size N is similar to (13.23), but we use different notation to indicate that we are dealing with the entire population.

TABLE 13.5

Sample Data for Example 13.5

STORE	x = NUMBER OF COMMERCIALS	y = SALES VOLUME ($100s)
1	2	24
2	5	28
3	1	22
4	3	26
5	4	25
6	1	24
7	5	26

$$\sigma_{xy} = \frac{\Sigma(x_i - \mu_x)(y_i - \mu_y)}{N}$$

(13.24)

where

μ_x = population mean for x

μ_y = population mean for y

In (13.24) we use the notation μ_x for the population mean of the variable x and μ_y for the population mean of the variable y. The sample covariance, s_{xy} is an estimate of the population covariance, σ_{xy} based upon a sample of size n.

Interpretation of the Covariance

To aid in the interpretation of the *sample covariance*, consider Figure 13.13. It is the same as the scatter diagram of Figure 13.12 with a vertical line at $x = 3$ (the value of $\bar{x}$) and a horizontal line at $y = 25$ (the value of $\bar{y}$). Four quadrants have been identified on the graph. Points that fall in quadrant I correspond to x_i values greater than $\bar{x}$ and y_i values greater than $\bar{y}$; points that fall in quadrant II correspond to x_i values less than $\bar{x}$ and y_i values greater than $\bar{y}$, and so on. Thus the value of $(x_i - \bar{x})(y_i - \bar{y})$ must be positive for points located in quadrant I, negative for points located in quadrant II, positive for points located in quadrant III and negative for points located in quadrant IV.

If the values of s_{xy} is positive, the points that have had the greatest effect on s_{xy} must lie in quadrants I and/or III. Hence, a positive value for s_{xy} is indicative of a positive linear association between x and y; that is, as the value of x increases, the value of y increases. If the value of s_{xy} is negative, however, the points that have had the greatest effect on s_{xy} lie in quadrants II and/or IV. Hence, a negative value for s_{xy} is indicative

TABLE 13.6

Calculations for the Sample Covariance

	x_i	y_i	$x_i - \bar{x}$	$y_i - \bar{y}$	$(x_i - \bar{x})(y_i - \bar{y})$
	2	24	−1	−1	1
	5	28	2	3	6
	1	22	−2	−3	6
	3	26	0	1	0
	4	25	1	0	0
	1	24	−2	−1	2
	5	26	2	1	2
Totals	21	175	0	0	17

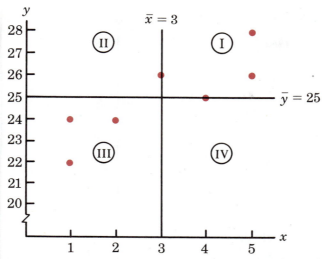

FIGURE 13.13 Quadrants I, II, III, and IV for the Stereo and Sound-Equipment Problem

of a negative linear association between x and y; that is, as the value of x increases, the value of y decreases. Finally, if the points are evenly distributed across all four quadrants, the value of s_{xy} will be close to 0, indicating no linear association between x and y. Figure 13.14 shows the values of s_{xy} that can be expected with these three different types of scatter diagrams.

From the previous discussion it might appear that a large positive value for the covariance is indicative of a strong positive linear relationship and that a large negative value is indicative of a strong negative linear relationship. However, one problem with using covariance as a measure of the strength of the linear relationship is that the value we obtain for the covariance depends upon the units of measurement for x and y. For example, suppose we were interested in the relationship between height (x) and weight (y) for individuals. If height is measured in inches, we will get much larger numerical values for $(x_i - \bar{x})$ than if it is measured in feet. Thus, with height measured in inches, we would obtain larger values for $\Sigma(x_i - \bar{x})(y_i - \bar{y})$—and hence a larger covariance—when, in fact, there is no difference in the relationship. A measure of relationship that avoids this difficulty is the *correlation coefficient*.

Correlation Coefficient

For sample data, the Pearson Product Moment correlation coefficient is defined as follows.

Equation (13.25) shows that the Pearson Product Moment correlation coefficient for sample data (commonly referred to more simply as the *sample correlation coefficient*) is computed by dividing the sample covariance by the product of the standard deviation of x and the standard deviation of y. Before we consider further interpretation of the sample correlation coefficient, let us consider the use of (13.25) for Example 13.5.

CHAPTER 13 LINEAR REGRESSION AND CORRELATION

FIGURE 13.14 Interpretation of Sample Covariance

**EXAMPLE 13.5
(continued)**

Using the data presented in Table 13.6, we can compute the sample correlation coefficient.

$$s_x = \sqrt{\frac{\Sigma(x_i - \bar{x})^2}{n-1}} = \sqrt{\frac{18}{6}} = 1.7321$$

$$s_y = \sqrt{\frac{\Sigma(y_i - \bar{y})^2}{n-1}} = \sqrt{\frac{22}{6}} = 1.9149$$

and, since $s_{xy} = 2.8333$, we have

$$r_{xy} = \frac{s_{xy}}{s_x s_y} = \frac{2.8333}{(1.7321)(1.9149)} = .854$$

    When using a calculator to compute the sample correlation coefficient, (13.26) is preferred since the computation of each deviation $x_i - \bar{x}$ and $y_i - \bar{y}$ is not necessary; thus, less round-off error is introduced.

**EXAMPLE 13.5
(continued)**

Algebraically, equations (13.25) and (13.26) are equivalent. In Table 13.7 we provide the calculations needed to use (13.26). Using these computations and (13.26), we obtain:

$$r_{xy} = \frac{542 - (21)(175)/7}{\sqrt{81 - (21)^2/7} \ \sqrt{4397 - (175)^2/7}} = \frac{17}{19.8997} = .854$$

Thus we see that the value obtained for r_{xy} using (13.26) is the same (to three decimal places) as the value obtained using (13.25).

◼

The formula for computing the correlation coefficient of a population, denoted by the Greek letter ρ_{xy} (rho, pronounced "row"), is as follows.

Pearson Product Moment Correlation Coefficient—Population Data

$$\rho_{xy} = \frac{\sigma_{xy}}{\sigma_x \sigma_y} \tag{13.27}$$

where

ρ_{xy} = population correlation coefficient

σ_{xy} = population covariance

σ_x = population standard deviation for x

σ_y = population standard deviation for y

The sample correlation coefficient, r_{xy}, is an estimate of the population correlation coefficient, ρ_{xy}.

First let us consider a simple example that illustrates the concept of perfect positive linear association.

EXAMPLE 13.6

The scatter diagram shown in Figure 13.15 depicts the relationship between the following $n = 3$ pairs of points:

TABLE 13.7

Computations for Using the Computational Formula for Computing r_{xy}

	x_i	y_i	$x_i y_i$	x_i^2	y_i^2
	2	24	48	4	576
	5	28	140	25	784
	1	22	22	1	484
	3	26	78	9	676
	4	25	100	16	625
	1	24	24	1	576
	5	26	130	25	676
Totals	21	175	542	81	4397

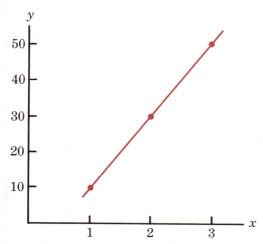

FIGURE 13.15 Scatter Diagram Depicting a Perfect Positive Linear Association

x_i	1	2	3
y_i	10	30	50

The straight line drawn through each of the 3 points shows that there is a perfect linear relationship between the two variables x and y. The calculations need to compute r_{xy} are shown in Table 13.8. Using the values in this table we obtain

$$r_{xy} = \frac{\Sigma x_i y_i - (\Sigma x_i \Sigma y_i)/n}{\sqrt{\Sigma x_i^2 - (\Sigma x_i)^2/n} \sqrt{\Sigma y_i^2 - (\Sigma y_i)^2/n}}$$

$$= \frac{220 - (6)(90)/3}{\sqrt{14 - (6)^2/3} \sqrt{3500 - (90)^2/3}} = \frac{40}{40} = 1$$

Thus we see that the value of the sample correlation coefficient for this data set is 1.

In general, it can be shown that if all the points in a data set fall on a straight line having positive slope, then the value of the sample correlation coefficient is $+1$; that is, a sample correlation coefficient of $+1$ corresponds to a perfect positive linear association

TABLE 13.8

Calculations for Computing r_{xy} for Example 13.6

	x_i	y_i	$x_i y_i$	x_i^2	y_i^2
	1	10	10	1	100
	2	30	60	4	900
	3	50	150	9	2500
Totals	6	90	220	14	3500

between x and y. Moreover, if the points in the data set fall on a straight line having negative slope, the value of the sample correlation coefficient is -1; that is, a sample correlation coefficient of -1 corresponds to a perfect negative linear association between x and y.

Let us now suppose that for a certain data set there is a positive linear association between x and y but that the relationship is not perfect. The value of r_{xy} will be less than $+1$, indicating that the points in the scatter diagram do not all fall on a straight line. As the points in a data set deviate more and more from a perfect positive linear association, the value of r_{xy} becomes smaller and smaller. A value of r_{xy} equal to 0 indicates no linear relationship between x and y and values of r_{xy} near zero indicate a weak relationship.

EXAMPLE 13.5 (continued)

Recall that, $r_{xy} = +.854$. Since $r_{xy} = +.854$ we conclude that there appears to be a positive linear association between the number of commercials and Saturday sales volume. More specifically, an increase in the number of commercials is associated with an increase in sales volume.

We have stated that values of r_{xy} near $+1$ indicate a strong linear association between two variables and values of r_{xy} near zero indicate little or no linear association between the variables. But we must be careful not to conclude that a value of r near zero means there is no relationship between the variables.

EXAMPLE 13.7

The scatter diagram in Figure 13.16 shows a case where $r_{xy} = 0$ and there is no linear relationship; however, in this case, there is a perfect curvilinear relationship between the

FIGURE 13.16 Even though $r = 0$, there is a Perfect Curvilinear Relationship for the Data

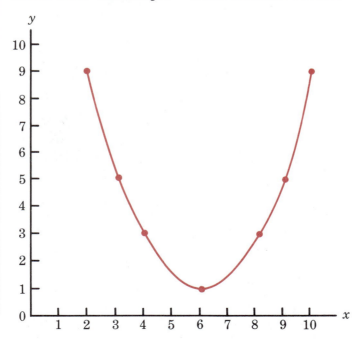

variables. Table 13.9 provides the calculations needed to compute r_{xy} for this example. Using these calculations, the computation of r_{xy} is as follows:

$$r_{xy} = \frac{210 - (42)(35)/7}{\sqrt{310 - (42)^2/7} \ \sqrt{231 - (35)^2/7}} = \frac{0}{56.9912} = 0$$

■

To reiterate, our last example illustrates an important concept regarding the proper interpretation of the sample correlation coefficient. The sample correlation coefficient measures only the degree of *linear association* between the two variables. A value of r_{xy} equal to 0 cannot be interpreted as implying that there is no relationship between the two variables. One should always look at the associated scatter diagram as well as the value of the sample correlation coefficient when attempting to determine if, and how, two variables are related.

In closing this part of the section we caution that while a correlation coefficient near ± 1 does imply a strong linear association between two variables, it does not imply a cause-effect relationship. For instance, it has been noted that as women's hemlines are raised, stock prices go up. There is a positive correlation, but it would be folly to infer a cause-effect relationship. No one truly believes that changes in the stock market are the result of changes in women's hemlines. Conclusions concerning cause-effect must be based on the judgment of the analyst.

Determining the Sample Correlation Coefficient from the Regression Analysis Output

In this discussion we will assume that the least squares estimated regression equation is $\hat{y} = b_0 + b_1 x$. In such cases the sample correlation coefficient can be computed using one of the following formulas:

Sample Correlation Coefficient

$$r_{xy} = (\text{sign of } b_1) \ \sqrt{\text{Coefficient of Determination}} = \pm \sqrt{r^2} \qquad (13.28)$$

$$r_{xy} = b_1 \left(\frac{s_x}{s_y} \right) \qquad (13.29)$$

where

$$s_x = \sqrt{\frac{\Sigma (x_i - \bar{x})^2}{n - 1}} \quad (\text{Sample standard deviation for } x)$$

$$s_y = \sqrt{\frac{\Sigma (y_i - \bar{y})^2}{n - 1}} \quad (\text{Sample standard deviation for } y)$$

Note that the sign of the sample correlation coefficient is the same as the sign of b_1, the slope of the estimated regression equation.

Testing for Significant Correlation

The sample correlation coefficient is really a point estimator of the population correlation coefficient, a measure of the actual linear relationship between x and y in the population. Recall from our previous discussion that ρ_{xy} denotes the population correlation coefficient. A statistical test for the significance of a linear association between x and y can be performed by testing the following hypotheses:

$$H_0: \quad \rho_{xy} = 0$$

$$H_a: \quad \rho_{xy} \neq 0$$

It can be shown that testing these hypotheses is equivalent to testing the hypotheses regarding the significance of β_1, the slope of the regression equation. Thus, the tests for significance introduced in Section 13.4 can be used to test for the significance of ρ_{xy}. To test for a significant correlation without performing a regression study, statisticians have shown that if H_0 is true, then the value of

$$r_{xy} \sqrt{\frac{n-2}{1-r_{xy}^2}} \tag{13.30}$$

has a t distribution with $n - 2$ degrees of freedom.

EXAMPLE 13.3
(continued)

Recall that $\hat{y} = 5.85 + 0.83x$ and $r^2 = .878$. Thus, using (13.28), we obtain

$$r_{xy} = +\sqrt{.878} = .937$$

With a sample correlation coefficient of $r_{xy} = .937$, the value of (13.30) is

$$.937 \sqrt{\frac{8}{1-.878}} = 7.59$$

TABLE 13.9

Calculations for Example 13.7 Illustrating $r_{xy} = 0$

	x_i	y_i	x_iy_i	x_i^2	y_i^2
	2	9	18	4	81
	3	5	15	9	25
	4	3	12	16	9
	6	1	6	36	1
	8	3	24	64	9
	9	5	45	81	25
	10	9	90	100	81
Totals	42	35	210	310	231

CHAPTER 13 LINEAR REGRESSION AND CORRELATION

For $\alpha = .05$ and $n - 2 = 10 - 2 = 8$ degrees of freedom, we see that the appropriate t value from Table 2 of Appendix B is 2.306. Thus, if the value of (13.30) exceeds 2.306 or is less than -2.306, we must reject the null hypothesis $H_0: \rho_{xy} = 0$. Since 7.59 exceeds the t value of 2.306, we reject H_0 and hence conclude that x and y have a significant correlation.

■

EXERCISES

37. Given are five observations taken for two variables.

x	4	6	11	3	16
y	50	50	40	60	30

a. Develop a scatter diagram with x on the horizontal axis.
b. What does the scatter diagram developed in (a) indicate about the relationship between the two variables?
c. Compute and interpret the sample covariance for these data.
d. Compute and interpret the sample correlation coefficient for these data.

38. Eight observations on two variables are given.

x	2	9	6	8	4	7	5	6
y	11	4	6	5	9	4	9	7

a. Develop a scatter diagram for these data with x on the horizontal axis.
b. What does the scatter diagram developed in (a) indicate about the relationship between the two variables?
c. Compute and interpret the sample covariance for these data.
d. Compute and interpret the sample correlation coefficient for these data.

39. The following data are shown for two variables x and y.

x	4	10	7	9	5	8	6	7
y	22	8	13	10	18	8	19	14

a. Compute and interpret the sample covariance for these data.
b. Compute and interpret the sample correlation coefficient for these data.

40. The following estimated regression line has been developed to estimate the relationship between x, the number of units produced per week, and y, the total weekly cost of production ($):

$$\hat{y} = 60 + 3.2x$$

The standard deviation of weekly production is 10 units, and the standard deviation of weekly cost is $35.00. Compute the sample correlation coefficient.

41. Eight observations on two random variables are given.

x_i	2	9	6	8	4	7	5	6
y_i	11	4	6	5	9	4	9	7

a. Compute r_{xy}.
b. Test the hypotheses

$$H_0: \quad \rho_{xy} = 0$$
$$H_a: \quad \rho_{xy} \neq 0$$

at the $\alpha = .01$ level of significance.

42. The data presented in Exercise 27 are shown.

Speed	30	50	40	55	30	25	60	25	50	55
Mileage	28	25	25	23	30	32	21	35	26	25

a. Compute the sample correlation coefficient for these data.
b. Test the hypotheses

$$H_0: \quad \rho_{xy} = 0$$
$$H_a: \quad \rho_{xy} \neq 0$$

at the $\alpha = .01$ level of significance.

43. As more U.S. households receive cable television, the advertising revenue has continued to increase. The following data show the 1988 and 1987 expenditures for the top ten cable television advertisers.

ADVERTISER	1988 EXPENDITURE ($1,000,000s)	1987 EXPENDITURE ($1,000,000s)
Procter & Gamble Co.	30.2	23.7
Philip Morris Cos.	23.1	20.6
Anheuser-Busch Cos.	21.4	22.9
Time Inc.	21.2	16.4
General Mills Inc.	20.0	18.6
RJR Nabisco Inc.	14.3	14.7
Eastman Kodak	11.0	2.6
Clorox Co.	10.1	6.9
Mars Inc.	10.0	14.9
Chrysler Corp.	9.5	6.1

a. Develop a scatter diagram for the data. Does it appear that the two variables are linearly related?
b. Compute the sample correlation coefficient for these data.
c. Test the hypotheses

$$H_0: \quad \rho_{xy} = 0$$
$$H_a: \quad \rho_{xy} \neq 0$$

at the $\alpha = .01$ level at significance.

In this chapter we introduced the topics of regression and correlation analysis. We discussed how regression analysis can be used to develop an equation showing how variables are related. For a given value of the independent variable, the estimated regression equation can be used to predict the value of the dependent variable. Correlation analysis can be used to determine the strength of the relationship between two variables but cannot be used for prediction. Before concluding our discussion, however, we would like to reemphasize a potential misinterpretation of these studies. Regression and correlation analyses can indicate only how or to what extent the variables are associated with each other. These techniques cannot be interpreted directly as showing cause-and-effect relationships. Cause-and-effect relationships should only be inferred if someone knowledgeable in the field concludes there is a good theoretical basis for such a relationship.

GLOSSARY

Note: The definitions here are all stated with the understanding that simple linear regression and correlation is being considered.

Simple linear regression   The simplest kind of regression, involving only two variables that are related by a straight line.

Dependent variable   The variable that is being predicted or explained. It is denoted by y in the regression equation.

Independent variable   The variable that is doing the predicting or explaining. It is denoted by x in the regression equation.

Scatter diagram   A graph of the available data in which the independent variable appears on the horizontal axis and the dependent variable appears on the vertical axis.

Least squares method   The approach used to develop the estimated regression equation which minimizes the sum of squares of the vertical distances from the points to the least squares fitted line.

Estimated regression equation line   The estimate of the regression equation obtained by the least squares method; that is, $\hat{y} = b_0 + b_1 x$.

Residual   The difference between the actual value of the dependent variable and the value predicted using the estimated regression equation; i.e., $y_i - \hat{y}_i$.

Coefficient of determination (r^2)   A measure of the variation explained by the estimated regression equation. It is a measure of how well the estimated regression equation fits the data.

Deterministic model   A relationship between an independent variable and a dependent variable in which specifying the value of the independent variable allows one to compute exactly the value of the dependent variable.

Probabilistic model   A relationship between an independent variable and a dependent variable in which specifying the value of the independent variable is not sufficient to allow determination of the value of the dependent variable.

Regression equation   The mathematical equation relating the independent variable to the expected value of the dependent variable; that is, $E(y) = \beta_0 + \beta_1 x$.

Sample correlation coefficient (r_{xy})   A statistical measure of the linear association between two variables.

ESTIMATED REGRESSION LINE

$$\hat{y} = b_0 + b_1 x \tag{13.1}$$

LEAST SQUARES OBJECTIVE

$$\min \Sigma (y_i - \hat{y}_i)^2 \tag{13.2}$$

SLOPE AND y-INTERCEPT FOR THE ESTIMATED REGRESSION LINE

$$b_1 = \frac{\Sigma x_i y_i - (\Sigma x_i \Sigma y_i)/n}{\Sigma x_i^2 - (\Sigma x_i)^2/n} \tag{13.3}$$

$$b_0 = \bar{y} - b_1 \bar{x} \tag{13.4}$$

SUM OF SQUARES DUE TO ERROR

$$SSE = \Sigma(y_i - \hat{y}_i)^2 \tag{13.5}$$

TOTAL SUM OF SQUARES

$$SST = \Sigma(y_i - \bar{y})^2 \tag{13.6}$$

SUM OF SQUARES DUE TO REGRESSION

$$SSR = \Sigma(\hat{y}_i - \bar{y})^2 \tag{13.7}$$

RELATIONSHIP AMONG SST, SSR, AND SSE

$$SST = SSR + SSE \tag{13.8}$$

COEFFICIENT OF DETERMINATION

$$r^2 = \frac{SSR}{SST} \tag{13.9}$$

COMPUTATIONAL FORMULA FOR SSR

$$SSR = \frac{[\Sigma x_i y_i - (\Sigma x_i \Sigma y_i)/n]^2}{\Sigma x_i^2 - (\Sigma x_i)^2/n} \tag{13.10}$$

COMPUTATIONAL FORMULA FOR SST

$$SST = \Sigma y_i^2 - \frac{(\Sigma y_i)^2}{n} \tag{13.11}$$

REGRESSION MODEL

$$y = \beta_0 + \beta_1 x + \epsilon \tag{13.12}$$

REGRESSION EQUATION

$$E(y) = \beta_0 + \beta_1 x \tag{13.13}$$

MEAN SQUARE ERROR (ESTIMATE OF σ^2)

$$s^2 = MSE = \frac{SSE}{n - 2} \tag{13.14}$$

ESTIMATE OF THE VARIANCE OF b_1

$$s_{b_1}^2 = s^2 \left(\frac{1}{\Sigma x_i^2 - (\Sigma x_i)^2/n} \right) \tag{13.16}$$

MEAN SQUARE DUE TO REGRESSION

$$\text{MSR} = \frac{\text{SSR}}{\text{Regression DF}} = \frac{\text{SSR}}{\text{Number of Independent Variables}} \qquad (13.17)$$

THE F STATISTIC

$$F = \frac{\text{MSR}}{\text{MSE}} \qquad (13.18)$$

ESTIMATED VARIANCE OF $\hat{y}_p$

$$s_{\hat{y}_p}^2 = s^2 \left(\frac{1}{n} + \frac{(x_p - \bar{x})^2}{\Sigma x_i^2 - (\Sigma x_i)^2/n} \right) \qquad (13.19)$$

CONFIDENCE INTERVAL ESTIMATE OF $E(y_p)$

$$\hat{y}_p \pm t_{\alpha/2} s_{\hat{y}_p} \qquad (13.20)$$

VARIANCE OF THE ESTIMATE OF AN INDIVIDUAL VALUE OF y_p

$$s_{\text{ind}}^2 = s^2 \left(1 + \frac{1}{n} + \frac{(x_p - \bar{x})^2}{\Sigma x_i^2 - (\Sigma x_i)^2/n} \right) \qquad (13.21)$$

PREDICTION INTERVAL ESTIMATE OF y_p

$$\hat{y}_p \pm t_{\alpha/2} s_{\text{ind}} \qquad (13.22)$$

SAMPLE COVARIANCE

$$s_{xy} = \frac{\Sigma(x_i - \bar{x})(y_i - \bar{y})}{n - 1} \qquad (13.23)$$

PEARSON PRODUCT MOMENT CORRELATION COEFFICIENT—SAMPLE DATA

$$r_{xy} = \frac{s_{xy}}{s_x s_y} \qquad (13.25)$$

PEARSON PRODUCT MOMENT CORRELATION COEFFICIENT—SAMPLE DATA, COMPUTATIONAL FORMULA

$$r = \frac{\Sigma x_i y_i - (\Sigma x_i \Sigma y_i)/n}{\sqrt{\Sigma x_i^2 - (\Sigma x_i)^2/n} \ \sqrt{\Sigma y_i^2 - (\Sigma y_i)^2/n}} \qquad (13.26)$$

DETERMINING THE SAMPLE CORRELATION COEFFICIENT FROM THE REGRESSION ANALYSIS

$$r_{xy} = (\text{sign of } b_1) \ \sqrt{\text{Coefficient of Determination}} \qquad (13.28)$$

or

$$r_{xy} = b_1 \left(\frac{s_x}{s_y} \right) \qquad (13.29)$$

TEST STATISTIC FOR SIGNIFICANT CORRELATION

$$r_{xy} \sqrt{\frac{n - 2}{1 - r_{xy}^2}} \qquad (13.30)$$

TRUE/FALSE

1. The least squares method is used to determine an estimated regression line that minimizes the squared deviations of the data values from the line.

2. The least squares method is applicable only in situations where the estimated regression line has a positive slope.

3. If the slope of the estimated regression line is positive, the correlation coefficient must be negative.

4. The slope of the estimated regression line (b_1) is a sample statistic, since, like other sample statistics, it is computed from the sample observations.

5. The coefficient of determination is the square root of the correlation coefficient.

6. The sum of squares due to regression (SSR) plus the sum of squares due to error (SSE) must equal the total sum of squares (SST).

7. A t test can be used to test whether or not there is a significant regression relationship.

8. The sampling distribution of b_1 is normal if the usual regression assumptions are satisfied.

9. An interval estimate for a particular value of the dependent variable yields a smaller interval than an interval estimate of the expected value of the dependent variable.

10. Minitab cannot be used for the computer solution of regression problems.

11. If two variables are perfectly linearly related, the sample correlation coefficient must equal -1 or 1.

12. The residual is the difference between the actual value of a dependent variable and the value predicted by the estimated regression line.

MULTIPLE CHOICE

13. If two variables x and y have a significant linear relationship, then
    a. there may or may not be any causal relationship between x and y
    b. x causes y to happen
    c. y causes x to happen

14. For the estimated regression line $\hat{y} = 3 - 10x$, the correlation coefficient r_{xy}
    c. equals 0
    b. is less than 0
    c. is greater than 0

15. If the correlation coefficient for two variables is $-.9$, the coefficient of determination is
    a. .9
    b. $-.81$
    c. .81

16. If a data set has SST = 200 and SSE = 150, then the coefficient of determination is
  a. .25
  b. .50
  c. .75
  d. 50

17. Compared to the prediction interval for a particular value of y, the confidence interval for a mean value of y will be
  a. narrower
  b. wider
  c. not enough information is given

18. A sample correlation coefficient is calculated from 15 pairs of x and y observations. The t distribution used to determine whether or not this coefficient is statistically significant will have how many degrees of freedom?
  a. 13
  b. 14
  c. 15
  d. 28

19. Which of the following is correct?
  a. SST = SSR − SSE
  b. SSE = SSR + SSE
  c. SSR = SSE + SST
  d. SSE = SST − SSR

20. The coefficient of determination is calculated as
  a. SST/SSE
  b. SSR/SST
  c. SSR/SSE
  d. SSE/SSR

21. Which of the following is an appropriate test statistic to test the null hypothesis that there is no linear relationship between x and y?
  a. SSR/SST
  b. MSE
  c. MSR/MSE
  d. MSE/MST

22. Which of the following points are *always* on the estimated linear regression line?
  a. $x = \bar{x}, y = \bar{y}$
  b. $x = 0, y = 0$
  c. $x = 1, y = 0$
  d. $x = 0, y = 1$

SUPPLEMENTARY EXERCISES

44. What is the difference between regression analysis and correlation analysis?

45. Does a high value of r^2 imply that two variables are causally related? Explain.

46. In your own words, explain the difference between an interval estimate of the mean value of y for a given x and an interval estimate for an individual value of y for a given x.

47. How do we measure how close the actual data points are to the estimated regression line? That is, how do we measure the goodness-of-fit of the regression line?

48. The following data show the percentage of women working in each company (x) and the percentage of management jobs held by women in that company (y); the data shown represent companies in retailing and trade *(Louis Rukeyser's Business Almanac)*.

COMPANY	x	y
Federated Department Stores	72	61
Kroger	47	16
Marriott	51	32
McDonald's	57	46
Sears, Roebuck	55	36

   a. Develop a scatter diagram for these data.
   b. What does the scatter diagram developed in part (a) indicate about the relationship between x and y?
   c. Use the method of least squares to develop an estimated regression equation for these data.
   d. Predict the percentage of management jobs held by women in a company that has 60% women employees.
   e. Use the estimated regression equation to predict the percentage of management jobs held by women in a company where 55% of the jobs are held by women. How does this predicted value compare to the 36% value observed for Sears, Roebuck, a company for which 55% of its employees are women?

49. A list of the best-selling cars for 1987 whose sales in units varied between 175,000 and 300,000 (rounded to the nearest thousand) is shown in the following table *(The World Almanac, 1989)*. The 1988 suggested retail price (in thousands of dollars, rounded to the nearest hundred dollars) is also shown.

MODEL	NUMBER SOLD (1000s)	1988 PRICE ($1000s)
Hyundai	264	5.4
Oldsmobile Ciera	245	11.4
Nissan Sentra	236	6.4
Ford Tempo	219	9.1
Chev. Corsica/Beretta	214	10.0
Pontiac Grand Am	211	10.3
Toyota Camry	187	11.2
Chevrolet Caprice	177	12.5

   a. Compute an estimate of σ^2.
   b. Compute an estimate of the variance of b_1.
   c. Use the t test to test the hypotheses ($\alpha = .05$).

$$H_0: \quad \beta_1 = 0$$

$$H_a: \quad \beta_1 \neq 0$$

   d. Use the F test to test the hypotheses in part *(c)* at the $\alpha = .05$ level of significance.

50. In a manufacturing process the assembly line speed (feet per minute) was thought to affect the number of defective parts found during the inspection process. To test this theory, management devised a situation where the same batch of parts was inspected visually at a variety of line speeds. The following data were collected.

LINE SPEED	NUMBER OF DEFECTIVE PARTS FOUND
20	21
20	19
40	15
30	16
60	14
40	17

a. Develop the estimated regression line that relates line speed to the number of defective parts found.

b. At the $\alpha = .05$ level of significance determine whether or not line speed and number of defective parts found are related.

c. Did the estimated regression line provide a good fit to the data?

d. Develop a 95% confidence interval to predict the mean number of defective parts for a line speed of 50 feet per minute.

51. The PJH&D Company is in the process of deciding whether or not to purchase a maintenance contract for its new word processing system. They feel that maintenance expense should be related to usage and have collected the following information on weekly usage (hours) and annual maintenance expense.

WEEKLY USAGE (HOURS)	ANNUAL MAINTENANCE EXPENSE ($100s)
13	17.0
10	22.0
20	30.0
28	37.0
32	47.0
17	30.5
24	32.5
31	39.0
40	51.5
38	40.0

a. Develop the estimated regression line that relates annual maintenance expense, in hundreds of dollars, to weekly usage.

b. Test the significance of the relationship in (a) at the $\alpha = .05$ level of significance.

c. PJH&D expects to operate the word processor 30 hours per week. Develop a 95% prediction interval for the company's annual maintenance expense.

d. If the maintenance contract costs $3000 per year, would you recommend purchasing it? Why or why not?

52. A sociologist was hired by a large city hospital to investigate the relationship between the number of unauthorized days that an employee is absent per year and the distance (miles) between home and work for employees. A sample of ten employees was chosen, and the following data were collected.

DISTANCE TO WORK (MILES)	NUMBER OF DAYS ABSENT
1	8
3	5
4	8
6	7
8	6
10	3
12	5
14	2
14	4
18	2

a. Develop a scatter diagram for the above data. Does a linear relationship appear reasonable? Explain.
b. Develop the least squares estimated regression line.
c. Is there a significant relationship between the two variables? Use $\alpha = .05$.
d. Did the estimated regression line provide a good fit? Explain.
e. Use the estimated regression line developed in (b) to develop a 95% confidence interval estimate of the expected number of days absent for employees living 5 miles from the company.

53. The regional transit authority for a major metropolitan area would like to determine if there is any relationship between the age of a bus and the annual maintenance cost. A sample of 10 buses resulted in the following data.

AGE OF BUS (YEARS)	MAINTENANCE COST ($)
1	350
2	370
2	480
2	520
2	590
3	550
4	750
4	800
5	790
5	950

a. Compute the sample correlation coefficient for the above data.
b. Using the sample correlation coefficient, test to see if the two variables are significantly related. Use $\alpha = .10$.

54. Reconsider the regional transit authority problem presented in Exercise 53.
a. Develop the least squares estimated regression line.
b. Test to see if the two variables are significantly related at $\alpha = .05$.
c. Did the least squares line provide a good fit to the observed data? Explain.
d. Develop a 90% confidence interval estimate of the maintenance cost for a specific bus that is 4 years old.

55. A psychology professor at Givens College is interested in the relation between time spent studying and total points earned in the course. Data collected on 10 students who took the course last quarter are given below.

TIME SPENT STUDYING	TOTAL POINTS EARNED
45	40
30	35
90	75
60	65
105	90
65	50
90	90
80	80
55	45
75	65

   a. Compute the sample correlation coefficient for the above data.
   b. Use the sample correlation coefficient and test to see if there is a significant relationship at the $\alpha = .05$ level.

56. Reconsider the Givens College data in Exercise 55.
   a. Develop an estimated regression line relating total points earned to hours spent studying.
   b. Test the significance of the model at the $\alpha = .05$ level.
   c. Predict the total points earned by Mark Sweeney. He spent 95 hours studying.
   d. Develop a 90% confidence interval for the total points earned by Mark Sweeney.

57. The "Statistics in the News" article at the beginning of this chapter described a system for developing a strength rating and ranking system for college basketball teams. It was stated that the strength ratings could be used to predict victory margins. Using the data in "Statistics in the News," answer the following questions.
   a. Compute the correlation coefficient between the predicted victory margin and the actual victory margin.
   b. Test for a significant relationship. Use $\alpha = .01$.
   c. Develop an estimated regression line, using the predicted victory margin as the independent variable and the actual victory margin as the dependent variable.
   d. What are the values of b_0 and b_1? Comment on what the values of β_0 and β_1 should be for an ideal system of predicting victory margins.

COMPUTER EXERCISE

As part of a study of transportation safety, a department of transportation collected data on the number of fatal accidents and the percentage of licensed drivers under the age of 21. The values for the two variables for a 1-year period were obtained from a sample of 42 cities. In order to account for differences in the population sizes of the various cities, the measure for the number of fatal accidents was defined in terms of the number of fatal accidents per 1000 licenses. The data are given.

PERCENT UNDER 21	NUMBER OF FATAL ACCIDENTS PER 1000 LICENSES	PERCENT UNDER 21	NUMBER OF FATAL ACCIDENTS PER 1000 LICENSES
13	2.962	17	4.100
12	0.708	8	2.190
8	0.885	16	3.623
12	1.652	15	2.623
11	2.091	9	0.835
17	2.627	8	0.820
18	3.830	14	2.890
8	0.368	8	1.267
13	1.142	15	3.224
8	0.645	10	1.014
9	1.028	10	0.493
16	2.801	14	1.443
12	1.405	18	3.614
9	1.433	10	1.926
10	0.039	14	1.643
9	0.338	16	2.943
11	1.849	12	1.913
12	2.246	15	2.814
14	2.855	13	2.634
14	2.352	9	0.926
11	1.294	17	3.256

QUESTIONS

1. Develop numerical and graphical measures to summarize the data.
2. Does there appear to be a relationship between the number of fatal accidents per 1000 licenses and the percentage of drivers under the age of 21? Explain.
3. What conclusions and recommendations can you derive from your analysis?

APPENDIX 13.1
Calculus-Based Derivation of Least Squares Formulas

As mentioned in the chapter, the least squares method is a procedure for determining the values of b_0 and b_1 that minimize the sum of squared residuals. The sum of squared residuals is given by

$$\Sigma(y_i - \hat{y}_i)^2$$

Substituting $\hat{y}_i = b_0 + b_1 x_i$, we get

$$\Sigma(y_i - b_0 - b_1 x_i)^2 \tag{13A.1}$$

as the expression that must be minimized.

To minimize (13A.1) we must take the partial derivatives with respect to b_0 and b_1, set them equal to zero, and solve. Doing so we get

$$\frac{\partial \Sigma(y_i - b_0 - b_1 x_i)^2}{\partial b_0} = -2\Sigma(y_i - b_0 - b_1 x_i) = 0 \tag{13A.2}$$

$$\frac{\partial \Sigma(y_i - b_0 - b_1 x_i)^2}{\partial b_1} = -2\Sigma x_i(y_i - b_0 - b_1 x_i) = 0 \tag{13A.3}$$

Dividing (13A.2) by 2 and summing each term individually yields

$$-\Sigma y_i + \Sigma b_0 + \Sigma b_1 x_i = 0$$

Bringing Σy_i to the other side of the equal sign and noting that $\Sigma b_0 = n b_0$, we obtain

$$n b_0 + (\Sigma x_i) b_1 = \Sigma y_i \tag{13A.4}$$

Similar algebraic simplification applied to (13A.3) yields

$$(\Sigma x_i) b_0 + (\Sigma x_i^2) b_1 = \Sigma x_i y_i \tag{13A.5}$$

(13A.4) and (13A.5) are known as the *normal equations*. Solving (13A.4) for b_0 yields

$$b_0 = \frac{\Sigma y_i}{n} - b_1 \frac{\Sigma x_i}{n} \tag{13A.6}$$

Using (13A.6) to substitute for b_0 in (13A.5) provides

$$\frac{\Sigma x_i \Sigma y_i}{n} - \frac{(\Sigma x_i)^2}{n} b_1 + (\Sigma x_i^2) b_1 = \Sigma x_i y_i \tag{13A.7}$$

Rearranging (13A.7), we obtain

$$b_1 = \frac{\Sigma x_i y_i - (\Sigma x_i \Sigma y_i)/n}{\Sigma x_i^2 - (\Sigma x_i)^2/n} \tag{13A.8}$$

Since $\bar{y} = \Sigma y_i/n$ and $\bar{x} = \Sigma x_i/n$, we can rewrite (13A.6):

$$b_0 = \bar{y} - b_1 \bar{x} \tag{13A.9}$$

Equations (13A.8) and (13A.9) are the formulas we used in the chapter to compute the coefficients in the estimated regression equation.

14 Multiple Regression

What You Will Learn in This Chapter:

- What multiple regression analysis is
- How to interpret the coefficients in a multiple regression model
- The important role computer packages play in performing multiple regression analysis
- How to use the t and F distributions to test for significant relationships in multiple regression analysis
- How to analyze the residuals in a multiple regression model

Contents

Father Versus Mother in Custody for Children of Divorce

Mothers have traditionally been awarded custody of children in divorce cases, due to the presumption by the courts that mothers were uniquely suited to raising children and also because of the powerful influence of our economic system, which has favored male employment. Thus children tend to go with the mother, whereas the father works to provide the financial support. However, the changing roles of both men and women and the various interrelated social changes have recently caused courts to consider fathers as custodians of children in divorce cases.

Although there are many variables, or factors, that need to be evaluated in making a custody decision, one consideration is that of providing the best environment for the children's academic achievement. Researchers Frederick Shilling and Patrick Lynch have investigated the effect that custody with the father rather than the mother has on the academic achievement of eighth grade children. Data for the Shilling-Lynch study were obtained from a sample of 3160 single-parent children in the eighth grade. In the sample, children living with fathers numbered 550, whereas 2610 children lived with the mothers.

Academic performance of the children was measured on reading, mathematics, and a composite of these two measures. Other data available included a measure of socioeconomic status based on parental occupation and education, sex of the child, sex of the single parent, residence (urban, suburban, rural) and a measure of the student's perception of parental interest in school. Using a method known as multiple linear regression, the researchers devel-

Jeffrey Grosscup

The courts have historically favored mother-custody for children of divorce.

oped an equation that used the socioeconomic status, sex of child, sex of parent, residence, and parental interest variables to predict academic performance scores. In terms of the sex of the single parent, the researchers found that verbal, mathematical, and overall academic achievement scores were better for children living with their mothers, other things being equal. The study can be construed as an encouragement to courts to continue assigning young children of broken families to the mothers instead of fathers, especially when academic achievement is an important factor in determining the best placement for the child.

Based on "Father Versus Mother Custody and Academic Achievement of Eighth Grade Children," by Frederick Shilling and Patrick D. Lynch, *Journal of Research and Development in Education* 18 (1985).

In Chapter 13 we discussed how simple linear regression can be used to develop a mathematical equation representing the relationship between two variables. Recall that the variable being predicted or explained by the mathematical equation is called the *dependent variable;* the variable being used to predict or explain the value of the dependent variable is called the *independent variable.* In this chapter we continue our study of regression analysis by considering situations that involve two or more independent variables. Regression analysis involving two or more independent variables is called *multiple regression analysis.* "Statistics in the News" reported on a multiple regression analysis using several independent variables to predict, or explain, the dependent variable of academic performance.

14.1 The Multiple Regression Model and Its Assumptions

To provide an introduction to multiple regression analysis, consider a simple example for which the techniques of multiple regression analysis can be applied.

EXAMPLE 14.1

Researchers studied the relationship of beer consumption, state alcohol policies and motor vehicle regulations to the number of fatal automobile accidents. One objective of the study was to develop an equation that could be used to predict the number of fatal accidents given the driving age (percent of drivers under 21) and the number of outlets selling alcohol for on-premises consumption. The number of fatal accidents would be the dependent variable, and the driving age and the number of outlets would be the independent variables.

In Example 14.1 we see that there are two independent variables. A notation that can be used to refer to these two independent variables is

$$x_1 = \text{driving age}$$

and

$$x_2 = \text{number of outlets}$$

The advantage of this notation is that with multiple regression problems involving more than two independent variables, we can continue to refer to each independent variable as x with an appropriate subscript. As in Chapter 13, we refer to the dependent variable with the letter y. Thus in Example 14.1 we denote the dependent variable as y = number of fatal accidents.

EXAMPLE 14.2

Butler Trucking Company is an independent trucking company located in southern California. A major portion of Butler's business involves deliveries throughout its local area. In an effort to develop better work schedules, the owner would like to develop an equation that could be used to help predict daily travel time for a truck. It is believed that the two most important predictors of daily travel time are the number of miles traveled and the number of deliveries made. Data collected for 10 days of operation are shown in Table 14.1.

The probabilistic model used for multiple regression analysis is a direct extension of the one introduced in the previous chapter for simple linear regression. For the case of two independent variables, the *multiple regression model* is given by

$$y = \beta_0 + \beta_1 x_1 + \beta_2 x_2 + \epsilon$$

Note that if $\beta_2 = 0$, then x_2 is not linearly related to y and hence the multiple regression model reduces to the one-independent-variable model discussed in Chapter 14; that is, $y = \beta_0 + \beta_1 x_1 + \epsilon$.

The multiple regression model shown can be extended to the case of p independent variables simply by adding more terms. Equation (14.1) shows the general case:

Multiple Regression Model

$$y = \beta_0 + \beta_1 x_1 + \beta_2 x_2 + \cdots + \beta_p x_p + \epsilon \qquad (14.1)$$

Note that if $\beta_3, \beta_4, \ldots, \beta_p$ all equal zero, (14.1) reduces to the two-independent-variable multiple regression model.

The assumptions made about the error term ϵ in Chapter 14 also apply for multiple regression analysis:

Assumptions About the Error Term ϵ

1. The error, ϵ, in the multiple regression model $y = \beta_0 + \beta_1 x_1 + \cdots + \beta_p x_p + \epsilon$ has a mean or expected value of 0; that is, $E(\epsilon) = 0$.
   Implication: For given values of $x_1, x_2, \ldots, x_p$, the average, or expected, value of y is given by

   $$E(y) = \beta_0 + \beta_1 x_1 + \beta_2 x_2 + \cdots + \beta_p x_p \qquad (14.2)$$

   Equation (14.2) is referred to as the *multiple regression equation*. In this equation $E(y)$ represents the average of all possible values of y that could occur for the given values of $x_1, x_2, \ldots, x_p$.
2. The variance of ϵ is denoted by σ^2 and is the same for all values of the independent variables $x_1, x_2, \ldots, x_p$.
   Implication: The variance of y equals σ^2 and is the same for all values of x_1, $x_2, \ldots, x_p$.
3. The values of ϵ are independent.
   Implication: The magnitude of the error for a given set of values for the independent variables is not related to the magnitude of the error for any other set of values.
4. The error ϵ is a normally distributed random variable reflecting the deviation between the actual y value and the expected value of y given by $\beta_0 + \beta_1 x_1 + \beta_2 x_2 + \cdots + \beta_p x_p$.
   Implication: Since $\beta_0, \beta_1, \ldots, \beta_p$ are constants, for the given values of x_1, $x_2, \ldots, x_p$ the dependent variable y is also a normally distributed random variable.

To obtain more insight into the form of the relationship for the two-independent variable case, refer to Figure 14.1 The graph of the multiple regression equation is a plane in three-dimensional space. In Figure 14.1 we show such a graph with x_1 and x_2 on the horizontal axis and y on the vertical axis. The specific data point shown in Figure 14.1 is for the case where $x_1 = x_1^*$ and $x_2 = x_2^*$.

**EXAMPLE 14.2
(continued)**

Let us return to the example involving the relationship between the travel time for a truck and the two independent variables, miles traveled and number of deliveries. Consider any truck that travels 50 miles ($x_1 = 50$) and makes two deliveries ($x_2 = 2$). When $x_1 = 50$ and $x_2 = 2$, the value for $E(y)$ in the multiple regression equation represents the average travel time for all trucks that travel 50 miles and make two deliveries. For these values of x_1 and x_2, the multiple regression assumptions imply that the values of y are normally distributed with a mean of $\beta_0 + \beta_1(50) + \beta_2(2)$ and a variance of σ^2; moreover, the values of travel time are assumed to be independent of one another.

■

In the previous chapter the least squares method was used to develop estimates of β_0 and β_1 for the one-independent-variable case. In multiple regression analysis, the least

TABLE 14.1

Data for the Butler Trucking Problem (Example 14.2)

DAY	x_1 = MILES TRAVELED	x_2 = NUMBER OF DELIVERIES	y = TRAVEL TIME (HOURS)
1	100	4	9.3
2	50	3	4.8
3	100	4	8.9
4	100	2	5.8
5	50	2	4.2
6	80	1	6.8
7	75	3	6.6
8	80	2	5.9
9	90	3	7.6
10	90	2	6.1

FIGURE 14.1 Graph of the Regression Equation for Multiple Regression Analysis involving Two Indpendent Variables

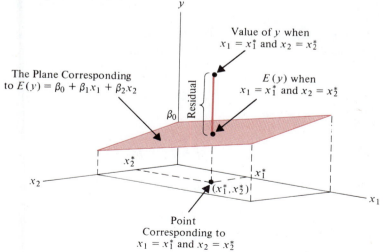

squares method is used in an analogous manner to develop estimates of the parameters β_0, β_1, β_2, $\cdots$, β_p. These estimates are denoted by b_0, b_1, b_2, $\cdots$, b_p; thus the corresponding estimated regression equation is written as follows.

Estimated Regression Equation

$$\hat{y} = b_0 + b_1 x_1 + b_2 x_2 + \cdots + b_p x_p \qquad (14.3)$$

In the two-independent-variables case, an estimate of the regression equation $E(y) = \beta_0 + \beta_1 x_1 + \beta_2 x_2$ is given by $\hat{y} = b_0 + b_1 x_1 + b_2 x_2$. Note that b_0 is the least squares estimate of β_0, b_1 is the least squares estimate of β_1, and b_2 is the least squares estimate of β_2.

At this point you should begin to see the similarity between the concepts of multiple regression analysis and those of simple linear regression. We have just extended the concepts of regression analysis involving one independent variable to the case involving two or more independent variables. In the next section we discuss how to develop the estimated regression equation for the data presented in Example 14.2.

14.2 Developing the Estimated Regression Equation

In Chapter 13 we presented formulas for estimating β_0 and β_1 for the regression line given by $E(y) = \beta_0 + \beta_1 x_1$. In the general multiple regression case, the usual presentation of formulas for computing the estimates of β_0, β_1, β_2, etc., involves the use of matrix algebra and is beyond the scope of this text. However, for the special case of two independent variables, we can show what is involved.

Recall from Chapter 13 that the least squares method chooses the values of the estimates of the population parameters that minimize the sum of squares of the deviations between the observed values of the dependent variable and the estimated values. For the case of two independent variables, the least squares method is a procedure for determining the values of b_0, b_1, and b_2 that minimize SSE, the sum of squares due to error. Recall that

$$\text{SSE} = \Sigma \, (y_i - \hat{y}_i)^2$$

Note that

$$\hat{y}_i = b_0 + b_1 x_{1i} + b_2 x_{2i}$$

where

$$x_{1i} = i\text{th value of } x_1$$

$$x_{2i} = i\text{th value of } x_2$$

$$\hat{y}_i = \text{predicted value of } y \text{ when } x_1 = x_{1i} \text{ and } x_2 = x_{2i}$$

TABLE 14.2

Calculation of Coefficients for Normal Equations

y_i	x_{1i}	x_{2i}	x_{1i}^2	x_{2i}^2	$x_{1i}x_{2i}$	$x_{1i}y_i$	$x_{2i}y_i$
9.3	100	4	10,000	16	400	930	37.2
4.8	50	3	2,500	9	150	240	14.4
8.9	100	4	10,000	16	400	890	35.6
5.8	100	2	10,000	4	200	580	11.6
4.2	50	2	2,500	4	100	210	8.4
6.8	80	1	6,400	1	80	544	6.8
6.6	75	3	5,625	9	225	495	19.8
5.9	80	2	6,400	4	160	472	11.8
7.6	90	3	8,100	9	270	684	22.8
6.1	90	2	8,100	4	180	549	12.2
66.0	815	26	69,625	76	2165	5594	180.6

Thus we can write the least squares criterion as follows.

$$\min \Sigma(y_i - b_0 - b_1x_{1i} - b_2x_{2i})^2 \tag{14.4}$$

It can be shown, using calculus, that the values of b_0, b_1, and b_2 that minimize (14.4) must satisfy the three equations below, called the *normal equations*.

Normal Equations—Two Independent Variables

$$nb_0 + (\Sigma x_{1i})b_1 + (\Sigma x_{2i})b_2 = \Sigma y_i \tag{14.5}$$

$$(\Sigma x_{1i})b_0 + (\Sigma x_{1j}^2)b_1 + (\Sigma x_{1i}x_{2i})b_2 = \Sigma x_{1i}y_i \tag{14.6}$$

$$(\Sigma x_{2i})b_0 + (\Sigma x_{1i}x_{2i})b_1 + (\Sigma x_{2i}^2)b_2 = \Sigma x_{2i}y_i \tag{14.7}$$

**EXAMPLE 14.2
(continued)**

In order to apply the normal equations for the Butler Trucking problem we must first find the values of the coefficients of b_0, b_1, and b_2 for these equations, as well as the right-hand side values. The necessary data for Butler Trucking are contained in Table 14.2.

Using the information in Table 14.2 we can substitute into the normal equations (14.5) to (14.7) to obtain the following normal equations for the Butler trucking problem.

$$10b_0 + 815b_1 + 26b_2 = 66.0 \tag{14.8}$$

$$815b_0 + 69,625b_1 + 2165b_2 = 5594.0 \tag{14.9}$$

$$26b_0 + 2165b_1 + 76b_2 = 180.6 \tag{14.10}$$

Since the least squares estimates must satisfy these three equations simultaneously, in order to obtain values for b_0, b_1, and b_2 we have to solve this system of three simultaneous linear equations in three variables. The solution* is given by $b_0 = .0367$, $b_1 = .0562$, and $b_2 = .7639$. Thus the estimated regression equation for Butler Trucking is

*In the chapter appendix we show how the solution is obtained.

$$\hat{y} = .0367 + .0562x_1 + .7639x_2$$

■

Note on Interpretation of Coefficients

Let us consider a modification of Example 14.2 in which we use only one independent variable, miles traveled, to predict the value of the dependent variable.

EXAMPLE 14.2 (continued)

Using the data in Table 14.1 we can use (13.3) and (13.4) to develop estimates of b_0 and b_1 for the estimated regression line $\hat{y} = b_0 + b_1x_1$; after rounding, these estimates are $b_0 = 1.13$ and $b_1 = .067$. Thus, using simple linear regression, the estimated regression line is $\hat{y} = 1.13 + .067x_1$.

One observation can be made concerning the relationship between the estimated regression equation with only miles traveled as an independent variable, and the one that includes the number of deliveries as a second independent variable. The value of b_1 is not the same in both cases. In the case where we use only one independent variable, we interpret $b_1 = .067$ as the amount of change in y for a 1-unit change in the independent variable. In the estimated multiple regression equation, this interpretation must be modified somewhat. That is, $b_1 = .0562$ represents the change in y corresponding to a 1-unit change in the independent variable x_1 when the other independent variable x_2 is held constant.

■

In general, in multiple regression analysis, we interpret each regression coefficient, b_i, as the change in y corresponding to a one unit change in x_i when all other independent variables are held constant.

Computer Solution

It can be shown that for multiple regression problems involving p independent variables, there are $p + 1$ normal equations that must be solved simultaneously for the estimated coefficients $b_0, b_1, b_2, \ldots, b_p$. The computational effort involved requires more sophisticated solution procedures than we have used in the solution of (14.8) to (14.10). Fortunately, computer software packages can be used to obtain these solutions with very little effort on the part of the user.

In Figure 14.2 we show the computer output from the Minitab computer package for the version of the Butler Trucking problem involving the two independent variables, miles traveled and number of deliveries. Note that in the column labeled "Coef," the values are the same as we obtained (except for rounding) for b_0, b_1, and b_2. Note also that the sum of squares due to regression (SSR = 18.9499) plus the sum of squares due to error (SSE = 5.0501) is equal to the total sum of squares (SST = 24.0). This relationship among SST, SSR, and SSE, which we introduced in Chapter 13 for simple linear regression analysis, also holds true in multiple regression analysis. We discuss the remainder of the computer output in the following sections.

```
The regression equation is
TIME = 0.04 + 0.0562 MILES + 0.764 DELIV

Predictor       Coef       Stdev     t-ratio        p
Constant       0.037       1.326       0.03      0.979
MILES        0.05616     0.01564       3.59      0.009
DELIV         0.7639      0.3053       2.50      0.041

s = 0.8494      R-sq = 79.0%     R-sq(adj) = 72.9%

Analysis of Variance

SOURCE        DF        SS         MS         F         p
Regression     2     18.9499     9.4749     13.13     0.004
Error          7      5.0501     0.7214
Total          9     24.0000
```

FIGURE 14.2 Minitab Regression Output for the Butler Trucking Problem with Two Independent Variables.

EXERCISES

1. Shown next is the estimated regression equation for a model involving 2 independent variables and 10 observations.

$$\hat{y} = 29.1270 + .5906x_1 + .4980x_2$$

Interpret the coefficients in this estimated regression equation.

2. The admissions' officer for Clearwater College developed the following estimated regression equation relating final college GPA to student's SAT mathematics scores and their high school GPA.

$$\hat{y} = -1.41 + .0235x_1 + .00486x_2$$

where

$$x_1 = \text{high school grade-point average}$$
$$x_2 = \text{SAT mathematics score}$$
$$y = \text{final college grade-point average}$$

Interpret the coefficients in this estimated regression equation.

3. The personnel director for Electronics Associates developed the following estimated regression equation relating an employee's score on a job satisfaction test to his or her length of service and wage rate.

$$\hat{y} = 14.4 - 8.69x_1 + 13.5x_2$$

where

$$x_1 = \text{length of service (years)}$$
$$x_2 = \text{wage rate (dollars)}$$
$$y = \text{job satisfaction test score (higher scores indicate more job satisfaction)}$$

Interpret the coefficients in this estimated regression equation.

4. A chemist has developed the following estimated regression equation relating the yield of a process (y) to the temperature (x_1) and the pressure (x_2).

$$\hat{y} = 69 + 1.17x_1 - 3.97x_2$$

where temperature is measured in degrees Fahrenheit and pressure in pounds per square inch. Interpret the coefficients in this estimated regression equation.

5. A shoe store has developed the following estimated regression equation relating sales to inventory investment and advertising expenditures.

$$\hat{y} = 25 + 10x_1 + 8x_2$$

where

$$x_1 = \text{inventory investment (1000s)}$$

$$x_2 = \text{advertising expenditures (\$1000s)}$$

$$y = \text{sales (\$1000s)}$$

Interpret the coefficients (b_1 and b_2) in this estimated regression equation.

6. (Computer recommended) The following data show the dollar value of prescriptions (in thousands) for 13 pharmacies in Iowa, the population of the city served by the given pharmacy, and the average prescription inventory value. ("The use of categorical variables in Data Envelopment Analysis," R.Banker and R. Morey, *Management Science*, December 1986).

PRESCRIPTIONS VALUE ($1000s)	POPULATION	AVERAGE INVENTORY VALUE ($)
61	1410	8,000
92	1523	9,000
93	1354	13,694
45	822	4,250
50	746	6,500
29	1281	7,000
56	1016	4,500
45	1070	5,000
183	1694	27,000
156	1910	21,560
120	1745	15,000
75	1353	8,500
122	1016	18,000

a. Determine the estimated regression equation that can be used to predict the dollar value of prescriptions (y) given the population size (x_1) and the average inventory value (x_2).
b. What other variables do you think might be useful in predicting y?

7. (Computer recommended) Two experts provided subjective lists of school districts that they think are among the best in the country. For each school district the average class size, the combined SAT score, and the percentage of students that attended a 4-year college were provided (*Wall Street Journal*, March 31, 1989).

DISTRICT	AVERAGE CLASS SIZE	COMBINED SAT SCORE	% ATTEND FOUR-YEAR COLLEGE
Blue Springs, Mo.	25	1083	74
Garden City, N.Y.	18	997	77
Indianapolis, Ind.	30	716	40
Newport Beach, Calif.	26	977	51
Novi, Mich.	20	980	53
Piedmont, Calif.	28	1042	75
Pittsburg, Pa.	21	983	66
Scarsdale, N.Y.	20	1110	87
Wayne, Pa.	22	1040	85
Weston, Mass.	21	1031	89
Farmingdale, N.Y.	22	947	81
Mamaroneck, N.Y.	20	1000	69
Mayfield, Ohio	24	1003	48
Morristown, N.J.	22	972	64
New Rochelle, N.Y.	23	1039	55
Newtown Square, Pa.	17	963	79
Omaha, Neb.	23	1059	81
Shaker Heights, Ohio	23	940	82

a. Using these data, develop an estimated regression equation relating the percentage of students that attend a 4-year college to the average class size and the combined SAT score.

b. Estimate the percentage of students that attend a 4-year college if the average class size is 20 and the combined SAT score is 1000.

14.3 Determining the Goodness of Fit

In Chapter 13 we used the coefficient of determination (r^2) to evaluate the goodness of fit for a regression model involving one independent variable. Recall that r^2 is computed as

$$r^2 = \frac{SSR}{SST}$$

In multiple regression analysis, we compute a similar quantity called the *multiple coefficient of determination*.

Multiple Coefficient of Determination

$$R^2 = \frac{SSR}{SST} \tag{14.11}$$

When multiplied by 100, the multiple coefficient of determination provides the percentage of variability in y that is explained by the estimated regression equation.

EXAMPLE 14.2 (continued)

In the case of Butler Trucking (refer to Figure 14.2 for SSR and SST), we find

$$R^2 = \frac{18.9499}{24.0000} = .7896$$

Therefore, 78.96% of the variability in y is explained by the relationship with miles traveled and number of deliveries. For the Butler Trucking problem, we showed in Chapter 13 that the regression model with only miles traveled (x_1) as the independent variable has an r^2 value of .60. Therefore, with the addition of the second independent variable, number of deliveries (x_2), the percentage of variability explained has increased from 60% to 78.96%.

EXAMPLE 14.1 (continued)

A study* by Israel Colon and Henry Cutter investigated factors related to motor vehicle fatalities based on data available from the United States Department of Transportation Fatal Accident Reporting System. In particular, data from all 50 states and the District of Columbia were examined in order to determine the relationship between alcohol availability and fatal motor vehicle accidents. Two multiple regression equations were developed: one with the number of fatal accidents as the dependent variable and one with the number of motor vehicle fatalities (potentially more than one per accident) as the dependent variable. In both cases, the seven independent or predictor variables considered were:

1. driving age (percent of drivers under 21)
2. beverage-purchase age
3. average beer consumption
4. number of outlets per million population selling alcohol for on-premises consumption
5. percentage of metropolitan residents
6. percentage of male drivers
7. mileage per driver per year

Results of the two regression analyses were similar, with the coefficient of determination slightly over .75 for each. The researchers concluded their study with interpretations that suggested possible strategies for reducing motor vehicle fatalities.

In general, it is always true that R^2 will increase as more independent variables are added to the regression equation because adding variables to the equation causes the prediction errors to be smaller, hence reducing SSE. Since SST = SSR + SSE, when SSE gets smaller SSR must get larger, causing R^2 = SSR/SST to increase.

Many analysts recommend adjusting R^2 for the number of independent variables in order to avoid overestimating the impact on the amount of explained variability of adding

*"The Relationship of Beer Consumption and State Alcohol and Motor Vehicle Policies to Fatal Accidents," by Israel Colon and Henry S. G. Cutter, *Journal of Safety Research* 14 (1983).

an independent variable. This *adjusted multiple coefficient of determination* is computed as follows.

Adjusted Multiple Coefficient of Determination

$$R_a^2 = 1 - (1 - R^2)\left(\frac{n-1}{n-p-1}\right) \tag{14.12}$$

**EXAMPLE 14.2
(continued)**

For the Butler Trucking problem, we obtain

$$R_a^2 = 1 - (1 - .7896)\,\frac{10 - 1}{10 - 2 - 1}$$

$$= 1 - (.2104)\,(1.2857)$$

$$= .7295$$

Note that both the adjusted and unadjusted coefficient of determination are provided by the Minitab output shown in Figure 14.2.

■

EXERCISES

8. In Exercise 1 the following estimated regression equation based on 10 observations was presented:

$$\hat{y} = 29.1270 + .5906x_1 + .4980x_2$$

The value of SST is 6,724.125 and SSR = 6,216.375
  a. Find SSE.
  b. Compute R^2.
  c. Compute R_a^2.
  d. Comment on the goodness of fit.

9. Shown next is a partial Minitab computer printout for a multiple regression problem involving two independent variables and 15 observations.
  a. Compute R^2.
  b. Compute R_a^2.

Analysis of Variance

SOURCE	DF	SS	MS
Regression	—	90.3	—
Error	12	——	—
Total	—	108.6	—

10. In Exercise 5 the following estimated regression equation for relating sales to inventory investment and advertising expenditure was given:

$$\hat{y} = 25 + 10x_1 + 8x_2$$

The data used to develop the model came from a survey of ten stores. In addition to the estimated regression equation, it was found as a result of a computer run that SST = 16,000 and SSR = 12,000.
  a. Compute R_a^2
  b. Comment on the goodness of fit.

11. Shown next is a partial Minitab computer printout from a regression analysis. There were 15 observations, and $R^2 = .923$.
  a. Compute SSE and SST.
  b. Compute R_a^2.
  c. Does the model appear to explain a large amount of the variability in the data?

```
Analysis of Variance

SOURCE        DF        SS        MS
Regression    __       1612       804
Error         12       ____      ____
Total         __       ____
```

12. In a regression analysis involving 18 observations and four independent variables, it was determined that SSR = 18,051.63 and SSE = 1,014.3.
  a. Compute SST.
  b. Compute R^2 and R_a^2.

13. In a regression analysis involving 30 observations, the following estimated regression equation was obtained.

$$\hat{y} = 17.6 + 3.8x_1 - 2.3x_2 + 7.6x_3 + 2.7x_4.$$

For this model SST = 1805 and SSR = 1760.
  a. Compute R^2.
  b. Compute R_a^2.

14.4 Testing for a Significant Relationship

Let us initially consider a multiple regression situation involving two independent variables. This implies an underlying regression equation of the form

$$E(y) = \beta_0 + \beta_1 x_1 + \beta_2 x_2$$

Therefore, the appropriate hypothesis test for determining whether or not there is a significant relationship of this form among x_1, x_2, and y is

$$H_0: \quad \beta_1 \text{ and } \beta_2 = 0$$

$$H_a: \quad \text{One or more coefficients is not zero}$$

If we reject H_0, we conclude that there is a significant relationship among x_1, x_2, and y.

F Test

In Chapter 13 we showed how the ANOVA table could provide a summary of the computational aspects associated with regression analysis. In multiple regression situations, we will use the ANOVA table. In Table 14.3 we show the general form of the ANOVA table for a multiple regression situation involving two independent variables.

Note that the number of degrees of freedom for SSR is 2. In regression analysis the number of degrees of freedom for SSR is always the number of independent variables. (It is 1 for the case of simple linear regression, since there is one independent variable.) Note also that the number of degrees of freedom for SSE is $n - 3$. The number of degrees of freedom for SSE is always equal to the number of observations minus the number of parameters estimated. In the two-independent-variable case we estimated three parameters (β_0, β_1, and β_2). As in the previous chapter the number of degrees of freedom for SST is $n - 1$; it is also the sum of the degrees of freedom for SSR and SSE.

The test concerning the significance of the relationship in multiple regression analysis is based upon the ratio of the two mean squares in the ANOVA table. Since

$$MSR = \frac{SSR}{2}$$

and

$$MSE = \frac{SSE}{n - 3}$$

the appropriate test statistic is given by

$$F = \frac{MSR}{MSE} \tag{14.13}$$

TABLE 14.3

General Form of the ANOVA Table for Multiple Regression Analysis Involving Two Independent Variables

SOURCE	SUM OF SQUARES	DEGREES OF FREEDOM	MEAN SQUARE
Regression	SSR	2	$MSR = \dfrac{SSR}{2}$
Error	SSE	$n - 3$	$MSE = \dfrac{SSE}{n - 3}$
Total	SST	$n - 1$	

If the null hypothesis is true (i.e., $\beta_1 = \beta_2 = 0$), the value that we obtain when computing this ratio should appear to be a value that would be obtained if we were randomly sampling values from an F distribution with 2 numerator degrees of freedom and $n - 3$ denominator degrees of freedom. In Table 4 of Appendix B, we provide critical values for the F distribution. For example, this table shows that for an F distribution with 2 numerator degrees of freedom and 7 denominator degrees of freedom, only 5% of the values obtained would be as large as or larger than the table value of 4.74. Thus if this were the appropriate F distribution for a multiple regression problem, we would compare the value of $F = $ MSR/MSE to 4.74; if this value exceeded 4.74, we would have sufficient statistical evidence to reject the null hypothesis at the 5% level of significance.

EXAMPLE 14.2
(continued)

In Figure 14.2 we showed the Minitab computer output for the Butler Trucking multiple regression problem. For this problem MSR $=$ 9.4749 and MSE $=$ .7214. Thus

$$F = \frac{\text{MSR}}{\text{MSE}} = \frac{9.4749}{.7214} = 13.13$$

Since the numerator in this ratio, MSR, has 2 degrees of freedom associated with it and the denominator, MSE, has 7 degrees of freedom, we refer to an F distribution with 2 numerator degrees of freedom and 7 denominator degrees of freedom. As we showed earlier, the critical value at the 5% level of significance in this case is 4.74. Thus we must reject the null hypothesis that $\beta_1 = \beta_2 = 0$ and conclude that there is a statistically significant relationship at the 5% level of significance. We can also note from the computer printout that the p-value associated with $F = $ 13.13 is .004. Thus, the null hypothesis would be rejected for any level of significance greater than .004.

The General ANOVA Table and *F* Test

Now that we know how the F test can be applied for a multiple regression situation involving two independent variables, let us generalize the test to the case involving p independent variables.

The appropriate hypotheses to test to determine if there is a significant linear relationship are as follows:

$$H_0: \quad \beta_1 = \beta_2 = \cdots = \beta_p = 0$$

$H_a:$ One or more of the p coefficients is not equal to zero

Again, if we reject H_0, we can conclude that there is a significant linear relationship.

The general form of the ANOVA table for the multiple regression case involving p independent variables is shown in Table 14.4. The only change from the two-independent-variable case is in the degrees of freedom corresponding to SSR and SSE. Here the sum of squares due to regression has p degrees of freedom corresponding to the p independent variables, and the degrees of freedom corresponding to error has $n - p - 1$ degrees of freedom. When looking up the critical value from the F distribution table, the numerator degrees of freedom are p and the denominator degrees of freedom are $n - p - 1$.

t Test for Significance of Individual Parameters

If, after using the F test, we conclude that the multiple regression relationship is significant (that is, we conclude that at least one of the β_i is not equal to 0), it is often of interest to conduct further tests to see which of the individual parameters, β_i, are significant. The t test that we introduced in Chapter 13 for the case of one independent variable can also be used to test for the significance of the individual parameters in multiple regression analysis.

The hypothesis test we wish to make is the same for the coefficient of each independent variable. It is stated as follows.

$$H_0: \quad \beta_i = 0$$

$$H_a: \quad \beta_i \neq 0$$

In Chapter 13 we learned how to conduct such a test for the case where there is only one independent variable. The hypotheses were:

$$H_0: \quad \beta_1 = 0$$

$$H_a: \quad \beta_1 \neq 0$$

To conduct the hypothesis test we computed the sample statistic b_1/s_{b_1}, where b_1 is the least squares estimator of β_1 and s_{b_1} is the estimator of the standard deviation of the sampling distribution of b_1. We learned that the sampling distribution of b_1/s_{b_1} follows a t distribution with $n - 2$ degrees of freedom. Thus to conduct the hypothesis test, we chose a value of α, computed b_1/s_{b_1} and used the following rejection rule:

$$\text{Reject } H_0 \text{ if } \quad \frac{b_1}{s_{b_1}} > t_{\alpha/2}$$

$$\text{or if } \quad \frac{b_1}{s_{b_1}} < -t_{\alpha/2}$$

The procedure for testing the individual parameters in the multiple regression case is essentially the same. The only differences are in the number of degrees of freedom for the appropriate t distribution in the formula for computing s_{b_1}. The number of degrees

TABLE 14.4

ANOVA Table for Multiple Regression Analysis Involving p Independent Variables

SOURCE	SUM OF SQUARES	DEGREES OF FREEDOM	MEAN SQUARE
Regression	SSR	p	$\text{MSR} = \dfrac{\text{SSR}}{p}$
Error	SSE	$n - p - 1$	$\text{MSE} = \dfrac{\text{SSE}}{n - p - 1}$
Total	SST	$n - 1$	

of freedom is the same as for the sum of squares due to error. Thus, we use $n - p - 1$ degrees of freedom, where p is the number of independent variables. (Note that for the case of one independent variable, this reduces to the $n - 2$ degrees of freedom used in Chapter 13.) The formula for s_{b_1} is more involved, and we do not present it here; however, s_{b_1} is calculated and printed by most computer software packages for multiple regression analysis.

EXAMPLE 14.2 (continued)

Let us return to the Butler Trucking problem to test the significance of the parameters β_1 and β_2. In the Minitab printout (Figure 14.2), the values of b_1, b_2, s_{b_1}, and s_{b_2} are given as

$$b_1 = .05616 \quad s_{b_1} = .01564$$

$$b_2 = .7639 \quad s_{b_2} = .3053$$

Therefore, for the parameters β_1 and β_2, we obtain

$$\frac{b_1}{s_{b_1}} = \frac{.05616}{.01564} = 3.59$$

$$\frac{b_2}{s_{b_2}} = \frac{.7639}{.3053} = 2.50$$

Note that both of these values were provided by the Minitab output of Figure 14.2 under the column labeled "t-ratio." Using $\alpha = .05$ and $10 - 2 - 1 = 7$ degrees of freedom, we can find the appropriate $t_{\alpha/2}$ value for the hypothesis tests in Table 2 of Appendix B. We obtain

$$t_{.025} = 2.365$$

Now, since $b_1/s_{b_1} = 3.59 > 2.365$, we reject the hypothesis that $\beta_1 = 0$. Furthermore, since $b_2/s_{b_2} = 2.50 > 2.365$, we reject the hypothesis that $\beta_2 = 0$. Note also that the p-values provided by the Minitab output permit us to perform these tests very easily. For instance, the p-value of .009 for "MILES" indicates that the null hypothesis H_0: $\beta_1 = 0$ should be rejected for all values of α larger than .009.

■

Multicollinearity

We have used the term independent variable in regression analysis to refer to any variable being used to predict or explain the value of the dependent variable. The term does not mean, however, that the independent variables themselves are independent in any statistical sense. Quite the contrary, most independent variables in a multiple regression problem are correlated to some degree with one another. For example, in the Butler Trucking problem involving the two independent variables x_1 (miles traveled) and x_2 (number of deliveries), we could treat the miles traveled as the dependent variable and the number of deliveries as the independent variable in order to determine if these two variables are themselves related. We could then compute the sample correlation coefficient $r_{x_1 x_2}$ to determine the extent to which these variables are related. We did so and obtained $r_{x_1x_2} = .28$. Thus there is some degree of linear association between the two independent variables. In multiple regression analysis we use the term *multicollinearity* to refer to the correlation among the independent variables.

To provide a better perspective of the potential problems of multicollinearity, let us consider a modification of the Butler Trucking problem. Instead of x_2 being the number of deliveries, let x_2 denote the number of gallons of gasoline consumed. Clearly, x_1 (the miles traveled) and x_2 are related; that is, we know that the number of gallons of gasoline used depends upon the number of miles traveled. Thus we would conclude logically that x_1 and x_2 are highly correlated independent variables.

Assume that we obtain the estimated regression equation $\hat{y} = b_0 + b_1x_1 + b_2x_2$ and find that the F test shows that the regression is significant. Then suppose that we were to conduct a t test on β_1 to determine if $\beta_1 \neq 0$, and we cannot reject H_0: $\beta_1 = 0$. Does this mean that travel time is not related to miles traveled? Not necessarily. What it probably means is that with x_2 already in the model, x_1 does not contribute a significant addition toward determining the value of y. This would seem to make sense here, since if we know the amount of gasoline consumed we do not gain much additional information useful in predicting y by knowing the miles traveled. Similarly, a t test might lead us to conclude $\beta_2 = 0$ on the grounds that with x_1 in the model knowledge of the amount of gasoline consumed does not add much.

To summarize, the difficulty caused by multicollinearity in conducting t tests for the significance of individual parameters is that it is possible to conclude that none of the individual parameters are significantly different from zero when an F test on the overall multiple regression equation indicates a significant relationship. This problem is avoided when there is very little correlation among the independent variables.

Ordinarily multicollinearity does not affect the way in which we perform the regression analysis or interpret the output from a study. However, when multicollinearity is severe—that is, when two or more of the independent variables are highly correlated with one another—we can run into difficulties interpreting the results of t tests on the individual parameters. In addition to the type of problem illustrated above, severe cases of multicollinearity have been shown to result in least squares estimates that even have the wrong sign. That is, in simulated studies where researchers created the underlying regression model and then applied the least squares technique to develop estimates of β_0, β_1, β_2, and so on, it has been shown that under conditions of high multicollinearity the least squares estimates can even have a sign opposite to that of the parameter being estimated. For example, β_2 might actually be $+10$ and b_2, its estimate, might turn out to be -2. Thus little faith can be placed in the individual coefficients themselves if multicollinearity is present to a high degree.

Statisticians have developed several tests for determining whether or not multicollinearity is high enough to cause these types of problems. One simple test, referred to as the "rule of thumb" test, says that multicollinearity is a potential problem if the absolute value of the sample correlation coefficient exceeds .7 for any two of the independent variables. The other types of tests are more advanced and beyond the scope of this text.

If possible, every attempt should be made to avoid including independent variables that are highly correlated. In practice, however, it is rarely possible to adhere to this policy strictly. Thus the data analyst should be warned that when there is reason to believe that substantial multicollinearity is present it is difficult to separate out the effect of the individual independent variables on the dependent variable.

14. In Exercise 1 the following estimated regression equation based on 10 observations was presented:

$$\hat{y} = 29.1270 + .5906x_1 + .4980x_2$$

Here SST $= 6{,}724.125$, SSR $= 6{,}216.375$, $s_{b_1} = .0813$, and $s_{b_2} = .0567$.

a. Compute MSR and MSE.
b. Compute F and perform the appropriate F test. Use $\alpha = .05$.
c. Perform a t test for the significance of β_1. Use $\alpha = .05$.
d. Perform a t test for the significance of β_2. Use $\alpha = .05$.

15. Refer again to Exercise 11.
a. Compute the entries in the DF, SS, and MS columns.
b. Compute the appropriate F statistic and test at the $\alpha = .05$ level for a significant relationship.

16. In Exercise 5 the following estimated regression equation for relating sales to inventory investment and advertising expenditures was given:

$$\hat{y} = 25 + 10x_1 + 8x_2$$

The data used to develop the model came from a survey of 10 stores. In addition to the estimated regression equation, it was found as a result of a computer run that SST $= 16{,}000$ and SSR $= 12{,}000$.

a. Compute SSE, MSE, and MSR.
b. Use an F test and an $\alpha = .05$ level of significance to determine if there is a significant relationship among the variables.

17. Refer to Exercise 9.
a. Find the appropriate values for regression and total degrees of freedom.
b. Find SSE, MSR, and MSE.
c. Compute F and test at the $\alpha = .01$ level whether a significant relationship exists or not.

18. The following estimated regression equation involving three independent variables has been developed:

$$\hat{y} = 18.31 + 8.12x_1 + 17.9x_2 - 3.6x_3$$

Computer output indicates that $s_{b_1} = 2.1$, $s_{b_2} = 9.72$, and $s_{b_3} = .71$. There were 15 observations in the study. Test each hypothesis at the $\alpha = .05$ level of significance.

a. $H_0: \beta_1 = 0$, $H_a: \beta_1 \neq 0$
b. $H_0: \beta_2 = 0$, $H_a: \beta_2 \neq 0$
c. $H_0: \beta_3 = 0$, $H_a: \beta_3 \neq 0$

19. (Computer recommended) The following data show the price-earnings (P/E) ratio, the net profit margin, and the growth rate for 19 companies listed in "The *Forbes* 500 on Wall Street." (*Forbes*, May 1, 1989)

FIRM	P/E RATIO	PROFIT MARGIN (%)	GROWTH RATE (%)
Exxon	11.3	6.5	10
Chevron	10.0	7.0	5

FIRM	P/E RATIO	PROFIT MARGIN (%)	GROWTH RATE (%)
Texaco	9.9	3.9	5
Mobil	9.7	4.3	7
Amoco	10.0	9.8	8
Pfizer	11.9	14.7	12
Bristol Meyers	16.2	13.9	14
Merck	21.0	20.3	16
American Home Products	13.3	16.9	11
Abbott Laboratories	15.5	15.2	18
Eli Lilly	18.9	18.7	11
Upjohn	14.6	12.8	10
Warner-Lambert	16.0	8.7	7
Amdahl	8.4	11.9	4
Digital	10.4	9.8	19
Hewlett-Packard	14.8	8.1	18
NCR	10.1	7.3	6
Unisys	7.0	6.9	6
IBM	11.8	9.2	6

a. Determine the estimated regression equation that can be used to predict the price-earnings ratio given the net profit margin and the growth rate.

b. At the $\alpha = .05$ level of significance, determine if there is a relationship among the variables.

c. Does there appear to be any multicollinearity present in the data? Explain.

20. The following estimated regression equation was developed for a model involving two independent variables:

$$\hat{y} = 40.7 + 8.63x_1 + 2.71x_2$$

After dropping x_2 from the model, the least squares method was used again to obtain an estimated regression equation involving only x_1 as an independent variable:

$$\hat{y} = 42.0 + 9.01x_1$$

a. Give an interpretation of the coefficient of x_1 in both models.

b. Could multicollinearity explain why the coefficient of x_1 differs in the two models? If so, how?

14.5   Estimation and Prediction

Estimating the mean value of y and predicting an individual value of y in multiple regression is similar to that for the case of regression analysis involving one independent variable. First, recall that in Chapter 13 we showed that the point estimate of the expected value of y for a given value of x was the same as the prediction of an individual value of y. In both cases we used $\hat{y} = b_0 + b_1x$ as the point estimate. In multiple regression we use the same procedure. That is, we substitute the given values of $x_1, x_2, \ldots, x_p$ into the estimated regression equation and use the corresponding value of $\hat{y}$ as the point estimate.

**EXAMPLE 14.2
(continued)**

Suppose that for the Butler Trucking problem, we wanted to use the estimated regression equation involving x_1 (miles traveled) and x_2 (number of deliveries) to do the following:

1. Estimate the mean value of travel time for all trucks that travel 50 miles and make two deliveries.
2. Predict the travel time for one specific truck that travels 50 miles and makes two deliveries.

Using the estimated regression equation $\hat{y} = .0367 + .0562x_1 + .7639x_2$ with $x_1 = 50$ and $x_2 = 2$, we obtain the following value of $\hat{y}$.

$$\hat{y} = .0367 + .0562(50) + .7639(2) = 4.3745$$

Hence the point estimate of travel time in both cases is approximately 4.4 hours.

■

To develop interval estimates for the mean value of y and for an individual value of y, we use a procedure similar to that for the case of regression analysis involving one independent variable. The formulas required, however, are complex and beyond the scope of the text. Nevertheless, computer packages for multiple regression analysis will often provide confidence intervals once the values of $x_1, x_2, \ldots, x_p$ are specified by the user.

**EXAMPLE 14.2
(continued)**

In Table 14.5 we show 95% confidence interval estimates for the Butler Trucking problem for selected values of x_1 and x_2. Note that the prediction interval for an individual estimate of y is wider than the interval estimate for the expected value of y. This simply reflects the fact that for given values of x_1 and x_2, we can predict the mean travel time for all trucks with more precision than we can the travel time for one specific truck.

■

TABLE 14.5

95% Confidence Interval and 95% Prediction Interval Estimates for the Butler Trucking Problem

VALUE OF x_1	VALUE OF x_2	EXPECTED VALUE OF y		INDIVIDUAL VALUE OF y	
		Lower Limit	Upper Limit	Lower Limit	Upper Limit
50	2	3.0841	5.6649	1.9869	6.7621
50	3	3.7127	6.5642	2.6750	7.6018
80	1	3.9907	6.7097	2.9006	7.6926
80	2	5.2984	6.8226	3.9120	8.2091
100	2	6.0774	8.2916	4.8908	9.4782
100	4	7.4853	9.9394	6.3584	11.0662

EXERCISES

21. For the regression model in Exercise 1, we obtained

$$\hat{y} = 29.1270 + .5906x_1 + .4980x_2$$

a. Estimate the average value of y when $x_1 = 180$ and $x_2 = 310$.

b. Predict the value of y when $x_1 = 190$ and $x_2 = 280$.

22. Reconsider Exercise 2, where the estimated regression equation relating the college GPA (y) to the high school GPA (x_1) and the SAT mathematics score (x_2) was

$$\hat{y} = -1.41 + .0235x_1 + .00486x_2$$

Estimate the final college GPA for a student who has a high school average of 84 and a score of 540 on the SAT mathematics test.

23. Reconsider Exercise 3, where the estimated regression equation developed was

$$\hat{y} = 14.4 - 8.69x_1 + 13.5x_2$$

where

$$x_1 = \text{length of service (years)}$$

$$x_2 = \text{wage rate (dollars)}$$

$$y = \text{job satisfaction test score (higher scores indicate more job satisfaction)}$$

Develop an estimate of the job satisfaction test score for an employee that has 4 years of service and makes $6.50 per hour.

24. In Exercise 4 we described a situation where a chemist developed the following estimated regression equation relating the yield of a process (y) to the temperature (x_1) and the pressure (x_2).

$$\hat{y} = 69 + 1.17x_1 - 3.97x_2$$

Temperature is measured in degrees Fahrenheit and pressure in pounds per square inch. Estimate the yield of the process corresponding to a temperature of 80° and a pressure of 16 pounds per square inch.

25. Recall Exercise 5, where we described a situation in which a shoe store had developed the following estimated regression equation relating sales ($1000s) to inventory investment ($1000s) and advertising expenditure ($1000s):

$$\hat{y} = 25 + 10x_1 + 8x_2$$

Estimate sales if there are a $15,000 investment in inventory and an advertising budget of $10,000.

14.6 Residual Analysis

In Chapter 13 we showed how residual plots could be used to determine whether or not the assumptions made regarding the error term and model form are valid. In the multiple regression case these same methods can be used to validate the assumptions made for the proposed model and provide additional information concerning the adequacy of the fitted least squares equation.

The statistical tests that we discussed in Section 14.4 are based on the assumptions presented in Section 14.1 for the error term, ϵ, and the assumption that the population regression equation has the linear form given by $E(y) = \beta_0 + \beta_1 x_1 + \cdots + \beta_p x_p$. Residual analysis will allow us to make a judgment about whether or not the model assumptions appear to be satisfied. In addition, residual analysis can often provide insight as to whether or not a different type of model—for example, one involving more variables or a different functional form—might better describe the observed relationship.

In simple linear regression we showed how a residual plot against the independent variable x could be used to determine if the assumptions regarding the error term are appropriate. In multiple regression analysis, there is more than one independent variable to consider. Thus, statisticians usually look first at a plot of the residuals versus the predicted values $\hat{y}$. The plots of the residuals against each independent variable can then be examined if additional information is desired.

Figure 14.3 shows three patterns that are often found in residual plots versus $\hat{y}$, the value of y predicted by the estimated regression equation. The plot in Panel A shows the type of pattern to expect when the model assumptions are satisfied.

The plot in Panel B shows a pattern that can be expected when the constant variance assumption is not satisfied. The error term gets larger as the value of $\hat{y}$ increases. Adding another independent variable will sometimes correct this problem.

Finally, the plot in Panel C shows a case where the errors are not independent. When $\hat{y}$ is small, the error term is positive; when $\hat{y}$ assumes intermediate values, the error term is negative; and when $\hat{y}$ is large, the error term is again positive. Often a curvilinear model is needed in this case.

EXAMPLE 14.2 (continued)

In the discussion of the interpretation of coefficients in Section 14.2, we noted that for Butler Trucking the estimated regression equation using the single independent variable miles traveled was

$$\hat{y} = 1.13 + .067x_1$$

where

$$\hat{y} = \text{predicted travel time}$$

$$x_1 = \text{miles traveled}$$

In Table 14.6 we show the computation of the residuals, and in Figure 14.4 we show a plot of the residuals versus $\hat{y}$ for this estimated regression equation. The residual plot has a pattern similar to that shown in Panel B of Figure 14.3. It appears that the assumption of constant variance (assumption 2 in Section 14.1) is not satisfied. The variability of the y values about the estimated regression line is increasing as $\hat{y}$ gets larger. This suggests we should question the adequacy of the model.

After adding a second independent variable, number of deliveries, we obtained the following estimated regression equation for Butler Trucking (see Section 14.2):

$$\hat{y} = .0367 + .0562x_1 + .7639x_2$$

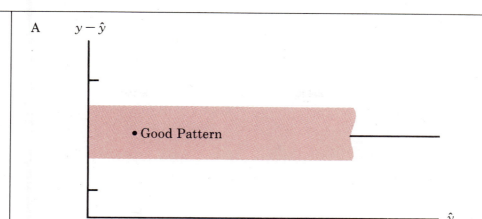

A $y - \hat{y}$

• Good Pattern

$\hat{y}$

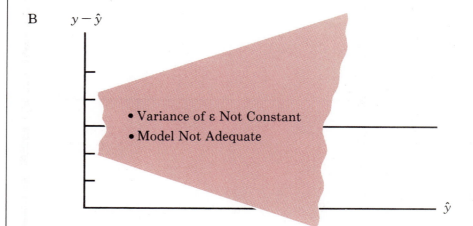

B $y - \hat{y}$

• Variance of ε Not Constant
• Model Not Adequate

$\hat{y}$

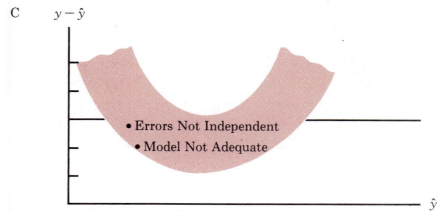

C $y - \hat{y}$

• Errors Not Independent
• Model Not Adequate

$\hat{y}$

FIGURE 14.3   Possible Residual Patterns and their Causes

where

$$\hat{y} = \text{predicted travel time}$$

$$x_1 = \text{miles traveled}$$

$$x_2 = \text{number of deliveries}$$

FIGURE 14.4 Residual Plot for the Butler Trucking Problem Corresponding to $\hat{y} = 1.13 +$ $.067x_1$

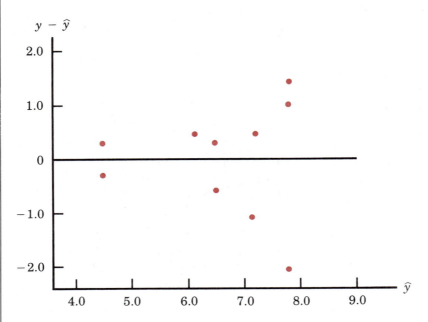

TABLE 14.6 **Computation of Residuals for Butler Trucking Using $\hat{y} = 1.13 + .067x_1$**

MILES TRAVELED $(x_{1\,i})$	TRAVEL TIME (y_i)	PREDICTED TRAVEL TIME $(\hat{y}_i)$	RESIDUAL $(y_i - \hat{y}_i)$
100	9.3	7.830	1.470
50	4.8	4.480	.320
100	8.9	7.830	1.070
100	5.8	7.830	−2.030
50	4.2	4.480	− .280
80	6.8	6.490	.310
75	6.6	6.155	.445
80	5.9	6.490	− .590
90	7.6	7.160	.440
90	6.1	7.160	−1.060

In Table 14.7 we show the computation of the residuals, and in Figure 14.5 we show a plot of the residuals versus $\hat{y}$ for this estimated regression equation. This residual plot does not exhibit any abnormalities. With the exception of one rather large residual, the plot fits the pattern shown in Panel A of Figure 14.3. Thus, we assume that the model assumptions are reasonable.

FIGURE 14.5 Residual Plot for the Butler Trucking Problem Corresponding to $\hat{y} = .0367 + .0562x_1 + .7639x_2$.

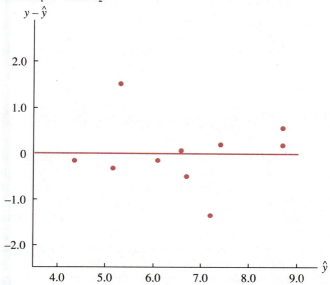

TABLE 14.7

Computation of Residuals for Butler Trucking Using $\hat{y} = .0367 + .0562x_1 + .7639x_2$

MILES TRAVELED ($x_{1\,i}$)	NUMBER OF DELIVERIES ($x_{2\,i}$)	TRAVEL TIME (y_i)	PREDICTED TRAVEL TIME ($\hat{y}_i$)	RESIDUAL ($y_i - \hat{y}_i$)
100	4	9.3	8.7084	.5916
50	3	4.8	5.1364	−.3364
100	4	8.9	8.7084	.1916
100	2	5.8	7.1807	−1.3807
50	2	4.2	4.3725	−.1725
80	1	6.8	5.2936	1.5064
75	3	6.6	6.5405	.0595
80	2	5.9	6.0574	−.1574
90	3	7.6	7.3829	.2171
90	2	6.1	6.6191	−.5191

26. (Computer recommended) Refer again to the data set in Exercise 6.
    a. Use the estimated regression equation developed in Exercise 6 to compute $\hat{y}_i$ for each observation.
    b. Compute the 13 residuals $(y_i - \hat{y}_i)$.
    c. Plot the residuals versus $\hat{y}$.
    d. Does the residual plot lead you to question the assumptions regarding ϵ?

27. (Computer recommended) Refer again to the data set in Exercise 7.
    a. Develop a plot of the residuals versus $\hat{y}$.
    b. Does the residual plot cause you to question any of the regression assumptions?

28. (Computer recommended) Refer again to the data set in Exercise 19.
    a. Develop a plot of the residuals versus $\hat{y}$.
    b. Does the residual plot cause you to question any of the assumptions regarding ϵ.

SUMMARY

In this chapter we have shown how extensions of the concepts of linear regression can be used to develop an estimated regression equation for predicting y that involves two or more independent variables. The interpretation of the coefficients has to be modified somewhat for this case. For instance, we interpret b_i as an estimate of the change in the dependent variable y that would result from a 1-unit change in independent variable x_i when the other independent variables are held constant.

A key part of any multiple regression study is the use of a computer software package for carrying out the computational work. Many excellent packages exist and, after a short learning period, can be used to develop the estimated regression equation, conduct the appropriate significance tests and prepare residual plots. We illustrated the use of the Minitab statistical package for the Butler Trucking problem. The Data Analyst can also be used for multiple regression problems.

GLOSSARY

Multiple regression analysis   Regression analysis involving two or more independent variables.

Normal equations   The equations that must be solved to compute b_0, b_1, b_2, . . . , b_p the least squares estimators of β_0, β_1, β_2, . . . , β_p

Multiple coefficient of determination (R^2)   When multiplied by 100, R^2 represents the percentage of variability in y that is explained by the regression relationship.

Adjusted multiple coefficient of determination (R_a^2)   The value of R^2 adjusted for the number of independent variables.

Multicollinearity   A term used to describe the case when the independent variables in a multiple regression model are correlated.

MULTIPLE REGRESSION MODEL

$$y = \beta_0 + \beta_1 x_1 + \beta_2 x_2 + \cdots + \beta_p x_p + \epsilon \qquad (14.1)$$

MULTIPLE REGRESSION EQUATION

$$E(y) = \beta_0 + \beta_1 x_1 + \cdots + \beta_p x_p \qquad (14.2)$$

ESTIMATED REGRESSION EQUATION

$$\hat{y} = b_0 + b_1 x_1 + \cdots + b_p x_p \qquad (14.3)$$

MULTIPLE COEFFICIENT OF DETERMINATION

$$R^2 = \frac{SSR}{SST} \qquad (14.11)$$

ADJUSTED MULTIPLE COEFFICIENT OF DETERMINATION

$$R_a^2 = 1 - (1 - R^2)\left(\frac{n-1}{n-p-1}\right) \qquad (14.12)$$

THE _F_ STATISTIC

$$F = \frac{MSR}{MSE} \qquad (14.13)$$

REVIEW QUIZ

TRUE/FALSE

1. In multiple regression analysis, there are two or more dependent variables.

2. The least squares method cannot be used to develop the coefficients of the estimated regression equation for multiple regression analysis.

3. The coefficient of x_1 in an estimated regression equation will be the same no matter how many independent variables are involved.

4. The computer solution of a multiple regression problem provides all the coefficients for the estimated regression equation.

5. An F test can be used to test for a significant relationship in multiple regression analysis.

6. The number of degrees of freedom for the error sum of squares in multiple regression analysis decreases as the number of independent variables increases.

7. The number of degrees of freedom in the t test for the significance of individual parameters in multiple regression analysis is always $n - 3$.

8. The adjusted multiple coefficient of determination will always be smaller than the unadjusted multiple coefficient of determination.

For question 9–12, consider a regression model in which two independent variables, x_1 and x_2, are used to explain the dependent variable, y.

9. In the test of the hypotheses $H_0: \beta_1 = \beta_2 = 0$ and H_a: either β_1 or β_2 or both $\neq 0$, the test statistic MSR/MSE has a sampling distribution that is the
   a. t distribution
   b. F distribution
   c. normal distribution
   d. chi-square distribution

10. The degrees of freedom for the sampling distribution in question 9 are
   a. $n - 2$
   b. $n - 3$
   c. 3 and $n - 2$
   d. 2 and $n - 3$

11. Which sampling distribution is used to test $H_0: \beta_2 = 0$, $H_a: \beta_2 \neq 0$?
   a. t distribution
   b. F distribution
   c. normal distribution
   d. chi-square distribution

12. The degrees of freedom for the distribution in question 11 are
   a. $n - 2$
   b. $n - 3$
   c. 3 and $n - 2$
   d. 2 and $n - 3$

SUPPLEMENTARY EXERCISES

29. Recall that in Exercise 2 the admissions' officer for Clearwater College developed the following estimated regression equation relating final college GPA to student's SAT mathematics scores and their high-school GPA.

$$\hat{y} = -1.41 + .0235x_1 + .00486x_2$$

where

$$x_1 = \text{high school grade-point average}$$

$$x_2 = \text{SAT mathematics score}$$

$$y = \text{final college grade-point average}$$

A portion of the Minitab computer output is shown.
   a. Complete the missing entries in this output.
   b. Compute F and test at the $\alpha = .05$ level whether a significant relationship exists or not.
   c. Did the estimated regression equation provide a good fit to the data? Explain.
   d. Use the t test and $\alpha = .05$ to test $H_0: \beta_1 = 0$ and $H_0: \beta_2 = 0$.

```
The regression equation is
Y = -1.41 + 0.0235 X1 + 0.00486 X2

Predictor        Coef        Stdev      t-ratio
Constant       -1.4053       0.4848      -2.90
X1            0.023467       0.008666    _____
X2            _____       0.001077    _____

s = 0.1298      R-sq = _____      R-sq(adj) = _____

Analysis of Variance

SOURCE          DF          SS          MS
Regression      __        1.76209       __
Error           __        _____      __
Total            9        1.88000
```

30. Recall that in Exercise 3 the personnel director for Electronics Associates developed the following estimated regression equation relating an employee's score on a job satisfaction test to their length of service and wage rate.

$$\hat{y} = 14.4 - 8.69x_1 + 13.5x_2$$

where

$$x_1 = \text{length of service (years)}$$

$$x_2 = \text{wage rate (dollars)}$$

$$y = \text{job satisfaction test score (higher scores indicate more job satisfaction)}$$

A portion of the Minitab computer output is shown:

```
The regression equation is
Y = 14.4 + -8.69 X1 + 13.5 X2

Predictor        Coef        Stdev      t-ratio
Constant       14.448        8.191       1.76
X1            _____       1.555      _____
X2            13.517         2.085      _____

s = 3.733       R-sq = _____      R-sq(adj) = _____

Analysis of Variance

SOURCE          DF          SS          MS
Regression       2        _____      __
Error           __         71.17        __
Total            7        720.00
```

a. Complete the missing entries in this output.
b. Compute F and test at the $\alpha = .05$ level whether a significant relationship exists or not.
c. Did the estimated regression equation provide a good fit to the data? Explain.
d. Use the t test and $\alpha = .05$ to test H_0: $\beta_1 = 0$ and H_0: $\beta_2 = 0$.

31. Recall that in Exercise 4 a chemist developed the following estimated regression equation relating the yield of a process (y) to the temperature (x_1) and the pressure (x_2).

$$\hat{y} = 69 + 1.17x_1 - 3.97x_2$$

where temperature is measured in degrees Fahrenheit and pressure in pounds per square inch. A portion of the Minitab computer output is shown.

```
The regression equation is
Y = 69.0 + 1.17 X1 - 3.97 X2

Predictor        Coef        Stdev      t-ratio
Constant        69.02        29.96        2.30
X1              1.1741       0.3396      _____
X2             -3.9667       0.9174      _____

s = 7.035       R-sq = _____      R-sq(adj) = _____

Analysis of Variance

SOURCE       DF        SS           MS
Regression   __      1333.60        __
Error        __       346.40        __
Total         9      _____
```

- **a.** Complete the missing entries in this output.
- **b.** Compute F and test at the $\alpha = .05$ level whether a significant relationship exists or not.
- **c.** Did the estimated regression equation provide a good fit to the data? Explain.
- **d.** Use the t test and $\alpha = .05$ to test $H_0: \beta_1 = 0$ and $H_0: \beta_2 = 0$.

32. Bauman Construction Company makes bids on a variety of projects. In an effort to estimate the bid to be made by one of its competitors, Bauman has obtained data on 15 previous bids and developed the following estimated regression equation.

$$\hat{y} = 80 + 45x_1 - 3x_2$$

where

$$y = \text{competitor's bid (\$1000s)}$$

$$x_1 = \text{square feet (1000s)}$$

$$x_2 = \text{local index of construction activity}$$

- **a.** Estimate the competitor's bid on a project involving 50,000 square feet and an index of construction activity of 70.
- **b.** If SSR = 19,780 and SST = 21,533, test at $\alpha = .01$ the significance of the relationship.

33. The following estimated regression equation was developed for a model involving two independent variables.

$$\hat{y} = 40.7 + 8.63x_1 + 2.71x_2$$

After dropping x_2 from the model, the least squares method was used again to obtain an estimated regression equation involving only x_1 as an independent variable.

$$\hat{y} = 42.0 + 9.01x_1$$

Give an interpretation of the coefficient of x_1 in each model.

34. The owner of TAI Movie Theaters, Inc. would like to investigate the effect of television advertising on weekly gross revenue for special promotion films. The following historical data were developed.

WEEKLY GROSS REVENUE ($1000s)	TELEVISION ADVERTISING ($1000s)
96	5.0
90	2.0
95	4.0
92	2.5
95	3.0
94	3.5
94	2.5
94	3.0

a. Using these data, develop an estimated regression equation relating weekly gross revenue to television advertising expenditure.
b. Estimate the weekly gross revenue in a week in which $3500 is spent on television advertising.

35. As an extension of Exercise 34, consider the possibility of incorporating the effect of newspaper advertising as well as television advertising on weekly gross revenue. The following data were developed from historical records.

WEEKLY GROSS REVENUE ($1000s)	NEWSPAPER ADVERTISING ($1000s)	TELEVISION ADVERTISING ($1000s)
96	1.5	5.0
90	2.0	2.0
95	1.5	4.0
92	2.5	2.5
95	3.3	3.0
94	2.3	3.5
94	4.2	2.5
94	2.5	3.0

Let

$$x_1 = \text{newspaper advertising (\$1000s)}$$

$$x_2 = \text{television advertising (\$1000s)}$$

$$y = \text{weekly gross revenue (\$1000s)}$$

Shown is the Minitab computer output for these data.

```
The regression equation is
Y = 83.2 + 1.30 X1 + 2.29 X2


Predictor        Coef       Stdev     t-ratio        p
Constant       83.230       1.574       52.88    0.000
X1             1.3010      0.3207       _____     0.010
X2             2.2902      0.3041       _____     0.001

s = 0.6426      R-sq = _____      R-sq(adj) = _____

Analysis of Variance

SOURCE         DF          SS         MS         F         p
Regression     __       23.435       __        __      0.002
Error          __       _____       __
Total          __       25.500
```

a. Complete the missing entries in this table.

b. Use the value of F to test at the $\alpha = .05$ level whether a significant relationship exists or not.

c. Did the estimated regression equation provide a good fit to the data? Explain.

d. Use the t test and $\alpha = .05$ to test $H_0: \beta_1 = 0$ and $H_0: \beta_2 = 0$.

e. Estimate the weekly gross revenue for a week in which $3500 is spent on television advertising and $1500 is spent on newspaper advertising.

f. Is the coefficient for television advertising expenditures the same as the coefficient obtained in Exercise 34? Interpret this coefficient in each case.

36. Heller Company believes that the quantity of lawnmowers sold depends on the price of its mower and the price of a competitor's mower. Let

$$y = \text{quantity sold (1000s)}$$

$$x_1 = \text{price of competitor's mower (dollars)}$$

$$x_2 = \text{price of Heller's mower (dollars)}$$

Management would like an estimated regression equation that relates quantity sold to the price of the Heller mower and the competitors mower. The following data are available concerning prices in ten different cities.

COMPETITOR'S PRICE (x_1)	HELLER'S PRICE (x_2)	QUANTITY SOLD (y)
120	100	102
140	110	100
190	90	120
130	150	77
155	210	46
175	150	93
125	250	26
145	270	69
180	300	65
150	250	85

The Minitab computer output for these data is shown.

```
The regression equation is
Y = 66.5 + 0.414 X1 - 0.270 X2

Predictor        Coef        Stdev      t-ratio        p
Constant        66.52        41.88        1.59      0.156
X1             0.4139       0.2604       _____      0.156
X2           -0.26978      0.08091       _____      0.013

s = 18.74       R-sq = _____        R-sq(adj) = _____

Analysis of Variance

SOURCE          DF           SS          MS         F         p
Regression      __        _____       __       6.58     0.025
Error           __         2457.3        __
Total           __         7076.1
```

    a.   Complete the missing entries in this table.

    b.   Use the value of F to test at the $\alpha = .05$ level whether a significant relationship exists or not.

    c.   Did the estimated regression equation provide a good fit to the data? Explain.

    d.   Use the t test and $\alpha = .05$ to test $H_0: \beta_1 = 0$ and $H_0: \beta_2 = 0$.

    e.   Predict the quantity sold in a city where Heller prices its mower at \$160 and the competitor prices its mower at \$170.

37. In a regression analysis involving 27 observations the following estimated regression equation was developed.

$$\hat{y} = 16.3 + 2.3x_1 + 12.1x_2 - 5.8x_3$$

Also, the following standard errors were obtained:

$$s_{b_1} = .53 \qquad s_{b_2} = 8.15 \qquad s_{b_3} = 1.30$$

At an $\alpha = .05$ level of significance conduct the following hypothesis tests.

    a. $H_0: \beta_1 = 0$ versus $H_a: \beta_1 \neq 0$

    b. $H_0: \beta_2 = 0$ versus $H_a: \beta_2 \neq 0$

    c. $H_0: \beta_3 = 0$ versus $H_a: \beta_3 \neq 0$

COMPUTER EXERCISE

CREDIT CARD CHARGES

A government agency conducted a study to determine the relationship between income, household size, and the amount charged in the last 12 months on credit cards. The objective of this study was to develop an equation that could be used to predict the amount charged on credit cards given the individual's income and household size. The following data were obtained.

INCOME ($1000s)	HOUSEHOLD SIZE	AMOUNT CHARGED	INCOME ($1000s)	HOUSEHOLD SIZE	AMOUNT CHARGED
33	3	2780	29	2	2491
32	2	2748	26	3	2300
35	1	3080	43	3	3690
34	2	2831	35	4	2935
35	2	3047	35	2	2946
37	1	3061	35	4	2670
38	2	3390	39	1	3485
35	2	3091	30	2	2722
41	2	3415	29	3	2620
47	1	4107	36	1	3126
42	1	3429	32	3	2726
33	5	2521	41	1	3486
38	1	3493	35	5	2720
38	3	3104	36	2	3172
34	3	2752	33	2	2769
30	4	2401	31	1	2804
27	2	2276	37	1	3170
34	2	2865	36	3	2923
29	3	2578	27	2	2270
37	1	3173	36	2	3053
38	5	3091	25	2	2266
33	3	2686	41	6	3187
38	5	2921	35	2	2947
28	4	2250	40	5	3224
35	5	2865	35	2	3105

QUESTIONS

1. Use numerical and graphical measures to summarize these data.

2. Does there appear to be any relationship between the amount charged and the individual's income? Between the amount charged and the household size?

3. Develop an estimated regression equation that can be used to predict the amount charged.

APPENDIX 14.1
Solving the Normal Equations for the Butler Trucking Company

In the chapter we developed an estimated regression equation for Butler Trucking involving two independent variables, miles traveled and number of deliveries. Substituting the data for this problem (see Section 14.2) into (14.5) to (14.7) provided the following normal equations:

$$10b_0 + 815b_1 + 26b_2 = 66.0 \qquad (14.8)$$

$$815b_0 + 69{,}625b_1 + 2165b_2 = 5594.0 \qquad (14.9)$$

$$26b_0 + 2165b_1 + 76b_2 = 180.6 \qquad (14.10)$$

By multiplying (14.8) by 81.5 and subtracting the result from (14.9) we can eliminate b_0 and obtain an equation involving b_1 and b_2 only:

$$
\begin{array}{r}
815b_0 + 69{,}625.0b_1 + 2165b_2 = 5594.0 \\
-815b_0 - 66{,}422.5b_1 - 2119b_2 = -5379.0 \\
\hline
3{,}202.5b_1 + 46b_2 = 215.0
\end{array}
\qquad (14A.1)
$$

Now multiply (14.8) by 2.6 and subtract the result from (14.10). This manipulation yields a second equation involving b_1 and b_2 only:

$$
\begin{array}{r}
26b_0 + 2165b_1 + 76.0b_2 = 180.6 \\
-26b_0 - 2119b_1 - 67.6b_2 = -171.6 \\
\hline
46b_1 + 8.4b_2 = 9.0
\end{array}
\qquad (14A.2)
$$

With equations (14A.1) and (14A.2) we can solve simultaneously for b_1 and b_2. Multiplying (14A.2) by 46/8.4 and subtracting the result from (14A.1) gives us an equation involving b_1 only:

$$
\begin{array}{r}
3202.5000b_1 + 46b_2 = 215.0000 \\
-251.9048b_1 - 46b_2 = -49.2857 \\
\hline
2950.5952b_1 = 165.7143
\end{array}
\qquad (14A.3)
$$

Using (14A.3) to solve for b_1 we get

$$b_1 = \frac{165.7143}{2950.5952} = .056163$$

Using this value for b_1 we can substitute into (14A.2) to solve for b_2:

$$46(.056163) + 8.4b_2 = 9$$
$$2.583498 + 8.4b_2 = 9$$
$$8.4b_2 = 6.416502$$
$$b_2 = .7638693$$

Now we can substitute the values obtained for b_1 and b_2 into (14.8), thus obtaining b_0:

$$10b_0 + 815(.056163) + 26(.7638693) = 66.0$$
$$10b_0 + 45.772845 \quad + 19.860602 \quad = 66.0$$
$$10b_0 = .366553$$
$$b_0 = .0366553$$

Rounding to four significant digits, we obtain the following estimated regression equation for the Butler Trucking problem:

$$\hat{y} = .0367 + .0562x_1 + .7639x_2 \tag{14A.4}$$

15 Regression Analysis: Model Building

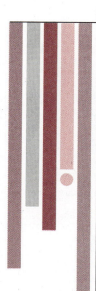

What You Will Learn in This Chapter:

- How the general linear model can be used to model curvilinear effects
- What interaction is and how it can be modeled
- How qualitative variables can be handled using dummy variables
- How the F test can be used to determine when to add or delete variables
- How variable selection procedures such as the stepwise procedure can be used to develop the best model
- How the analysis of residuals can be used to identify outliers and influential observations
- How the Durbin-Watson test can be used to identify if autocorrelation is present

Contents

Working Wives and Expenditures on Time-Saving Services

The dramatic rise of married women in the labor force has increased the number of families where both spouses are wage earners. Since working wives have fewer hours per week to devote to housework than do full-time homemakers, how does the housework get done when both husband and wife are in the work force? Perhaps the husband or other members of the family take on more of the household duties. However, research has shown that, in general, this is not the case. Thus, a working wife with less time at home must find other ways to successfully balance her role in the home with her role at the office.

Don Bellante and Ann Foster, researchers at Auburn University, have studied how much working wives spend on time-saving goods and services, such as the expenditure on food away from home, child care, domestic services, clothing care, and personal care, in order to become more efficient in meeting their household responsibilities. Specifically, the researchers used regression analysis to study the relationship between these time-saving services and factors such as family income, weeks worked per year, hours worked per week, family size, the number of children under the age of six, wife's education, and homeownership status.

The research findings showed that the more income the family has the more the family will spend on time-saving services. In addition, it was found that spending increases when the number of hours worked per week increases, when there are more children under the age of six, when the wife has

Nancy Brown/The Image Bank

Time-saving services help working wives balance their home, family, and career responsibilities.

more education, and when the family owns a home. The research supports the notion that when the job at home gets bigger, the families with working wives tend to spend more on time-saving services. The challenge of balancing a career, a home, and a family is an every day battle for the modern working wife.

Based on "Working Wives and Expenditures on Services," by Don Bellante and Ann C. Foster, *Journal of Consumer Research*, September, 1984.

Model building in regression analysis is the process of developing the regression model that best describes the relationship among the independent and dependent variables. The major issues are determining the proper form of the relationship (linear or curvilinear) and variable selection. Variable selection involves determining which of a candidate set of independent variables should be included in the regression model.

In Chapters 13 and 14, we introduced and worked with two regression models:

Simple Linear Regression: $\quad y = \beta_0 + \beta_1 x + \epsilon$

Multiple Regression: $\quad y = \beta_0 + \beta_1 x_1 + \beta_2 x_2 + \cdots + \beta_p x_p + \epsilon$

Both of these models specify a linear relationship between the independent variable(s), the x's, and the dependent variable, y.

The primary method introduced in Chapters 13 and 14 for determining whether or not a regression model was adequate was residual analysis. Essentially, if the residual plot looked like a horizontal band, we concluded that the assumptions concerning model form and the error term were satisfied. Otherwise, we concluded that the model was inadequate. When the regression model is judged inadequate, it may be because we have chosen the wrong functional form for the model (for example, the actual relationship is curvilinear) and/or the proper independent variables have not been included. In Chapter 14, we saw that when the residual plot for the Butler Trucking problem showed a nonconstant variance, a model that corrected this deficiency was created by adding a second independent variable.

In this chapter we focus on the model-building issues of identifying the proper model form and variable selection. Section 15.1, which establishes the framework for model building, introduces the concept of a *general linear model*. Surprisingly the general linear model makes it possible for us to accommodate curvilinear relationships between the independent variables and the dependent variable with no more computational difficulty than that involved in multiple regression. Section 15.2 extends our modeling capability by showing how qualitative independent variables may be incorporated into a regression model.

Section 15.3 provides the foundation for the more sophisticated computer-based procedures for variable selection. The issue of when one variable or a group of variables should be added to a regression model is examined. We show that the general approach to determining when to add or delete variables is based on the value of an F statistic. In Section 15.4, a large problem of the type often encountered in practice is introduced. It involves 25 observations on 8 independent variables. This larger problem provides an illustration for the computer-based variable selection procedures of Section 15.5. The stepwise, forward-selection, and backward-elimination procedures are explained.

In Section 15.6, we introduce some additional diagnostic procedures based on analysis of residuals. We show how the residuals can be used to identify potentially troublesome observations: outliers and observations that exert an unusually large amount of influence on the regression results. The Durbin-Watson test is also introduced as a diagnostic for detecting serial or autocorrelation. The chapter concludes with a discussion of how regression analysis can be used for solving analysis of variance problems.

A caveat is in order before beginning this chapter. The "truly correct model" will never be found for real data. Several models may do an adequate job; the model-builder's goal is to find the best from a set of acceptable models.

15.1 The General Linear Model

Suppose we were faced with a situation in which the following regression model was appropriate:

$$y = \beta_0 + \beta_1 x^2 + \epsilon \qquad (15.1)$$

Since x is squared, the relationship between y and x is said to be curvilinear. At first glance, it might not appear that the regression techniques developed in Chapters 13 and 14 are applicable to this model. But they are!

By making the substitution $z = x^2$, the preceding model can be converted into the following simple linear regression model with z as the independent variable:

$$y = \beta_0 + \beta_1 z + \epsilon \qquad (15.2)$$

If there are n observations, each z_i is given by x_i^2. Using the data for y and the computed data for z, the least squares method can then be used to compute b_0 and b_1. Let us see how the techniques we have already developed can be used with the type of curvilinear regression model in (15.1) by considering the problem faced by Nugent Industries.

EXAMPLE 15.1

Management of Nugent Industries has been investigating the relationship between each customer's annual order size and the number of sales calls made per year. A random sample of seven customer accounts was obtained; the data are presented in Table 15.1 and a scatter diagram is shown in Figure 15.1.

From the scatter diagram in Figure 15.1, it appears that a curvilinear relationship might be a better model of the relationship than a straight line. Note that the curved line in the figure appears to provide a very good fit to the data. Suppose, after viewing the scatter diagram, we hypothesize that y is related to x by the following regression model:

$$y = \beta_0 + \beta_1 x^2 + \epsilon$$

Using the approach noted before we can make the substitution $z = x^2$ to obtain a simple linear regression model with the independent variable denoted by z:

$$y = \beta_0 + \beta_1 z + \epsilon$$

To develop estimates of β_0 and β_1 in this model, we simply substitute z_i for x_i in the least squares formulas for b_0 and b_1. Doing so, we obtain

$$b_1 = \frac{\sum z_i y_i - (\sum z_i \sum y_i)/n}{\sum z_i^2 - (\sum z_i)^2/n}$$

$$b_0 = \bar{y} - b_1 \bar{z}.$$

Computing the values of b_1 and b_0 in these formulas requires that we substitute the value of x_i^2 every place z_i appears. The calculations of b_0 and b_1 for this problem are

TABLE 15.1

Annual Order Size and Number of Sales Calls for Nugent Industries

NUMBER OF SALES CALLS (x)	ANNUAL SALES ($1000s) (y)
2	12
3	17
4	16
5	24
6	26
7	34
8	46

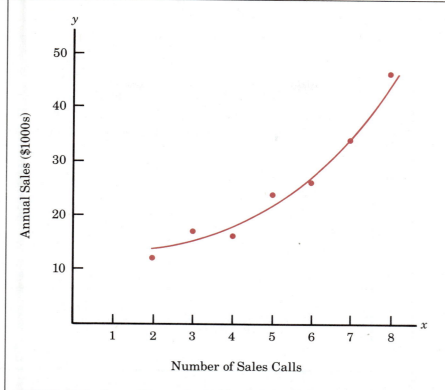

FIGURE 15.1 Scatter Diagram for the Nugent Industries Problem Data

summarized in Table 15.2. Rounding to two decimal places and substituting x^2 for z leads to the estimated regression equation

$$\hat{y} = 9.64 + .53x^2$$

This equation represents a curvilinear (quadratic) relationship between x and y.

A residual plot of $y_i - \hat{y}_i$ versus $\hat{y}_i$ corresponding to the estimated regression equation $\hat{y} = 9.64 + .53x^2$ is shown in Figure 15.2. The pattern observed does not suggest any assumptions have been violated. In Figure 15.3, we show the graph of the estimated regression equation; it clearly fits the data well. It turns out that the relationship between y and x^2 is statistically significant and that the value of r^2 for this equation is .97. On the basis of this analysis, we recommend that $\hat{y} = 9.64 + .53x^2$ be used for developing predictions of annual sales given the number of sales calls per year. However, we caution that this model should not be used to make predictions outside the range of x values observed. Obviously one would expect diminishing returns per sales call beyond some point.

The method by which we developed a curvilinear relationship between x and y for the Nugent Industries problem can be generalized to handle curvilinear types of relationships for the multiple regression case. To show how this can be done, consider the following *general linear model* involving p independent variables.

FIGURE 15.2 Residual Plot for the Estimated Regression Equation $\hat{y} = 9.64 + .53x^2$

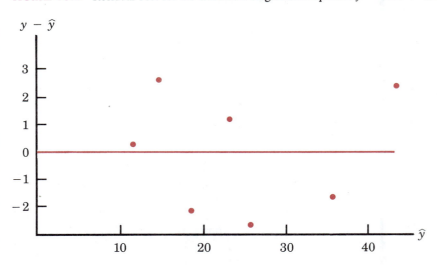

TABLE 15.2

Calculations for the Estimated Regression Equation $\hat{y} = b_0 + b_1 x^2$

CUSTOMER (i)	x_i	$z_i = x_i^2$	y_i	$z_i y_i$	z_i^2
1	2	4	12	48	16
2	3	9	17	153	81
3	4	16	16	256	256
4	5	25	24	600	625
5	6	36	26	936	1296
6	7	49	34	1666	2401
7	8	64	46	2944	4096
Totals	35	203	175	6603	8771

$$\bar{z} = \frac{\sum z_i}{n} = \frac{203}{7} = 29$$

$$\bar{y} = \frac{\sum y_i}{n} = \frac{175}{7} = 25$$

$$b_1 = \frac{\sum z_i y_i - (\sum z_i \sum y_i)/n}{\sum z_i^2 - (\sum z_i)^2/n} = \frac{(6603) - (203)(175)/7}{(8771) - (203)^2/7} = .5298$$

$$b_0 = \bar{y} - b_1 \bar{z} = 25 - .5298(29) = 9.6358$$

$$\hat{y} = 9.6358 + .5298z$$

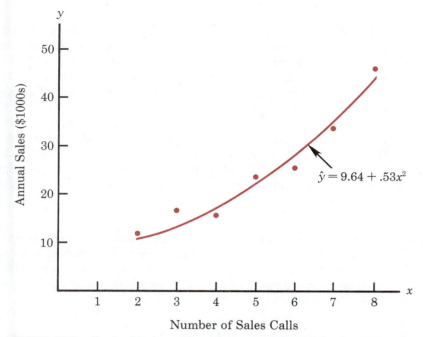

FIGURE 15.3 Graph of Estimated Regression Equation for the Nugent Industries Problem

In (15.3) each of the independent variables z_j, $j = 1, 2, \ldots, p$ is a function of $x_1, x_2, \ldots, x_k$, the variables for which data have been collected. In some cases, each z_j may be a function of only one x-variable. The simplest case occurs in which we have collected data for just one variable, x_1, and want to estimate y using a straight-line relationship. In this case $z_1 = x_1$ and (15.3) becomes

$$y = \beta_0 + \beta_1 x_1 + \epsilon$$

Note that this is just the simple linear regression model introduced in Chapter 13 with the exception that the independent variable is labeled x_1 instead of x. In the statistical literature this model is referred to as a *simple first-order model with one predictor variable*.

More complex types of relationships can be modeled easily with (15.3). For example, if $p = 1$ and $z_1 = x_1^2$, we have the model used earlier in this section (with x_1 replacing x) for the Nugent Industries problem (Example 15.1). In situations where the relationship between y and x_1 is better described by a combination of a linear and a quadratic effect, we could set $z_1 = x_1$ and $z_2 = x_1^2$ in (15.3) to obtain the model

$$y = \beta_0 + \beta_1 x_1 + \beta_2 x_1^2 + \epsilon$$

This model is referred to as a *second-order model with one predictor variable*.

If the original data set consists of observations for y and two independent variables, x_1 and x_2, we can develop a *second-order model with two predictor variables* by setting $z_1 = x_1$, $z_2 = x_2$, $z_3 = x_1^2$, $z_4 = x_2^2$, and $z_5 = x_1 x_2$ in the general linear model of (15.3). The model obtained is

$$y = \beta_0 + \beta_1 x_1 + \beta_2 x_2 + \beta_3 x_1^2 + \beta_4 x_2^2 + \beta_5 x_1 x_2 + \epsilon$$

In this second-order model, the variable $z_5 = x_1 x_2$ is added to account for the potential effects of the two variables acting together. This type of effect is called *interaction*.

It should be apparent that many types of relationships can be modeled by (15.3). Each different function of the x variables generates a separate term, denoted by z_i. Thus the regression techniques with which we have been working are definitely not limited to linear, or straight-line, relationships.

In regression analysis (15.3) is referred to as the *general linear model*. The term linear in this context can be confusing, since it really refers only to the fact that $\beta_0, \beta_1, \ldots, \beta_p$ all have exponents of 1; it does not imply that the relationship between y and the x_is is linear. Indeed, in this section we have seen one example where (15.3) can be used to model a curvilinear relationship. Let us now review an example involving interaction.

EXAMPLE 15.2

To understand better how interaction is treated with the general linear model, let us review the regression study conducted by Tyler Personal Care for one of its shampoo products. Two factors believed to have the most influence on sales were unit selling price and advertising expenditures. The following regression model was used.

$$y = \beta_0 + \beta_1 x_1 + \beta_2 x_2 + \beta_3 x_1 x_2 + \epsilon$$

where

$$y = \text{unit sales (1000s)}$$

$$x_1 = \text{selling price (\$)}$$

$$x_2 = \text{advertising expenditure (\$1000s)}$$

This model reflected Tyler's belief that sales depended linearly on selling price and advertising expenditures (the $\beta_1 x_1$ and $\beta_2 x_2$ terms) but that there was some interaction between the two (the $\beta_3 x_1 x_2$ term). Indeed, Tyler believed that at higher unit selling prices, the effect of increased advertising would diminish.

In order to develop an estimated regression equation, a general linear model involving three independent variables (z_1, z_2, and z_3) was used.

$$y = \beta_0 + \beta_1 z_1 + \beta_2 z_2 + \beta_3 z_3 + \epsilon$$

where

$$z_1 = x_1$$

$$z_2 = x_2$$

$$z_3 = x_1 x_2$$

The computer solution of the least squares equations for this model yielded $b_0 = 200$, $b_1 = -50$, $b_2 = 10$, and $b_3 = -3$. Substituting for z_1, z_2, and z_3 in the general linear model provided the following estimated regression equation for Tyler's shampoo product:

$$\hat{y} = 200 - 50x_1 + 10x_2 - 3x_1x_2$$

The term x_1x_2 is called a cross-product term; it accounts for the combined effect of x_1 and x_2 in predicting y. To understand better the interaction effect modeled by the cross-product term, consider the following values of $\hat{y}$ for this equation.

SELLING PRICE x_1	ADVERTISING x_2	PREDICTED UNIT SALES $\hat{y}$
1.50	50	400
1.50	100	675
2.00	50	300
2.00	100	500

Holding price constant at $1.50 the difference in unit sales between an advertising expenditure of $100,000 and an advertising expenditure of $50,000 is 675,000 − 400,000 = 275,000. Holding price constant at $2.00, the difference in unit sales between an advertising expenditure of $100,000 and an advertising expenditure of $50,000 is 500,000 − 300,000 = 200,000. Thus, the regression results do show that the effect of advertising on unit sales depends on the level of the selling price.

∎

The general linear model of (15.3) can be used to accommodate any situation in which $z_1, z_2, \ldots, z_p$ are functions of the x variables. However, it cannot be used in cases where the parameters, β_i, appear nonlinearly. Sometimes, however, a transformation can be applied to allow even these cases to be handled with a general linear model. One such transformation, the logarithmic transformation, is discussed briefly here.

The Logarithmic Transformation

Models in which the parameters $(\beta_0, \beta_1, \ldots, \beta_p)$ have exponents other than 1 are referred to as nonlinear models. However, for the case of the exponential model, it is possible to perform a transformation of variables that will permit us to perform regression analysis using (15.3), the general linear model. The exponential model involves the following regression equation:

$$E(y) = \beta_0\beta_1^x \qquad (15.4)$$

This model is appropriate in cases where the dependent variable y increases or decreases by a constant percentage, instead of by a fixed amount, as x increases.

EXAMPLE 15.3

Suppose that sales for a product (y) were related to advertising expenditure x (in $1000s) according to the following exponential model:

$$E(y) = 500(1.2)^x$$

EXAMPLE 15.3
(continued)

Thus, for $x = 1$, $E(y) = 500(1.2)^1 = 600$; for $x = 2$, $E(y) = 500(1.2)^2 = 720$; and for $x = 3$, $E(y) = 500(1.2)^3 = 864$. Note that $E(y)$ is not increasing by a constant amount in this case, but by a constant percentage; the percentage increase is 20%.

We can transform the nonlinear model in Example 15.3 to a linear model by taking the logarithm of both sides of (15.4):

$$\log E(y) = \log \beta_0 + x \log \beta_1 \tag{15.5}$$

Now if we let $y' = \log E(y)$, $\beta_0^1 = \log \beta_0$, and $\beta_1^1 = \log \beta_1$, we can rewrite (15.5) as

$$y' = \beta_0^1 + \beta_1^1 x$$

It is clear that the formulas for simple linear regression can now be used to develop estimates of β_0^1 and β_1^1. Denoting the estimates as b_0^1 and b_1^1 leads to the following estimated regression equation:

$$\hat{y}' = b_0^1 + b_1^1 x \tag{15.6}$$

To obtain predictions of the original dependent variable y given a value of x, we would first substitute the value of x into (15.6) and compute $\hat{y}'$. The antilog of $\hat{y}'$ would be the point estimate of y, or the expected value of y.

We should make it clear that there are many nonlinear models that cannot be transformed into an equivalent linear model. Furthermore, the mathematical background needed for study of such models is beyond the scope of this text.

EXERCISES

1. The highway department is doing a study on the relationship between traffic flow and speed. The following model has been hypothesized:

$$y = \beta_0 + \beta_1 x + \epsilon$$

where

$y =$ traffic flow in vehicles per hour

$x =$ vehicle speed in miles per hour

The following data have been collected during rush hour for six highways leading out of the city:

TRAFFIC FLOW (y)	VEHICLE SPEED (x)
1256	35
1329	40
1226	30

TRAFFIC FLOW (y)	VEHICLE SPEED (x)
1335	45
1349	50
1124	25

a. Develop an estimated regression equation for these data.
b. Using $\alpha = .01$ test for a significant relationship.

2. In working further with the data of Exercise 1 statisticians suggested the use of the following curvilinear estimated regression equation:

$$\hat{y} = b_0 + b_1 x + b_2 x^2$$

a. Use the data of Exercise 1 to estimate the parameters of this estimated regression equation.
b. Using $\alpha = .01$ test for a significant relationship.
c. Estimate the traffic flow in vehicles per hour at speeds of 38 miles per hour.

3. The following estimated regression equation has been developed for the relationship between y, sales ($1000s), and x, store size (square feet $\times$ 10,000):

$$\hat{y} = 150 + 100x - 10x^2$$

Ten stores were included in the sample. Values of SST $= 168,000$ and SSR $= 140,000$ were obtained.
a. Compute R^2 and R_a^2.
b. Using $\alpha = .05$ test for a significant relationship.

4. A study of emergency service facilities investigated the relationship between the number of facilities and the average distance travelled to provide the emergency service *(Management Science,* July 1988). The following data were collected:

NUMBER OF FACILITIES	AVERAGE TRAVEL DISTANCE (MILES)
9	1.66
11	1.12
16	0.83
21	0.62
27	0.51
30	0.47

a. Develop a scatter diagram for these data treating average travel distance as the dependent variable.
b. Does a simple linear model appear to be appropriate? Explain.
c. Fit a model to the data that you believe will best explain the relationship between these two variables.

15.2 The Use of Qualitative Variables

So far the variables that we have used to build regression models have been quantitative variables; that is, variables that are measured in terms of how much or how many.

Frequently, however, we will need to use variables that are not measured in these terms. We refer to such variables as *qualitative variables*. For instance, suppose that we were interested in predicting sales for a product which was available in either bottles or cans. Clearly the independent variable "container type" could influence the dependent variable "sales"—but container type is a qualitative, not a quantitative, variable. The distinguishing feature of qualitative variables is that there is no natural measure of "how much" or "how many"; these variables are used to refer to attributes that are either present or not present. Let us see how qualitative variables might be used in the context of the Butler Trucking problem introduced in Chapter 14.

EXAMPLE 15.4

In Chapter 14 we concluded that the travel time (y) could be predicted using the number of miles traveled (x_1) and the number of deliveries (x_2). The estimated regression equation we developed was $\hat{y} = .0367 + .0562x_1 + .7639x_2$. Suppose that management felt that the type of truck should also be considered in attempting to predict total travel time. Butler Trucking has only two types of trucks: pickups and vans. Thus truck type is an example of a qualitative variable. Table 15.3 shows the expanded data set for the Butler Trucking problem with the addition of truck type as a third independent variable.

To incorporate the effect of truck type into a model to predict total travel time, we define the following variable:

$$x_3 = \begin{cases} 0 & \text{if the truck is a pickup} \\ 1 & \text{if the truck is a van} \end{cases}$$

When preparing the data, whenever an observation involves a pickup truck we will set $x_3 = 0$; whenever an observation involves a van we will set $x_3 = 1$. In regression analysis this type of variable is commonly referred to as a *dummy*, or *indicator*, variable.

Adding this dummy variable to the previous regression equation for predicting travel time results in the following:

$$E(y) = \beta_0 + \beta_1 x_1 + \beta_2 x_2 + \beta_3 x_3 \qquad (15.7)$$

TABLE 15.3

Data for the Butler Trucking Problem, Including Truck Type

DAY (I)	MILES TRAVELED (x_{1i})	NUMBER OF DELIVERIES (x_{2i})	TRUCK TYPE (x_{3i})		TRAVEL TIME (HOURS) y_i
1	100	4	Van	1	9.3
2	50	3	Pickup	0	4.8
3	100	4	Van	1	8.9
4	100	2	Pickup	0	5.8
5	50	2	Pickup	0	4.2
6	80	1	Van	1	6.8
7	75	3	Van	1	6.6
8	80	2	Pickup	0	5.9
9	90	3	Pickup	0	7.6
10	90	2	Van	1	6.1

We can see that when $x_3 = 0$, and hence the truck is a pickup, the regression equation reduces to

$$E(y) = \beta_0 + \beta_1 x_1 + \beta_2 x_2 + \beta_3(0)$$

$$= \beta_0 + \beta_1 x_1 + \beta_2 x_2 \qquad (15.8)$$

However, if we want to predict y when a van is used, $x_3 = 1$ and the regression equation becomes

$$E(y) = \beta_0 + \beta_1 x_1 + \beta_2 x_2 + \beta_3(1)$$

$$= \beta_0 + \beta_1 x_1 + \beta_2 x_2 + \beta_3 \qquad (15.9)$$

If we subtract (15.8), the expected travel time for a pickup, from (15.9), the expected travel time for a van, we obtain

$$\underbrace{(\beta_0 + \beta_1 x_1 + \beta_2 x_2 + \beta_3)}_{\substack{\text{Expected travel} \\ \text{time for a van}}} - \underbrace{(\beta_0 + \beta_1 x_1 + \beta_2 x_2)}_{\substack{\text{Expected travel} \\ \text{time for a pickup}}} = \beta_3$$

Thus β_3 can be interpreted as the difference in the expected travel time between a van and a pickup truck.

When we fit an estimated regression equation using the least squares method and incorporate the possible effect of truck type we obtain the following estimated regression equation:

$$\hat{y} = b_0 + b_1 x_1 + b_2 x_2 + b_3 x_3$$

As usual, b_3 turns out to be the least squares estimate of β_3 and hence the best estimate of the effect of truck type.

Computer Solution

In Figure 15.4 we show the computer output obtained using the Minitab statistical computer software system. The estimated regression equation is

$$\hat{y} = .522 + .0464 x_1 + .7102 x_2 + .9000 x_3$$

Thus we see that $b_3 = .9000$. Hence the best estimate of the difference in the expected travel time when a van is used instead of a pickup truck is .9 hours, or 54 minutes.

To test for the significance of x_3 given that x_1 and x_2 are in the model, the appropriate hypotheses are

$$H_0: \quad \beta_3 = 0$$

$$H_a: \quad \beta_3 \neq 0$$

Using $\alpha = .05$ and the p-value of .139, we cannot reject H_0 and must conclude that truck type is not a significant factor in predicting travel time once the effects of miles traveled and number of deliveries have been accounted for. Note that we concluded not

```
The regression equation is
TIME = 0.52 + 0.0464 MILES + 0.710 DELIV + 0.900 TYPE

Predictor        Coef        Stdev      t-ratio         p
Constant        0.522       1.210         0.43       0.681
MILES         0.04640     0.01500         3.09       0.021
DELIV          0.7102      0.2725         2.61       0.040
TYPE           0.9000      0.5281         1.70       0.139

s = 0.7531      R-sq = 85.8%      R-sq(adj) = 78.7%

Analysis of Variance

SOURCE        DF        SS          MS         F          p
Regression     3     20.5969     6.8656     12.10      0.006
Error          6      3.4031     0.5672
Total          9     24.0000
```

FIGURE 15.4 Minitab Output for the Butler Trucking Problem

that truck type is not significant, but that it is not significant once the effect of x_1 and x_2 have been accounted for.

◼

In closing this section we should also point out that the concept of interaction and the method for dealing with it also applies when working with qualitative variables. All that needs to be done is to add cross-product terms of the form $x_1 x_2$, where x_1, x_2, or both may be qualitative variables.

EXERCISES

5. The following regression model has been proposed to predict sales at a fast-food outlet:

$$y = \beta_0 + \beta_1 x_1 + \beta_2 x_2 + \beta_3 x_3 + \epsilon$$

where

x_1 = number of competitors within 1 mile

x_2 = population within 1 mile (1000s)

$x_3 = \begin{cases} 1 & \text{if drive-up window present} \\ 0 & \text{otherwise} \end{cases}$

y = sales ($1000s)

The following estimated regression equation was developed after 20 outlets were surveyed:

$$\hat{y} = 10.1 - 4.2x_1 + 6.8x_2 + 15.3x_3$$

a. What is the expected amount of sales attributable to the drive-up window?
b. Predict sales for a store with two competitors, a population of 8000 within 1 mile, and no drive-up window.

c. Predict sales for a store with one competitor, a population within 1 mile of 3000, and a drive-up window.

6. In order to investigate the relationship among the service time to repair a machine and (1) the number of months since the machine was serviced and (2) whether a mechanical failure or an electrical failure had occurred, the following data were obtained:

REPAIR TIME (HOURS)	TIME SINCE PREVIOUS SERVICE-CALL (MONTHS)	TYPE OF FAILURE
2.9	2	Electrical
3.0	6	Mechanical
4.8	8	Electrical
1.8	3	Mechanical
2.9	2	Electrical
4.9	7	Electrical
4.2	9	Mechanical
4.8	8	Mechanical
4.4	4	Electrical
4.5	6	Electrical

a. Ignore for now the type of failure associated with the machine. Develop the estimated simple linear regression equation to predict the repair time given the number of months since the previous service call.

b. Does the equation that you developed in (a) provide a good fit for the observed data? Explain.

7. This problem is an extension of the situation described in exercise 6. In an attempt to incorporate the possible effect of the type of failure the following dummy variable was added to the regression model:

$$x_2 = \begin{cases} 1 & \text{if the failure was electrical} \\ 0 & \text{if the failure was mechanical} \end{cases}$$

With the addition of this variable, the following regression equation was proposed:

$$E(y) = \beta_0 + \beta_1 x_1 + \beta_2 x_2$$

where

$$x_1 = \text{number of months since the previous service call}$$

$$y = \text{repair time (hours)}$$

a. What is the interpretation of β_2 in this regression equation?

b. Develop the estimated regression equation using both the number of months since the previous service call (x_1) and the type of failure associated with the machine (x_2).

c. At the $\alpha = .05$ level of significance, test whether or not the estimated regression equation developed in (b) represents a significant relationship between the independent variables and the dependent variable.

d. Does the estimated regression equation developed in (b) provide a better fit than the equation developed in exercise 6? Explain.

e. Use the estimated regression equation developed in (b) to determine on the average how much longer it takes to service a machine involving an electrical failure than one with a mechanical failure.

8. The following data show the price-earnings ratio, the net profit margin, and the growth rate for 19 companies listed in "The *Forbes* 500 on Wall Street." (*Forbes*, May 1, 1989) The data in the column labeled "Industry" are simply codes used to define the industry for each company: 1 = energy-international oil; 2 = health-drugs, and 3 = electronics-computers.

FIRM	P/E RATIO	PROFIT MARGIN	GROWTH RATE	INDUSTRY
Exxon	11.3	6.5	10	1
Chevron	10.0	7.0	5	1
Texaco	9.9	3.9	5	1
Mobil	9.7	4.3	7	1
Amoco	10.0	9.8	8	1
Pfizer	11.9	14.7	12	2
Bristol Meyers	16.2	13.9	14	2
Merck	21.0	20.3	16	2
American Home Products	13.3	16.9	11	2
Abbott Laboratories	15.5	15.2	18	2
Eli Lilly	18.9	18.7	11	2
Upjohn	14.6	12.8	10	2
Warner-Lambert	16.0	8.7	7	2
Amdahl	8.4	11.9	4	3
Digital	10.4	9.8	19	3
Hewlett-Packard	14.8	8.1	18	3
NCR	10.1	7.3	6	3
Unisys	7.0	6.9	6	3
IBM	11.8	9.2	6	3

a. Develop the estimated regression equation that can be used to predict the price-earnings ratio given the profit margin, growth rate, and type of industry.

b. Consider modifying the model developed in (a) to account for possible interaction involving profit margin and the type of industry. Is there significant interaction involving these variables? Explain.

15.3   Determining When to Add or Delete Variables

In this section we will show how an F test can be used to determine whether or not it is advantageous to add one variable—or a group of variables—to a multiple regression model. This test is based on a determination of the amount of reduction in the error sum of squares resulting from adding one or more independent variables to the model. We will first illustrate how the test might be used in the context of the Butler Trucking problem.

EXAMPLE 15.4 (continued)

With miles traveled (x_1) as the only independent variable, the least squares procedure provided the following estimated regression equation:

$$\hat{y} = 1.13 + .067x_1$$

In Chapter 14 we showed that the error sum of squares for this model was SSE = 9.5669. When x_2, the number of deliveries, was added as a second independent variable we obtained the following estimated regression equation:

$$\hat{y} = .0367 + .0562x_1 + .7639x_2$$

The error sum of squares for this model was 5.0501. Clearly, adding x_2 resulted in a reduction of SSE. The question we want to answer is: Does adding the variable x_2 lead to a *significant* reduction in SSE?

We will use the notation SSE(x_1) to denote the error sum of squares when x_1 is the only independent variable, SSE(x_1, x_2) the error sum of squares when x_1 and x_2 are both independent variables, and so on. Hence the reduction in SSE resulting from adding x_2 to the model involving just x_1 is

$$\text{SSE}(x_1) - \text{SSE}(x_1, x_2) = 9.5669 - 5.0501 = 4.5168$$

An F test is conducted to determine whether or not this reduction is significant.

The numerator of the F statistic is the reduction in SSE divided by the number of variables added to the original model. Here only one variable, x_2, has been added; thus the numerator is

$$\frac{\text{SSE}(x_1) - \text{SSE}(x_1, x_2)}{1} = 4.5168$$

The numerator is a measure of the reduction in SSE per variable added to the model. The denominator of the F statistic is the mean square error for the model that includes all of the variables. For Butler Trucking this corresponds to the model containing both x_1 and x_2; thus $p = 2$ and hence

$$\text{MSE} = \frac{\text{SSE}(x_1, x_2)}{n - p - 1} = \frac{5.0501}{7} = .7214$$

The following F statistic provides the basis for testing whether or not the addition of x_2 is statistically significant:

$$F = \frac{\dfrac{\text{SSE}(x_1) - \text{SSE}(x_1, x_2)}{1}}{\dfrac{\text{SSE}(x_1, x_2)}{n - p - 1}} \qquad (15.10)$$

The numerator degrees of freedom for this F test equal the number of variables added to the model, and the denominator degrees of freedom equal $n - p - 1$.

For the Butler Trucking problem, we obtain

$$F = \frac{\dfrac{4.5168}{1}}{\dfrac{5.0501}{7}} = \frac{4.5168}{.7214} = 6.26$$

Refer to Table 4 of Appendix B. We find that for a level of significance of $\alpha = .05$,

$$F_{.05} = 5.59$$

Since

$$F = 6.26 > F_{.05} = 5.59$$

we reject the null hypothesis that x_2 is not statistically significant; in other words, adding x_2 to the model involving only x_1 results in a significant reduction in the error sum of squares.

■

When we want to test for the significance of adding only one additional independent variable to an existing model, the result found with the F test just described could also be obtained by using the t test for the significance of an individual parameter (described in Section 14.5). Indeed, the F statistic we just computed is the square of the t statistic used to test the hypothesis that an individual parameter is zero.

Since the t test is equivalent to the F test when only one variable is being added to the model, we can now further clarify the proper use of the t test on the individual parameters. If an individual parameter is not significant, the corresponding variable can be dropped from the model. However, no more than one variable can ever be dropped from a model on the basis of a t test; if one variable is dropped, a second variable that was not significant initially might become significant.

We now turn briefly to a consideration of whether or not the addition of more than one variable—as a set—results in a significant reduction in the error sum of squares.

The General Case

Consider the following multiple regression model involving q independent variables, where $q < p$:

$$y = \beta_0 + \beta_1 x_1 + \beta_2 x_2 + \cdots + \beta_q x_q + \epsilon \qquad (15.11)$$

If we add variables $x_{q+1}, x_{q+2}, \cdots, x_p$ to this model, we obtain a model involving p independent variables:

$$
\begin{aligned}
y = \ &\beta_0 + \beta_1 x_1 + \beta_2 x_2 + \cdots + \beta_q x_q \\
&+ \beta_{q+1} x_{q+1} + \beta_{q+2} x_{q+2} + \cdots + \beta_p x_p + \epsilon
\end{aligned} \qquad (15.12)
$$

To test whether or not the addition of $x_{q+1}, x_{q+2}, \cdots, x_p$ is statistically significant, the null and alternative hypotheses can be stated as follows:

$$H_0: \quad \beta_{q+1} = \beta_{q+2} = \cdots = \beta_p = 0$$

$$H_a: \quad \text{One or more of the coefficients is not equal to zero}$$

The following F statistic provides the basis for testing whether or not the additional variables are statistically significant:

$$F = \frac{\dfrac{\text{SSE}(x_1, x_2, \cdots, x_q) - \text{SSE}(x_1, x_2, \cdots, x_q, x_{q+1}, \cdots, x_p)}{p - q}}{\dfrac{\text{SSE}(x_1, x_2, \cdots, x_q, x_{q+1}, \cdots, x_p)}{n - p - 1}} \qquad (15.13)$$

This computed F value is then compared with F_α, the table value with $p - q$ numerator degrees of freedom and $n - p - 1$ denominator degrees of freedom. If $F > F_\alpha$, we reject H_0 and conclude that the set of additional variables is statistically significant. Note that for the special case where $q = 1$ and $p = 2$, (15.13) reduces to (15.10).

EXERCISES

9. In a regression analysis involving 27 observations, the following estimated regression equation was developed:

$$\hat{y} = 16.3 + 2.3x_1 + 12.1x_2 - 5.8x_3$$

Also, the following standard errors were obtained:

$$s_{b_1} = .53 \quad s_{b_2} = 8.15 \quad s_{b_3} = 1.30$$

At an $\alpha = .05$ level of significance conduct the following hypothesis tests:
   a. $H_0\colon \beta_1 = 0$ versus $H_a\colon \beta_1 \neq 0$.
   b. $H_0\colon \beta_2 = 0$ versus $H_a\colon \beta_2 \neq 0$.
   c. $H_0\colon \beta_3 = 0$ versus $H_a\colon \beta_3 \neq 0$.
   d. Can any of the variables be dropped from the model? Why or why not?

10. In a regression analysis involving 30 observations the following estimated regression equation was obtained:

$$\hat{y} = 17.6 + 3.8x_1 - 2.3x_2 + 7.6x_3 + 2.7x_4$$

For this model SST $= 1805$ and SSR $= 1760$.
   a. Compute R^2.
   b. Compute R_a^2.
   c. At $\alpha = .05$ test the significance of the relationship among the variables.

11. Refer again to exercise 10. Variables x_1 and x_4 were dropped from the model, and the following estimated regression equation was obtained:

$$\hat{y} = 11.1 - 3.6x_2 + 8.1x_3$$

For this model SST $= 1805$ and SSR $= 1705$.
   a. Compute SSE(x_1, x_2, x_3, x_4).
   b. Compute SSE(x_2, x_3).
   c. Use an F test and an $\alpha = .05$ level of significance to determine if x_1 and x_4 contribute significantly to the model.

TABLE 15.4

The Cravens Data

SALES	TIME	POTEN	ADV	SHARE	CHANGE	ACCTS	WORK	RATING
3,669.88	43.10	74,065.1	4,582.9	2.51	0.34	74.86	15.05	4.9
3,473.95	108.13	58,117.3	5,539.8	5.51	0.15	107.32	19.97	5.1
2,295.10	13.82	21,118.5	2,950.4	10.91	−0.72	96.75	17.34	2.9
4,675.56	186.18	68,521.3	2,243.1	8.27	0.17	195.12	13.40	3.4
6,125.96	161.79	57,805.1	7,747.1	9.15	0.50	180.44	17.64	4.6
2,134.94	8.94	37,806.9	402.4	5.51	0.15	104.88	16.22	4.5
5,031.66	365.04	50,935.3	3,140.6	8.54	0.55	256.10	18.80	4.6
3,367.45	220.32	35,602.1	2,086.2	7.07	−0.49	126.83	19.86	2.3
6,519.45	127.64	46,176.8	8,846.2	12.54	1.24	203.25	17.42	4.9
4,876.37	105.69	42,053.2	5,673.1	8.85	0.31	119.51	21.41	2.8
2,468.27	57.72	36,829.7	2,761.8	5.38	0.37	116.26	16.32	3.1
2,533.31	23.58	33,612.7	1,991.8	5.43	−0.65	142.28	14.51	4.2
2,408.11	13.82	21,412.8	1,971.5	8.48	0.64	89.43	19.35	4.3
2,337.38	13.82	20,416.9	1,737.4	7.80	1.01	84.55	20.02	4.2
4,586.95	86.99	36,272.0	10,694.2	10.34	0.11	119.51	15.26	5.5
2,729.24	165.85	23,093.3	8,618.6	5.15	0.04	80.49	15.87	3.6
3,289.40	116.26	26,878.6	7,747.9	6.64	0.68	136.58	7.81	3.4
2,800.78	42.28	39,572.0	4,565.8	5.45	0.66	78.86	16.00	4.2
3,264.20	52.84	51,866.1	6,022.7	6.31	−0.10	136.58	17.44	3.6
3,453.62	165.04	58,749.8	3,721.1	6.35	−0.03	138.21	17.98	3.1
1,741.45	10.57	23,990.8	861.0	7.37	−1.63	75.61	20.99	1.6
2,035.75	13.82	25,694.9	3,571.5	8.39	−0.43	102.44	21.66	3.4
1,578.00	8.13	23,736.3	2,845.5	5.15	0.04	76.42	21.46	2.7
4,167.44	58.44	34,314.3	5,060.1	12.88	0.22	136.58	24.78	2.8
2,799.97	21.14	22,809.5	3,552.0	9.14	−0.74	88.62	24.96	3.9

15.4   First Steps in the Analysis of a Larger Problem: The Cravens Data

In introducing multiple regression analysis, we utilized the Butler Trucking problem extensively. Although the small size of this problem was an advantage when exploring introductory concepts, the limited size of the problem makes it difficult to illustrate some of the variable-selection issues involved in model building. To provide an illustration of the variable-selection procedures discussed in the next section, we now introduce a data set consisting of 25 observations on 8 independent variables. Permission to use these data was provided by Dr. David W. Cravens of the Department of Marketing at Texas Christian University. Consequently, we refer to the data set as the Cravens data.*

The Cravens data involve a company that sells products in a number of sales territories, each of which is assigned to a single sales representative. It was desired to conduct a regression analysis to determine if sales in each territory could be explained using a variety of predictor (independent) variables. A random sample of 25 sales territories resulted in the data shown in Table 15.4; the variable definitions are shown in Table 15.5.

*For details see David W. Cravens, Robert B. Woodruff, and Joe C. Stamper, "An Analytical Approach for Evaluating Sales Territory Performance," *Journal of Marketing* 36 (January 1972): 31-37.

As a preliminary step, let us consider the sample correlation coefficients between each pair of variables. Figure 15.5 shows the correlation matrix obtained using the Minitab correlation command. Note that the sample correlation coefficient between SALES and TIME is .623, between SALES and POTEN is .598, and so on.

Looking at the sample correlation coefficients between the independent variables, we see that the correlation between TIME and ACCTS is .758; thus, if ACCTS is used as an independent variable, TIME would not be able to provide much more explanatory power to the model. Recall from the discussion of multicollinearity in Section 14.5 that the rule-of-thumb test says that multicollinearity can cause problems if the absolute value of the sample correlation coefficient exceeds .7 for any two of the independent variables. If possible, then, we should avoid including both TIME and ACCTS in the same regression model. The sample correlation coefficient of .549 between CHANGE and RATING is also quite high and may warrant further consideration.

Looking at the sample correlation coefficients between SALES and each of the independent variables can provide us with a quick indication of which independent variables are, by themselves, good predictors. We see that the single best predictor of SALES is ACCTS, since it has the highest sample correlation coefficient (.754). Recall that for the case of one independent variable, the square of the sample correlation coefficient is the coefficient of determination. Thus, ACCTS can explain $(.754)^2(100)$, or 56.85%, of the variability in SALES. The next most important independent variables are TIME, POTEN, and ADV, each with a sample correlation coefficient of approximately .6.

Although there are potential multicollinearity problems, let us for the moment consider developing an estimated regression equation using all eight independent variables.

FIGURE 15.5 Sample Correlation Coefficients for the Cravens Data (as Printed by Minitab)

	SALES	TIME	POTEN	ADV	SHARE	CHANGE	ACCTS	WORK
TIME	0.623							
POTEN	0.598	0.454						
ADV	0.596	0.249	0.174					
SHARE	0.484	0.106	-0.211	0.264				
CHANGE	0.489	0.251	0.268	0.377	0.085			
ACCTS	0.754	0.758	0.479	0.200	0.403	0.327		
WORK	-0.117	-0.179	-0.259	-0.272	0.349	-0.288	-0.199	
RATING	0.402	0.101	0.359	0.411	-0.024	0.549	0.229	-0.277

TABLE 15.5

Minitab Variable Names for the Craven's Data

VARIABLE	DEFINITION
SALES	Total sales in units credited to the sales representative
TIME	Length in time employed in months
POTEN	Market potential; total industry sales in units of the sales territory*
ADV	Advertising expenditure in the sales territory
SHARE	Market share; weighted average for the past 4 years
CHANGE	Change in the market share over the previous 4 years
ACCTS	Number of accounts assigned to the sales representative*
WORK	Work load; a weighted index based on annual purchases and concentrations of accounts
RATING	Sales representative overall rating on eight performance dimensions; an aggregate rating on a 1-7 scale

*These data were coded to preserve confidentiality.

Using the Minitab computer package provided the results shown in Figure 15.6. The eight-variable multiple regression model has an adjusted coefficient of determination of 88.3%, which is very high for real data. Note, however, that the "p" column (the p-values for the t tests of individual parameters) shows that only POTEN, ADV, and SHARE are significant at the $\alpha = .05$ level, given the effect of all the others. Thus we might be inclined to investigate the results that would be obtained if we used just these three variables. Figure 15.7 shows the Minitab results obtained for the model that uses just these three variables. We see that the model using just three independent variables has an adjusted coefficient of determination of 82.7%, which, although not quite as good as for the eight-independent-variable model, is still very high.

How can we find a model that will do the best job given the data available? One approach sometimes advocated for determining the best model is to compute all possible regressions. That is, one could develop 8 one-variable models (each of which corresponds to one of the independent variables), 28 two-variable models (the number of combinations of 8 variables taken 2 at a time), and so on. In all, for the Cravens data, there are 255

FIGURE 15.6 Minitab Output for the Model involving all Eight Independent Variables

```
The regression equation is
SALES = - 1508 + 2.01 TIME + 0.0372 POTEN + 0.151 ADV + 199 SHARE + 291 CHANGE
        + 5.55 ACCTS + 19.8 WORK + 8 RATING
```

Predictor	Coef	Stdev	t-ratio	p
Constant	1507.8	778.6	-1.94	0.071
TIME	2.010	1.931	1.04	0.313
POTEN	0.037205	0.008202	4.54	0.000
ADV	0.15099	0.04711	3.21	0.006
SHARE	199.02	67.03	2.97	0.009
CHANGE	290.9	186.8	1.56	0.139
ACCTS	5.551	4.776	1.16	0.262
WORK	19.79	33.68	0.59	0.565
RATING	8.2	128.5	0.06	0.950

s = 449.0   R-sq = 92.2%   R-sq(adj) = 88.3%

Analysis of Variance

SOURCE	DF	SS	MS	F	p
Regression	8	38153568	4769196	23.65	0.000
Error	16	3225984	201624		
Total	24	41379552			

FIGURE 15.7 Minitab Output for the Model involving POTEN, ADV, and SHARE

```
The regression equation is
SALES = - 1604 + 0.0543 POTEN + 0.167 ADV + 283 SHARE
```

Predictor	Coef	Stdev	t-ratio	p
Constant	-1603.6	505.6	-3.17	0.005
POTEN	0.054286	0.007474	7.26	0.000
ADV	0.16748	0.04427	3.78	0.001
SHARE	282.75	48.76	5.80	0.000

s = 545.5   R-sq = 84.9%   R-sq(adj) = 82.7%

Analysis of Variance

SOURCE	DF	SS	MS	F	p
Regression	3	35130240	11710080	39.35	0.000
Error	21	6249310	297586		
Total	24	41379552			

different models involving one or more independent variables that would have to be fitted to the data.

With the excellent computer packages available today, it is possible to compute all possible regressions. But, doing so involves a great deal of computation and requires the model-builder to review a great deal of computer output, much of which is associated with obviously poor models. Statisticians usually prefer a more systematic approach to selecting the subset of independent variables providing the best model. In the next section, we introduce some of the more popular approaches.

15.5   Variable-Selection Procedures

In this section, we discuss three computer-based variable-selection procedures for selecting the independent variables in a regression model: stepwise regression, forward selection, and backward elimination. Given a data set involving several possible independent variables (the x_i's), these methods can be used to identify which independent variables provide the best model. All three methods are iterative; at each step a single variable is added or deleted and the new model is evaluated. The process continues until a stopping criterion indicates that the procedure cannot find a better model.

The criterion for selecting an independent variable to add or delete from the model at each step is based on the F statistic introduced in Section 15.3. Suppose, for instance, that we were considering adding x_3 to a model involving x_1 or deleting x_3 from a model involving x_1 and x_3. In Section 15.3, we showed that

$$F = \frac{\dfrac{\text{SSE}\,(x_1) - \text{SSE}\,(x_1, x_3)}{1}}{\dfrac{\text{SSE}\,(x_1, x_3)}{n - p - 1}}$$

can be used as a criterion for determining whether or not the presence of x_3 in the model causes a significant reduction in the error sum of squares. The value of this F statistic is the criterion used by all three methods to determine whether or not a variable should be added to or deleted from the regression model at each step. It is also used to indicate when the iterative procedure should stop. All three procedures stop when no more significant reduction in the error sum of squares can be obtained. As also noted in Section 15.3, when only one variable at a time is to be added or deleted, the t statistic (recall that $t^2 = F$) provides the same criterion.

With the stepwise regression procedure, a variable may be added or deleted at each step. The procedure stops when no more improvement can be obtained by adding or deleting a variable. With the forward-selection procedure, a variable is added at each step, but variables are never deleted. The procedure stops when no more improvement can be obtained by adding a variable. With the backward-elimination procedure, the procedure starts with a model involving all the possible independent variables. At each step a variable is eliminated. The procedure stops when no more improvement can be obtained by deleting a variable.

Stepwise Regression

We will illustrate the stepwise regression procedure using the Cravens data. To see how a step of the procedure is performed, suppose that after three steps the following three independent variables have been selected: ACCTS, ADV, and POTEN. At the next step, the procedure first determines if any of the variables *already in the model* should be deleted. It does so by first determining which of the three variables is the least significant addition in moving from a two- to three-independent-variable model. To determine this an F statistic is computed for each of the three variables. The F statistic for ACCTS enables us to test whether or not adding ACCTS to a model that already includes ADV and POTEN leads to a significant reduction in SSE. If not, the stepwise procedure will consider dropping ACCTS from the model. Before doing so, however, the same F statistic is computed for each of the other two variables. The one that makes the least significant addition in moving from a two- to three-independent-variable regression model becomes a candidate for deletion. If any variable is to be deleted, it will be the one chosen.

We will denote by FMIN the smallest of the F statistics for all variables in the regression model at the beginning of a new step. The variable with the smallest F statistic is the least significant addition to the model. If the value of FMIN is small enough not to be significant, the corresponding variable is deleted from the model. On the other hand, if FMIN is large enough to be significant, none of the variables are deleted from the model (none of the other variables can have smaller F statistics).

The user of a computer-based stepwise regression procedure must specify a cutoff value for the F statistic so that the method can determine when FMIN is large enough to be significant. With the Minitab package, the smallest significant F value is denoted by FREMOVE. If the user does not specify a value for FREMOVE, it is automatically set equal to 4 by Minitab. Anytime FMIN < FREMOVE, the stepwise procedure of Minitab will delete the corresponding variable from the model. If FMIN ≥ FREMOVE, no variable is deleted at that step of the procedure.

If no variable can be removed from the model, the stepwise procedure next checks to see if adding a variable can improve the model. For each variable *not in the model,* an F statistic is computed. The largest of these F statistics corresponds to the variable that will cause the largest reduction in SSE. This variable then becomes a candidate for inclusion in the model. We will denote the largest F statistic for variables not currently in the model by FMAX. Again a cutoff value for the F statistic must be used to determine if FMAX is large enough for the corresponding variable to make a significant improvement in the model.

The cutoff value for determining when to add a variable is denoted by FENTER in the Minitab computer package. The user of the package may specify a cutoff value for FENTER; if the user does not, Minitab will automatically set FENTER = 4. If FMAX > FENTER, the corresponding variable is added to the model, and the stepwise regression procedure goes on to the next step. The procedure stops when no variables can be deleted and no variables can be added.

In summary, at each step of the stepwise regression procedure, the first consideration is to see if any variable can be removed. If none of the variables can be removed, the procedure then checks to see if any variables can be added. Because of the nature of the stepwise procedure, it is possible for a variable to enter the model at one step, be deleted at a subsequent step, and then reenter the model at a later step. The procedure stops when FMIN ≥ FREMOVE (no variables can be deleted) and FMAX ≤ FENTER (no variables can be added).

```
STEPWISE REGRESSION OF SALES ON 8 PREDICTORS, WITH N = 25

    STEP        1        2        3        4
CONSTANT    709.32    50.30  -327.23  -1441.93

ACCTS         21.7     19.0     15.6      9.2
T-RATIO       5.50     6.41     5.19     3.22

ADV                   0.227    0.216    0.195
T-RATIO                4.50     4.77     4.74

POTEN                          0.0219   0.0382
T-RATIO                         2.53     4.79

SHARE                                    190
T-RATIO                                  3.82

S              881      650      583      454
R-SQ         56.85    77.51    82.77    90.04
```

FIGURE 15.8   Minitab Output using Stepwise Regression for the Cravens Data

Figure 15.8 shows the result of using the Minitab stepwise regression procedure for the Cravens data. As we noted in Section 15.3, when only one variable is being added, the t statistic provides the same criterion as the F. One can show that $F = t^2$. The entries in the "T-RATIO" row are the t statistics. The values of FREMOVE and FENTER were both automatically set equal to 4. At step 1, there are no variables to consider for deletion. The variable providing the largest value for the F statistic is ACCTS, with $F = t^2 = (5.5)^2 = 30.25$. Since $30.25 > 4$, ACCTS is added to the model. On the next three steps, ADV, POTEN, and SHARE are added to the model. After step 4, an F statistic was computed for each of the four variables in the model. The values of the F statistics were $t^2 = (3.22)^2 = 10.37$, $t^2 = 22.47$, $t^2 = 22.94$, and $t^2 = 14.59$ for ACCTS, ADV, POTEN, and SHARE, respectively. Thus, FMIN $= 10.37$, and the corresponding variable in ACCTS. Since $10.37 > 4$, no variable is dropped from the model.

An F statistic was then computed for each of the other four variables not in the model. The stepwise procedure stopped at this point; no variables could be deleted and none could be added to improve the model. The results shown in Figure 15.8 were printed at this point. The estimated regression equation identified by the Minitab stepwise regression procedure is

$$\hat{y} = -1441.93 + 9.2 \text{ ACCTS} + 0.175 \text{ ADV} + .0382 \text{ POTEN} + 190 \text{ SHARE}$$

Note also in Figure 15.8 that, with the error sum of squares being reduced at each step, $s = \sqrt{\text{MSE}}$ has been reduced from 881 with the best one-variable model to 454 after four steps. The value of r^2 has been increased from 56.85 to 90.04.

Forward Selection

Forward selection is another computer-based procedure for variable selection. It is similar to the stepwise regression procedure except that it does not permit a variable to be deleted from the model once it has been added. The forward-selection procedure starts out with no independent variables. Then it adds variables, one at a time, as long as a significant reduction in the error sum of squares (SSE) can be achieved. When no variable can be added that will cause a further significant reduction in SSE, the procedure stops and prints out the results. For the Cravens data, the stepwise regression procedure added one variable at each step and did not delete any variables. Thus, for the Cravens' data, the forward-selection procedure leads to the same model as that provided by the stepwise procedure.

Backward Elimination

The backward-elimination procedure begins with a model including all the independent variables the model builder wants considered. (Figure 15.6 shows a regression model involving all eight independent variables for the Cravens data.) It then deletes one variable at a time using the same criterion as that for removing variables using the stepwise regression procedure. The variable with the smallest F statistic is deleted, provided F is less than the preestablished cutoff criterion (FREMOVE for Minitab). The major difference between the backward-elimination procedure and the stepwise procedure is that once a variable has been removed from the model, it cannot reenter at a later step.

Making the Final Choice

The analysis performed on the Cravens data to this point is good preparation for choosing a final model. But, more analysis should be conducted before making the final choice. As we noted in Chapters 13 and 14, a careful analysis of the residuals should be made. We want the residual plot for the model chosen to resemble approximately a horizontal band. More will be said on analysis of residuals in the next section. For now let us assume that there is no difficulty with the residuals and that we want to utilize the results of the stepwise regression procedure to help choose the model to be used.

The stepwise regression procedure has shown us that if we consider adding or deleting only one variable at a time, the best model utilizes four independent variables: ACCTS, ADV, POTEN, and SHARE. However, it may be that another model is better or that a simpler model using fewer variables will be preferred by the person using it. Table 15.6 is helpful in making the final choice. It shows several possible models consisting of some or all of these four independent variables.

From Table 15.6, we see that the model involving just ACCTS and ADV is pretty good. The adjusted R^2 is 75.5%, whereas the model using all four variables only provides a 12.6 increase. The simpler two-variable model might be preferred if, for instance, it is difficult to measure market potential (POTEN). On the other hand, if the data are readily available, the model builder would clearly prefer the model involving all four variables if highly accurate predictions of sales are needed.

TABLE 15.6

Selected Models Involving ACCTS, ADV, POTEN, and SHARE

MODEL	INDEPENDENT VARIABLES	ADJUSTED R^2
1	ACCTS	55.0
2	ACCTS, ADV	75.5
3	POTEN, SHARE	72.3
4	ACCTS, ADV, POTEN	80.3
5	ADV, POTEN, SHARE	82.7
6	ACCTS, ADV, POTEN, SHARE	88.1

EXERCISES

12. Two experts provided subjective lists of school districts that they think are among the best in the country. For each school district, the following data were obtained: average class size, instructional spending per student, average teacher salary, combined SAT score, percent of students taking the SAT, and the percentage of students that attended a 4-year college (*The Wall Street Journal,* March 31, 1989).

	AVERAGE CLASS SIZE	INSTRUCTIONAL SPENDING PER STUDENT	AVERAGE TEACHER SALARY	COMBINED SAT SCORE (% TAKING TEST)	ATTEND FOUR-YEAR COLLEGE
Blue Springs, Mo.	25	$3,060	$29,359	1083 (8%)	74%
Garden City, N.Y.	18	$9,700	$51,000	997 (99%)	77%
Indianapolis, Ind.	30	$3,222	$30,482	716 (42%)	40%
Newport Beach, Calif.	26	$4,028	$37,043	977 (46%)	51%
Novi, Mich.	20	$3,067	$39,797	980 (15%)	53%
Piedmont, Calif.	28	$4,208	$37,274	1042 (91%)	75%
Pittsburgh, Pa.	21	$4,884	$37,156	983 (80%)	66%
Scarsdale, N.Y.	20	$9,853	$31,555	1110 (98%)	87%
Wayne, Pa.	22	$5,022	$40,406	1040 (95%)	85%
Weston, Mass.	21	$4,680	$39,800	1031 (99%)	89%

	AVERAGE CLASS SIZE	INSTRUCTIONAL SPENDING PER STUDENT	AVERAGE TEACHER SALARY	COMBINED SAT SCORE (% TAKING TEST)	ATTEND FOUR-YEAR COLLEGE
Farmingdale, N.Y.	22	$6,729	$45,846	947 (75%)	81%
Mamaroneck, N.Y.	20	$10,405	$49,625	1000 (90%)	69%
Mayfield, Ohio	24	$5,881	$36,228	1003 (25%)	48%
Morristown, N.J.	22	$6,300	$37,000	972 (80%)	64%
New Rochelle, N.Y.	23	$8,875	$41,650	1039 (80%)	55%
Newton Square, Pa.	17	$5,313	$38,000	963 (75%)	79%
Omaha, Neb.	23	$4,815	$32,500	1059 (31%)	81%
Shaker Heights, Ohio	23	$4,370	$38,639	940 (56%)	82%

Let the dependent variable be the percentage of students that attend a four-year college.
   a. Develop the best one-variable model.
   b. Use the stepwise procedure to develop the best model.
   c. Use the backward-elimination procedure to develop the best model.

13. Refer to exercise 12. Let the dependent variable be the combined SAT score.
   a. Develop the best one-variable model.
   b. Use the stepwise procedure to develop the best model.
   c. Use the backward elimination procedure to develop the best model.

14. The following table shows some of the data available for 14 teams in the National Football League at the end of week 15 for the 1988 season.

					INTERCEPTIONS	
TEAM	WON-LOST	TOTAL POINTS	RUSHING YARDS	PASSING YARDS	Made by Team	Made by Opponent
Atlanta	5-10	305	1907	2473	19	23
Chicago	12-3	187	2134	2718	14	24
Dallas	3-12	358	1858	3386	24	10
Detroit	4-11	292	1184	1971	15	12
Green Bay	3-12	298	1274	3046	22	20
L.A. Rams	9-6	277	1882	3604	17	22
Minnesota	10-5	206	1744	3633	16	35
N. Orleans	9-6	274	1843	2963	15	17
N.Y. Giants	10-5	277	1492	3096	14	15
Philadelphia	9-6	312	1812	3247	17	29
Phoenix	7-8	372	1909	3633	19	14
San Fran.	10-5	256	2453	3131	14	21
Tampa Bay	4-11	340	1650	3169	33	18
Washington	7-8	367	1377	3930	24	14

Let the dependent variable be the number of wins
   a. Develop the best one-variable model.
   b. Use the stepwise procedure to develop the best model.
   c. Use the backward-elimination procedure to develop the best model.

15. Refer to exercise 14. Let the dependent variable be the total points scored.
   a. Develop the best one-variable model.
   b. Use the stepwise procedure to develop the best model.
   c. Use the backward-elimination procedure to develop the best model.

16. Refer to exercise 8.
  a. Use the stepwise procedure to develop a model that can be used to predict the price-earnings ratio.
  b. Use the backwards elimination procedure to develop a model that can be used to predict the price-earnings ratio.

Residual Analysis

In Chapters 13 and 14 we showed how residual plots could be used to determine when violations of assumptions concerning the regression model had occurred. We looked for violations of assumptions concerning the error term (ϵ) and the assumed functional form of the model. Some of the actions that can be taken when such violations are detected have been discussed in this chapter. When a different functional form is needed, curvilinear and interaction terms can be included through the use of the general linear model. When additional variables need to be considered, some of the variable-selection procedures of the preceding section may be appropriate.

In this section, we discuss how residual analysis can be used to identify observations that can be classified as outliers or as being especially influential in determining the estimated regression equation. Some steps that should be taken when such observations have been found are noted. Also, in many regression studies involving economic data, a special type of correlation involving the error terms can cause problems; it is called *autocorrelation,* or *serial correlation*. We show how residual analysis using the Durbin-Watson test can be used to determine when autocorrelation is a problem.

Detecting Outliers

In regression analysis, *outliers* are data points (observations) that do not fit the trend. Figure 15.9 shows an outlier for a data set involving a single independent variable. Outliers represent observations that are suspect and warrant careful examination. They may represent erroneous data; if so, the data should be corrected. They may signal a

FIGURE 15.9 A Data Set involving an Outlier

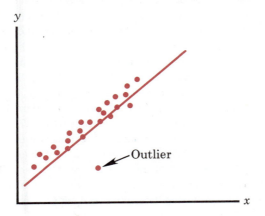

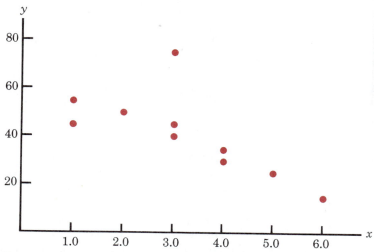

FIGURE 15.10 Scatter Diagram for the Data Set of Table 15.7

violation of model assumptions; if so, other models should be considered. And, finally, they may simply be unusual values that have occurred by chance. In this case they should be retained.

To illustrate the process of detecting outliers, consider the data set shown in Table 15.7; a scatter diagram is shown in Figure 15.10. Except for observation 4 ($x = 3$, $y = 75$), a pattern suggesting a negative linear relationship is apparent. Indeed, given the pattern of the rest of the data, we would have expected the y value for observation 4 to be much smaller and thus would identify the observation as an outlier. For the case of simple linear regression, one can usually detect outliers by simply examining the scatter diagram. In multiple regression models, the situation is a bit more difficult.

Many of the residual plots provided by computer software packages use a standardized version of the residuals. As we have seen in earlier chapters, a random variable is standardized by subtracting its mean and dividing the result by its standard deviation. With the least squares method, the mean of the residuals is zero. Thus, simply dividing each residual by its standard deviation provides the standardized residual. More advanced texts on regression analysis discuss how the standard deviation of each residual can be computed. It is interesting to note, however, that the standard deviation of the ith residual depends upon $s = \sqrt{\text{MSE}}$ and the corresponding values of the independent variables.

In the multiple regression case, we can use the standardized residuals to identify outliers. If the value of y for a particular x is unusually large or small (does not seem to follow the trend of the rest of the data) the corresponding standardized residual will be large in absolute value. Many computer packages automatically identify observations with standardized residuals that are large in absolute value. In Figure 15.11 we show the Minitab output from a regression analysis of the data in Table 15.7. The last line of the output (labeled "ROW 4" shows that the standardized residual for observation 4 is 2.67. Minitab considers a standardized residual less than -2 or greater than $+2$ to be an outlier; in such cases the observation is printed on a separate line with an R next to the standardized residual, as shown in Figure 15.11. Assuming normally distributed errors, standardized residuals should fall outside these limits only approximately 5% of the time.

In deciding how to handle an outlier, we should first check to see if it is a valid observation. Perhaps an error has been made in initially recording the data or in entering the data into the computer system. For example, suppose that in checking the data for the outlier in Table 15.7, we find that an error has been made and that the correct value for observation 4 is $x = 3$, $y = 30$. Figure 15.12 shows the Minitab output obtained after correcting the value of y_4. We see that the effect of using an incorrect value for the dependent variable had a substantial effect on the goodness of fit. With the correct data, the value of r^2 has increased from 49.7% to 83.8% and the value of b_0 has decreased from 64.958 to 59.237. The slope of the line, however, has changed only from -7.331 to -6.949.

FIGURE 15.11 Minitab Output for Regression Analysis of the Data Set with an Outlier (Table 15.7)

```
The regression equation is
Y = 65.0 - 7.33 X

Predictor       Coef      Stdev     t-ratio        p
Constant      64.958      9.258        7.02    0.000
X             -7.331      2.608       -2.81    0.023

s = 12.67      R-sq = 49.7%     R-sq(adj) = 43.4%

Analysis of Variance

SOURCE        DF          SS         MS        F        p
Regression     1       1268.2     1268.2     7.90    0.023
Error          8       1284.3      160.5
Total          9       2552.5

Unusual Observations
Obs.      X          Y       Fit Stdev.Fit  Residual   St.Resid
  4     3.00      75.00     42.97     4.04     32.03       2.67R

R denotes an obs. with a large st. resid.
```

TABLE 15.7

Data Set Illustrating The Effect Of An Outlier

x	y
1	45
1	55
2	50
3	75
3	40
3	45
4	30
4	35
5	25
6	15

```
The regression equation is
Y = 59.2 - 6.95 X

Predictor        Coef       Stdev     t-ratio        p
Constant       59.237       3.835       15.45    0.000
X              -6.949       1.080       -6.43    0.000

s = 5.248      R-sq = 83.8%      R-sq(adj) = 81.8%

Analysis of Variance

SOURCE         DF         SS          MS        F         p
Regression      1      1139.7      1139.7    41.38    0.000
Error           8       220.3        27.5
Total           9      1360.0
```

FIGURE 15.12   Minitab Output for the Revised Data Set in Table 15.7

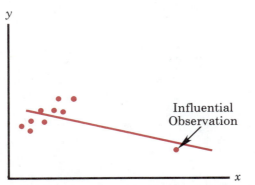

FIGURE 15.13   A Data Set with an Influential Observation

Detection of Influential Observations

In regression analysis, it sometimes happens that one or more observations have a strong influence on the results obtained. Figure 15.13 shows an example of an *influential observation* in simple linear regression. The estimated regression line has a negative slope. But, if the influential observation is dropped from the data set, the slope of the estimated regression line would change from negative to positive, and the y-intercept would be smaller. Clearly this one observation is much more influential in determining the estimated regression line than any of the others; dropping one of the other observations from the data set would have very little effect on the estimated regression equation.

Since influential observations can have such a dramatic effect on the estimated regression equation, it is important that they be examined carefully. We should first check to make sure that no error has been made in collecting or recording the data. If an error has occurred, it can be corrected and a new estimated regression equation can be developed. On the other hand, if the observation is valid, we might consider ourselves fortunate to have it. Such a point, if valid, can contribute to a better understanding of the appropriate model and can lead to a better estimated regression equation. The presence of the influential observation in Figure 15.13, if valid, would suggest trying to obtain data on intermediate values of x to understand better the relationship between x and y.

Influential observations can be identified from a scatter diagram when only one independent variable is present. An influential observation may be an outlier (an observation with a y value that deviates substantially from the trend); it may correspond to an x value far away from its mean (see, for example, Figure 15.13); or it may be caused by a combination of the two (a somewhat off-trend y value and a somewhat extreme x value). With multiple regression it is not as easy to identify influential observations. However, most computer packages offer diagnostics that help.

We have already seen how outliers are identified using Minitab. They are observations with large standardized residuals. Observations with extreme values for the independent variables are called high leverage points. The influential observation in Figure 15.13, caused by an extreme value of x, is a point with high leverage. The *leverage* of an observation is determined by how far the values of the independent variables are from their mean values. For the single-independent-variable case, the leverage of the ith observation, denoted by h_i, can be computed using (15.14).

Leverage of Observation i

$$h_i = \frac{1}{n} + \frac{(x_i - \bar{x})^2}{\Sigma(x_i - \bar{x})^2} \qquad (15.14)$$

From the formula, it is clear that the farther x_i is from its mean $\bar{x}$, the higher the leverage of observation i.

Many computer packages automatically identify observations with high leverage and large standardized residuals as part of the standard regression output. We have already seen how Minitab identifies observations with large standardized residuals. To provide an illustration of how computer packages identify points with high leverage, let us consider the data set presented in Table 15.8.

A scatter diagram for the data set in Table 15.8 is shown in Figure 15.14. From the scatter diagram, it is clear that observation 7 ($x = 70$, $y = 100$) is an observation with an extreme value of x. Thus we would expect it to be identified as a point with high leverage. For this observation, the leverage is computed using (15.14) as follows:

$$h_7 = \frac{1}{n} + \frac{(x_7 - \bar{x})^2}{\Sigma(x_i - \bar{x})^2} = \frac{1}{7} + \frac{(70 - 24.286)^2}{2621.43} = .94$$

TABLE 15.8

Data Set Containing an Observation with High Leverage

x	y
10	125
10	130
15	120
20	115
20	120
25	110
70	100

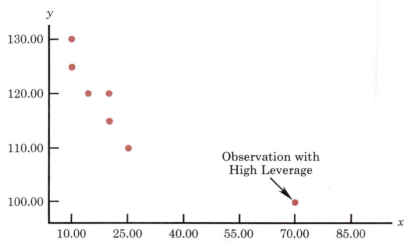

FIGURE 15.14 Scatter Diagram for the Data Set in Table 15.8

```
The regression equation is
Y = 127 -0.425 X

Predictor        Coef        Stdev      t-ratio          p
Constant       127.466        2.961        43.04      0.000
X             -0.42507      0.09537        -4.46      0.007

s = 4.883        R-sq = 79.9%      R-sq(adj) = 75.9%

Analysis of Variance

SOURCE        DF           SS          MS          F          p
Regression     1       473.65      473.65      19.87      0.007
Error          5       119.21       23.84
Total          6       592.86

Unusual Observations
Obs.      X          Y        Fit Stdev.Fit   Residual    St.Resid
  7     70.0     100.00      97.71      4.73       2.29       1.91 X

X denotes an obs. whose X value gives it large influence.
```

FIGURE 15.15 Minitab Output for the Data Set in Table 15.8

The Minitab computer package identifies observations as having high leverage if $h_i > 3(p + 1)/n$, where p is the number of independent variables and n is the number of observations. For the data set in Table 15.8,

$$\frac{3(p + 1)}{n} = \frac{3(1 + 1)}{7} = \frac{6}{7} = .86$$

Since $h_7 = .94 > .86$, Minitab will identify observation 7 as a high leverage point. Figure 15.15 shows the Minitab output for a regression analysis of this data set. Observation 7 ($x = 70$, $y = 100$) is identified as having high leverage; it is printed on a separate line at the bottom, with an "X" in the right-hand margin.

Influential observations are caused by an interaction of large residuals and high leverage. Diagnostic procedures are available that take both into account in determining when an observation is influential. One such measure is called Cook's D statistic. Cook's D statistic and other measures of influence are discussed in texts on regression analysis (see Appendix A). For our purposes, we shall simply note that potentially influential observations can be identified as the observations with large standardized residuals and/ or high leverage in the output of many computer regression packages. Such observations should be carefully reviewed to establish their validity and to evaluate their influence on the estimated regression equation.

Autocorrelation and the Durbin-Watson Test

The data used for regression studies in business and economics often have been collected over time. In such cases it is not uncommon for the value of y at time t, denoted by y_t, to be related to the value of y at previous time periods. When this occurs, we say autocorrelation (also called serial correlation) is present in the data. If the value of y in time period t is related to its value in time period $t - 1$, we say first-order autocorrelation is present. If the value of y in time period t is related to the value of y in time period $t - 2$, we say second-order autocorrelation is present, and so on.

When autocorrelation is present, one of the assumptions of the regression model is violated: the error terms are not independent. In the case of first-order autocorrelation, the error at time t, denoted by ϵ_t, will be related to the error at time period $t - 1$, denoted by ϵ_{t-1}. Two cases of first-order autocorrelation are shown in Figure 15.16. Panel A illustrates the case of positive autocorrelation, and Panel B illustrates the case of negative autocorrelation. With positive autocorrelation we expect a positive residual in one period to be followed by a positive residual in the next period, a negative residual in one period to be followed by a negative residual in the next period, and so on. With negative autocorrelation, we expect a positive residual in one period to be followed by a negative residual in the next period, then a positive residual, and so on.

When autocorrelation is present, serious errors can be made in statistical inferences about the regression model. Thus it is important to be able to detect autocorrelation and

FIGURE 15.16 Two Data Sets with First-Order Autocorrelation

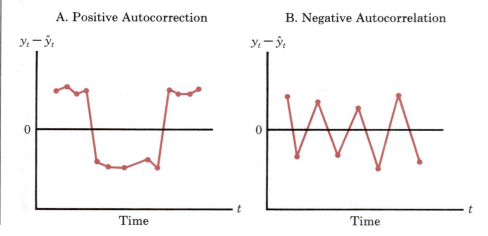

A. Positive Autocorrection

B. Negative Autocorrelation

take corrective action. We will show how the Durbin-Watson statistic can be used to detect first-order autocorrelation.

Suppose that the values of ϵ are not independent but are related in the following manner:

$$\epsilon_t = \rho\epsilon_{t-1} + z_t \tag{15.15}$$

where ρ is a parameter with an absolute value less than 1 and z_t is a normally and independently distributed random variable with mean 0 and variance σ^2. From (15.15) we see that if $\rho = 0$, then the error terms are not related, and each has a mean of 0 and a variance of σ^2. In this case, there is no autocorrelation and the regression assumptions are satisfied. If $\rho > 0$, we have positive autocorrelation; if $\rho < 0$, we have negative autocorrelation. In either of these cases the regression assumptions concerning the error term are violated.

The *Durbin-Watson test* for autocorrelation uses the residuals to determine whether or not $\rho = 0$. To simplify the notation for the Durbin-Watson statistic, we shall denote the ith residual by $e_i = y_i - \hat{y}_i$. The Durbin-Watson statistic, denoted by d, is given by the following.

Durbin-Watson Statistic

$$d = \frac{\sum_{t=2}^{n}(e_t - e_{t-1})^2}{\sum_{t=1}^{n}e_t^2} \tag{15.16}$$

If successive values of the residuals are close together (positive autocorrelation), the Durbin-Watson statistic will be small. If successive values of the residuals are far apart (negative autocorrelation), the Durbin-Watson statistic will tend to be large.

The Durbin-Watson statistic ranges in value between 0 and 4, with a value of 2 indicating no autocorrelation is present. Durbin and Watson have developed tables that can be used to determine when their test statistic indicates the presence of autocorrelation. Table 15.9 shows lower and upper bounds (d_L and d_U) for hypothesis tests using $\alpha = .05$, $\alpha = .025$, and $\alpha = .01$; in the table, n denotes the number of observations and k is the number of independent variables in the model. The null hypothesis to be tested is always taken to be one of no autocorrelation:

$$H_0: \quad \rho = 0$$

The alternative hypothesis to test for positive autocorrelation is

$$H_a: \quad \rho > 0$$

The alternative hypothesis to test for negative autocorrelation is

$$H_a: \quad \rho < 0$$

A. Test for Positive Autocorrelation

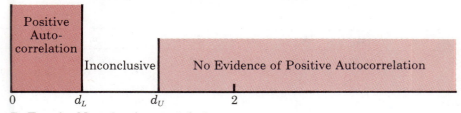

B. Test for Negative Autocorrelation

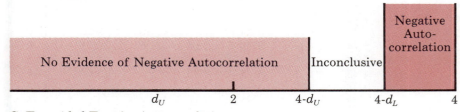

C. Two-sided Test for Autocorrelation

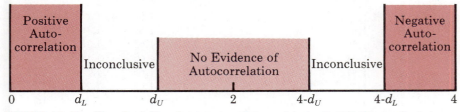

FIGURE 15.17 Hypothesis Test for Autocorrelation using the Durbin-Watson Statistic

A two-sided test is also possible. In this case the alternative hypothesis is

$$H_a: \quad \rho \neq 0$$

Figure 15.17 shows how the values of d_L and d_U in Table 15.9 are to be used to test for autocorrelation. Panel A illustrates the test for positive autocorrelation. If $d < d_L$, we conclude positive autocorrelation is present. If $d_L \leq d \leq d_U$, we say the test is inconclusive. If $d > d_U$, we conclude there is no evidence of positive autocorrelation.

Panel B illustrates the test of negative autocorrelation. If $d > 4 - d_L$, we conclude negative autocorrelation is present. If $4 - d_U \leq d \leq 4 - d_L$, we say the test is inconclusive. If $d < 4 - d_U$, we conclude there is no evidence of negative autocorrelation.

Panel C illustrates the two-sided test. If $d < d_L$ or $d > 4 - d_L$, we reject H_0 and conclude autocorrelation is present. If $d_L \leq d \leq d_U$ or $4 - d_U \leq d \leq 4 - d_L$, we say the test is inconclusive. If $d_U < d < 4 - d_U$, we conclude there is no evidence of autocorrelation.

If significant autocorrelation is identified, we should investigate whether we have omitted one or more key independent variables that have time-ordered effects on the dependent variable. If no such variables can be identified, including an independent variable that measures the time of the observation (for instance, the value of this variable could be 1 for the first observation, 2 for the second observation, and so on) will sometimes

TABLE 15.9   Critical Values for the Durbin-Watson Test for Autocorrelation

SIGNIFICANCE POINTS OF d_L AND d_U: $\alpha = .05$
NUMBER OF INDEPENDENT VARIABLES (k)

n	1 d_L	1 d_U	2 d_L	2 d_U	3 d_L	3 d_U	4 d_L	4 d_U	5 d_L	5 d_U
15	1.08	1.36	0.95	1.54	0.82	1.75	0.69	1.97	0.56	2.21
16	1.10	1.37	0.98	1.54	0.86	1.73	0.74	1.93	0.62	2.15
17	1.13	1.38	1.02	1.54	0.90	1.71	0.78	1.90	0.67	2.10
18	1.16	1.39	1.05	1.53	0.93	1.69	0.82	1.87	0.71	2.06
19	1.18	1.40	1.08	1.53	0.97	1.68	0.86	1.85	0.75	2.02
20	1.20	1.41	1.10	1.54	1.00	1.68	0.90	1.83	0.79	1.99
21	1.22	1.42	1.13	1.54	1.03	1.67	0.93	1.81	0.83	1.96
22	1.24	1.43	1.15	1.54	1.05	1.66	0.96	1.80	0.86	1.94
23	1.26	1.44	1.17	1.54	1.08	1.66	0.99	1.79	0.90	1.92
24	1.27	1.45	1.19	1.55	1.10	1.66	1.01	1.78	0.93	1.90
25	1.29	1.45	1.21	1.55	1.12	1.66	1.04	1.77	0.95	1.89
26	1.30	1.46	1.22	1.55	1.14	1.65	1.06	1.76	0.98	1.88
27	1.32	1.47	1.24	1.56	1.16	1.65	1.08	1.76	1.01	1.86
28	1.33	1.48	1.26	1.56	1.18	1.65	1.10	1.75	1.03	1.85
29	1.34	1.48	1.27	1.56	1.20	1.65	1.12	1.74	1.05	1.84
30	1.35	1.49	1.28	1.57	1.21	1.65	1.14	1.74	1.07	1.83
31	1.36	1.50	1.30	1.57	1.23	1.65	1.16	1.74	1.09	1.83
32	1.37	1.50	1.31	1.57	1.24	1.65	1.18	1.73	1.11	1.82
33	1.38	1.51	1.32	1.58	1.26	1.65	1.19	1.73	1.13	1.81
34	1.39	1.51	1.33	1.58	1.27	1.65	1.21	1.73	1.15	1.81
35	1.40	1.52	1.34	1.58	1.28	1.65	1.22	1.73	1.16	1.80
36	1.41	1.52	1.35	1.59	1.29	1.65	1.24	1.73	1.18	1.80
37	1.42	1.53	1.36	1.59	1.31	1.66	1.25	1.72	1.19	1.80
38	1.43	1.54	1.37	1.59	1.32	1.66	1.26	1.72	1.21	1.79
39	1.43	1.54	1.38	1.60	1.33	1.66	1.27	1.72	1.22	1.79
40	1.44	1.54	1.39	1.60	1.34	1.66	1.29	1.72	1.23	1.79
45	1.48	1.57	1.43	1.62	1.38	1.67	1.34	1.72	1.29	1.78
50	1.50	1.59	1.46	1.63	1.42	1.67	1.38	1.72	1.34	1.77
55	1.53	1.60	1.49	1.64	1.45	1.68	1.41	1.72	1.38	1.77
60	1.55	1.62	1.51	1.65	1.48	1.69	1.44	1.73	1.41	1.77
65	1.57	1.63	1.54	1.66	1.50	1.70	1.47	1.73	1.44	1.77
70	1.58	1.64	1.55	1.67	1.52	1.70	1.49	1.74	1.46	1.77
75	1.60	1.65	1.57	1.68	1.54	1.71	1.51	1.74	1.49	1.77
80	1.61	1.66	1.59	1.69	1.56	1.72	1.53	1.74	1.51	1.77
85	1.62	1.67	1.60	1.70	1.57	1.72	1.55	1.75	1.52	1.77
90	1.63	1.68	1.61	1.70	1.59	1.73	1.57	1.75	1.54	1.78
95	1.64	1.69	1.62	1.71	1.60	1.73	1.58	1.75	1.56	1.78
100	1.65	1.69	1.63	1.72	1.61	1.74	1.59	1.76	1.57	1.78

Source: J. Durbin and G.S. Watson, "Testing for serial correlation in least square regression II," *Biometrika,* 38, 1951, 159-178.

Entries in the table give the critical values for a one-tailed Durbin-Watson test for autocorrelation. For a two-tailed test, the level of significance is doubled.

TABLE 15.9 (Continued)

SIGNIFICANCE POINTS OF d_L AND d_U: $\alpha = .025$
NUMBER OF INDEPENDENT VARIABLES (k)

n	1		2		3		4		5	
	d_L	d_U	d_L	d_U	d_L	d_U	d_L	d_U	d_L	d_U
15	0.95	1.23	0.83	1.40	0.71	1.61	0.59	1.84	0.48	2.09
16	0.98	1.24	0.86	1.40	0.75	1.59	0.64	1.80	0.53	2.03
17	1.01	1.25	0.90	1.40	0.79	1.58	0.68	1.77	0.57	1.98
18	1.03	1.26	0.93	1.40	0.82	1.56	0.72	1.74	0.62	1.93
19	1.06	1.28	0.96	1.41	0.86	1.55	0.76	1.72	0.66	1.90
20	1.08	1.28	0.99	1.41	0.89	1.55	0.79	1.70	0.70	1.87
21	1.10	1.30	1.01	1.41	0.92	1.54	0.83	1.69	0.73	1.84
22	1.12	1.31	1.04	1.42	0.95	1.54	0.86	1.68	0.77	1.82
23	1.14	1.32	1.06	1.42	0.97	1.54	0.89	1.67	0.80	1.80
24	1.16	1.33	1.08	1.43	1.00	1.54	0.91	1.66	0.83	1.79
25	1.18	1.34	1.10	1.43	1.02	1.54	0.94	1.65	0.86	1.77
26	1.19	1.35	1.12	1.44	1.04	1.54	0.96	1.65	0.88	1.76
27	1.21	1.36	1.13	1.44	1.06	1.54	0.99	1.64	0.91	1.75
28	1.22	1.37	1.15	1.45	1.08	1.54	1.01	1.64	0.93	1.74
29	1.24	1.38	1.17	1.45	1.10	1.54	1.03	1.63	0.96	1.73
30	1.25	1.38	1.18	1.46	1.12	1.54	1.05	1.63	0.98	1.73
31	1.26	1.39	1.20	1.47	1.13	1.55	1.07	1.63	1.00	1.72
32	1.27	1.40	1.21	1.47	1.15	1.55	1.08	1.63	1.02	1.71
33	1.28	1.41	1.22	1.48	1.16	1.55	1.10	1.63	1.04	1.71
34	1.29	1.41	1.24	1.48	1.17	1.55	1.12	1.63	1.06	1.70
35	1.30	1.42	1.25	1.48	1.19	1.55	1.13	1.63	1.07	1.70
36	1.31	1.43	1.26	1.49	1.20	1.56	1.15	1.63	1.09	1.70
37	1.32	1.43	1.27	1.49	1.21	1.56	1.16	1.62	1.10	1.70
38	1.33	1.44	1.28	1.50	1.23	1.56	1.17	1.62	1.12	1.70
39	1.34	1.44	1.29	1.50	1.24	1.56	1.19	1.63	1.13	1.69
40	1.35	1.45	1.30	1.51	1.25	1.57	1.20	1.63	1.15	1.69
45	1.39	1.48	1.34	1.53	1.30	1.58	1.25	1.63	1.21	1.69
50	1.42	1.50	1.38	1.54	1.34	1.59	1.30	1.64	1.26	1.69
55	1.45	1.52	1.41	1.56	1.37	1.60	1.33	1.64	1.30	1.69
60	1.47	1.54	1.44	1.57	1.40	1.61	1.37	1.65	1.33	1.69
65	1.49	1.55	1.46	1.59	1.43	1.62	1.40	1.66	1.36	1.69
70	1.51	1.57	1.48	1.60	1.45	1.63	1.42	1.66	1.39	1.70
75	1.53	1.58	1.50	1.61	1.47	1.64	1.45	1.67	1.42	1.70
80	1.54	1.59	1.52	1.62	1.49	1.65	1.47	1.67	1.44	1.70
85	1.56	1.60	1.53	1.63	1.51	1.65	1.49	1.68	1.46	1.71
90	1.57	1.61	1.55	1.64	1.53	1.66	1.50	1.69	1.48	1.71
95	1.58	1.62	1.56	1.65	1.54	1.67	1.52	1.69	1.50	1.71
100	1.59	1.63	1.57	1.65	1.55	1.67	1.53	1.70	1.51	1.72

Source: J. Durbin and G.S. Watson, "Testing for serial correlation in least squares regression II," *Biometrika,* 38, 1951, 159-178.

TABLE 15.9 (Continued)

	SIGNIFICANCE POINTS OF d_L AND d_U: $\alpha = .01$									
	NUMBER OF INDEPENDENT VARIABLES (k)									
	1		**2**		**3**		**4**		**5**	
n	d_L	d_U	d_L	d_U	d_L	d_U	d_L	d_U	d_L	d_U
15	0.81	1.07	0.70	1.25	0.59	1.46	0.49	1.70	0.39	1.96
16	0.84	1.09	0.74	1.25	0.63	1.44	0.53	1.66	0.44	1.90
17	0.87	1.10	0.77	1.25	0.67	1.43	0.57	1.63	0.48	1.85
18	0.90	1.12	0.80	1.26	0.71	1.42	0.61	1.60	0.52	1.80
19	0.93	1.13	0.83	1.26	0.74	1.41	0.65	1.58	0.56	1.77
20	0.95	1.15	0.86	1.27	0.77	1.41	0.68	1.57	1.60	1.74
21	0.97	1.16	0.89	1.27	0.80	1.41	0.72	1.55	0.63	1.71
22	1.00	1.17	0.91	1.28	0.83	1.40	0.75	1.54	0.66	1.69
23	1.02	1.19	0.94	1.29	0.86	1.40	0.77	1.53	0.70	1.67
24	1.04	1.20	0.96	1.30	0.88	1.41	0.80	1.53	0.72	1.66
25	1.05	1.21	0.98	1.30	0.90	1.41	0.83	1.52	0.75	1.65
26	1.07	1.22	1.00	1.31	0.93	1.41	0.85	1.52	0.78	1.64
27	1.09	1.23	1.02	1.32	0.95	1.41	0.88	1.51	0.81	1.63
28	1.10	1.24	1.04	1.32	0.97	1.41	0.90	1.51	0.83	1.62
29	1.12	1.25	1.05	1.33	0.99	1.42	0.92	1.51	0.85	1.61
30	1.13	1.26	1.07	1.34	1.01	1.42	0.94	1.51	0.88	1.61
31	1.15	1.27	1.08	1.34	1.02	1.42	0.96	1.51	0.90	1.60
32	1.16	1.28	1.10	1.35	1.04	1.43	0.98	1.51	0.92	1.60
33	1.17	1.29	1.11	1.36	1.05	1.43	1.00	1.51	0.94	1.59
34	1.18	1.30	1.13	1.36	1.07	1.43	1.01	1.51	0.95	1.59
35	1.19	1.31	1.14	1.37	1.08	1.44	1.03	1.51	0.97	1.59
36	1.21	1.32	1.15	1.38	1.10	1.44	1.04	1.51	0.99	1.59
37	1.22	1.32	1.16	1.38	1.11	1.45	1.06	1.51	1.00	1.59
38	1.23	1.33	1.18	1.39	1.12	1.45	1.07	1.52	1.02	1.58
39	1.24	1.34	1.19	1.39	1.14	1.45	1.09	1.52	1.03	1.58
40	1.25	1.34	1.20	1.40	1.15	1.46	1.10	1.52	1.05	1.58
45	1.29	1.38	1.24	1.42	1.20	1.48	1.16	1.53	1.11	1.58
50	1.32	1.40	1.28	1.45	1.24	1.49	1.20	1.54	1.16	1.59
55	1.36	1.43	1.32	1.47	1.28	1.51	1.25	1.55	1.21	1.59
60	1.38	1.45	1.35	1.48	1.32	1.52	1.28	1.56	1.25	1.60
65	1.41	1.47	1.38	1.50	1.35	1.53	1.31	1.57	1.28	1.61
70	1.43	1.49	1.40	1.52	1.37	1.55	1.34	1.58	1.31	1.61
75	1.45	1.50	1.42	1.53	1.39	1.56	1.37	1.59	1.34	1.62
80	1.47	1.52	1.44	1.54	1.42	1.57	1.39	1.60	1.36	1.62
85	1.48	1.53	1.46	1.55	1.43	1.58	1.41	1.60	1.39	1.63
90	1.50	1.54	1.47	1.56	1.45	1.59	1.43	1.61	1.41	1.64
95	1.51	1.55	1.49	1.57	1.47	1.60	1.45	1.62	1.42	1.64
100	1.52	1.56	1.50	1.58	1.48	1.60	1.46	1.63	1.44	1.65

Source: J. Durbin and G.S. Watson, "Testing for serial correlation in least squares regression II," *Biometrika,* 38, 1951, 159-178.

eliminate or reduce the autocorrelation. When these attempts to reduce or remove autocorrelation do not work, transformations on the dependent or independent variables can prove helpful; a discussion of such transformations can be found in more advanced texts on regression analysis.

In closing this section we note that the Durbin-Watson tables list the smallest sample size as 15. The reason for this is that the test is generally inconclusive for small sample sizes. A rule of thumb suggested by statisticians is that the sample size should be at least 50 in order for the test to produce worthwhile results.

Notes and Comments

Once an observation has been identified as potentially influential because of a large residual or high leverage, its impact on the estimated regression equation should be evaluated. More advanced texts discuss diagnostics for doing so. However, if one is not familiar with the more advanced material, a simple procedure is to run the regression analysis with and without the observation. Although more time consuming, this approach will reveal the influence of the observation on the results.

EXERCISES

17. Refer to the NFL data set presented in exercise 14. Let y, the dependent variable, be the number of wins.
  a. Develop a model that can be used to predict y given the number of interceptions made by the team.
  b. Develop a residual plot for the model developed in (a). Does the pattern of the residual plot appear acceptable? Explain.
  c. Are there any outliers?
  d. Are there any influential observations? If so, what effect do they have on the model?

18. Refer to the NFL data set presented in exercise 14. Let y, the dependent variable, be the number of wins.
  a. Develop a model that can be used to predict y given the number of interceptions made by the opponents.
  b. Develop a residual plot for the model developed in (a). Does the pattern of the residual plot appear acceptable? Explain.
  c. Are there any outliers?
  d. Are there any influential observations? If so, what effect do they have on the model?

19. Consider the data set presented in exercise 12. Let y denote the percentage of students that attend a four-year college and x denote the combined SAT score.
  a. Develop a scatter diagram showing y given x.
  b. Develop a model that can be used to predict y given x.
  c. Refer to the scatter diagram developed in (a). Are there any influential observations in the data set? If so, determine their effect on the model developed in (b).

20. Refer to the Cravens data set presented in Table 15.4.
  a. In Section 15.5 we indicated that the model developed using ACCTS and ADV did a good job in predicting SALES. For this model are there any influential observations? If so, what effect do they have on the fitted model?

b. Consider a model using only ACCTS. For this model are there any influential observations? If so, what effect do they have on the fitted model?

c. Consider a model using only ADV. For this model are there any influential observations? If so, what effect do they have on the fitted model?

21. Consider the data set presented in exercise 8.

a. Develop the estimated regression equation which can be used to predict the price-earnings ratio given the profit margin.

b. Plot the residuals obtained from the model developed in (a) as a function of the order in which the data are presented. Does there appear to be any autocorrelation present in the data? Explain.

c. At the 5% level of significance, test for any positive autocorrelation in the data.

22. Refer to the Cravens data set presented in Table 15.4. In Section 15.5 we showed that the model involving ACCTS, ADV, POTEN, and SHARE had an adjusted R^2 of 88.1%. At the 5% level of significance, use the Durbin-Watson test to determine if positive autocorrelation is present.

15.7 Multiple Regression Approach to Analysis of Variance

In Section 15.2 we discussed the use of dummy variables in multiple regression analysis. In this section we show how the use of dummy variables in a multiple regression equation can provide another approach to solving analysis of variance problems. We will demonstrate the multiple regression approach to analysis of variance by applying it to the reading-competency experiment introduced in Chapter 12.

EXAMPLE 15.5

Recall that the reading competency experiment (Example 12.1) involved a school district that had developed an 80-question reading-competency examination. For students who answered fewer than 55 questions correctly on this examination, three remedial programs were proposed:

Program 1: A review class that meets 1 hour each week for 10 weeks

Program 2: A personalized system of instruction using microcomputers

Program 3: A formal 10-week class that meets for 1 hour each day

The objective of the experiment was to determine if the three programs were different in terms of their effect on retest scores. Table 15.10 shows the data collected for the 15 students who were retested.

We begin the regression approach to this problem by defining two dummy variables that will be used to indicate the program associated with each sample observation. Since

TABLE 15.10 **Reading Competency Examination Scores for the 15 Students Who Were Retested**

		OBSERVATION				
	Program 1	48	54	57	54	62
Treatment	Program 2	73	63	66	64	74
	Program 3	51	63	61	54	56

there are three programs in the reading-competency experiment, we need two dummy variables. In general, if there are k populations we need to define $k - 1$ dummy variables. For the reading-competency experiment we define x_1 and x_2 as shown in Table 15.11.

We can use the dummy variables x_1 and x_2 to relate the retest score of each student (y) to the type of remedial program:

$$E(y) = \text{expected value of } y = \text{mean exam score}$$
$$= \beta_0 + \beta_1 x_1 + \beta_2 x_2$$

Thus if we are interested in the expected value of y for a student who completed Program 1, our procedure for assigning numerical values to the dummy variables x_1 and x_2 would result in setting $x_1 = x_2 = 0$. The multiple regression equation then reduces to

$$E(y) = \beta_0 + \beta_1(0) + \beta_2(0) = \beta_0$$

Thus we can interpret β_0 as the mean exam score for students who complete Program 1.

Next let us consider the forms of the multiple regression equation for each of the other programs. For program 2, $x_1 = 1$ and $x_2 = 0$ and

$$E(y) = \beta_0 + \beta_1(1) + \beta_2(0) = \beta_0 + \beta_1$$

For Program 3, $x_1 = 0$ and $x_2 = 1$ and

$$E(y) = \beta_0 + \beta_1(0) + \beta_2(1) = \beta_0 + \beta_2$$

We see that $\beta_0 + \beta_1$ represents the mean exam score for students who complete Program 2, and $\beta_0 + \beta_2$ represents the mean score for students who complete Program 3.

We now want to estimate the coefficients β_0, β_1, and β_2 and hence develop estimates of mean exam score for each program. The sample consisting of 15 observations of x_1, x_2, and y was entered into Minitab. The actual input data and the output from Minitab are shown in Table 15.12 and Figure 15.18, respectively.

Refer to Figure 15.18. We see that the estimates of β_0, β_1, and β_2 are $b_0 = 55$, $b_1 = 13$, and $b_2 = 2$. Thus the best estimate of the mean exam score for each type of program is as follows:

TYPE OF PROGRAM	ESTIMATE OF $E(y)$
Program 1	$b_0 = 55$
Program 2	$b_0 + b_1 = 55 + 13 = 68$
Program 3	$b_0 + b_2 = 55 + 2 = 57$

TABLE 15.11

Dummy Variables for Example 15.5

x_1	x_2	These Values Are Used Whenever
0	0	Observation is associated with Program 1
1	0	Observation is associated with Program 2
0	1	Observation is associated with Program 3

Note that the best estimate of the expected value of the retest score for each program obtained from the regression analysis is the same as the sample means found earlier when applying the ANOVA procedure. That is, $\bar{x}_1 = 55$, $\bar{x}_2 = 68$, and $\bar{x}_3 = 57$.

Now let us see how we can use the output from the multiple regression package in order to test if the means for the three programs are equal. First, we observe that if there is no difference in the means, then

$$E(y) \text{ for Program 2} - E(y) \text{ for Program 1} = 0$$

$$E(y) \text{ for Program 3} - E(y) \text{ for Program 1} = 0$$

Since β_0 equals $E(y)$ for Program 1 and $\beta_0 + \beta_1$ equals $E(y)$ for Program 2, the first difference above is equal to $(\beta_0 + \beta_1) - \beta_0 = \beta_1$. Moreover, since $\beta_0 + \beta_2$ equals $E(y)$ for Program 3, the second difference above is equal to $(\beta_0 + \beta_2) - \beta_0 = \beta_2$.

FIGURE 15.18 Multiple Regression Output for Example 15.5

```
The regression equation is
Y = 55.0 + 13.0 X1 + 2.00 X2

Predictor      Coef        Stdev      t-ratio         p
Constant     55.000        2.266       24.28     0.000
X1           13.000        3.204        4.06     0.002
X2            2.000        3.204        0.62     0.544

s = 5.066      R-sq = 61.4%    R-sq(adj) = 55.0%

Analysis of Variance

SOURCE         DF          SS          MS         F         p
Regression      2      490.00      245.00      9.55     0.003
Error          12      308.00       25.67
Total          14      798.00
```

TABLE 15.12 **Input Data for Example 12**

OBSERVATIONS CORRESPOND TO	x_1	x_2	y
	0	0	48
	0	0	54
Program 1	0	0	57
	0	0	54
	0	0	62
	1	0	73
	1	0	63
Program 2	1	0	66
	1	0	64
	1	0	74
	0	1	51
	0	1	63
Program 3	0	1	61
	0	1	54
	0	1	56

Hence we would conclude that there is no difference in the three means if $\beta_1 = 0$ and $\beta_2 = 0$. Thus the null hypothesis for a test for difference of means can be stated as

$$H_0: \quad \beta_1 = \beta_2 = 0$$

Recall that in order to test this type of null hypothesis about the significance of the regression relationship, we must compare the value of MSR/MSE to the critical value from an F distribution with numerator and denominator degrees of freedom equal to the degrees of freedom for the regression sum of squares and the error sum of squares, respectively. In the current problem the regression sum of squares has 2 degrees of freedom and the error sum of squares has 12 degrees of freedom. Thus the values for MSR and MSE are

$$\text{MSR} = \frac{\text{SSR}}{2} = \frac{490}{2} = 245$$

$$\text{MSE} = \frac{\text{SSE}}{12} = \frac{308}{12} = 25.67$$

Hence the computed F value is

$$F = \frac{\text{MSR}}{\text{MSE}} = \frac{245}{25.67} = 9.55$$

At the $\alpha = .05$ level of significance, the critical value of F with 2 numerator degrees of freedom and 12 denominator degrees of freedom is 3.89. Since the observed value of F is greater than the critical value of 3.89, we reject the null hypothesis $H_0: \beta_1 = \beta_2 = 0$. Hence we conclude that the means for the three remedial programs are not all equal.

■

EXERCISES

23. The Jacobs Chemical Company wants to estimate the mean time (minutes) required to mix a batch of material on machines produced by three different manufacturers. In order to limit the cost of testing, four batches of material were mixed on machines produced by each of the three manufacturers. The times needed to mix the material were recorded and are as follows:

MANUFACTURER 1	MANUFACTURER 2	MANUFACTURER 3
20	28	20
26	26	19
24	31	23
22	27	22

a. Write a multiple regression equation that can be used to analyze the data.
b. What are the best estimates of the coefficients in your regression equation?

c. In terms of the regression equation coefficients, what hypotheses do we have to test to see if the mean time to mix a batch of material is the same for each manufacturer?

d. For the $\alpha = .05$ level of significance what conclusion should be drawn?

24. Four different paints are advertised as having the same drying time. In order to check the manufacturers' claim, five paint samples were tested for each brand of paint. The time in minutes until the paint was dry enough for a second coat to be applied was recorded for each sample. The following data were obtained:

PAINT 1	PAINT 2	PAINT 3	PAINT 4
128	144	133	150
137	133	143	142
135	142	137	135
124	146	136	140
141	130	131	153

a. For an $\alpha = .05$ level of significance, test for a difference in mean drying times of the paints.

b. What is your estimate of mean drying time for paint 2? How is it obtained from the computer output?

SUMMARY

In this chapter we discussed several concepts used by statisticians in identifying the best estimated regression equation. First, we introduced the concept of a general linear model in order to show how the methods discussed in Chapters 13 and 14 could be extended to handle curvilinear relationships and interaction effects. Then, we discussed how dummy variables could be used in model building to account for the effect of qualitative variables.

In applications of regression analysis to real problems, there are usually a large number of potential independent variables to consider. We presented a general approach, based on an F statistic for adding or deleting variables from a regression model. We then introduced a larger problem involving 25 observations and 8 independent variables. We saw that one issue encountered when solving larger problems is finding the best subset of the possible independent variables. To help in this regard, we discussed several variable-selection procedures, including stepwise regression, forward selection, and backward elimination.

In Section 15.6, we extended the applications of residual analysis to identify potentially troublesome observations (outliers and influential observations) and to show the Durbin-Watson test for autocorrelation. The chapter concluded with a discussion of how multiple regression models could be developed in order to provide another approach for solving analysis of variance problems.

GLOSSARY

General linear model A model of the form $y = \beta_0 + \beta_1 z_1 + \beta_2 z_2 + \cdots + \beta_p z_p + \epsilon$, where each of the independent variables z_j, $j = 1, 2, \ldots, p$, is a function of $x_1, x_2, \ldots, x_k$, the variables for which data has been collected.

Interaction The joint effect of two variables acting together.

Qualitative variable A variable that is not measured in terms of how much or how many, but instead is assigned values to represent categories.

Dummy variable A variable that takes on the values 0 or 1 and is used to incorporate the effects of qualitative variables in a regression model.

Variable-selection procedures Computer-based methods for selecting a subset of the potential independent variables for a regression model.

Autocorrelation Correlation in the errors that arises when the error terms at successive points in time are related. First-order autocorrelation is when ϵ_t and ϵ_{t-1} are related, second-order is when ϵ_t and ϵ_{t-2} are related, and so on.

Serial correlation Same as autocorrelation.

Outlier An observation with a residual that is far greater in magnitude than the rest of the residual values.

Influential observation An observation that has a great deal of influence in determining the estimated regression equation.

Leverage A measure designed to indicate how far an observation is from the others in terms of the values of the independent variables.

Durbin-Watson test A test to determine whether or not first-order autocorrelation is present.

KEY FORMULAS

GENERAL LINEAR MODEL

$$y = \beta_0 + \beta_1 z_1 + \beta_2 z_2 + \cdots + \beta_p z_p + \epsilon \tag{15.3}$$

GENERAL F TEST FOR ADDING OR DELETING $p - q$ VARIABLES

$$F = \frac{\dfrac{\text{SSE}(x_1, x_2, \cdots, x_q) - \text{SSE}(x_1, x_2, \cdots, x_q, x_{q+1}, \cdots, x_p)}{p - q}}{\dfrac{\text{SSE}(x_1, x_2, \cdots, x_q, x_{q+1}, \cdots, x_p)}{n - p - 1}} \tag{15.13}$$

AUTOCORRELATED ERROR TERMS

$$\epsilon_t = \rho \epsilon_{t-1} + z_t \tag{15.15}$$

DURBIN-WATSON STATISTIC

$$d = \frac{\sum\limits_{t=2}^{n}(e_t - e_{t-1})^2}{\sum\limits_{t=1}^{n}e_t^2} \tag{15.16}$$

REVIEW QUIZ

TRUE/FALSE

1. Model building is solely concerned with identifying which independent variables should be included in the model.

2. Since the "truly correct model" can never be found for real data, the model-builder's goal is to find the best model from a set of acceptable models.

3. Models in which the independent variables have exponents other than 1 are referred to as nonlinear models.

4. All nonlinear models can be transformed into equivalent linear models using the logarithmic transformation.

5. Interaction (and the method for dealing with it) does not apply when working with qualitative variables.

6. The t test for determining the significance of adding only one additional independent variable to an existing model is equivalent to the F test.

7. With the stepwise regression procedure, a variable is added at each step, but variables are never deleted.

8. At each step of the stepwise regression procedure, the first consideration is to see if any variable can be added.

9. In the stepwise procedure, FENTER cannot be set greater than FREMOVE.

10. Outliers represent erroneous data and thus should be dropped from the data set.

11. Influential observations can be identified from a scatter diagram when only one independent variable is present.

12. When autocorrelation is present, at least one of the assumptions of the regression model is violated.

13. The Durbin-Watson test is generally inconclusive for small sample sizes.

MULTIPLE CHOICE

14. The model $y = \beta_0 + \beta_1 x_1 + \beta_2 x_1^2 + \epsilon$ is called a
    a. simple nonlinear model with one predictor variable
    b. simple first-order model with one predictor variable
    c. second-order model with one predictor variable
    d. second-order model with two predictor variables

15. Interaction terms are added to a regression model in order to
    a. study the joint effects of quantitative and qualitative variables.
    b. account for curvilinear effects in the data.
    c. transform a nonlinear model into an equivalent linear model.
    d. account for the combined effect of two independent variables.

Questions 16–18 involve the following problem. A car-rental company wanted to predict the annual operating cost (y) of its cars using the number of miles a car was driven (x_1) and the size of the car (subcompact, compact, midsize, and full size). To incorporate the effect of the size of the car, the following dummy variables were defined:

x_2	x_3	x_4	CAR SIZE
0	0	0	sub-compact
1	0	0	compact
0	1	0	mid-size
0	0	1	full-size

16. The mean annual operating cost for a compact car is
   a. β_0
   b. $\beta_0 + \beta_1 x_1 + \beta_2 x_2$
   c. $\beta_0 + \beta_1 x_1 + \beta_2$
   d. $\beta_0 + \beta_1 x_1 + \beta_3$
   e. $\beta_0 + \beta_1 x_1 + \beta_4$

17. The mean annual operating cost for a subcompact car is
   a. β_0
   b. $\beta_0 + \beta_1 x_1$
   c. $\beta_0 + \beta_1 x_1 + \beta_2$
   d. $\beta_0 + \beta_1 x_1 + \beta_3$
   e. $\beta_0 + \beta_1 x_1 + \beta_4$

18. The difference in mean annual operating costs between a full-size car and a subcompact car is
   a. β_0
   b. β_4
   c. $\beta_0 + \beta_4$
   d. $\beta_0 + \beta_1 x_1 + \beta_4$

19. The F test for determining whether to add two variables to an existing model is based on a determination of
   a. the increase in SSE resulting from adding the two variables
   b. the decrease in SSR resulting from adding the two variables
   c. the change in SST resulting from adding the two variables
   d. the decrease in SSE resulting from adding the two variables
   e. none of the above

20. The forward selection procedure
   a. starts out with all the independent variables
   b. adds variables, one at a time, until all the independent variables have entered the model
   c. allows a variable that entered at a previous step to be deleted at a later step
   d. none of the above

21. In the stepwise procedure, which of the following are acceptable values for FENTER AND FREMOVE?
   a. FENTER = 2 and FREMOVE = 3
   b. FENTER = 3 AND FREMOVE = 2
   c. none of the above combinations are acceptable

22. Observations with extreme values for the independent variables are called
   a. outliers
   b. influential observations
   c. high leverage points
   d. unusual observations

23. If significant autocorrelation is identified, we should
   a. try to remove some of the independent variables that are highly correlated with one another
   b. investigate whether we have omitted one or more key independent variables that have time-ordered effects on the dependent variable

c. recommend caution in using the model, since one or more of the model assumptions have been violated

d. all of the above

25. Refer to the Cravens data set presented in Table 15.4.

  a. Develop a scatter diagram showing SALES as a function of TIME.

  b. Does a linear relationship between SALES and TIME appear to be appropriate? Explain.

  c. Develop a model that can be used to predict SALES using just TIME or some appropriate function of TIME.

26. A study reported in the *Journal of Accounting Research* (Autumn 1987) investigated the relationship between audit delay (AUDELAY), the length of time from a company's fiscal year-end to the date of the auditor's report, and variables that describe the client and the auditor. Some of the independent variables that were included in this study were

INDUS A dummy variable coded as 1 if the firm was an industrial company or 0 if the firm was a bank, savings and loan, or insurance company.

PUBLIC A dummy variable coded as 1 if the company was traded on an organized exchange or over the counter; otherwise coded 0.

ICQUAL A measure of overall quality of internal controls, as judged by the auditor, using a five-point scale ranging from "virtually none" (1) to "excellent" (5).

INTFIN A measure ranging from 1 to 4, as judged by the auditor, where 1 indicates "all work performed subsequent to year-end" and 4 indicates "most work performed prior to year-end."

Suppose that in a similar study a sample of 40 companies provided the following data:

AUDELAY	INDUS	PUBLIC	ICQUAL	INTFIN
62	0	0	3	1
45	0	1	3	3
54	0	0	2	2
71	0	1	1	2
91	0	0	1	1
62	0	0	4	4
61	0	0	3	2
69	0	1	5	2
80	0	0	1	1
52	0	0	5	3
47	0	0	3	2
65	0	1	2	3
60	0	0	1	3
81	1	0	1	2
73	1	0	2	2
89	1	0	2	1
71	1	0	5	4
76	1	0	2	2
68	1	0	1	2
68	1	0	5	2
86	1	0	2	2

AUDELAY	INDUS	PUBLIC	ICQUAL	INTFIN
76	1	1	3	1
67	1	0	2	3
57	1	0	4	2
55	1	1	3	2
54	1	0	5	2
69	1	0	3	3
82	1	0	5	1
94	1	0	1	1
74	1	1	5	2
75	1	1	4	3
69	1	0	2	2
71	1	0	4	4
79	1	0	5	2
80	1	0	1	4
91	1	0	4	1
92	1	0	1	4
46	1	1	4	3
72	1	0	5	2
85	1	0	5	1

a. Develop the estimated regression equation using all of the independent variables.
b. Did the model developed in (a) provide a good fit? Explain.
c. Develop a scatter diagram which shows AUDELAY as a function of INTFIN. What does this scatter diagram indicate about the relationship between AUDELAY and INTFIN?
d. Based upon your observations regarding the relationship between AUDELAY and INTFIN, develop an alternative model to the one developed in (a) in order to explain as much of the variability in AUDELAY as possible.

27. The following data set, reported in *Louis Rukeyser's Business Almanac* (1988 Simon and Schuster, p. 47), shows the percentage of management jobs held by women in various companies and the percentage of women in each company.

INDUSTRY/COMPANY	PERCENTAGE OF MANAGEMENT JOBS HELD BY WOMEN	PERCENTAGE OF WOMEN
Industrial		
du Pont	7	22
Exxon	8	27
General Motors	6	19
Goodyear Tire and Rubber	25	39
Technology		
AT&T	32	48
General Electric	6	26
IBM	16	28
Xerox	23	38
Consumer products		
Johnson & Johnson	18	47
PepsiCo	28	46
Phillip Morris		
(excluding General Foods)	14	31
Proctor & Gamble	17	28

INDUSTRY/COMPANY	PERCENTAGE OF MANAGEMENT JOBS HELD BY WOMEN	PERCENTAGE OF WOMEN
Retailing and trade		
Federated Department Stores	61	72
Kroger	16	47
Marriott	32	51
McDonald's	46	57
Sears, Roebuck	36	55
Media		
ABC (excluding Capital Cities)	36	43
Time	46	54
Times Mirror	27	37
Financial Services		
American Express	37	57
Bank America	64	72
Chemical Bank	34	57
Prudential Life Insurance	32	53
Wells Fargo Bank	58	71

a. Fit a simple linear regression model which can be used to predict the percentage of management jobs held by women given the percentage of women employed by the company.

b. Did the model developed in (a) provide a good fit to the data? Explain.

c. Use dummy variables to develop a model which relates the percentage of management jobs held by women to the type of industry (industrial, technology, and so on).

d. What conclusions can you reach based upon the model developed in (c)?

e. Develop a model which can be used to predict the percentage of management jobs held by women using the percentage of women employed by the company and the type of industry.

f. What final conclusions can be made regarding the percentage of management jobs held by women based upon your analyses?

28. Refer to the data set in exercise 26.
   a. Develop the best one-variable model which can be used to predict AUDELAY.
   b. Use the stepwise procedure to develop the best model.
   c. Use the backward-elimination procedure to develop the best model.

29. Refer to the data set in exercise 27. In addition to the percentage of women employed in the company, create additional independent variables by using dummy variables to account for the type of industry.
   a. Use the stepwise procedure to develop the best model.
   b. Use the backward-elimination procedure to develop the best model.

30. Refer to exercise 26.
   a. Develop a residual plot for the multiple regression model developed in (a) of exercise 26. Does the pattern of the residual plot appear acceptable? Explain.
   b. Are there any outliers?
   c. Are there any influential observations? If so, what effect do they have on the model?

31. Refer to the NFL data set presented in exercise 14. Let y, the dependent variable, be the number of points scored.
   a. Develop a model that can be used to predict y given the number of interceptions made by the team.
   b. Develop a residual plot for the model developed in (a). Does the pattern of the residual plot appear acceptable? Explain.

c. Are there any outliers?

d. Are there any influential observations? If so, what effect do they have on the model?

32. Refer to the NFL data set presented in exercise 14. Let *y*, the dependent variable, be the number of points scored.

    a. Develop a model that can be used to predict *y* given the number of interceptions made by the opponents.

    b. Develop a residual plot for the model developed in (a). Does the pattern of the residual plot appear acceptable? Explain.

    c. Are there any outliers?

    d. Are there any influential observations? If so, what effect do they have on the model?

33. Refer to the data in exercise 26. Consider a model in which only INDUS is used to predict AUDELAY. At the $\alpha = .01$ level of significance, test for any positive autocorrelation in the data.

34. Refer to the data in exercise 26.

    a. Develop an estimated regression equation which can be used to predict AUDELAY using INDUS and ICQUAL.

    b. Plot the residuals obtained from the model developed in (a) as a function of the order in which the data are presented. Does there appear to be any autocorrelation present in the data? Explain.

    c. At the 5% level of significance, test for any positive autocorrelation in the data.

35. Refer to the data in exercise 27.

    a. Plot the residuals obtained from the model developed in 27 (a) as a function of the order in which the data are presented. Does there appear to be any autocorrelation present in the data? Explain.

    b. At the 5% level of significance, test for any positive autocorrelation in the data.

36. A study was conducted to investigate the browsing activity by shoppers (*Journal of the Academy of Marketing Science,* Winter 1989). Shoppers were classified as nonbrowsers, light browsers, and heavy browsers. For each shopper in the study a measure was obtained to determine how comfortable the shopper was in the store. Higher scores indicated greater comfort. Assume that the following data is from this study. Use a .05 level of significance to test for differences between comfort levels for the three types of browsers.

NONBROWSER	LIGHT BROWSER	HEAVY BROWSER
4	5	5
5	6	7
6	5	5
3	4	7
3	7	4
4	4	6
5	6	5
4	5	7

COMPUTER EXERCISE

An article in the *Industrial and Labor Relations Review* (Vol. 41, No. 3, April 1988) reported the results of an investigation of factors related to the number of weeks a manufacturing worker was jobless since displacement (WEEKS). Some of the factors that were considered in this study were

	AGE	The age of the worker					
	EDUC	The number of years of education					
	MARRIED	A dummy variable; 1 if married, 0 otherwise					
	HEAD	A dummy variable; 1 if the head-of-household, 0 otherwise					
	TENURE	The number of years on the old job					
	MGT	A dummy variable; 1 if management occupation, 0 otherwise					
	SALES	A dummy variable; 1 if sales occupation, 0 otherwise					

Suppose that in a related study of 50 displaced workers that the following data were obtained.

WEEKS	AGE	EDUC	MARRIED	HEAD	TENURE	MGT	SALES
37	30	14	1	1	1	0	0
62	27	14	1	0	6	0	0
49	32	10	0	1	11	0	0
73	44	11	1	0	2	0	0
8	21	14	1	1	2	0	0
15	26	13	1	0	7	1	0
52	26	15	1	0	6	0	0
72	33	13	0	1	6	0	0
11	27	12	1	1	8	0	0
13	33	12	0	1	2	0	0
39	20	11	1	0	1	0	0
59	35	7	1	1	6	0	0
39	36	17	0	1	9	1	0
44	26	12	1	1	8	0	0
56	36	15	0	1	8	0	0
31	38	16	1	1	11	0	1
62	34	13	0	1	13	0	0
25	27	19	1	0	8	0	0
72	44	13	1	0	22	0	0
65	45	15	1	1	6	0	0
44	28	17	0	1	3	0	1
49	25	10	1	1	1	0	0
80	31	15	1	0	12	0	0
7	23	15	1	0	2	0	0
14	24	13	1	1	7	0	0
94	62	13	0	1	8	0	0
48	31	16	1	0	11	0	0
82	48	18	0	1	30	0	0
50	35	18	1	1	5	0	0
37	33	14	0	1	6	0	1
62	46	15	0	1	6	0	0
37	35	8	0	1	6	0	0
40	32	9	1	1	13	0	0
16	40	17	1	0	8	1	0
34	23	12	1	1	1	0	0
4	36	16	0	1	8	0	1
55	33	12	1	0	10	0	1
39	32	16	0	1	11	0	0
80	62	15	1	0	16	0	1
19	29	14	1	1	12	0	0
98	45	12	1	0	17	0	0
30	38	15	0	1	6	0	1

WEEKS	AGE	EDUC	MARRIED	HEAD	TENURE	MGT	SALES
22	40	8	1	1	16	0	1
57	42	13	1	0	2	1	0
64	45	16	1	1	22	0	0
22	39	11	1	1	4	0	0
27	27	15	1	0	10	0	1
20	42	14	1	1	6	1	0
30	31	10	1	1	8	0	0
23	33	13	1	1	8	0	0

Use the methods presented in this chapter to analyze this data set. Present your findings, conclusions, and recommendations in the form of brief report. Include in an appendix to your report any technical material, such as computer output, residual plots, etc., that you feel are appropriate.

16

Tests of Goodness of Fit and Independence

What You Will Learn in This Chapter:

- What a goodness of fit test is

- What a contingency table is

- How to use the chi-square distribution to conduct tests of goodness of fit and independence

Contents

Do Soccer Injuries Differ Between Game and Training Sessions?

A study of injuries occurring in three Swedish semiprofessional soccer teams showed that 75% of the players sustained one or more injuries each year. The study showed an average of 13 injuries per 1000 hours of game conditions and 3 injuries per 1000 hours during training. Each injury was classified as minor, moderate, or major. Minor injuries were those in which the player was absent from training and/or games for less than 1 week, moderate injuries were those in which the player was absent from 1 week to 1 month, and major injuries resulted in absences of more than 1 month. To better understand the relationship between the severity of an injury and when it occurred, training versus game conditions, the researchers collected the following data:

These data show that 40/63 or 63.5% of the injuries occurred during game conditions and 36.5% of the injuries occurred during training. Moreover, for injuries sustained during a game, 15/40 or 38% of the injuries were classified as minor, 13/40 or 33% were classified as moderate, and 12/40 or 30% were classified as major. Similarly, 26% of the injuries sustained during training were minor, 39% were moderate, and 35% were major. Based upon these data, can we conclude that the severity of an injury is independent of the playing conditions under which the injury occurred? Using a statistical test for independence, it was concluded that the severity of the injury is independent of the playing conditions.

	Minor	Moderate	Major	Totals
Game	15	13	12	40
Training	6	9	8	23
Totals	21	22	20	63

The risk of injury is high for today's soccer players.

© Adam Stoltman/DUOMO 1990

The results from this study also showed that the position of the player within the team did not influence the injury rate, and that there was no significant difference between ankle sprain rates in players using ankle tape versus players not using ankle tape. The fact that 20/63 = 32% of the players sustained a major injury indicates that soccer at this level is a relatively dangerous sport resulting in a high incidence of injury.

Based on "Does a major knee injury definitely sideline an elite soccer player?" by Björn Engström, Magnus Forssblad, Christer Johansson, and Hans Törnkvist, *The American Journal of Sports Medicine* (January, 1990)

In Chapter 11 we introduced the chi-square distribution and illustrated how it could be used in interval estimation and hypothesis testing. In this chapter we introduce two more hypothesis-testing procedures that are based on the use of the chi-square distribution. As with other hypothesis-testing procedures, these tests compare observed sample results with those that are expected when the null hypothesis is true. The acceptance or rejection of the null hypothesis is based upon how "close" the sample or observed results are to the expected results.

In the following section we introduce a goodness-of-fit test involving a multinomial population. Later we discuss a test for independence using contingency tables.

16.1   Goodness of Fit Test—A Multinomial Population

In this section we consider the situation in which each element of a population is assigned to one and only one of several classes or categories. Such a population is described by a *multinomial probability distribution*. Example 16.1 describes a multinomial distribution with three classes, or categories.

EXAMPLE 16.1

Patients that arrive for treatment at the emergency room of a large metropolitan hospital are assigned to one of the following three categories based upon the seriousness of their condition.

> Category 1: Patient condition is stable; immediate treatment by a physician is not required.

> Category 2: Patient condition is serious; immediate treatment is not required, but patient should be monitored for vital signs until a physician is available.

> Category 3: Patient condition is critical; the patient's life will be endangered without immediate treatment.

In this example the population of interest is a multinomial population since the condition of each patient is classified into one and only one of three classifications or categories: stable, serious, and critical.

In general, a multinomial population involves k categories. A goodness-of-fit test can be used to determine if a hypothesized probability distribution provides a good description of a particular population of interest. To perform such a test we must first formulate a null hypothesis concerning the particular multinomial probability distribution.

**EXAMPLE 16.1
(continued)**

Over the past year the hospital's records show that 50% of the patients that arrived for treatment were classified as stable, 30% were classified as serious, and 20% were classified as critical. The hospital's reputation for providing superior emergency room treatment, however, has resulted in an increased volume for the emergency room. The director is concerned that the percentage of patients classified as having stable, serious, or critical conditions may have also changed. Let us define the following notation:

$$p_1 = \text{percentage of patients classified as stable}$$

$$p_2 = \text{percentage of patients classified as serious}$$

$$p_3 = \text{percentage of patients classified as critical}$$

Based upon the assumption that the increase in volume for the emergency room has not altered the distribution of patients among the categories, the null and alternative hypotheses would be stated as follows:

$$H_0: p_1 = .50, \ p_2 = .30, \ \text{and} \ p_3 = .20$$

$$H_a: \text{The population proportions are not}$$

$$p_1 = .50, \ p_2 = .30, \ \text{and} \ p_3 = .20$$

Once the hypotheses have been formulated we must obtain a simple random sample of n items from the population in order to conduct the test. Using the sample of n items, we record the observed frequencies for each of the k classes or categories. Then, given the usual hypothesis-testing assumption that the null hypothesis is true, we determine the expected frequency for each category. The expected frequency for each category is found by multiplying the sample size n by the proportion assumed to be in that category under the null hypothesis.

EXAMPLE 16.1 (continued)

Suppose the hospital has selected a sample of 200 patients who have been treated since the volume increased in the emergency room. The observed frequencies for this group are summarized as follows.

STABLE CONDITION	SERIOUS CONDITION	CRITICAL CONDITION
98	48	54

The next step is to compute the expected frequencies for the 200 patients under the null hypothesis assumption that $p_1 = .50$, $p_2 = .30$, and $p_3 = .20$. Doing so provides the following expected frequencies:

STABLE CONDITION	SERIOUS CONDITION	CRITICAL CONDITION
200(.50) = 100	200(.30) = 60	200(.20) = 40

Note that the expected frequency for each category is found by multiplying the sample size of 200 by the hypothesized proportion for the category.

The goodness-of-fit test now focuses on the differences between the observed and expected frequencies. One or more large differences between observed and expected frequencies casts doubt on the assumption that the hypothesized proportions are correct. However, small differences between observed and expected frequencies do not provide sufficient evidence to reject the null hypothesis. In the case where observed and expected frequencies are equal, the sample data provide a perfect fit to the hypothesized distribution.

Let

$$O_i = \text{observed frequency for category i}$$

$$E_i = \text{expected frequency for category i based on the}$$
$$\text{assumption that the null hypothesis is true,}$$

$$k = \text{the number of categories.}$$

If the null hypothesis is true and if the sample size is large, then the quantity

$$\chi^2 = \sum_{i=1}^{k} \frac{(O_i - E_i)^2}{E_i} \tag{16.1}$$

has a chi-square distribution. From the numerator of (16.1), we see that larger differences between observed and expected frequencies result in larger values of χ^2, and vice versa. For the case where the null hypothesis involves proportions for k categories of a multinomial population, the appropriate chi-square distribution has $k - 1$ degrees of freedom. The requirement of a *large* sample size is satisfied whenever the expected frequency for each category is 5 or more.

EXAMPLE 16.1
(continued)

With expected frequencies $E_i \geq 5$ for all three categories, the large sample size requirement is satisfied, and we can proceed with the computation of the chi-square (χ^2) value in (16.1), as follows:

$$\chi^2 = \frac{(98 - 100)^2}{100} + \frac{(48 - 60)^2}{60} + \frac{(54 - 40)^2}{40}$$

$$= \frac{4}{100} + \frac{144}{60} + \frac{196}{40}$$

$$= .04 + 2.40 + 4.90 = 7.34$$

We will reject the null hypothesis only if the differences between observed and expected cell frequencies are *large*. Thus the larger the value of χ^2, the more likely it is we will reject the null hypothesis. Using $\alpha = .05$, we will place a rejection area of .05 in the upper tail of the chi-square distribution. Checking the chi-square distribution table (Table 3 of the Appendix B), we find that with $k - 1 = 3 - 1 = 2$ degrees of freedom $\chi^2_{.05} = 5.99147$. Thus, as with similar one-tailed tests, we will reject H_0 if the computed chi-square value exceeds the critical value of 5.99147.

Since $\chi^2 = 7.34$ is larger than the critical value of 5.99147, we reject the null hypothesis. In rejecting H_0 we conclude that the increase in volume for the emergency room has altered the percentages of patients whose conditions are stable, serious, or critical. While the goodness-of-fit test itself permits no further conclusions, we can informally compare the observed and expected frequencies to obtain an idea of where the significant differences are. For instance, considering the critical-condition category, we find that the observed frequency of 54 is larger than the expected frequency of 40. Since the expected frequency was based upon the historical percentage observed, the larger observed frequency suggests that associated with the increase in volume for the emergency room has been an increase in the percentage of patients whose conditions are classified as critical. Similarly, there has been a decrease in the percentage of patients whose conditions are classified as serious.

■

As illustrated in Example 16.1, the goodness-of-fit test uses the chi-square distribution to determine if a hypothesized probability distribution for a population provides a good fit. Acceptance or rejection of the hypothesized population distribution is based upon differences between observed frequencies in a sample and the expected frequencies based on the assumed population distribution. Let us outline the general steps that can be used to conduct a goodness-of-fit test for any hypothesized population distribution:

1. Formulate a null hypothesis indicating a hypothesized distribution for k classes or categories of a population.
2. Select a simple random sample of n items and record the observed frequencies for each of the k classes or categories.
3. Based upon the assumption that the null hypothesis is true, determine the expected frequencies for each category.
4. Use the observed and expected frequencies and (16.1) to compute a value of χ^2 for the test.
5. Rejection rule:

$$\text{Reject } H_0 \text{ if } \chi^2 > \chi_\alpha^2$$

where α is the level of significance for the test.

EXERCISES

1. During the first 13 weeks of the television season, the Saturday evening 8:00 P.M. to 9:00 P.M. audience proportions were recorded as: ABC, 29%; CBS, 28%; NBC, 25%; and independents 18%. A sample of 300 homes two weeks after a Saturday night schedule revision showed the following viewing audience data: ABC 95 homes, CBS 70 homes, NBC 89 homes, and independents 46 homes. Test with $\alpha = .05$ to determine if the viewing audience proportions have changed.

2. A new container design has been adopted by a manufacturer. Color preferences indicated in a sample of 150 individuals are as follows:

RED	BLUE	GREEN
40	64	46

Test using $\alpha = .10$ to see if the color preferences are the same. (*Hint:* Formulate the null hypothesis as H_0: $p_1 = p_2 = p_3 = \frac{1}{3}$.)

3. Grade distribution guidelines for a statistics course at a major university are as follows: 10% A, 30% B, 40% C, 15% D, and 5% F. A sample of 120 statistics grades at the end of a semester showed 18 A's, 30 B's, 40 C's, 22 D's, and 10 F's. Use $\alpha = .05$ and test to see if the actual grades differ significantly from the grade distribution guidelines.

4. An October 1988 poll sponsored by the *Cincinnati Post* used a random sample of registered voters throughout the state of Ohio to determine how Ohioans rate their local schools. The following rating categories and results were reported:

SCHOOL RATING CATEGORY	FREQUENCY
Excellent	155
Good	234
Fair	129
Poor	98

Assume Ohio school administrators had expected rating percentages of 20% excellent, 40% good, 25% fair, and 15% poor for the population of Ohio registered voters. Use the chi-square test and the actual survey data to determine whether or not the survey is consistent with the administrators expectations.

5. At Ontario University entering freshmen have historically selected the following colleges:

COLLEGE	PERCENTAGE
Business	15%
Education	20%
Engineering	30%
Liberal Arts	25%
Science	10%

Data obtained for the most recent class show that 73 students selected business, 105 selected education, 150 selected engineering, 124 chose liberal arts, and 47 selected science. Use $\alpha = .10$ and test whether or not the historical percentages have changed.

16.2   Test of Independence—Contingency Tables

Another important application of the chi-square distribution involves using sample data to test for the independence of two variables. To illustrate the test of independence, we consider the study conducted by Alber's Brewery.

EXAMPLE 16.2

Alber's Brewery manufactures and distributes three types of beers: a low-calorie light beer, a regular beer, and a dark beer. In an analysis of the market segments for the three beers, the firm's market research group has raised the question of whether or not preferences for the three beers differ between male and female beer drinkers. If beer preference is independent of the sex of the beer drinker, one advertising campaign will be initiated for all of Alber's beers. However, if beer preference depends upon the sex of the beer drinker, the firm will tailor its promotions toward different target markets.

A test of independence addresses the question of whether or not the beer preference (light, regular, or dark) is independent of the sex of the beer drinker (male, female). The hypotheses for this test of independence are as follows:

H_0: Beer preference is independent of the sex of the beer drinker

H_a: Beer preference is not independent of the sex of the beer drinker (i.e., males and females differ in their preferences)

Table 16.1 can be used to describe the situation being studied. By identifying the population as all male and female beer drinkers, a sample can be selected and each individual asked to state his or her preference for the three Alber's beers. Every individual in the sample will be placed in one of the six cells in the table. For example, an individual may be a male preferring regular beer (cell 2), a female preferring light beer (cell 4), a female preferring dark beer (cell 6), and so on. Since we have listed all possible combinations of beer preference and sex—or, in other words, listed all possible contingencies—Table 16.1 is called a *contingency table*. The test of indpendence makes use of the contingency table format and for this reason is sometimes referred to as a *contingency table test*.

Suppose that a simple random sample of 150 beer drinkers has been selected. After taste-testing the three beers, the individuals in the sample are asked to state their preference, or first choice. The contingency table in Table 16.2 summarizes the responses to the study. As we see in the contingency table, the data for the test of independence are collected in terms of counts, or frequencies, for each cell or category. Thus of the 150 individuals in the sample, 20 were men who favored light beer, 40 were men who favored regular beer, 20 were men who favored dark beer, and so on.

The data in Table 16.2 contain the sample, or observed, frequencies for each of six classes or categories. If we can determine the expected frequenices under the assumption of independence between beer preference and sex of the beer drinker, we can use the chi-square distribution, just as we did in the previous section, to determine whether or not there is a significant difference between observed and expected frequencies.

Expected frequencies for the cells of the contingency table are based on the following rationale: First, we assume that the null hypothesis of independence between beer pref-

TABLE 16.1

Contingency Table—Beer Preference and Sex of Beer Drinkers

SEX	BEER PREFERENCE		
	Light	Regular	Dark
Male	(Cell 1)	(Cell 2)	(Cell 3)
Female	(Cell 4)	(Cell 5)	(Cell 6)

TABLE 16.2

Sample Results of Beer Preferences for Male and Female Beer Drinkers (Observed Frequencies)

SEX	BEER PREFERENCE			Totals
	Light	Regular	Dark	
Male	20	40	20	80
Female	30	30	10	70
Totals	50	70	30	150

erence and sex of the beer drinker is true. Then we note that the sample of 150 beer drinkers showed a total of 50 preferring light beer, 70 preferring regular beer, and 30 preferring dark beer. In terms of fractions, we conclude that $50/150 = 1/3$ of the beer drinkers prefer light beer, $70/150 = 7/15$ prefer regular beer, and $30/150 = 1/5$ prefer dark beer. If the *independence* assumption is valid, we argue that these same fractions must be applicable to both male and female beer drinkers. Thus under the assumption of independence, we would expect the 80 male beer drinkers to show that $(1/3)80 = 26.67$ prefer light beer, $(7/15)80 = 37.33$ prefer regular beer, and $(1/5)80 = 16$ prefer dark beer. Application of these same fractions to the 70 female beer drinkers provides the expected frequencies as shown in Table 16.3.

Let E_{ij} stand for the expected frequency for the contingency table category in row i and column j. With this notation let us reconsider the expected frequency calculation for males (row $i = 1$) who prefer regular beer (column $j = 2$)—that is, expected frequency E_{12}. Following our previous argument for the computation of expected frequencies, we showed that

$$E_{12} = \left(\frac{7}{15}\right) 80 = 37.33$$

Writing this slightly differently, we find

$$E_{12} = \left(\frac{7}{15}\right) 80 = \left(\frac{70}{150}\right) 80 = \frac{(80)(70)}{150} = 37.33$$

Note that the 80 in the above expression is the total number of males (row 1 total), the 70 is the total number preferring regular beer (column 2 total), and the 150 is the total sample size. Thus we see that

$$E_{12} = \frac{(\text{Row 1 Total})(\text{Column 2 Total})}{\text{Sample Size}}$$

■

Generalization of this last expression shows that the following formula provides the expected frequencies for a contingency table in the test for independence.

TABLE 16.3

Expected Frequencies if Beer Preference is Independent of the Sex of the Beer Drinker

SEX	BEER PREFERENCE			Totals
	Light	Regular	Dark	
Male	26.67	37.33	16.00	80
Female	23.33	32.67	14.00	70
Totals	50	70	30	150

$$E_{ij} = \frac{(\text{Row } i \text{ Total})(\text{Column } j \text{ Total})}{\text{Sample Size}} \qquad (16.2)$$

The test procedure for comparing the observed frequencies of Table 16.2 with the expected frequencies of Table 16.3 is similar to the goodness-of-fit calculations made in the previous section. Specifically, the value of χ^2, based on the differences between the observed and expected frequencies, is computed as follows.

$$\chi^2 = \sum_i \sum_j \frac{(O_{ij} - E_{ij})^2}{E_{ij}} \qquad (16.3)$$

where

O_{ij} = observed frequency for contingency table category
in row i and column j,

E_{ij} = expected frequency for contingency table category
in row i and column j.

The double summation in (16.3) indicates that the values must be summed for each cell in the contingency table.

Before we apply (16.3), we must check to see that the expected frequencies in each cell are at least 5. This is the same check that we used in the previous section to determine whether or not the sample size was large enough for the chi-square distribution assumption to be made.

EXAMPLE 16.2 (continued)

Since all expected frequencies in Table 16.3 are at least 5, we can conclude that the sample size is adequate. The resulting value of χ^2 is found as follows:

$$\chi^2 = \frac{(20 - 26.67)^2}{26.67} + \frac{(40 - 37.33)^2}{37.33} + \cdots + \frac{(10 - 14.00)^2}{14.00}$$

$$= 1.67 + .19 + \cdots + 1.14 = 6.13$$

The number of degrees of freedom for the appropriate chi-square distribution is computed by multiplying the *number of rows minus* 1 times the *number of columns minus* 1. With two rows and three columns, we have $(2 - 1)(3 - 1) = (1)(2) = 2$ degrees of freedom for the test of independence of beer preference and the sex of the beer drinker. Using $\alpha = .05$ for the level of significance of the test, Table 3 of the Appendix shows an upper-tail value of $\chi^2_{.05} = 5.99147$. Again we are using the upper-tail value because we will reject the null hypothesis only if the differences in observed and expected frequencies provide a large value of χ^2. Since $\chi^2 = 6.13$ is greater than the critical value of $\chi^2_{.05}$

= 5.99147, we reject the null hypothesis of independence and conclude that the preference for the beers is not independent of the sex of the beer drinkers.

Although the test for independence allows only the above conclusion, we can again informally compare the observed and expected frequencies in order to obtain an idea of how the dependence between the beer preference and sex of the beer drinker comes about. (Refer to Tables 16.2 and 16.3.) We see that male beer drinkers have higher observed than expected frequencies for both regular and dark beers, while female beer drinkers have a higher observed than expected frequency for only the light beer. These observations give us an insight into the differing preferences between male and female beer drinkers. This information can be used by the company in targeting its promotions for the different beers.

■

EXERCISES

6. The number of units of three different products sold by three salespersons over a 3-month period are shown.

	PRODUCT		
SALESPERSON	A	B	C
Troutman	14	12	4
Kempton	21	16	8
McChristian	15	5	10

Use $\alpha = .05$ and test for the independence of salesperson and type of product sold.

7. The Graduate Management Admission Council (GMAC) sponsored a survey of MBA students to learn about characteristics of the population of students interested in graduate education in business administration. The following table was published in the *GMAC Occasional Papers* (March 1988). Do the data suggest males and females differ in the reason for application to MBA programs? Explain. Use $\alpha = .05$.

	PRIMARY REASON FOR APPLICATION		
	Program Quality	Convenience/Cost	Other
Male	519	599	86
Female	298	390	36

8. The *GMAC Occasional Papers* (March 1988) study referred to in Exercise 7 also provided data on the reason for application to MBA schools by full-time and part-time students. Do the data suggest full-time and part-time students differ in the reason for application to MBA programs? Explain. Use $\alpha = .05$.

	PRIMARY REASON FOR APPLICATION		
	Program Quality	Convenience/Cost	Other
Full-time	421	393	76
Part-time	400	593	46

9. Two hundred graduates were classified by their major and the industry in which they obtained employment upon graduation.

DEGREE MAJOR	INDUSTRY			
	Oil	Chemical	Electrical	Computer
Business	30	15	15	40
Engineering	30	30	20	20

Use $\alpha = .01$ and test for independence of degree major and industry type.

10. A sport preference poll shows the following data for men and women.

SEX	FAVORITE SPORT		
	Baseball	Basketball	Football
Men	19	15	24
Women	16	18	16

Use $\alpha = .05$ and test for similar sport preferences by men and women. What is your conclusion?

SUMMARY

In this chapter we introduced the goodness-of-fit test and the test of independence procedures, both of which are based on the chi-square distribution. The purpose of the goodness-of-fit test is to determine whether or not a hypothesized probability distribution provides a good description of a particular population of interest. The computations for conducting the goodness-of-fit test involve comparing observed frequencies from a sample with expected frequencies when the hypothesized probability distribution is assumed true. A chi-square distribution is used to determine if the differences in observed and expected frequencies are large enough to reject the hypothesized probability distribution. We illustrated the goodness-of-fit test for an assumed multinomial probability distribution.

A test of independence for two variables is a straightforward extension of the methodology employed in the goodness-of-fit test for a multinomial population. A contingency table is used to display the observed and expected frequencies. Then a chi-square value is computed. Large chi-square values, caused by large differences between observed and expected frequencies, lead to the rejection of the null hypothesis of independence.

GLOSSARY

Goodness-of-fit test   A statistical test conducted to determine whether to accept or reject a hypothesized probability distribution for a population.

Contingency table   A table used to summarize observed and expected frequencies for a test of independence of two variables associated with a population.

GOODNESS-OF-FIT TEST

$$\chi^2 = \sum_{i=1}^{k} \frac{(O_i - E_i)^2}{E_i} \qquad (16.1)$$

EXPECTED FREQUENCIES FOR CONTINGENCY TABLES UNDER THE ASSUMPTION OF INDEPENDENCE

$$E_{ij} = \frac{(\text{Row } i \text{ Total})(\text{Column } j \text{ Total})}{\text{Sample Size}} \qquad (16.2)$$

CONTINGENCY TABLE TEST

$$\chi^2 = \sum_{i}\sum_{j} \frac{(O_{ij} - E_{ij})^2}{E_{ij}} \qquad (16.3)$$

REVIEW QUIZ

TRUE/FALSE

1. A goodness-of-fit test can be used to determine if a particular multinomial distribution provides a good description of a population.

2. The chi-square probability distribution should not be used for a goodness-of-fit test.

3. In computing the expected frequencies for a goodness-of-fit test, it is not proper to assume the null hypothesis is true.

4. In conducting a goodness-of-fit test, the expected frequency for each category must be greater than or equal to 10.

5. In conducting a goodness-of-fit test, the observed frequency for one or more categories may be less than 5.

6. The chi-square distribution is used in conducting a contingency table test.

7. In a contingency table test of independence, the number of rows must be equal to the number of columns.

8. In conducting a contingency table test for independence, the expected frequency for each cell must be at least 5.

9. In conducting either a goodness-of-fit or contingency table test, the larger the differences between the observed and expected frequencies, the more likely it is that the null hypothesis will be rejected.

10. The appropriate number of degrees of freedom for a contingency table test is given by the product of the number of rows times the number of columns.

MULTIPLE CHOICE

11. The sampling distribution for a goodness of fit test is the
  a. chi-square distribution
  b. t distribution
  c. F distribution
  d. normal distribution

Use the following information for questions 12-15. A firm that manufactures kitchen appliances takes a random sample of 300 families to check whether or not there is any significant difference in color preferences for appliances. The results are as follows.

COLOR PREFERRED	NUMBER OF FAMILIES
White	65
Coppertone	89
Avocado	72
Harvest gold	74
	300

12. The critical value for the multinomial goodness-of-fit test with a .01 level of significance is
  a. 6.25139
  b. 7.77944
  c. 11.3449
  d. 13.2767

13. The expected frequency is
  a. 4 for each color
  b. 50 for each color
  c. 75 for each color
  d. different for each color

14. The calculated value of the test statistic for this chi-square goodness-of-fit test is
  a. 4.08
  b. 8.52
  c. 11.3
  d. 306

15. The result of the test at the .01 level of significance is which of the following?
  a. There is a significant difference in color preference.
  b. There is no significant difference in color preference.

16. The value of χ^2 that cuts off an area in the upper tail of .01 with 8 degrees of freedom is closest to
  a. 14
  b. 19
  c. 22
  d. 25

17. The value of χ^2 that cuts off an area in the upper tail of .025 with 70 degrees of freedom is closest to
  a. 20
  b. 50
  c. 75
  d. 100

SUPPLEMENTARY EXERCISES

11. A regional transit authority was concerned about the number of riders on one of their routes. In setting up this route, the assumption was that the number of riders was uniformly distributed from Monday through Friday. The following historical data were obtained:

DAY	NUMBER OF RIDERS
Monday	13
Tuesday	16
Wednesday	28
Thursday	17
Friday	16

Test using $\alpha = .05$ to determine if the transit authority's assumption appears to be justified.

12. An automobile dealer sells three models of a certain make of pickup truck. Over the most recent sales period the dealer sold 27 units of model 1, 39 units of model 2, and 30 units of model 3. Using $\alpha = .05$, test whether or not consumer preferences vary among the three models.

13. In setting sales quotas, the marketing manager makes the assumption that order potentials are the same in each of four sales territories. A sample of 200 sales shows the following number of orders from each region.

SALES TERRITORIES			
I	II	III	IV
60	45	59	36

Do these data support the manager's assumptions? Use $\alpha = .05$.

14. A sample of parts provided the following contingency table data concerning part quality and production shift.

SHIFT	NUMBER GOOD	NUMBER DEFECTIVE
First	368	32
Second	285	15
Third	176	24

Use $\alpha = .05$ and test the assumption that part quality is independent of the production shift. What is your conclusion?

15. An analysis of attendance records and performance on the final examination was made for a freshman mathematics course. The following results were obtained.

NUMBER OF CLASSES MISSED	GRADE ON FINAL			
	A	B	C	D/F
None	18	11	5	1
1-5	10	14	16	8
More than 5	3	8	20	12

a. Compute the expected frequencies. Are the assumptions required for using the chi-square test for independence satisified for this data set? Explain and, if necessary, combine classes to meet the expected frequency requirement.

b. Use $\alpha = .05$ and test for independence between number of classes missed and the grade on the final examination. What do the results of your analysis tell the professor in charge of the course?

16. The October 1988 poll of registered voters in Ohio referred to in Exercise 4 provided the following tabular results on whether voters were for or against a proposed 1% increase in the state sales tax in order to finance public schools.

	VOTER POLITICAL AFFILIATION		
	Republican	Democrat	Independent
For sales tax	135	154	50
Against sales tax	122	120	35

	VOTER AGE		
	Under 30	**30-45**	**Over 45**
For sales tax	110	126	103
Against sales tax	65	112	100

a. What is the overall percentage of voters favoring the sales tax increase for public schools?

b. Is being for or against the sales tax independent of political party affiliation? voter age? Present statistical evidence and discuss. Use $\alpha = .05$.

17. As part of the standard course evaluation, students are asked to rate the course as either poor, good, or excellent. The course evaluation form also asks students to indicate whether or not the course taken was a required part of their academic program or was taken as an elective. The dean of the college is interested in determining if the rating of the course is independent of the reason for taking the course. The following results were obtained.

REASON FOR TAKING THE COURSE	RATING		
	Poor	Good	Excellent
Required	16	38	16
Elective	4	10	16

How would you respond to the dean? Use $\alpha = .01$.

18. The director of placement for Southwestern University has been studying the records of engineering students who have interviewed with Computer Systems, Inc., one of the major employers over the past 5 years. One interest was to see whether or not receiving a job offer was dependent upon the student's grade-point average. The following results were obtained.

GRADE POINT AVERAGE	COMPANY DECISION	
	Offer Made	No Offer Made
Below 2.5	5	15
2.5-3.5	7	8
Above 3.5	18	12

At the $\alpha = .10$ level, determine whether or not the company decision to make a job offer is independent of the student's grade-point average.

19. A lending institution shows the following data regarding loan approvals by four different loan officers.

LOAN OFFICER	LOAN APPROVAL DECISION	
	Approved	Rejected
Miller	24	16
McMahon	17	13
Games	35	15
Runk	11	9

Use $\alpha = .05$ and test to determine if the loan approval decision is independent of the loan officer reviewing the loan application.

20. The United Way in Rochester, New York collected data on the characteristics of donors and their perception about the United Way campaign. A portion of the data collected is shown below. Analyze the data to determine if perception of United Way administrative expenses is independent of occupation of the donor. Use $\alpha = .05$. Discuss the results.

OCCUPATION	ESTIMATED PERCENT OF FUNDS GOING TO UNITED WAY ADMINISTRATIVE EXPENSES		
	Up to 10%	11–20%	More than 20%
Production	6	6	22
Clerical	15	27	35
Sales	6	16	16
Managers	37	13	7

The computer exercise in Chapter 2 described a study conducted by Consolidated Foods, Inc. The company had taken a sample of 100 customers in order to learn about how customers were paying for their food purchases. The data collected for each customer included how much was spent on the purchase and how the customer paid for the purchase. The alternative payment methods included cash, an approved check, or a credit card. In addition, data were also collected on the sex of the customer. The data obtained for the 100 customers are shown below.

AMOUNT SPENT ($)	SEX	METHOD OF PAYMENT	AMOUNT SPENT ($)	SEX	METHOD OF PAYMENT
84.12	Male	Check	86.34	Female	Check
34.66	Male	Credit card	20.23	Female	Credit card
37.27	Female	Credit card	108.70	Female	Check
38.82	Female	Credit card	45.36	Female	Credit card

AMOUNT SPENT ($)	SEX	METHOD OF PAYMENT	AMOUNT SPENT ($)	SEX	METHOD OF PAYMENT
46.50	Female	Credit card	83.31	Male	Check
99.67	Female	Check	64.45	Male	Credit card
70.18	Female	Check	54.33	Female	Credit card
99.21	Male	Check	16.78	Female	Cash
138.42	Female	Check	115.96	Male	Check
93.68	Female	Check	95.83	Female	Check
120.89	Female	Check	19.76	Female	Cash
10.14	Female	Cash	35.37	Male	Cash
74.51	Male	Check	111.98	Female	Check
17.91	Male	Check	103.95	Female	Check
49.59	Male	Check	90.40	Male	Credit card
4.74	Male	Cash	6.68	Male	Cash
48.14	Male	Cash	32.09	Male	Credit card
65.67	Male	Credit card	79.70	Male	Credit card
89.66	Female	Check	96.08	Male	Credit card
96.40	Female	Check	20.60	Male	Cash
54.16	Female	Credit card	78.81	Female	Check
79.55	Female	Check	123.62	Female	Check
67.95	Female	Check	125.01	Female	Check
30.69	Male	Cash	41.58	Male	Credit card
151.89	Female	Check	36.73	Male	Credit card
130.41	Female	Check	52.07	Female	Credit card
98.80	Female	Check	19.78	Male	Cash
23.59	Female	Cash	66.44	Female	Check
104.67	Female	Check	5.08	Male	Cash
90.04	Female	Check	50.15	Male	Credit card
77.62	Female	Check	114.42	Female	Check
36.01	Male	Cash	97.26	Male	Credit card
88.17	Female	Check	22.75	Male	Cash
66.76	Female	Credit card	53.63	Female	Credit card
23.50	Male	Cash	132.31	Female	Check
127.34	Female	Check	105.54	Male	Check
26.02	Male	Cash	66.09	Male	Check
79.77	Male	Check	62.24	Female	Check
29.35	Male	Check	97.93	Female	Check
71.31	Female	Credit card	10.57	Female	Cash
43.57	Female	Credit card	51.21	Male	Credit card
76.18	Female	Credit card	90.17	Female	Check
59.38	Male	Credit card	24.08	Male	Credit card
72.99	Male	Credit card	42.72	Male	Cash
19.24	Male	Cash	97.72	Female	Check
80.20	Female	Check	112.67	Female	Check
55.79	Female	Cash	14.30	Female	Cash
134.27	Female	Check	28.76	Male	Credit card
64.68	Female	Credit card	81.85	Female	Check
75.54	Female	Check	56.84	Female	Credit card

QUESTIONS

1. Develop a contingency table showing the sex of the customer and the method of payment.

2. Is method of payment independent of the sex of the customer? Explain.

17 Nonparametric Methods

What You Will Learn in This Chapter:

- When nonparametric methods are applicable
- The advantages of nonparametric methods
- How to use and interpret the Sign test
- How to use the Wilcoxon Signed-Rank test
- How to use the Mann-Whitney-Wilcoxon test for differences between two populations
- How to use the Kruskal-Wallis test for k populations
- How to compute the Spearman rank correlation coefficient

Contents

Physical Contact in the Family

Physical contact is an important means of connecting people with one another. The touch, the handshake, the hug, the embrace, and the kiss reflect different levels of physical intimacy and commitment. Physical contact is a nonverbal means of interpersonal bonding and an important aspect of family activity. Children require the physical contact of parents, especially during infancy, if they are to thrive. Although physical contact is an important aspect of close relationships, it has been analyzed infrequently by social scientists because of its delicacy and related measurement difficulties.

A study by Oscar Grusky, Phillip Bonacich, and Mark Peyrot attempts to measure and test hypotheses about physical contact within the family unit. Using 48 two-parent, two-child families, the researchers asked each family to pose for a family photograph. In this setting, the researchers observed who touched whom and the frequency of the physical contacts.

The researchers hypothesized that higher-status persons in the family unit would touch those of lower status more often than vice versa. The touches were expected to come from father to mother, from parent to child, and/or from older child to younger child. Using a technique from nonparametric statistics known as the sign test, the researchers found support for this hypothesis. They noted that fathers were significantly more likely to touch mothers than vice versa and that parents were more likely to touch children than vice versa.

The researchers found that children were more likely to be touched than parents and that mothers were significantly more likely to be touched than

Physical contact among father, mother and children helps provide a close family relationship.

fathers. Younger children were also more likely to be touched than older children. Fathers were touched less than any other member of the family and experienced the least physical contact of any family member.

The researchers conclude by noting that family physical contact is a complex multidimensional phenomenon. On one level it signifies status and power differential between members; on another level it reflects warmth, closeness, support and expressiveness. Clearly touching and physical contact are important aspects in the health and well-being of a family unit.

Based on "Physical Contact in the Family," by Oscar Grusky, Phillip Bonacich, and Mark Peyrot, *Journal of Marriage and the Family* (August 1984).

The statistical methods presented thus far in the text are generally referred to as *parametric methods*. In this chapter we introduce several statistical methods that are referred to as *nonparametric methods*. Nonparametric methods are often applicable in situations where parametric methods of the preceding chapters are not. They typically require less restrictive assumptions concerning the level of data measurement and/or fewer assumptions concerning the form of the probability distributions generating the data.

One consideration used to determine whether a parametric or a nonparametric method should be used is the scale of measurement for the data. As noted in Chapter 2, there are four scales of measurement: nominal, ordinal, interval, and ratio. All data are generated by one of these four scales; thus all statistical analyses are conducted with either nominal, ordinal, interval, or ratio data. Let us briefly review the four scales of measurement.

1. *Nominal scale*. The scale of measurement is nominal when the data is simply a label used to identify an attribute of the item.
2. *Ordinal scale*. The scale of measurement is ordinal when the data have all the properties of nominal data and the data can be ranked or ordered with respect to some criterion.
3. *Interval scale*. The scale of measurement is interval when the data have all the properties of ordinal data and the distance between values can be expressed in terms of a fixed unit of measure.
4. *Ratio scale*. The scale of measurement is ratio when the data have all the properties of interval data and a zero value is inherently defined for the measurement scale. With this scale, the ratio of data values is meaningful.

Most of the statistical methods referred to as parametric methods require the use of interval or ratio scaled data. With these scales of measurement, means, variances, standard deviations, and so on, can be computed, interpreted, and used in the analysis. Generally, with nominal or ordinal data, parametric methods cannot be used. Thus whenever the data are nominal or ordinal, nonparametric methods are often the only way to analyze the data and draw statistical conclusions.

Another consideration used to determine whether a parametric method or a nonparametric method should be employed is the assumption concerning the population generating the data. For example, a parametric procedure for testing a hypothesis about the difference between the means of two populations was presented in Chapter 10. In the small-sample case, the *t* distribution can be used for this test provided we are willing to assume that both populations are normally distributed with equal variances. If this assumption about the populations is not appropriate, the parametric method based on the use of the *t* distribution should not be used. However, nonparametric methods, which require no assumptions about the population probability distributions, are available for testing for differences between two populations. Because of this, and other cases in which no population assumptions are required, nonparametric methods are often referred to as *distribution-free* methods.

In general, for a statistical method to be classified as nonparametric it must satisfy at least one of the following conditions:*

1. The method may be used with nominal data.
2. The method may be used with ordinal data.
3. The method may be used with interval or ratio data when no assumption can be made about the population probability distribution.

If the scale of measurement is interval or ratio and if the necessary probability distribution assumptions for the population are appropriate, parametric methods provide a more powerful or more discerning statistical procedure. In many cases where a nonparametric method as well as a parametric method can be applied, the nonparametric method is almost as good as or almost as powerful as the parametric method. In cases where the data are nominal or ordinal or in cases where the assumptions required by parametric methods are inappropriate, only nonparametric methods are available. Because of the less restrictive data-measurement requirements and the fewer assumptions needed concerning the population distribution, nonparametric methods are regarded as more

*See W. J. Conover, *Practical Nonparametric Statistics,* 2d ed. (New York: John Wiley and Sons, 1980).

generally applicable than parametric methods. The sign test, the Wilcoxon signed-rank test, the Mann-Whitney-Wilcoxon test, the Kruskal-Wallis test, and Spearman rank correlation are the nonparametric methods that are presented in this chapter.

Sign Test

The *sign test,* one of the oldest nonparametric methods, can be used with either nominal or ordinal data. A common market-research application of the sign test involves using a sample of n potential customers to identify preferences for 2 brands of a product such as coffee, soft drinks, detergents, and so on. Given these data the objective is to determine whether or not a difference in preference exists. As we will see, the sign test is a nonparametric statistical procedure that can be used to help answer this question.

Small-Sample Case

The small-sample case for the sign test is appropriate whenever $n \leq 20$. Let us illustrate the use of the sign test for the small-sample case by considering a study conducted for Sun Coast Farms.

EXAMPLE 17.1

Sun Coast Farms produces an orange juice product marketed under the name "Citrus Valley." A competitor of Sun Coast Farms has begun producing a new orange juice product known as "Tropical Orange." In a study of consumer preferences for the two brands, 12 individuals were given unmarked samples of the two brands of orange juice. The brand each individual tasted first was randomly selected. After tasting the two products, the individuals were asked to state a preference for one of the two brands. The purpose of the study is to determine whether or not preferences for the two products are equal. Letting p indicate the proportion of the population of consumers favoring Citrus Valley, Sun Coast Farms would like to test the following hypotheses:

Hypothesis	Conclusion
$H_0: p = .50$	No difference in preference for one product over the other exists
$H_a: p \neq .50$	A difference in preference for one product over the other exists

In recording the preference data for the 12 individuals participating in the study, a " $+$ " sign will be recorded if the individual expresses a preference for Citrus Valley and a " $-$ " will be recorded if the individual expresses a preference for Tropical Orange. Using this procedure the data will be recorded in terms of the $+$ or $-$ signs; thus the nonparametric test is referred to as the sign test.

Under the assumption that H_0 is true ($p = .50$), the number of $+$ signs follows a binomial probability distribution with $p = .50$. With a sample size of $n = 12$, Table 5 in Appendix B shows the following probabilities for the binomial probability distribution with $p = .50$:

NUMBER OF + SIGNS	BINOMIAL PROBABILITY
0	.0002
1	.0029
2	.0161
3	.0537
4	.1208
5	.1934
6	.2256
7	.1934
8	.1208
9	.0537
10	.0161
11	.0029
12	.0002

A graphical representation of this binomial probability distribution is shown in Figure 17.1. This probability distribution shows the probability of the number of + signs under the assumption that H_0 is true. We will use this sampling distribution to determine a rejection rule. The approach will be similar to the method we used to develop rejection rules for hypothesis testing in Chapter 8. For example, using $\alpha = .05$, we would place a rejection region or area of approximately .025 in each tail of the distribution shown in Figure 17.1. Starting at the lower end of the distribution, we see that the probability of obtaining 0, 1, or 2 + signs is $.0002 + .0029 + .0161 = .0192$. Note that we stop at

FIGURE 17.1   Binomial Probabilities for the Number of + Signs when $n = 12$ and $p = .50$

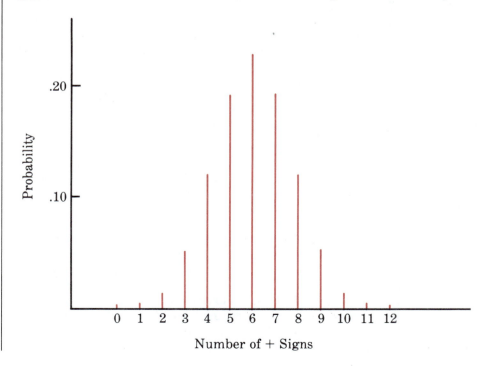

2 + signs because adding the probability of 3 + signs would make the area in the lower tail equal to .0192 + .0537 = .0729, which substantially exceeds the desired area of .025. At the upper end of the distribution, we find the same probability of .0192 corresponding to 10, 11, or 12 + signs. Thus, the closest we can come to $\alpha = .05$ is .0192 + .0192 = .0384. As a result, we adopt the following rejection rule:

Reject H_0 if the number of + signs is less than 3 or greater than 9

The preference data that were obtained for the Sun Coast Farms example are shown in Table 17.1. Since only 2 + signs were observed, the null hypothesis is rejected. There is evidence from this study that consumers preference differs for the two brands of orange juice. We would advise Sun Coast Farms that consumers appear to prefer the competitor's Tropical Orange brand.

■

In the Sun Coast Farm example, all 12 individuals in the study were able to state a preference. In many situations, one or more individuals in the sample may not be able to state a definite preference. In such cases the individual's response of no preference can be removed from the study and the analysis conducted with a smaller sample size.

The binomial probability distribution shown in Table 5 of Appendix B can be used to determine the rejection rule for a sign test up to a sample size of $n = 20$. In addition, by considering the probabilities in only the lower or upper tail of the binomial probability distribution, rejection rules can also be developed for one-tailed tests. Appendix B does not provide binomial probability distribution tables for sample sizes greater than 20. In these cases we can use the large-sample normal approximation of binomial probabilities to determine the appropriate rejection rule for the sign test.

Large-Sample Case

Using the null hypothesis H_0: $p = .50$ and a sample size of $n > 20$, the normal approximation of the sampling distribution for the number of + signs is as follows:

TABLE 17.1

Preference Data for the Sun Coast Farms' Taste Test

INDIVIDUAL	BRAND PREFERENCE	RECORDED DATA
1	Tropical Orange	−
2	Tropical Orange	−
3	Citrus Valley	+
4	Tropical Orange	−
5	Tropical Orange	−
6	Tropical Orange	−
7	Tropical Orange	−
8	Tropical Orange	−
9	Citrus Valley	+
10	Tropical Orange	−
11	Tropical Orange	−
12	Tropical Orange	−

EXAMPLE 17.2

A poll taken during a recent presidential election campaign asked 200 registered voters to rate the Democratic and Republican candidates in terms of best overall foreign policy. Results of the poll showed that 72 rated the Democratic candidate higher, 103 rated the Republican candidate higher, and 25 indicated no difference between the candidates. Does the poll indicate that there is a significant difference between public perception of the foreign policies of the two candidates?

Using the sign test, we see that $n = 200 - 25 = 175$ individuals were able to indicate the candidate they believed had the best overall foreign policy. Using (17.1) and (17.2), the sampling distribution of the number of plus signs has the following properties.

$$\mu = .50n = .50(175) = 87.5$$
$$\sigma = \sqrt{.25n} = \sqrt{.25(175)} = 6.6$$

In addition, with $n = 175$ we can assume that the sampling distribution is approximately normal. This distribution is shown in Figure 17.2. Since the distribution is approximately normal, we can use the table of areas for the standard normal probability distribution to develop the rejection rule for the sign test. Using Table 1 of Appendix B, we find that for a two-tailed test with $\alpha = .05$, the area in each tail is .025; the corresponding z

FIGURE 17.2 Probability Distribution of the Number of + Signs for a Sign Test with $n = 175$

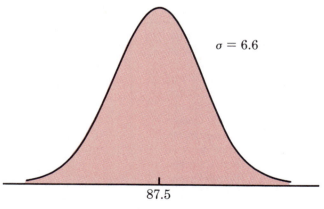

$\sigma = 6.6$

87.5

Number of + Signs

values are -1.96 and $+1.96$. Thus the rejection rule for this sign test can be written as follows:

$$\text{Reject } H_0 \text{ if } z < -1.96 \text{ or if } z > +1.96$$

Using the number of times the Democratic candidate received the higher foreign policy rating as the number of plus signs (72), we have the following value of the standardized test statistic:

$$z = \frac{72 - 87.5}{6.6} = -2.35$$

Note that $z = -2.35$ is less than -1.96. Thus the hypothesis of no difference in foreign policy for the two candidates should be rejected at the .05 level of significance. Based on this study the Republican candidate is perceived to have the higher-rated foreign policy.

■

Hypothesis Tests About a Median

In earlier chapters we described statistical tests that can be used to make inferences about population means. We now show how the sign test can be used to conduct hypothesis tests about the value of a population median. Recall that the median splits a population such that 50% of the values fall at the median or above and 50% fall at the median or below. We can apply the sign test to conduct a hypothesis test about the valve of a median by using a plus sign whenever the data value in the sample is above the median and a minus sign whenever the data value in the sample is below the median. Any data value exactly equal to the hypothesized value of the median should be discarded. The computations for the sign test are done in exactly the same manner as before.

EXAMPLE 17.3

The following hypothesis test is being conducted on the median price of new homes in St. Louis, Missouri.

$$H_0: \text{Median} = \$75,000$$

$$H_a: \text{Median} \neq \$75,000$$

From a sample of 62 sales of new homes, 37 had prices above $75,000, 23 had prices below $75,000, and 2 had prices of exactly $75,000.

Using (17.1) and (17.2) for the $n = 60$ homes with prices different than $75,000, we have

$$\mu = .50(60) = 30$$

$$\sigma = \sqrt{.25(60)} = 3.87$$

Using 37 as the number of plus signs, the value of z becomes

$$z = \frac{37 - 30}{3.87} = 1.81$$

With a level of significance of $\alpha = .05$, we reject H_0 if $z < -1.96$ or if $z > +1.96$. Since $z = 1.81$, we are unable to reject H_0.

■

EXERCISES

1. The following data show the preferences indicated by 10 individuals in taste tests involving 2 brands of coffee.

INDIVIDUAL	BRAND A VERSUS BRAND B	INDIVIDUAL	BRAND A VERSUS BRAND B
1	+	6	+
2	+	7	−
3	+	8	+
4	−	9	−
5	+	10	+

Using $\alpha = .05$, test for a significant difference between the preferences for the 2 brands. A plus indicates a preference for brand A over brand B.

2. Researchers studied physical contact between parents and children in the same family (*Journal of Marriage and the Family*, August 1984). Based on this study, assume that a sample of interactions for 20 different families with two children showed 14 cases where the mother was touched more than the father, 4 cases where the father was touched more than the mother, and 2 cases where the touching was judged equal.
   a. What are the null and alternative hypotheses for the sign test?
   b. Using $\alpha = .05$, what is the rejection rule?
   c. What is your conclusion?

3. In a television preference poll, 180 individuals were asked to state a preference for one of the two shows aired at the same time on Friday evenings. "Big Town Detective" was favored by 100, 65 favored "The Friday Variety Special," and 15 were unable to state a preference for one over the other. Is there evidence of a significant difference between the preferences for the two shows? Use $\alpha = .05$ for the test.

4. Menu planning at the Hampshire House Restaurant involves the question of customer preferences for steak and seafood. In a sample, 250 customers were asked to state a preference for the two menu items. A preference for steak was stated by 140, and 110 stated a preference for seafood. Use $\alpha = .05$ and test for a difference between the preference for the two menu items.

5. The nationwide median hourly wage for a particular labor group is $14.50 per hour. A sample of 200 individuals in this labor group showed that 134 individuals had a wage rate less than $14.50 per hour, 54 individuals had a wage rate greater than $14.50 per hour, and 12 individuals had a wage rate of $14.50. Test the null hypothesis that the median hourly wage in this city is the same as the nationwide median hourly wage. Use a .02 level of significance.

6. In a sample of 150 college basketball games, it was found that the home team won 98 games. Test to see if this data supports the claim that there is a home-team advantage in college basketball. Use a .05 level of significance. What is your conclusion?

7. The median number of part-time employees at fast-food restaurants in a particular city was known to be 15 last year. City officials think the use of part-time employees may be increasing. A sample of 9 fast-food restaurants showed that there were more than 15 part-time employees at 7 of the restaurants, 1 restaurant had exactly 15 part-time employees, and 1 had fewer than 15 part-time employees. Test at $\alpha = .05$ whether or not there has been an increase in the median number of part-time employees.

8. *The Wall Street Journal,* October 22, 1988, reported that the median age at first marriage for men is 25.9 years and for women is 23.6 years. Suppose a sample of 225 first marriages in a certain Ohio county showed 122 cases where men were less than 25.9 years of age and 103 cases where men were more than 25.9 years of age. Test the hypothesis that the median age at first marriage for men in the sampled county is the same as the reported 25.9 years. Use $\alpha = .05$. What is your conclusion?

17.2 Wilcoxon Signed-Rank Test

The Wilcoxon signed-rank test is the nonparametric alternative to the parametric matched-sample test presented in Chapter 10. In the matched-sample situation, each experimental unit generates two paired or matched observations, one from population 1 and one from population 2. The differences between the matched observations provide insight concerning the differences between the two populations.

The methodology of the parametric matched-sample analysis (the *t* test for paired differences) requires interval data and the assumption that the population of differences between the pairs of observations is *normally distributed.* With this assumption, the *t* distribution can be used to test the hypothesis of no difference between population means. If some question exists concerning the appropriateness of the assumption of normally distributed differences, the nonparametric Wilcoxon signed-rank test can be used. We illustrate this nonparametric test for data used to compare the effectiveness of two production methods.

EXAMPLE 17.4

A manufacturing firm is attempting to determine if a difference between task-completion times exists for two production methods. A sample of 11 workers was selected, and each worker completed a production task using both production methods. The production method that each worker used first was selected randomly. Thus each worker in the sample provided a pair of observations shown in Table 17.2. A positive difference in task completion times indicates that method 1 required more time and a negative difference in times indicates that method 2 required more time. Do the data indicate that the methods are significantly different in terms of task-completion times?

The question raised is whether or not the two methods provide differences in task-completion times. In effect, we have two populations of task-completion times, one population associated with each method. The hypotheses that will be tested are

H_0: The populations are identical

H_a: The populations are not identical

If H_0 can be rejected, we will conclude that the two methods differ in terms of task-completion times.

The first step of the Wilcoxon signed-rank test requires us to rank the *absolute value* of the differences in the two methods. To do this we first discard any differences of zero and then rank the remaining absolute differences from lowest to highest. Tied differences are assigned average rank values. The ranking of the absolute values of differences is shown in the fourth column of Table 17.3. Note that the difference of 0 for worker 8 is discarded from the rankings; then the smallest absolute difference of .1 is assigned the rank of 1. This ranking of absolute differences continues with the largest absolute difference of .9 assigned the rank of 10. The absolute differences of .4 for workers 3 and 5 are assigned the average rank of 3.5, while the absolute differences of .5 for workers 4 and 10 are assigned the average rank of 5.5.

Once the ranks of the absolute differences have been determined, the ranks are given the sign of the original difference in the data. For example, the .1 difference for worker 7, which was assigned the rank of 1, is given the value of $+1$ because the observed difference between the two methods was positive. The .2 difference, which was assigned the rank of 2, is given the value of -2 because the observed difference between the two

TABLE 17.2

Production Task Completion Times (minutes)

WORKER	METHOD 1	METHOD 2	DIFFERENCE
1	10.2	9.5	.7
2	9.6	9.8	− .2
3	9.2	8.8	.4
4	10.6	10.1	.5
5	9.9	10.3	− .4
6	10.2	9.3	.9
7	10.6	10.5	.1
8	10.0	10.0	.0
9	11.2	10.6	.6
10	10.7	10.2	.5
11	10.6	9.8	.8

TABLE 17.3

Ranking of Absolute Differences for the Production Task Completion Time Example

WORKER	DIFFERENCE	ABSOLUTE VALUE OF DIFFERENCE	RANK	SIGNED RANK
1	.7	.7	8	+ 8
2	− .2	.2	2	− 2
3	.4	.4	3.5	+ 3.5
4	.5	.5	5.5	+ 5.5
5	− .4	.4	3.5	− 3.5
6	.9	.9	10	+10
7	.1	.1	1	+ 1
8	0	0	—	—
9	.6	.6	7	+ 7
10	.5	.5	5.5	+ 5.5
11	.8	.8	9	+ 9
			Sum of signed ranks	+44

methods was negative. The complete list of signed ranks, together with their sum, is shown in the last column of Table 17.3.

Let us return to the original hypothesis of identical population task-completion times for the two methods. If the populations representing task-completion times for each of the two methods are identical, we would expect the positive ranks and the negative ranks to cancel each other, so that the sum of the signed rank values would be approximately 0. Thus the test for significance under the Wilcoxon signed-rank test involves determining whether or not the computed sum of signed ranks (+44 in our example) is significantly different from 0.

■

Let T denote the sum of the signed-rank values in a Wilcoxon signed-rank test. It can be shown that if the two populations are identical and the number of matched pairs of data is 10 or more, the sampling distribution of T can be approximated as follows.

Sampling Distribution of T for Identical Populations

$$\text{Mean: } \mu_T = 0 \tag{17.3}$$

$$\text{Standard Deviation: } \sigma_T = \sqrt{\frac{n(n + 1)(2n + 1)}{6}} \tag{17.4}$$

Distribution Form: Approximately normal provided $n \geq 10$

EXAMPLE 17.4 (continued)

Since we discarded the observation with the difference of 0 (worker 8), n = 10. Thus using (17.4), we have

$$\sigma_T = \sqrt{\frac{10(11)(21)}{6}} = 19.62$$

The sampling distribution of T under the assumption of identical populations is shown in Figure 17.3.

FIGURE 17.3 Sampling Distribution of the Wilcoxon T for the Production Task Completion Time Example

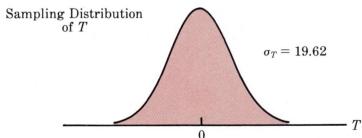

Sampling Distribution of T

$\sigma_T = 19.62$

0

T

The value of z is as follows.

$$z = \frac{T - \mu_T}{\sigma_T} = \frac{44 - 0}{19.62} = 2.24$$

Testing the null hypothesis of no difference using a level of significance of $\alpha = .05$, the rejection region will be $z < -1.96$ or $z > +1.96$. With the value of $z = 2.24$, we reject H_0 and conclude that the two populations are not identical in terms of task-completion times. Although we have determined that a difference between populations exists, the Wilcoxon signed-rank test does not enable us to conclude in what ways the populations differ. However, the fact that method 2 showed the shorter completion times for 8 of the 11 workers would lead us to conclude that differences between the two populations indicate method 2 to be the better production method.

■

EXERCISES

9. A sample of 10 individuals was used in a study to test the effects of a relaxant on the time required to fall asleep for male adults. Data for 10 subjects showing the number of minutes required to fall asleep with and without the relaxant are given. Use a .05 level of significance to determine if the relaxant reduces the time required to fall asleep. What is your conclusion?

SUBJECT	WITHOUT RELAXANT	WITH RELAXANT
1	15	10
2	12	10
3	22	12
4	8	11
5	10	9
6	7	5
7	8	10
8	10	7
9	14	11
10	9	6

10. Shown below are the number of baggage-related complaints per 1000 passengers for ten airlines during the months of December 1988 and January 1989 (*U.S. Department of Transportation*, March 1989). Use $\alpha = .05$ and the Wilcoxon signed-rank test to determine if the data indicate the number of baggage related complaints for the airline industry has *decreased* over the two months studied. What is your conclusion?

AIRLINE	DECEMBER COMPLAINTS	JANUARY COMPLAINTS
American	8.9	8.0
Delta	8.2	7.9
Continental	7.9	8.2
Eastern	7.5	7.8
Northwest	9.6	6.5
Pan American	5.0	5.1
Piedmont	12.3	11.0

AIRLINE	DECEMBER COMPLAINTS	JANUARY COMPLAINTS
TWA	11.2	10.9
United	7.7	7.4
USAir	8.6	7.7

11. A test was conducted comparing two overnight mail-delivery services. Two samples of identical deliveries were set up such that both delivery services were notified of the need for a delivery at the same time. The number of hours required to make the delivery was recorded for each service. Do the data shown suggest a difference in the delivery times for the two services? Use a .05 level of significance for the test.

DELIVERY	SERVICE 1	SERVICE 2
1	24.5	28.0
2	26.0	25.5
3	28.0	32.0
4	21.0	20.0
5	18.0	19.5
6	36.0	28.0
7	25.0	29.0
8	21.0	22.0
9	24.0	23.5
10	26.0	29.5
11	31.0	30.0

12. Harding Investors, Inc. provides a 6-week training program for newly hired management trainees. As part of the program evaluation procedure, the firm gives each trainee a pretest and post-test. Use a one-tailed test with $\alpha = .05$ and analyze the following data as part of the evaluation of the firm's management training program. What is your conclusion?

TRAINEE	PRETEST SCORE	POST-TEST SCORE
1	45	65
2	60	70
3	65	63
4	60	67
5	52	60
6	62	58
7	57	70
8	70	65
9	72	80
10	66	88
11	78	74

13. Ten test-market cities were selected as part of a market research study designed to evaluate the effectiveness of a particular advertising campaign. The sales dollars for each city were recorded for the week prior to the promotional program. Then the campaign was conducted for 2 weeks, with new sales data collected for the week immediately following the campaign. The resulting data with sales in thousands of dollars are shown.

CITY	PRECAMPAIGN SALES	POSTCAMPAIGN SALES
Kansas City	130	160
Dayton	100	105

CITY	PRECAMPAIGN SALES	POSTCAMPAIGN SALES
Cincinnati	120	140
Columbus	95	90
Cleveland	140	130
Indianapolis	80	82
Louisville	65	55
St. Louis	90	105
Pittsburgh	140	152
Peoria	125	140

Use $\alpha = .05$. What conclusion would you draw concerning the value of the advertising program?

17.3 Mann-Whitney-Wilcoxon Test

In this section we present another nonparametric method that can be used to determine if there is a difference between two populations. This test, unlike the signed-rank test, is not based on a matched sample. Two independent samples, one from each population, are used. This test, developed jointly by Mann, Whitney, and Wilcoxon, is sometimes referred to as the *Mann-Whitney test* and is sometimes referred to as the *Wilcoxon rank-sum test*. Both the Mann-Whitney and Wilcoxon versions of this test are equivalent; thus we refer to it as the *Mann-Whitney-Wilcoxon (MWW) test*.

The MWW test is based upon independent random samples from each population. Recall that in Chapter 10 we conducted a parametric test for the difference between the means of two populations. The hypotheses tested were as follows:

$$H_0: \mu_1 - \mu_2 = 0$$
$$H_a: \mu_1 - \mu_2 \neq 0$$

In the small-sample case, the parametric method used was based on two assumptions:

1. Both populations are normally distributed.
2. The variance of the two populations are equal.

The nonparametric MWW test does not require either of the above assumptions. The only requirement of the MWW test is that the measurement scale for the data generated by the two independent random samples is at least ordinal. Instead of testing for the difference between the means of the two populations, the MWW test is to determine whether or not the two populations are identical. The hypotheses for the Mann-Whitney-Wilcoxon test are as follows:

$$H_0: \text{The two populations are identical}$$

$$H_a: \text{The two populations are not identical}$$

We first demonstrate how the MWW test can be applied by showing an application for the small-sample case.

Small-Sample Case

The small-sample size case for the MWW test is appropriate whenever the sample sizes for both populations are less than or equal to 10. We illustrate the use of the MWW test for the small-sample case by considering the academic achievement of students attending Johnston High School.

EXAMPLE 17.5

The majority of students attending Johnston High School previously attended either Garfield Junior High School or Mulberry Junior High School. The question raised by school administrators was whether or not the population of students that attended Garfield were identical to the population of students that attended Mulberry in terms of academic potential. The hypotheses under consideration were expressed as follows:

H_0: The two populations are identical in terms of academic potential

H_a: The two populations are not identical in terms of academic potential

Using high school records, Johnston High School administrators selected a random sample of 4 high school students who had attended Garfield Junior High and another random sample of 5 students who had attended Mulberry Junior High. The current high school class standing was recorded for each of the 9 students used in the study. The ordinal class standings for the 9 students are shown in Table 17.4.

The first step in the MWW procedure is to rank the *combined* data from the two samples from low to high. The lowest value (class standing 8) receives a rank of 1 and the highest value (classing standing 202) receives a rank of 9. The complete ranking of the 9 students is as follows:

STUDENT	CLASS STANDING	COMBINED SAMPLE RANK
Fields	8	1
Tibbs	21	2
Clark	52	3
Hart	70	4
Jones	112	5
Kirkwood	144	6
Guest	146	7
Abbott	175	8
Phipps	202	9

TABLE 17.4

High School Class-Standing Data

GARFIELD STUDENTS		MULBERRY STUDENTS	
Student	Class Standing	Student	Class Standing
Fields	8	Hart	70
Clark	52	Phipps	202
Jones	112	Kirkwood	144
Tibbs	21	Abbott	175
		Guest	146

TABLE 17.5

Rank Sums for High School Students from Each Junior High School

	GARFIELD STUDENTS			MULBERRY STUDENTS	
Student	Class Standing	Sample Rank	Student	Class Standing	Sample Rank
Fields	8	1	Hart	70	4
Clark	52	3	Phipps	202	9
Jones	112	5	Kirkwood	144	6
Tibbs	21	2	Abbott	175	8
			Guest	146	7
Sum of ranks		11			34

EXAMPLE 17.5 (continued)

The next step is to sum the ranks for each sample separately. This calculation is shown in Table 17.5 The MWW procedure may utilize the sum of the ranks for either sample. In the following discussion we use the sum of the ranks for the sample of 4 students from Garfield. We denote this sum by the symbol T. Thus for our example, $T = 11$.

Let us consider for a moment the properties of the sum of the ranks for the Garfield sample. Since there are 4 students in the sample, Garfield could have the top 4 ranking students in the study. If this were the case, $T = 1 + 2 + 3 + 4 = 10$ would be the smallest value possible for the rank sum T. On the other hand, Garfield could have the bottom 4 ranking students, in which case $T = 6 + 7 + 8 + 9 = 30$ would be the largest value possible for T. Thus, T for the Garfield sample must take on a value between 10 and 30.

Note that values of T near 10 imply Garfield has the significantly better, or higher-ranking, students, whereas values of T near 30 imply Garfield has the significantly weaker, or lower-ranking, students. Thus, if the two populations of students were identical in terms of academic potential, we would expect the value of T to be near the average of the above two values, or $(10 + 30)/2 = 20$.

■

Critical values of the MWW T statistic are provided in Table 9 of Appendix B for cases where both sample sizes are less than or equal to 10.* In these tables, n_1 refers to the sample size corresponding to the sample whose rank sum is being used in the test. The value of T_L is read directly from the tables and the value of T_U is computed from (17.5).

$$T_U = n_1(n_1 + n_2 + 1) - T_L \qquad (17.5)$$

The null hypothesis of identical populations should be rejected if T is *strictly less than* T_L or *strictly greater than* T_U.

EXAMPLE 17.5 (continued)

For example, using Table 9 of Appendix B with a .05 level of significance we see that the lower-tail critical value for the MWW statistic with $n_1 = 4$ (Garfield) and $n_2 = 5$ (Mulberry) is $T_L = 12$. The upper-tail critical value for the MWW statistic is computed using (17.5) as follows:

*A more complete table of critical values for the Mann-Whitney-Wilcoxon test can be found in *Practical Nonparametric Statistics,* by W. J. Conover.

$$T_U = 4(4 + 5 + 1) - 12 = 28$$

Thus the MWW decision rule indicates that the null hypothesis of identical populations cannot be rejected if the sum of the ranks for the first sample (Garfield) is between 12 and 28, inclusively. The rejection rule for the example can be written as follows:

Reject H_0 if $T < 12$ or if $T > 28$

Referring to Table 17.5, we see that $T = 11$. Thus, the null hypothesis H_0 is rejected and we can conclude that the population of students at Garfield differs from the population of students at Mulberry in terms of academic potential. The higher class ranking obtained by the sample of Garfield students indicates that Garfield students appear to be better prepared for high school than the Mulberry students.

■

Large-Sample Case

In the case where both sample sizes are greater than or equal to 10, a normal approximation of the distribution of T can be used to conduct the analysis for the MWW test. We show an example of the large-sample case by considering an application at the Third National Bank.

EXAMPLE 17.6

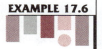

The Third National Bank has two branch offices. Data collected from two independent simple random samples, one from each branch, are shown in Table 17.6. Do the data indicate that the populations of checking account balances at the two branch banks are not identical?

The first step in the MWW test is to rank the *combined* data from the lowest to the highest values. Using the combined set of 22 observations shown in Table 17.6, we find the lowest data value of $750 (sixth item of sample 2) and assign to it a rank of 1. Continuing the ranking, we have the results shown on the next page.

TABLE 17.6

Account Balances for Two Branches of the Third National Bank

BRANCH 1		BRANCH 2	
Sampled Account	Account Balance ($)	Sampled Account	Account Balance ($)
1	1095	1	885
2	955	2	850
3	1200	3	915
4	1195	4	950
5	925	5	800
6	950	6	750
7	805	7	865
8	945	8	1000
9	875	9	1050
10	1055	10	935
11	1025		
12	975		

ACCOUNT BALANCE	ITEM	ASSIGNED RANK
$ 750	6th of sample 2	1
$ 800	5th of sample 2	2
$ 805	7th of sample 1	3
$ 850	2nd of sample 2	4
•	•	•
•	•	•
•	•	•
$1195	4th of sample 1	21
$1200	3rd of sample 1	22

In ranking the combined data, we may find that 2 or more data values are the same. In this case, these same values are given the *average* ranking of their positions in the combined data set. This situation of *ties* occurs with the ranking of the 22 account balances from the two branch banks. For example, the balance of $945 (eighth item of sample 1) will be assigned the rank of 11. However, the next 2 values in the data set are tied with values of $950 (see the sixth item of sample 1 and the fourth item of sample 2). Since these two values will be considered for assigned ranks of 12 and 13, they are both given the assigned rank of 12.5. At the next highest data value of $955, we continue the ranking process by assigning $955 the rank of 14. Table 17.7 shows the entire data set with the assigned rank of each observation.

The next step in the MWW test is to sum the ranks for each sample. These sums are shown in Table 17.7. The test procedure can be based upon the sum of the ranks for either sample. We use the sum of the ranks for the sample from Branch 1. Thus for this example, $T = 169.5$.

Given that the sample sizes are $n_1 = 12$ and $n_2 = 10$, we can use the normal approximation to the sampling distribution of the rank sum, T. The appropriate sampling distribution is given by (17.6) and (17.7).

TABLE 17.7

Combined Ranking of the Data in the Two Samples from the Third National Bank

BRANCH 1			BRANCH 2		
Sampled Account	Account Balance ($)	Rank	Sampled Account	Account Balance ($)	Rank
1	1095	20	1	885	7
2	955	14	2	850	4
3	1200	22	3	915	8
4	1195	21	4	950	12.5
5	925	9	5	800	2
6	950	12.5	6	750	1
7	805	3	7	865	5
8	945	11	8	1000	16
9	875	6	9	1050	18
10	1055	19	10	935	10
11	1025	17		Sum of ranks	83.5
12	975	15			
	Sum of ranks	169.5			

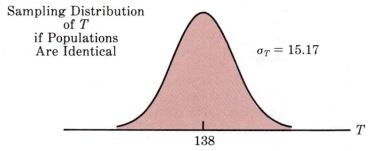

FIGURE 17.4 Sampling Distribution of T for the Third National Bank Example

<table>
<tr><td colspan="3">Sampling Distribution of T for Identical Populations</td></tr>
<tr><td>Mean:</td><td>$\mu_T = \frac{1}{2} n_1(n_1 + n_2 + 1)$</td><td>(17.6)</td></tr>
<tr><td>Standard Deviation:</td><td>$\sigma_T = \sqrt{\frac{1}{12} n_1 n_2(n_1 + n_2 + 1)}$</td><td>(17.7)</td></tr>
<tr><td colspan="3">Distribution Form: Approximately normal provided $n_1 \geq 10$ and $n_2 \geq 10$</td></tr>
</table>

For Branch 1, we have

$$\mu_T = \frac{1}{2}\, 12(12 + 10 + 1) = 138$$

$$\sigma_T = \sqrt{\frac{1}{12}\, 12(10)(12 + 10 + 1)} = 15.17$$

The sampling distribution of T is shown in Figure 17.4. Following the usual hypothesis-testing procedure, we compute the standardized test statistic z to determine if the observed value of T appears to be from the sampling distribution of Figure 17.4. If T appears to be from this distribution, cannot reject the hypothesis that the two populations are identical. However, if T does not appear to be from this distribution, we will reject the null hypothesis and conclude that the populations are not identical. Computing the value of z we have

$$z = \frac{T - \mu_T}{\sigma_T} = \frac{169.5 - 138}{15.17} = 2.08$$

At an $\alpha = .05$ level of significance, we know that we can reject H_0 if $z < -1.96$ or if $z > 1.96$. With $z = 2.08$, we reject H_0 and conclude that the two populations are not identical. That is, the populations of account balances at the two branches are not the same.

■

In summary, the Mann-Whitney-Wilcoxon rank-sum test follows the steps outlined below in order to determine if two independent random samples are selected from identical populations.

1. Rank the combined sample observations from lowest to highest, with tied values being assigned the average of the tied rankings.
2. Compute T, the sum of the ranks for the first sample.
3. In the large-samples case, make the test for significant differences between the two populations by using the observed value of T and comparing it to the sampling distribution of T for identical populations (see equations (17.6) and (17.7)). The value of the standardized test statistic z will provide the basis deciding whether or not to reject the null hypothesis of identical populations. In the small sample case, use Table 9 in Appendix B to find the critical values for the test.

NOTES AND COMMENTS

The nonparametric test discussed in this section is used to determine whether or not two populations are identical. Parametric statistical tests, such as the t test described in Chapter 10, test the equality of two population means. When we reject the hypothesis that the means are equal, we conclude that the populations differ only in their means. When we reject the hypothesis that the populations are identical using the MWW test, we cannot state how they differ. The populations could have different means, different variances, and/or different forms. Nonetheless, if we believe that the populations are approximately the same in every way except for the means, a rejection of H_0 using the nonparametric method implies that the means differ. The major advantages of the MWW test, compared to the parametric t test, are that it does not require any assumptions about the form of the probability distribution from which the measurements come and the test may be used with ordinal data.

EXERCISES

14. Anderson Company has sent two groups of employees to a privately run program providing word-processing training. One group was from the data-processing department; the other was from the typing pool. At the completion of the program, Anderson Company received a report showing the class rank for each of its employees. Of the 70 persons finishing the program, the class ranks of the 13 employees of Anderson Company are given (by group).

DATA-PROCESSING DEPARTMENT	TYPING POOL
1	17
12	26
15	29
23	33
30	45
33	51
	62

Using $\alpha = .10$, test to see whether or not there is a performance difference between the two groups in the word-processing program.

15. Mileage performance tests were conducted for two models of automobiles. Twelve automobiles of each model were randomly selected and a miles-per-gallon rating for each model was developed based upon 1000 miles of highway driving. The data are shown.

MODEL 1		MODEL 2	
Automobile	Miles per Gallon	Automobile	Miles per Gallon
1	20.6	1	21.3
2	19.9	2	17.6
3	18.6	3	17.4
4	18.9	4	18.5
5	18.8	5	19.7
6	20.2	6	21.1
7	21.0	7	17.3
8	20.5	8	18.8
9	19.8	9	17.8
10	19.8	10	16.9
11	19.2	11	18.0
12	20.5	12	20.1

Using $\alpha = .10$, test for a significant difference in the populations of miles-per-gallon ratings for the two models.

16. Insurance costs for men and women were reported in *Newsweek*, May 8, 1989. Assume that the following data show the annual cost of $100,000, 5-year term insurance policies for nonsmoking men and nonsmoking women. Use the MWW to test for a significant difference between the costs for men and women. Use $\alpha = .05$.

MEN	WOMEN
167	146
175	162
160	164
165	148
172	166
180	158
185	150
170	150
163	140
184	142

17. The following data from police records show the number of daily crime reports from a sample of days during the winter months and a sample of days during the summer months. Using a .05 level of significance, determine if there is a significant difference between the number of crime reports in winter and summer months.

WINTER	SUMMER
18	28
20	18
15	24
16	32
21	18
20	29
12	23

WINTER	SUMMER
16	38
19	28
20	18

18. A certain brand of microwave oven was priced at 10 stores in Dallas and 13 stores in San Antonio. The data are as follows.

DALLAS	SAN ANTONIO
445	460
489	451
405	435
485	479
439	475
449	445
436	429
420	434
430	410
405	422
	425
	459
	430

Use a .05 level of significance to test whether or not prices for the microwave oven are the same in the two cities.

17.4   The Kruskal-Wallis Test

The MWW test in Section 17.3 can be used to test whether or not two populations are identical. This test has been extended to the case of three or more populations by Kruskal and Wallis. The hypotheses for the *Kruskal-Wallis test* with $k \geq 3$ populations can be written as follows:

$$H_0: \text{All } k \text{ populations are identical}$$

$$H_a: \text{Not all populations are identical}$$

The Kruskal-Wallis test is based on the analysis of independent random samples from each of the k populations.

In Chapter 12 we introduced the completely randomized experimental design as a procedure that could be used to test for the equality of means among 3 or more populations. The parametric procedure known as analysis of variance (ANOVA) was used to analyze the data and conduct the test. The ANOVA procedure requires interval or ratio data and normally distributed populations with equal variances.

The nonparametric Kruskal-Wallis test can be used with ordinal data as well as with interval or ratio data. In addition, the Kruskal-Wallis test does not require the assumptions of normality and equal variances that are required by the parametric ANOVA procedure. Thus, whenever the data from $k \geq 3$ independent random samples is ordinal or whenever the assumptions of normality and equal variances are questionable, the Kruskal-Wallis test provides an alternate statistical procedure for testing whether or not the populations are identical. Let us demonstrate the Kruskal-Wallis test by showing how it can be used in an employee-selection application.

EXAMPLE 17.7

Williams Manufacturing Company hires employees for its management staff from three area colleges. Recently the company's personnel department has been collecting and reviewing annual performance ratings in an attempt to determine if there are differences in performance among the managers hired from the three area colleges. Performance-rating data are available from independent samples of 7 employees from college A, 6 employees from college B, and 7 employees from college C. These data are summarized in Table 17.8; the overall performance rating of each manager is given on a 0 to 100 scale, with 100 being the highest possible performance evaluation. Suppose that we are interested in testing whether or not the three populations are identical with respect to performance evaluations.

■

The Kruskal-Wallis test statistic, which is based on the sum of ranks for each of the samples, can be computed as follows:

Kruskal-Wallis Test Statistic

$$W = \frac{12}{n_T(n_T + 1)} \left[\sum_{i=1}^{k} \frac{R_i^2}{n_i} \right] - 3(n_T + 1) \tag{17.8}$$

where

k = the number of populations

n_i = the number of items in sample i

$n_T = \Sigma\, n_i$ = total number of items in all samples

R_i = sum of the ranks for sample i

TABLE 17.8

Performance Evaluation Ratings for 20 Employees

COLLEGE A	COLLEGE B	COLLEGE C
25	60	50
70	20	70
60	30	60
85	15	80
95	40	90
90	35	70
80		75

TABLE 17.9

Rankings for the 20 Employees

COLLEGE A	RANK	COLLEGE B	RANK	COLLEGE C	RANK
25	3	60	9	50	7
70	12	20	2	70	12
60	9	30	4	60	9
85	17	15	1	80	15.5
95	20	40	6	90	18.5
90	18.5	35	5	70	12
80	15.5			75	14
Sum of ranks	95		27		88

Kruskal and Wallis were able to show that under the null hypothesis that the populations are identical, the sampling distribution of W can be approximated by a chi-square distribution with $k - 1$ degrees of freedom. This approximation is generally acceptable if each of the sample sizes is greater than or equal to 5.

**EXAMPLE 17.7
(continued)**

In order to compute the W statistic we must first rank-order all 20 data items. The lowest data value of 15 from the college B sample receives a rank of 1, whereas the highest data value of 95 from the college A sample receives a rank of 20. The data values, their associated ranks and the sum of the ranks for the three samples are shown in Table 17.9. Note that we assign the average rank to tied items*; for example the data values of 60, 70, 80, and 90 had ties.

The sample sizes are:

$$n_1 = 7 \quad n_2 = 6 \quad n_3 = 7$$

and

$$n_T = \Sigma n_i = 7 + 6 + 7 = 20$$

Using (17.8) the W statistic is computed as follows:

$$W = \frac{12}{20(21)} \left[\frac{(95)^2}{7} + \frac{(27)^2}{6} + \frac{(88)^2}{7} \right] - 3(20 + 1) = 8.92$$

The chi-square distribution table (Table 3 of Appendix B) shows that with $k - 1 = 2$ degrees of freedom and $\alpha = .05$ in the upper tail of the distribution, the critical chi-square value is $\chi^2 = 5.99147$. Since the test statistic $W = 8.92$ is greater than 5.99147, we reject the null hypothesis that the 3 populations are identical. As a result we conclude that manager performance differs significantly depending on the college attended. Furthermore, since the performance ratings were the lowest for college B, it would appear reasonable for the company to either cut back on its recruiting from college B or at least do a more thorough evaluation of graduates from this college.

*If numerous tied ranks are observed, (17.8) must be modified; the modified formula can be found in *Practical Nonparametric Statistics,* by W. J. Conover.

We note that the Kruskal-Wallis procedure illustrated in Example 17.7 began with the collection of interval-scaled data showing employee-performance evaluation ratings. However, the procedure would have also worked had the original data been the ordinal rankings of the 20 employees. In this case the Kruskal-Wallis test could have been applied directly to the original data; the step of constructing the rank orderings from the performance evaluation ratings could have been omitted.

EXERCISES

19. Three products received the following performance ratings by a panel of 15 consumers.

PERFORMANCE RATINGS

Product A	50	62	75	48	65
Product B	80	95	98	87	90
Product C	60	45	30	58	57

Use the Kruskal-Wallis test and $\alpha = .05$ to determine if there is a significant difference in the performance evaluations.

20. Three different admission-test-preparation programs are being evaluated. The scores obtained by a sample of 20 people utilizing the test-preparation programs yielded these results.

PROGRAM A	PROGRAM B	PROGRAM C
540	450	600
400	540	630
490	400	580
530	410	490
490	480	590
610	370	620
	550	570

Use the Kruskal-Wallis test to determine if there is a significant difference in the three test preparation programs. Use $\alpha = .01$.

21. In Chapter 12, the ANOVA procedure was used to test for significant differences in gas mileage for three types of automobiles. The data obtained from tests on 5 automobiles of each type are shown again.

MILES PER GALLON

Automobile A	19	21	20	19	21
Automobile B	19	20	22	21	23
Automobile C	24	26	23	25	27

a. Use the Kruskal-Wallis test with $\alpha = .05$ to determine if there is a significant difference in the gasoline mileage for the 3 automobiles.

b. What information available in the data is used by the ANOVA procedure and not the Kruskal-Wallis test.

22. A large corporation has been sending many of its first-level managers to an off-site supervisory skills course. Four different management-development centers offer this course and the cor-

poration is interested in determining if there are differences in the quality of training provided. A sample of 20 employees who have attended these programs has been obtained and the employees ranked with respect to supervisory skills. The results are shown:

PROGRAM	SUPERVISORY SKILLS RANK				
1	3	14	10	12	13
2	2	7	1	5	11
3	19	16	9	18	17
4	20	4	15	6	8

Note that the top-ranked supervisor attended supervisory skills course 2 and the lowest-ranked supervisor attended course 4. Use $\alpha = .05$ and test to see if there is a significant difference in the training provided by the four programs.

17.5   Rank Correlation

Correlation was introduced in Chapter 13 as a measure of the linear association between two variables for which interval or ratio data are available. In this section we consider measures of association between two variables when only rank-order or ordinal data are available. The *Spearman rank-correlation coefficient, r_s*, has been developed for this purpose.

The formula for the Spearman rank-correlation coefficient is as follows.

Spearman Rank-Correlation Coefficient

$$r_s = 1 - \frac{6\Sigma\, d_i^2}{n(n^2 - 1)} \tag{17.9}$$

where

$\quad n = $ the number of items or individuals being ranked

$\quad x_i = $ the rank of item i with respect to one variable

$\quad y_i = $ the rank of item i with respect to a second variable

$\quad d_i = x_i - y_i$

EXAMPLE 17.8

A company wants to determine if individuals who were expected at the time of employment to be better salespersons actually turn out to have better sales records. To investigate this question, the vice president in charge of personnel has carefully reviewed the original job interview summaries, academic records, and letters of recommendations for 10 current members of the firm's sales force. Based on the review of this information, the vice president ranked the 10 individuals in terms of their potential for success, basing the assessment solely upon the information available at the time of employment. Then a list was obtained of the number of units sold by each salesperson over the first 2 years. Based on actual sales performance, a second ranking of the 10 salespersons was carried out. Table 17.10 shows the relevant data and the two rankings. The statistical question involves determining whether or not there is agreement between the ranking of potential at the time of employment and the ranking based upon the actual sales performance over the first 2 years. Let us compute the Spearman rank-correlation coefficient for the data in Table 17.10. The computations for the rank-correlation coefficient are summarized in Table 17.11. Here we see that the rank-correlation coefficient is a positive .73. The Spearman rank-correlation coefficient ranges from -1.0 to $+1.0$, with an interpretation similar to the sample correlation coefficient in that positive values near 1.0 indicate a strong association between the rankings; as one rank increases, the other rank increases. On the other hand, rank correlations near -1.0 indicate a strong negative association in

TABLE 17.10

Sales Potential and Actual 2-Year Sales in Units for 10 Salespersons

SALESPERSONS	RANKING OF POTENTIAL	2-YEAR SALES (UNITS)	RANKING ACCORDING TO 2-YEAR SALES
A	2	400	1
B	4	360	3
C	7	300	5
D	1	295	6
E	6	280	7
F	3	350	4
G	10	200	10
H	9	260	8
I	8	220	9
J	5	385	2

TABLE 17.11

Computation of the Spearman Rank-Correlation Coefficient for Sales Potential and Sales Performance

SALESPERSON	x_i = RANKING OF POTENTIAL	y_i = RANKING OF SALES PERFORMANCE	$d_i = x_i - y_i$	d_i^2
A	2	1	1	1
B	4	3	1	1
C	7	5	2	4
D	1	6	-5	25
E	6	7	-1	1
F	3	4	-1	1
G	10	10	0	0
H	9	8	1	1
I	8	9	-1	1
J	5	2	3	9

$$\Sigma\, d_i^2 = 44$$

$$r_s = 1 - \frac{6\Sigma\, d_i^2}{n(n^2 - 1)} = 1 - \frac{6(44)}{10(100 - 1)} = .73$$

the ranks (as one rank increases, the other rank decreases). The value $r_s = .73$ indicates a positive correlation between potential and actual performance. Individuals ranked high on potential tend to rank high on performance.

◼

A Test for Significant Rank Correlation

At this point, we have seen how sample results can be used to compute the sample rank-correlation coefficient. As with many other statistical procedures, we may wish to use the sample results to make an inference about the population rank correlation, ρ_s, between two variables. In our example, the population rank-correlation coefficient could be obtained by making the rank-correlation coefficient computations for all members of the sales force. However, we would like to avoid all this data collection and make an inference about the population rank-correlation based on the sample rank-correlation coefficient, r_s. To make this inference, we must test the following hypotheses:

$$H_0: \rho_s = 0$$

$$H_a: \rho_s \neq 0$$

Under the null hypothesis of no rank correlation ($\rho_s = 0$), the rankings are independent and the sampling distribution of r_s is as follows.

Sampling Distribution of r_s

Mean: $\mu_{r_s} = 0$ (17.10)

Standard deviation: $\sigma_{r_s} = \sqrt{\dfrac{1}{(n-1)}}$ (17.11)

Form: Approximately normal provided $n \geq 10$

EXAMPLE 17.8 (continued)

The sample rank-correlation coefficient for sales potential and sales performance was $r_s = .73$. Use this value to test for a significant rank correlation.

From (17.10) we have $\mu_{r_s} = 0$, and from (17.11) we have $\sigma_{r_s} = \sqrt{1/(10 - 1)} = .33$. Thus we have

$$z = \frac{r_s - \mu_{r_s}}{\sigma_{r_s}} = \frac{.73 - 0}{.33} = 2.21$$

Using a level of significance of $\alpha = .05$, the null hypothesis of no correlation can be rejected if $z < -1.96$ or if $z > 1.96$. Since $z = 2.21 > 1.96$, we reject the hypothesis of no rank correlation. Thus we can conclude that a significant rank correlation exists between sales potential and sales performance.

◼

23. Consider the following two sets of rankings for 6 items.

	CASE ONE			CASE TWO	
Item	First Ranking	Second Ranking	Item	First Ranking	Second Ranking
A	1	1	A	1	6
B	2	2	B	2	5
C	3	3	C	3	4
D	4	4	D	4	3
E	5	5	E	5	2
F	6	6	F	6	1

Note that in the first case the rankings are identical, whereas in the second case the rankings are exactly opposite. What value should you expect for the Spearman rank-correlation coefficient for each of these cases? Explain. Calculate the Spearman rank-correlation coefficient for each case.

24. Airline passengers file complaints about lost, stolen, damaged, and delayed baggage. How do the airlines compare? Using the U.S. Department of Transportation (March 1989) data in Exercise 10, ten airlines can be ranked from fewest to most complaints per 1000 passengers. Shown below are the rankings of the airlines for December 1988 and January 1989.

DECEMBER 1988	JANUARY 1989
Pan American	Pan American
Eastern	Northwest
United	United
Continental	USAir
Delta	Eastern
USAir	Delta
American	American
Northwest	Continental
TWA	TWA
Piedmont	Piedmont

a. Compute the rank correlation for the airlines for the two months of data.
b. Test for significant rank correlation using $\alpha = .05$. What is your conclusion?

25. In the baseball draft 8 players are ranked by a scout in terms of speed and then in terms of power hitting.

PLAYER	SPEED RANKING	POWER-HITTING RANKING
A	1	8
B	2	5
C	3	6
D	4	7
E	5	2
F	6	3
G	7	4
H	8	1

Use the Spearman rank-correlation coefficient to measure the association between speed and power. Use $\alpha = .05$ and test for the significance of this correlation coefficient.

26. In a poll of men and women television viewers, preferences for the top 10 shows led to the following rankings. Is there a relationship between the rankings provided for the two groups? Use $\alpha = .10$.

TELEVISION SHOW	RANKING BY MEN	RANKING BY WOMEN
1	1	5
2	5	10
3	8	6
4	7	4
5	2	7
6	3	2
7	10	9
8	4	8
9	6	1
10	9	3

27. A student organization surveyed both recent graduates and current students in an attempt to obtain information on the quality of teaching at a particular university. An analysis of the responses provided the following rankings for 10 professors on the basis of teaching ability.

PROFESSOR	RANKING BY CURRENT STUDENTS	RANKING BY RECENT GRADUATES
1	4	6
2	6	8
3	8	5
4	3	1
5	1	2
6	2	3
7	5	7
8	10	9
9	7	4
10	9	10

Do the rankings given by the current students agree with the rankings given by the recent graduates? Use $\alpha = .10$ and test for a significant rank correlation.

SUMMARY

In this chapter we have presented several statistical procedures that are classified as nonparametric methods. The parametric methods of the earlier chapters generally required interval or ratio data and were often based on assumptions such as the population probability distribution is normal.

Since nonparametric methods can be applied to nominal and ordinal data as well as interval and ratio data and since nonparametric methods do not require population-distribution assumptions, nonparametric methods expand the class of problems that can be subjected to statistical analysis.

The sign test provides a nonparametric procedure for identifying differences between two populations. In the small-sample case, the binominal probability distribution can be used to determine the critical values for the sign test; in the large-sample case, a normal approximation may be used. The Wilcoxon signed-rank test provides a procedure for analyzing matched sample data whenever interval or ratio scaled data are available for each matched pair. The Wilcoxon procedure tests the hypothesis that the two populations are identical.

The Mann-Whitney-Wilcoxon test provides a nonparametric method for testing for a difference between two populations based on two independent random samples. Tables were presented for the small-sample case and a normal approximation was provided for the large-sample case. The Kruskal-Wallis test extended the Mann-Whitney-Wilcoxon test to the case of 3 or more populations. The Kruskal-Wallis test is the nonparametric version of the parametric ANOVA test for differences among population means.

In the last section of this chapter we introduced the Spearman rank-correlation coefficient as a measure of association for two ordinal or rank-ordered sets of items.

GLOSSARY

Nonparametric methods   A collection of statistical methods that generally require very few, if any, assumptions about the population distributions and the level of measurement. These methods can be applied when nominal or ordinal data are available.

Distribution-free methods   Another name for nonparametric statistical methods suggested by the lack of assumptions required concerning the population distribution.

Sign test   A nonparametric statistical test for identifying differences between two populations based on the analysis of two matched or paired samples.

Wilcoxon signed-rank test   A nonparametric statistical test for identifying differences between two populations based on the analysis of two matched or paired samples.

Mann-Whitney-Wilcoxon (MWW) test   A nonparametric statistical test for identifying differences between two populations based on the analysis of two independent samples.

Kruskal-Wallis test   A nonparametric test for identifying differences among 3 or more populations.

Spearman rank-correlation coefficient   A correlation measure based on rank-order data for two variables.

KEY FORMULAS

SIGN TEST (LARGE SAMPLE CASE-NORMAL APPROXIMATION)

$$\text{Mean:} \qquad \mu = .50n \qquad (17.1)$$

$$\text{Standard deviation:} \quad \sigma = \sqrt{.25n} \qquad (17.2)$$

WILCOXON SIGNED-RANK TEST

$$\mu_T = 0 \qquad (17.3)$$

$$\sigma_T = \sqrt{\frac{n(n + 1)(2n + 1)}{6}} \qquad (17.4)$$

MANN-WHITNEY-WILCOXON (NORMAL APPROXIMATION)

$$\mu_T = \tfrac{1}{2} n_1(n_1 + n_2 + 1) \tag{17.6}$$

$$\sigma_T = \sqrt{\tfrac{1}{12} n_1 n_2(n_1 + n_2 + 1)} \tag{17.7}$$

KRUSKAL-WALLIS TEST STATISTIC

$$W = \frac{12}{n_T(n_T + 1)} \left[\sum_{i=1}^{k} \frac{R_i^2}{n_i} \right] - 3(n_T + 1) \tag{17.8}$$

SPEARMAN RANK CORRELATION COEFFICIENT

$$r_s = 1 - \frac{6\Sigma d_i^2}{n(n^2 - 1)} \tag{17.9}$$

REVIEW QUIZ

TRUE/FALSE

1. The sign test can be used to test whether or not individuals prefer one item over another.

2. Nonparametric methods are often referred to as distribution-free methods.

3. Nonparametric statistical methods are not applicable to ordinal data.

4. With the Wilcoxon signed-rank test we must assume the populations sampled from are normal.

5. If ties occur in ranking the data from the two samples, the Mann-Whitney-Wilcoxon test cannot be used.

6. With the Mann-Whitney-Wilcoxon test, we need to know only the two sample sizes in order to compute the standard deviation of the rank sum.

7. With the Mann-Whitney-Wilcoxon test, we need to know only the two sample sizes in order to compute the expected value of the rank sum.

8. The Wilcoxon signed-rank test uses paired differences to test whether or not two populations are identical.

9. With the Wilcoxon signed-rank test, the expected value of the sum of signed ranks is equal to the sample size.

10. The Kruskal-Wallis test cannot be used unless the populations sampled from are approximately normal.

11. The Spearman rank correlation coefficient can be computed only if quantitative data is available for the variables.

MULTIPLE CHOICE

12. A municipal transit system has collected data on which of two seating configurations is preferred by the passengers on its buses. In a sign test to

determine if one seating arrangement is significantly preferred, the null hypothesis would be

a. H_0: $\mu = 0$
b. H_0: $\mu = .5$
c. H_0: $p = 0$
d. H_0: $p = .5$

13. A nonparametric test for the equivalence of two populations would be used instead of a parametric test for the equivalence of the population parameters if

a. the samples are very large
b. the samples are not independent
c. no information about the populations is available
d. the parametric test is always used in this situation

14. A variable assumes the following values.

$$10, 12, 15, 15, 16, 18$$

What rank is assigned to 15 in a rank-sum test?

a. 3.5 for both values of 15
b. 3 for one 15 and 4 for the other 15
c. only one value of 15 should be used and given the rank 3
d. no rank needed because ties should be omitted from the rankings

15. For a Mann-Whitney-Wilcoxon test, the first and second samples have sizes 15 and 24, respectively. The expected value of T is

a. 180
b. 300
c. 0
d. not enough information given

16. In a Wilcoxon signed-rank test, the two samples each have $n = 17$. The expected value of T is

a. 0
b. 145
c. 298
d. 300

SUPPLEMENTARY EXERCISES

28. Mueller Beverage Products of Milwaukee, Wisconsin, has conducted a market research study designed to determine if there is a consumer preference for Mueller's Old Brew Beer over the individual consumer's usual beer. Each individual participating in the test was provided with a glass of his or her usual beer and a glass of Mueller's Old Brew. The two glasses were not labeled, and thus the individuals had no way of knowing beforehand which of the two glasses was Mueller's Old Brew and which was the individual's usual brand. The glass that each individual tasted first was randomly selected. After tasting the beer in each glass, the individuals were asked to indicate their *preferred* beer. The test results from a sample of 24 individuals are shown:

INDIVIDUAL	BRAND PREFERRED	VALUE RECORDED
1	Old Brew	+
2	Old Brew	+
3	Usual Brand	−
4	Old Brew	+
5	Usual Brand	−
6	Old Brew	+
7	Usual Brand	−
8	Old Brew	+
9	Old Brew	+
10	Usual Brand	−
11	Old Brew	+
12	Usual Brand	−
13	Usual Brand	−
14	Usual Brand	−
15	Old Brew	+
16	Usual Brand	−
17	Old Brew	+
18	Old Brew	+
19	Old Brew	+
20	Usual Brand	−
21	Old Brew	+
22	Old Brew	+
23	Usual Brand	−
24	Old Brew	+

If an individual selected Mueller's Old Brew as the preferred beer, a plus sign was recorded. On the other hand, if the individual stated a preference for his or her usual brand, a minus sign was recorded. Do the data for the 24 individuals indicate a significant difference in the preferences for the beers? Use $\alpha = .05$.

29. Two pilots for a prime-time television show (a western and a mystery show) are being tested. Both have been shown to a group of 12 viewers. The viewer preferences are shown.

VIEWER	PREFERENCE	VIEWER	PREFERENCE
1	Mystery	7	Mystery
2	Mystery	8	Western
3	Mystery	9	Mystery
4	Western	10	Mystery
5	Mystery	11	Mystery
6	Western	12	Mystery

Using $\alpha = .05$, test to see if there is a significant difference in preferences.

30. In a soft-drink taste test, 48 individuals stated a preference for one of two well-known brands. Results showed 28 favoring brand A, 16 favoring brand B, and 4 undecided. Use the sign test with $\alpha = .10$ to determine whether or not there is a significant difference in the preferences for the two brands of soft-drinks.

31. Use the sign test to determine whether or not the task-completion times for the following two production methods differ. Use $\alpha = .05$.

WORKER	METHOD 1 (Minutes)	METHOD 2 (Minutes)
1	10.2	9.5
2	9.6	9.8
3	9.2	8.8
4	10.6	10.1
5	9.9	10.3
6	10.2	9.3
7	10.6	10.5
8	10.0	10.0
9	11.2	10.6
10	10.7	10.2
11	10.6	9.8

32. The national median price of new homes for 1988 was $123,500 (*U.S. News & World Report*, September 19, 1988). Assume that data on the prices of new homes were obtained from samples of loans recorded in Chicago and Dallas-Fort Worth. Use the data to test the hypothesis that the median price of homes in each of the two cities is the same as the national median price. Use $\alpha = .05$. State your conclusion for each city.

	GREATER THAN $123,500	EQUAL TO $123,500	LESS THAN $123,500
Chicago	55	6	28
Dallas-Fort Worth	42	3	36

33. Mayfield Products, Inc. has collected data on preferences of 12 individuals concerning cleaning power of 2 brands of detergent. The individuals and their preferences are shown below. A plus indicates a preference for brand A.

INDIVIDUAL	BRAND A VERSUS BRAND B	INDIVIDUAL	BRAND A VERSUS BRAND B
1	−	7	−
2	+	8	+
3	+	9	+
4	+	10	−
5	−	11	+
6	+	12	+

Using $\alpha = .10$, test for a significant difference in the preference for the 2 brands.

34. Twelve homemakers were asked to estimate the retail selling price of two models of refrigerators. The estimates of selling price provided by the homemakers are shown.

HOMEMAKER	MODEL 1	MODEL 2
1	$650	$ 900
2	760	720
3	740	690
4	700	850
5	590	920
6	620	800
7	700	890

HOMEMAKER	MODEL 1	MODEL 2
8	690	920
9	900	1000
10	500	690
11	610	700
12	720	700

At the .05 level of significance determine if there is a difference between the homemaker's perception of selling price for the two models.

35. A study was designed to evaluate the weight-gain potential of a new poultry feed. A sample of 12 chickens was used in a 6-week study. The weight of each chicken was recorded before and after the 6-week test period. The difference between the before and after weights of each chicken are as follows: 1.5, 1.2, −.2, .0, .5, .7, .8, 1.0, .0, .6, .2, −.01. A negative value indicates a weight loss during the test period, whereas .0 indicates no weight change over the period. Use a .05 level of significance to determine if the new feed appears to provide a weight gain for the chickens.

36. The following data show product weights for items produced on two production lines. Test for a difference between the product weights for the two lines. Use $\alpha = .10$.

PRODUCTION LINE 1	PRODUCTION LINE 2
13.6	13.7
13.8	14.1
14.0	14.2
13.9	14.0
13.4	14.6
13.2	13.5
13.3	14.4
13.6	14.8
12.9	14.5
14.4	14.3
	15.0
	14.9

37. It is desired to determine if there is a significant difference in the time required to complete a program evaluation with the 3 different methods that are in common use. The times (in hours) required for each of 18 evaluators to conduct a program evaluation are given below.

METHOD 1	METHOD 2	METHOD 3
68	62	58
74	73	67
65	75	69
76	68	57
77	72	59
72	70	62

Using $\alpha = .05$, test to see if there is a significant difference in the time required by the 3 methods.

38. A sample of 20 engineers who have been with a company for 3 years was taken and rank-ordered with respect to managerial potential. Some of the engineers have attended the company's management-development course, others have attended an off-site management-development program at a local university, and the remainder have not attended any program. Use the rankings given and $\alpha = .025$ to test for a significant difference in the managerial potential of these three groups.

NO PROGRAM	COMPANY PROGRAM	OFF-SITE PROGRAM
16	12	7
9	20	1
10	17	4
15	19	2
11	6	3
13	18	8
	14	5

39. Shown below are course evaluation ratings for four instructors. Use $\alpha = .05$ and the Kruskal-Wallis procedure to test for a significant difference in teaching abilities.

INSTRUCTOR	COURSE-EVALUATION RATING							
Black	88	80	79	68	96	69			
Jennings	87	97	78	85	99	99	85	94	
Swanson	88	76	68	82	85	82	84	83	81
Wilson	80	85	56	71	89	87			

40. A group of investment analysts ranked 12 companies, first with respect to book value and then with respect to growth potential.

COMPANY	RANKING OF BOOK VALUE	RANKING OF GROWTH POTENTIAL
1	12	2
2	2	9
3	8	6
4	1	11
5	9	4
6	7	5
7	3	12
8	11	1
9	4	7
10	5	10
11	6	8
12	10	3

For these data does a relationship exist between the companies' book values and growth potentials? Use $\alpha = .05$.

41. Two individuals provided the following preference rankings of seven soft drinks. For $\alpha = .05$, is there a significant rank correlation for the 2 individuals?

SOFT DRINK	RANKING BY INDIVIDUAL 1	RANKING BY INDIVIDUAL 2
A	1	3
B	3	2
C	5	5

SOFT DRINK	RANKING BY INDIVIDUAL 1	RANKING BY INDIVIDUAL 2
D	6	7
E	7	6
F	4	1
G	2	4

42. A sample of 15 students obtained the following rankings on midterm and final examinations in a statistics course.

MIDTERM RANK	FINAL RANK
1	4
2	7
3	1
4	3
5	8
6	2
7	5
8	12
9	6
10	9
11	14
12	15
13	11
14	10
15	13

Compute the Spearman rank-correlation coefficient for the data and test for a significant correlation using $\alpha = .10$.

Appendixes

A References and Bibliography

General

Freedman, D., R. Pisani, and R. Purves, *Statistics*, New York, W. W. Norton, 1978.

Freund, J. E., and R. E. Walpole, *Mathematical Statistics*, 4th ed., Englewood Cliffs, N.J., Prentice-Hall, 1987.

Hogg, R. V., and A. T. Craig, *Introduction to Mathematical Statistics*, 4th ed., New York, Macmillan, 1978.

Mood, A. M., F. A. Graybill, and D. C. Boes, *Introduction to the Theory of Statistics*, 3d ed., New York, McGraw-Hill, 1974.

Neter, J., W. Wasserman, and G. W. Whitmore, *Applied Statistics*, 3d. ed., Boston, Allyn & Bacon, 1988.

Ryan, T. A., B. L. Joiner, and B. F. Ryan, *Minitab Handbook*, 2d ed., Boston, PWS-Kent, 1985.

Tanur, J. M., ed., *Statistics: A Guide to the Unknown*, San Francisco, Holden-Day, 1972.

Exploratory Data Analysis

Hartwig, F., and B. E. Dearing, *Exploratory Data Analysis*, Beverly Hills, Sage Publications, 1979.

Hoaglin, D. C., F. Mosteller, and J. W. Tukey, *Understanding Robust and Exploratory Data Analysis*, New York, John Wiley & Sons, 1983.

Tukey, J. W., *Exploratory Data Analysis*, Menlo Park, Calif., Addison-Wesley, 1977.

Vellerman, P. F., and D. C. Hoaglin, *Applications, Basics, and Computing of Exploratory Data Analysis*, Boston, Duxbury Press, 1981.

Probability

Barr, D. R., and P. W. Zehna, *Probability: Modeling Uncertainty*, Reading, Mass., Addison-Wesley, 1983.

Feller, W., *An Introduction to Probability Theory and Its Applications*, Vol. I, 3d ed., New York, John Wiley & Sons, 1968.

Feller, W., *An Introduction to Probability Theory and Its Applications*, Vol. II, 2d ed., New York, John Wiley & Sons, 1971.

Hoel, P. G., S. C. Port, and C. J. Stone, *Introduction to Probability Theory*, Boston, Houghton Mifflin, 1971.

Mendenhall, W., R. L. Scheaffer, and D. Wackerly, *Mathematical Statistics with Applications*, 3d ed., Boston, PWS-Kent, 1986.

Parzen, E., *Modern Probability Theory and Its Applications*, New York, John Wiley & Sons, Inc. 1960.

Zehna, P. W., *Probability Distributions and Statistics*, Boston, Allyn & Bacon, 1970.

Sampling Methods

Cochran, W. G., *Sampling Techniques*, 3d ed., New York, John Wiley & Sons, 1977.

Kish, L., *Survey Sampling*, New York, John Wiley & Sons, 1965.

Scheaffer, R. L., W. Mendenhall, and L. Ott, *Elementary Survey Sampling*, 2d ed., North Scituate, Mass., Duxbury Press, 1979.

Williams, B., *A Sampler on Sampling*, New York, John Wiley & Sons, 1978.

Analysis of Variance and Experimental Design

Anderson, V. L., and R. A. McLean, *Design of Experiments: A Realistic Approach*, New York, Marcel Dekker, 1974.

Box, G. E. P., W. G. Hunter and J. S. Hunter, *Statistics for Experimenters*, New York, John Wiley & Sons, 1978.

Cochran, W. G., and G. M. Cox, *Experimental Designs*, 2d ed., New York, John Wiley & Sons, 1957.

Hicks, C. R., *Fundamental Concepts in the Design of Experiments*, 2d. ed., New York, Holt, Rinehart and Winston 1973.

Mendenhall, W., *Introduction to Linear Models and the Design and Analysis of Experiments*, Belmont, Calif., Duxbury Press, 1968.

Montgomery, D. C., *Design and Analysis of Experiments*, 2d ed., New York, John Wiley & Sons, 1983.

Winer, B. J., *Statistical Principles in Experimental Design*, 2d ed., New York, McGraw-Hill, 1971.

Regression Analysis

Chatterjee, S., and R. Price, *Regression Analysis by Example*, New York, John Wiley & Sons, 1977.

Daniel, C., and F. Wood, *Fitting Equations to Data*, 2d ed., New York, John Wiley & Sons, 1980.

Draper, N. R., and H. Smith, *Applied Regression Analysis*, 2d ed., New York, John Wiley & Sons, 1981.

Gunst, R. F., and R. L. Mason, *Regression Analysis and Its Application: a Data-Oriented Approach*, New York, Marcel Dekker, 1980.

Kleinbaum, D. G., and L. L. Kupper, *Applied Regression Analysis and Other Multivariable Methods*, North Scituate, Mass., Duxbury Press, 1978.

Mosteller, F., and J. W. Tukey, *Data Analysis and Regression: A Second Course in Statistics*, Reading, Mass., Addison-Wesley, 1977.

Neter, J., W. Wasserman and M. Kutner, *Applied Linear Statistical Models*, 2d ed. Homewood, Ill., Richard D. Irwin, 1985.

Neter, J., W. Wasserman and M. Kutner, *Applied Linear Regression Models*, Homewood, Ill., Richard D. Irwin, 1983.

Weidberg, S., *Applied Linear Regression*, 2d ed., New York, John Wiley & Sons, 1985.

Wonnacott, T. H., and R. J. Wonnacott, *Regression: A Second Course in Statistics*, New York, John Wiley & Sons, 1981.

Nonparametric Methods

Conover, W. J., *Practical Nonparametric Statistics*, 2d ed., New York, John Wiley & Sons, 1980.

Gibbons, J. D. I. Olkin, and M. Sobel, *Selecting and Ordering Populations: A New Statistical Methodology*, New York, John Wiley & Sons, 1977.

Hollander, M., and D. A. Wolfe, *Nonparametric Statistical Methods*, New York, John Wiley & Sons, 1973.

Mosteller, F., and R. E. K. Rourke, *Sturdy Statistics*, Reading, Mass., Addison-Wesley, 1973.

Siegel, S., *Nonparametric Statistics for the Behavioral Sciences*, New York, McGraw-Hill, 1956.

B Tables

TABLE 1 **Standard Normal Distribution**

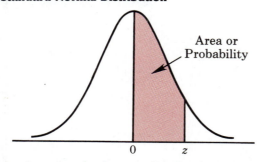

Area or Probability

Entries in the table give the area under the curve between the mean and z standard deviations above the mean. For example, for $z = 1.25$ the area under the curve between the mean and z is .3944.

z	.00	.01	.02	.03	.04	.05	.06	.07	.08	.09
.0	.0000	.0040	.0080	.0120	.0160	.0199	.0239	.0279	.0319	.0359
.1	.0398	.0438	.0478	.0517	.0557	.0596	.0636	.0675	.0714	.0753
.2	.0793	.0832	.0871	.0910	.0948	.0987	.1026	.1064	.1103	.1141
.3	.1179	.1217	.1255	.1293	.1331	.1368	.1406	.1443	.1480	.1517
.4	.1554	.1591	.1628	.1664	.1700	.1736	.1772	.1808	.1844	.1879
.5	.1915	.1950	.1985	.2019	.2054	.2088	.2123	.2157	.2190	.2224
.6	.2257	.2291	.2324	.2357	.2389	.2422	.2454	.2486	.2518	.2549
.7	.2580	.2612	.2642	.2673	.2704	.2734	.2764	.2794	.2823	.2852
.8	.2881	.2910	.2939	.2967	.2995	.3023	.3051	.3078	.3106	.3133
.9	.3159	.3186	.3212	.3238	.3264	.3289	.3315	.3340	.3365	.3389
1.0	.3413	.3438	.3461	.3485	.3508	.3531	.3554	.3577	.3599	.3621
1.1	.3643	.3665	.3686	.3708	.3729	.3749	.3770	.3790	.3810	.3830
1.2	.3849	.3869	.3888	.3907	.3925	.3944	.3962	.3980	.3997	.4015
1.3	.4032	.4049	.4066	.4082	.4099	.4115	.4131	.4147	.4162	.4177
1.4	.4192	.4207	.4222	.4236	.4251	.4265	.4279	.4292	.4306	.4319
1.5	.4332	.4345	.4357	.4370	.4382	.4394	.4406	.4418	.4429	.4441
1.6	.4452	.4463	.4474	.4484	.4495	.4505	.4515	.4525	.4535	.4545
1.7	.4554	.4564	.4573	.4582	.4591	.4599	.4608	.4616	.4625	.4633
1.8	.4641	.4649	.4656	.4664	.4671	.4678	.4686	.4693	.4699	.4706
1.9	.4713	.4719	.4726	.4732	.4738	.4744	.4750	.4756	.4761	.4767
2.0	.4772	.4778	.4783	.4788	.4793	.4798	.4803	.4808	.4812	.4817
2.1	.4821	.4826	.4830	.4834	.4838	.4842	.4846	.4850	.4854	.4857
2.2	.4861	.4864	.4868	.4871	.4875	.4878	.4881	.4884	.4887	.4890

z	.00	.01	.02	.03	.04	.05	.06	.07	.08	.09
2.3	.4893	.4896	.4898	.4901	.4904	.4906	.4909	.4911	.4913	.4916
2.4	.4918	.4920	.4922	.4925	.4927	.4929	.4931	.4932	.4934	.4936
2.5	.4938	.4940	.4941	.4943	.4945	.4946	.4948	.4949	.4951	.4952
2.6	.4953	.4955	.4956	.4957	.4959	.4960	.4961	.4962	.4963	.4964
2.7	.4965	.4966	.4967	.4968	.4969	.4970	.4971	.4972	.4973	.4974
2.8	.4974	.4975	.4976	.4977	.4977	.4978	.4979	.4979	.4980	.4981
2.9	.4981	.4982	.4982	.4983	.4984	.4984	.4985	.4985	.4986	.4986
3.0	.4986	.4987	.4987	.4988	.4988	.4989	.4989	.4989	.4990	.4990

TABLE 2

t Distribution

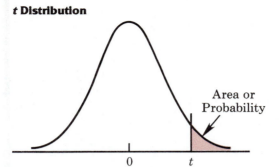

Area or
Probability

0   *t*

Entries in the table give t_α values, where α is the area of probability in the upper tail of the *t* distribution. For example, with 10 degrees of freedom and a .05 area in the upper tail, $t_{.05} = 1.812$.

DEGREES OF FREEDOM	AREA IN UPPER TAIL				
	.10	**.05**	**.025**	**.01**	**.005**
1	3.078	6.314	12.706	31.821	63.657
2	1.886	2.920	4.303	6.965	9.925
3	1.638	2.353	3.182	4.541	5.841
4	1.533	2.132	2.776	3.747	4.604
5	1.476	2.015	2.571	3.365	4.032
6	1.440	1.943	2.447	3.143	3.707
7	1.415	1.895	2.365	2.998	3.499
8	1.397	1.860	2.306	2.896	3.355
9	1.383	1.833	2.262	2.821	3.250
10	1.372	1.812	2.228	2.764	3.169
11	1.363	1.796	2.201	2.718	3.106
12	1.356	1.782	2.179	2.681	3.055
13	1.350	1.771	2.160	2.650	3.012
14	1.345	1.761	2.145	2.624	2.977
15	1.341	1.753	2.131	2.602	2.947
16	1.337	1.746	2.120	2.583	2.921
17	1.333	1.740	2.110	2.567	2.898
18	1.330	1.734	2.101	2.552	2.878
19	1.328	1.729	2.093	2.539	2.861
20	1.325	1.725	2.086	2.528	2.845
21	1.323	1.721	2.080	2.518	2.831
22	1.321	1.717	2.074	2.508	2.819
23	1.319	1.714	2.069	2.500	2.807
24	1.318	1.711	2.064	2.492	2.797
25	1.316	1.708	2.060	2.485	2.787
26	1.315	1.706	2.056	2.479	2.779
27	1.314	1.703	2.052	2.473	2.771
28	1.313	1.701	2.048	2.467	2.763
29	1.311	1.699	2.045	2.462	2.756
30	1.310	1.697	2.042	2.457	2.750
40	1.303	1.684	2.021	2.423	2.704
60	1.296	1.671	2.000	2.390	2.660
120	1.289	1.658	1.980	2.358	2.617
∞	1.282	1.645	1.960	2.326	2.576

Reprinted by permission of Biometrika Trustees from Table 12, Percentage Points of the *t* Distribution, 3rd. Edition, 1966. E. S. Pearson and H. O. Hartley, *Biometrika Tables for Statisticians*, Vol. I.

TABLE 3   **Chi-Square Distribution**

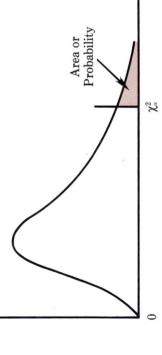

Area or Probability

Entries in the table give χ_α^2 values, where α is the area of probability in the χ_α^2 upper tail of the chi-square distribution. For example, with 10 degrees of freedom and a .01 area in the upper tail, $\chi_{.01}^2 = 23.2093$.

DEGREES OF FREEDOM	AREA IN UPPER TAIL									
	.995	.99	.975	.95	.90	.10	.05	.025	.01	.005
1	$392,704 \times 10^{-10}$	$157,088 \times 10^{-9}$	$982,069 \times 10^{-9}$	$393,214 \times 10^{-8}$	.0157908	2.70554	3.84146	5.02389	6.63490	7.87944
2	.0100251	.0201007	.0506356	.102587	.210720	4.60517	5.99147	7.37776	9.21034	10.5966
3	.0717212	.114832	.215795	.351846	.584375	6.25139	7.81473	9.34840	11.3449	12.8381
4	.206990	.297110	.484419	.710721	1.063623	7.77944	9.48773	11.1433	13.2767	14.8602
5	.411740	.554300	.831211	1.145476	1.61031	9.23635	11.0705	12.8325	15.0863	16.7496
6	.675727	.872085	1.237347	1.63539	2.20413	10.6446	12.5916	14.4494	16.8119	18.5476
7	.989265	1.239043	1.68987	2.16735	2.83311	12.0170	14.0671	16.0128	18.4753	20.2777
8	1.344419	1.646482	2.17973	2.73264	3.48954	13.3616	15.5073	17.5346	20.0902	21.9550
9	1.734926	2.087912	2.70039	3.32511	4.16816	14.6837	16.9190	19.0228	21.6660	23.5893
10	2.15585	2.55821	3.24697	3.94030	4.86518	15.9871	18.3070	20.4831	23.2093	25.1882
11	2.60321	3.05347	3.81575	4.57481	5.57779	17.2750	19.6751	21.9200	24.7250	26.7569
12	3.07382	3.57056	4.40379	5.22603	6.30380	18.5494	21.0261	23.3367	26.2170	28.2995
13	3.56503	4.10691	5.00874	5.89186	7.04150	19.8119	22.3621	24.7356	27.6883	29.8194
14	4.07468	4.66043	5.62872	6.57063	7.78953	21.0642	23.6848	26.1190	29.1413	31.3193

df										
15	4.60094	5.22935	6.26214	7.26094	8.54675	22.3072	24.9958	27.4884	30.5779	32.8013
16	5.14224	5.81221	6.90766	7.96164	9.31223	23.5418	26.2962	28.8454	31.9999	34.2672
17	5.69724	6.40776	7.56418	8.67176	10.0852	24.7690	27.5871	30.1910	33.4087	35.7185
18	6.26481	7.01491	8.23075	9.39046	10.8649	25.9894	28.8693	31.5264	34.8053	37.1564
19	6.84398	7.63273	8.90655	10.1170	11.6509	27.2036	30.1435	32.8523	36.1908	38.5822
20	7.43386	8.26040	9.59083	10.8508	12.4426	28.4120	31.4104	34.1696	37.5662	39.9968
21	8.03366	8.89720	10.28293	11.5913	13.2396	29.6151	32.6705	35.4789	38.9321	41.4010
22	8.64272	9.54249	10.9823	12.3380	14.0415	30.8133	33.9244	36.7807	40.2894	42.7958
23	9.26042	10.19567	11.6885	13.0905	14.8479	32.0069	35.1725	38.0757	41.6384	44.1813
24	9.88623	10.8564	12.4011	13.8484	15.6587	33.1963	36.4151	39.3641	42.9798	45.5585
25	10.5197	11.5240	13.1197	14.6114	16.4734	34.3816	37.6525	40.6465	44.3141	46.9278
26	11.1603	12.1981	13.8439	15.3791	17.2919	35.5631	38.8852	41.9232	45.6417	48.2899
27	11.8076	12.8786	14.5733	16.1513	18.1138	36.7412	40.1133	43.1944	46.9630	49.6449
28	12.4613	13.5648	15.3079	16.9279	18.9392	37.9159	41.3372	44.4607	48.2782	50.9933
29	13.1211	14.2565	16.0471	17.7083	19.7677	39.0875	42.5569	45.7222	49.5879	52.3356
30	13.7867	14.9535	16.7908	18.4926	20.5992	40.2560	43.7729	46.9792	50.8922	53.6720
40	20.7065	22.1643	24.4331	26.5093	29.0505	51.8050	55.7585	59.3417	63.6907	66.7659
50	27.9907	29.7067	32.3574	34.7642	37.6886	61.1671	67.5048	71.4202	76.1539	79.4900
60	35.5346	37.4848	40.4817	43.1879	46.4589	74.3970	79.0819	83.2976	88.3794	91.9517
70	43.2752	45.4418	48.7576	51.7393	55.3290	85.5271	90.5312	95.0231	100.425	104.215
80	51.1720	53.5400	57.1532	60.3915	64.2778	96.5782	101.879	106.629	112.329	116.321
90	59.1963	61.7541	65.6466	69.1260	73.2912	107.565	113.145	118.136	124.116	128.299
100	67.3276	70.0648	74.2219	77.9295	82.3581	118.498	124.342	129.561	135.807	140.169

Reprinted by permission of Biometrika Trustees from Table 8, Percentage Points of the χ^2 Distribution by E. S. Pearson and H. O. Hartley, *Biometrika Tables for Statisticians*, Vol. I, 3rd Edition, 1966.

TABLE 4

F Distribution

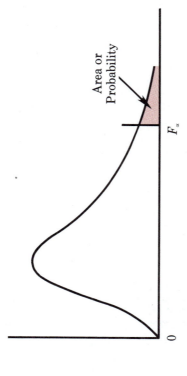

Area or Probability

F_α

Entries in the table give F_α values, where α is the area of probability in the upper tail of the F distribution. For example, with 12 numerator degrees of freedom, 15 denominator degrees of freedom, and a .05 area in the upper tail, $F_{.05} = 2.48$.

TABLE OF $F_{.05}$ VALUES

Denominator Degrees of Freedom	Numerator Degrees of Freedom																		
	1	2	3	4	5	6	7	8	9	10	12	15	20	24	30	40	60	120	∞
1	161.4	199.5	215.7	224.6	230.2	234.0	236.8	238.9	240.5	241.9	243.9	245.9	248.0	249.1	250.1	251.1	252.2	253.3	254.3
2	18.51	19.00	19.16	19.25	19.30	19.33	19.35	19.37	19.38	19.40	19.41	19.43	19.45	19.45	19.46	19.47	19.48	19.49	19.50
3	10.13	9.55	9.28	9.12	9.01	8.94	8.89	8.85	8.81	8.79	8.74	8.70	8.66	8.64	8.62	8.59	8.57	8.55	8.53
4	7.71	6.94	6.59	6.39	6.26	6.16	6.09	6.04	6.00	5.96	5.91	5.86	5.80	5.77	5.75	5.72	5.69	5.66	5.63
5	6.61	5.79	5.41	5.19	5.05	4.95	4.88	4.82	4.77	4.74	4.68	4.62	4.56	4.53	4.50	4.46	4.43	4.40	4.36
6	5.99	5.14	4.76	4.53	4.39	4.28	4.21	4.15	4.10	4.06	4.00	3.94	3.87	3.84	3.81	3.77	3.74	3.70	3.67
7	5.59	4.74	4.35	4.12	3.97	3.87	3.79	3.73	3.68	3.64	3.57	3.51	3.44	3.41	3.38	3.34	3.30	3.27	3.23
8	5.32	4.46	4.07	3.84	3.69	3.58	3.50	3.44	3.39	3.35	3.28	3.22	3.15	3.12	3.08	3.04	3.01	2.97	2.93
9	5.12	4.26	3.86	3.63	3.48	3.37	3.29	3.23	3.18	3.14	3.07	3.01	2.94	2.90	2.86	2.83	2.79	2.75	2.71
10	4.96	4.10	3.71	3.48	3.33	3.22	3.14	3.07	3.02	2.98	2.91	2.85	2.77	2.74	2.70	2.66	2.62	2.58	2.54
11	4.84	3.98	3.59	3.36	3.20	3.09	3.01	2.95	2.90	2.85	2.79	2.72	2.65	2.61	2.57	2.53	2.49	2.45	2.40
12	4.75	3.89	3.49	3.26	3.11	3.00	2.91	2.85	2.80	2.75	2.69	2.62	2.54	2.51	2.47	2.43	2.38	2.34	2.30
13	4.67	3.81	3.41	3.18	3.03	2.92	2.83	2.77	2.71	2.67	2.60	2.53	2.46	2.42	2.38	2.34	2.30	2.25	2.21
14	4.60	3.74	3.34	3.11	2.96	2.85	2.76	2.70	2.65	2.60	2.53	2.46	2.39	2.35	2.31	2.27	2.22	2.18	2.13

15	4.54	3.68	3.29	3.06	2.90	2.79	2.71	2.64	2.59	2.54	2.48	2.40	2.33	2.29	2.25	2.20	2.16	2.11	2.07
16	4.49	3.63	3.24	3.01	2.85	2.74	2.66	2.59	2.54	2.49	2.42	2.35	2.28	2.24	2.19	2.15	2.11	2.06	2.01
17	4.45	3.59	3.20	2.96	2.81	2.70	2.61	2.55	2.49	2.45	2.38	2.31	2.23	2.19	2.15	2.10	2.06	2.01	1.96
18	4.41	3.55	3.16	2.93	2.77	2.66	2.58	2.51	2.46	2.41	2.34	2.27	2.19	2.15	2.11	2.06	2.02	1.97	1.92
19	4.38	3.52	3.13	2.90	2.74	2.63	2.54	2.48	2.42	2.38	2.31	2.23	2.16	2.11	2.07	2.03	1.98	1.93	1.88
20	4.35	3.49	3.10	2.87	2.71	2.60	2.51	2.45	2.39	2.35	2.28	2.20	2.12	2.08	2.04	1.99	1.95	1.90	1.84
21	4.32	3.47	3.07	2.84	2.68	2.57	2.49	2.42	2.37	2.32	2.25	2.18	2.10	2.05	2.01	1.96	1.92	1.87	1.81
22	4.30	3.44	3.05	2.82	2.66	2.55	2.46	2.40	2.34	2.30	2.23	2.15	2.07	2.03	1.98	1.94	1.89	1.84	1.78
23	4.28	3.42	3.03	2.80	2.64	2.53	2.44	2.37	2.32	2.27	2.20	2.13	2.05	2.01	1.96	1.91	1.86	1.81	1.76
24	4.26	3.40	3.01	2.78	2.62	2.51	2.42	2.36	2.30	2.25	2.18	2.11	2.03	1.98	1.94	1.89	1.84	1.79	1.73
25	4.24	3.39	2.99	2.76	2.60	2.49	2.40	2.34	2.28	2.24	2.16	2.09	2.01	1.96	1.92	1.87	1.82	1.77	1.71
26	4.23	3.37	2.98	2.74	2.59	2.47	2.39	2.32	2.27	2.22	2.15	2.07	1.99	1.95	1.90	1.85	1.80	1.75	1.69
27	4.21	3.35	2.96	2.73	2.57	2.46	2.37	2.31	2.25	2.20	2.13	2.06	1.97	1.93	1.88	1.84	1.79	1.73	1.67
28	4.20	3.34	2.95	2.71	2.56	2.45	2.36	2.29	2.24	2.19	2.12	2.04	1.96	1.91	1.87	1.82	1.77	1.71	1.65
29	4.18	3.33	2.93	2.70	2.55	2.43	2.35	2.28	2.22	2.18	2.10	2.03	1.94	1.90	1.85	1.81	1.75	1.70	1.64
30	4.17	3.32	2.92	2.69	2.53	2.42	2.33	2.27	2.21	2.16	2.09	2.01	1.93	1.89	1.84	1.79	1.74	1.68	1.62
40	4.08	3.23	2.84	2.61	2.45	2.34	2.25	2.18	2.12	2.08	2.00	1.92	1.84	1.79	1.74	1.69	1.64	1.58	1.51
60	4.00	3.15	2.76	2.53	2.37	2.25	2.17	2.10	2.04	1.99	1.92	1.84	1.75	1.70	1.65	1.59	1.53	1.47	1.39
120	3.92	3.07	2.68	2.45	2.29	2.17	2.09	2.02	1.96	1.91	1.83	1.75	1.66	1.61	1.55	1.50	1.43	1.35	1.25
∞	3.84	3.00	2.60	2.37	2.21	2.10	2.01	1.94	1.88	1.83	1.75	1.67	1.57	1.52	1.46	1.39	1.32	1.22	1.00

TABLE OF $F_{.025}$ VALUES

Denominator Degrees of Freedom	Numerator Degrees of Freedom																		
	1	2	3	4	5	6	7	8	9	10	12	15	20	24	30	40	60	120	∞
1	647.8	799.5	864.2	899.6	921.8	937.1	948.2	956.7	963.3	968.6	976.7	984.9	993.1	997.2	1,001	1,006	1,010	1,014	1,018
2	38.51	39.00	39.17	39.25	39.30	39.33	39.36	39.37	39.39	39.40	39.41	39.43	39.45	39.46	39.46	39.47	39.48	39.49	39.50
3	17.44	16.04	15.44	15.10	14.88	14.73	14.62	14.54	14.47	14.42	14.34	14.25	14.17	14.12	14.08	14.04	13.99	13.95	13.90
4	12.22	10.65	9.98	9.60	9.36	9.20	9.07	8.98	8.90	8.84	8.75	8.66	8.56	8.51	8.46	8.41	8.36	8.31	8.26
5	10.01	8.43	7.76	7.39	7.15	6.98	6.85	6.76	6.68	6.62	6.52	6.43	6.33	6.28	6.23	6.18	6.12	6.07	6.02
6	8.81	7.26	6.60	6.23	5.99	5.82	5.70	5.60	5.52	5.46	5.37	5.27	5.17	5.12	5.07	5.01	4.96	4.90	4.85
7	8.07	6.54	5.89	5.52	5.29	5.12	4.99	4.90	4.82	4.76	4.67	4.57	4.47	4.42	4.36	4.31	4.25	4.20	4.14
8	7.57	6.06	5.42	5.05	4.82	4.65	4.53	4.43	4.36	4.30	4.20	4.10	4.00	3.95	3.89	3.84	3.78	3.73	3.67
9	7.21	5.71	5.08	4.72	4.48	4.32	4.20	4.10	4.03	3.96	3.87	3.77	3.67	3.61	3.56	3.51	3.45	3.39	3.33
10	6.94	5.46	4.83	4.47	4.24	4.07	3.95	3.85	3.78	3.72	3.62	3.52	3.42	3.37	3.31	3.26	3.20	3.14	3.08
11	6.72	5.26	4.63	4.28	4.04	3.88	3.76	3.66	3.59	3.53	3.43	3.33	3.23	3.17	3.12	3.06	3.00	2.94	2.88
12	6.55	5.10	4.47	4.12	3.89	3.73	3.61	3.51	3.44	3.37	3.28	3.18	3.07	3.02	2.96	2.91	2.85	2.79	2.72
13	6.41	4.97	4.35	4.00	3.77	3.60	3.48	3.39	3.31	3.25	3.15	3.05	2.95	2.89	2.84	2.78	2.72	2.66	2.60
14	6.30	4.86	4.24	3.89	3.66	3.50	3.38	3.29	3.21	3.15	3.05	2.95	2.84	2.79	2.73	2.67	2.61	2.55	2.49
15	6.20	4.77	4.15	3.80	3.58	3.41	3.29	3.20	3.12	3.06	2.96	2.86	2.76	2.70	2.64	2.59	2.52	2.46	2.40
16	6.12	4.69	4.08	3.73	3.50	3.34	3.22	3.12	3.05	2.99	2.89	2.79	2.68	2.63	2.57	2.51	2.45	2.38	2.32
17	6.04	4.62	4.01	3.66	3.44	3.28	3.16	3.06	2.98	2.92	2.82	2.72	2.62	2.56	2.50	2.44	2.38	2.32	2.25
18	5.98	4.56	3.95	3.61	3.38	3.22	3.10	3.01	2.93	2.87	2.77	2.67	2.56	2.50	2.44	2.38	2.32	2.26	2.19
19	5.92	4.51	3.90	3.56	3.33	3.17	3.05	2.96	2.88	2.82	2.72	2.62	2.51	2.45	2.39	2.33	2.27	2.20	2.13
20	5.87	4.46	3.86	3.51	3.29	3.13	3.01	2.91	2.84	2.77	2.68	2.57	2.46	2.41	2.35	2.29	2.22	2.16	2.09
21	5.83	4.42	3.82	3.48	3.25	3.09	2.97	2.87	2.80	2.73	2.64	2.53	2.42	2.37	2.31	2.25	2.18	2.11	2.04
22	5.79	4.38	3.78	3.44	3.22	3.05	2.93	2.84	2.76	2.70	2.60	2.50	2.39	2.33	2.27	2.21	2.14	2.08	2.00
23	5.75	4.35	3.75	3.41	3.18	3.02	2.90	2.81	2.73	2.67	2.57	2.47	2.36	2.30	2.24	2.18	2.11	2.04	1.97
24	5.72	4.32	3.72	3.38	3.15	2.99	2.87	2.78	2.70	2.64	2.54	2.44	2.33	2.27	2.21	2.15	2.08	2.01	1.94
25	5.69	4.29	3.69	3.35	3.13	2.97	2.85	2.75	2.68	2.61	2.51	2.41	2.30	2.24	2.18	2.12	2.05	1.98	1.91
26	5.66	4.27	3.67	3.33	3.10	2.94	2.82	2.73	2.65	2.59	2.49	2.39	2.28	2.22	2.16	2.09	2.03	1.95	1.88
27	5.63	4.24	3.65	3.31	3.08	2.92	2.80	2.71	2.63	2.57	2.47	2.36	2.25	2.19	2.13	2.07	2.00	1.93	1.85
28	5.61	4.22	3.63	3.29	3.06	2.90	2.78	2.69	2.61	2.55	2.45	2.34	2.23	2.17	2.11	2.05	1.98	1.91	1.83
29	5.59	4.20	3.61	3.27	3.04	2.88	2.76	2.67	2.59	2.53	2.43	2.32	2.21	2.15	2.09	2.03	1.96	1.89	1.81
30	5.57	4.18	3.59	3.25	3.03	2.87	2.75	2.65	2.57	2.51	2.41	2.31	2.20	2.14	2.07	2.01	1.94	1.87	1.79
40	5.42	4.05	3.46	3.13	2.90	2.74	2.62	2.53	2.45	2.39	2.29	2.18	2.07	2.01	1.94	1.88	1.80	1.72	1.64
60	5.29	3.93	3.34	3.01	2.79	2.63	2.51	2.41	2.33	2.27	2.17	2.06	1.94	1.88	1.82	1.74	1.67	1.58	1.48
120	5.15	3.80	3.23	2.89	2.67	2.52	2.39	2.30	2.22	2.16	2.05	1.94	1.82	1.76	1.69	1.61	1.53	1.43	1.31
∞	5.02	3.69	3.12	2.79	2.57	2.41	2.29	2.19	2.11	2.05	1.94	1.83	1.71	1.64	1.57	1.48	1.39	1.27	1.00

TABLE OF $F_{.01}$ VALUES

Numerator Degrees of Freedom

Denominator Degrees of Freedom	1	2	3	4	5	6	7	8	9	10	12	15	20	24	30	40	60	120	∞
1	4,052	4,999.5	5,403	5,625	5,764	5,859	5,928	5,982	6,022	6,056	6,106	6,157	6,209	6,235	6,261	6,287	6,313	6,339	6,366
2	98.50	99.00	99.17	99.25	99.30	99.33	99.36	99.37	99.39	99.40	99.42	99.43	99.45	99.46	99.47	99.47	99.48	99.49	99.50
3	34.12	30.82	29.46	28.71	28.24	27.91	27.67	27.49	27.35	27.23	27.05	26.87	26.69	26.60	26.50	26.41	26.32	26.22	26.13
4	21.20	18.00	16.69	15.98	15.52	15.21	14.98	14.80	14.66	14.55	14.37	14.20	14.02	13.93	13.84	13.75	13.65	13.56	13.46
5	16.26	13.27	12.06	11.39	10.97	10.67	10.46	10.29	10.16	10.05	9.89	9.72	9.55	9.47	9.38	9.29	9.20	9.11	9.06
6	13.75	10.92	9.78	9.15	8.75	8.47	8.26	8.10	7.98	7.87	7.72	7.56	7.40	7.31	7.23	7.14	7.06	6.97	6.88
7	12.25	9.55	8.45	7.85	7.46	7.19	6.99	6.84	6.72	6.62	6.47	6.31	6.16	6.07	5.99	5.91	5.82	5.74	5.65
8	11.26	8.65	7.59	7.01	6.63	6.37	6.18	6.03	5.91	5.81	5.67	5.52	5.36	5.28	5.20	5.12	5.03	4.95	4.86
9	10.56	8.02	6.99	6.42	6.06	5.80	5.61	5.47	5.35	5.26	5.11	4.96	4.81	4.73	4.65	4.57	4.48	4.40	4.31
10	10.04	7.56	6.55	5.99	5.64	5.39	5.20	5.06	4.94	4.85	4.71	4.56	4.41	4.33	4.25	4.17	4.08	4.00	3.91
11	9.65	7.21	6.22	5.67	5.32	5.07	4.89	4.74	4.63	4.54	4.40	4.25	4.10	4.02	3.94	3.86	3.78	3.69	3.60
12	9.33	6.93	5.95	5.41	5.06	4.82	4.64	4.50	4.39	4.30	4.16	4.01	3.86	3.78	3.70	3.62	3.54	3.45	3.36
13	9.07	6.70	5.74	5.21	4.86	4.62	4.44	4.30	4.19	4.10	3.96	3.82	3.66	3.59	3.51	3.43	3.34	3.25	3.17
14	8.86	6.51	5.56	5.04	4.69	4.46	4.28	4.14	4.03	3.94	3.80	3.66	3.51	3.43	3.35	3.27	3.18	3.09	3.00
15	8.68	6.36	5.42	4.89	4.56	4.32	4.14	4.00	3.89	3.80	3.67	3.52	3.37	3.29	3.21	3.13	3.05	2.96	2.87
16	8.53	6.23	5.29	4.77	4.44	4.20	4.03	3.89	3.78	3.69	3.55	3.41	3.26	3.18	3.10	3.02	2.93	2.84	2.75
17	8.40	6.11	5.18	4.67	4.34	4.10	3.93	3.79	3.68	3.59	3.46	3.31	3.16	3.08	3.00	2.92	2.83	2.75	2.65
18	8.29	6.01	5.09	4.58	4.25	4.01	3.84	3.71	3.60	3.51	3.37	3.23	3.08	3.00	2.92	2.84	2.75	2.66	2.57
19	8.18	5.93	5.01	4.50	4.17	3.94	3.77	3.63	3.52	3.43	3.30	3.15	3.00	2.92	2.84	2.76	2.67	2.58	2.49
20	8.10	5.85	4.94	4.43	4.10	3.87	3.70	3.56	3.46	3.37	3.23	3.09	2.94	2.86	2.78	2.69	2.61	2.52	2.42
21	8.02	5.78	4.87	4.37	4.04	3.81	3.64	3.51	3.40	3.31	3.17	3.03	2.88	2.80	2.72	2.64	2.55	2.46	2.36
22	7.95	5.72	4.82	4.31	3.99	3.76	3.59	3.45	3.35	3.26	3.12	2.98	2.83	2.75	2.67	2.58	2.50	2.40	2.31
23	7.88	5.66	4.76	4.26	3.94	3.71	3.54	3.41	3.30	3.21	3.07	2.93	2.78	2.70	2.62	2.54	2.45	2.35	2.26
24	7.82	5.61	4.72	4.22	3.90	3.67	3.50	3.36	3.26	3.17	3.03	2.89	2.74	2.66	2.58	2.49	2.40	2.31	2.21
25	7.77	5.57	4.68	4.18	3.85	3.63	3.46	3.32	3.22	3.13	2.99	2.85	2.70	2.62	2.54	2.45	2.36	2.27	2.17
26	7.72	5.53	4.64	4.14	3.82	3.59	3.42	3.29	3.18	3.09	2.96	2.81	2.66	2.58	2.50	2.42	2.33	2.23	2.13
27	7.68	5.49	4.60	4.11	3.78	3.56	3.39	3.26	3.15	3.06	2.93	2.78	2.63	2.55	2.47	2.38	2.29	2.20	2.10
28	7.64	5.45	4.57	4.07	3.75	3.53	3.36	3.23	3.12	3.03	2.90	2.75	2.60	2.52	2.44	2.35	2.26	2.17	2.06
29	7.60	5.42	4.54	4.04	3.73	3.50	3.33	3.20	3.09	3.00	2.87	2.73	2.57	2.49	2.41	2.33	2.23	2.14	2.03
30	7.56	5.39	4.51	4.02	3.70	3.47	3.30	3.17	3.07	2.98	2.84	2.70	2.55	2.47	2.39	2.30	2.21	2.11	2.01
40	7.31	5.18	4.31	3.83	3.51	3.29	3.12	2.99	2.89	2.80	2.66	2.52	2.37	2.29	2.20	2.11	2.02	1.92	1.80
60	7.08	4.98	4.13	3.65	3.34	3.12	2.95	2.82	2.72	2.63	2.50	2.35	2.20	2.12	2.03	1.94	1.84	1.73	1.60
120	6.85	4.79	3.95	3.48	3.17	2.96	2.79	2.66	2.56	2.47	2.34	2.19	2.03	1.95	1.86	1.76	1.66	1.53	1.38
∞	6.63	4.61	3.78	3.32	3.02	2.80	2.64	2.51	2.41	2.32	2.18	2.04	1.88	1.79	1.70	1.59	1.47	1.32	1.00

TABLE 5   Binomial Probabilities

Entries in the table give the probability of x successes in n trials of a binomial experiment, where p is the probability of a success on one trial. For example, with six trials and $p = .40$, the probability of two successes is .3110.

n	x	.05	.10	.15	.20	p .25	.30	.35	.40	.45	.50
1	0	.9500	.9000	.8500	.8000	.7500	.7000	.6500	.6000	.5500	.5000
	1	.0500	.1000	.1500	.2000	.2500	.3000	.3500	.4000	.4500	.5000
2	0	.9025	.8100	.7225	.6400	.5625	.4900	.4225	.3600	.3025	.2500
	1	.0950	.1800	.2550	.3200	.3750	.4200	.4550	.4800	.4950	.5000
	2	.0025	.0100	.0225	.0400	.0625	.0900	.1225	.1600	.2025	.2500
3	0	.8574	.7290	.6141	.5120	.4219	.3430	.2746	.2160	.1664	.1250
	1	.1354	.2430	.3251	.3840	.4219	.4410	.4436	.4320	.4084	.3750
	2	.0071	.0270	.0574	.0960	.1406	.1890	.2389	.2880	.3341	.3750
	3	.0001	.0010	.0034	.0080	.0156	.0270	.0429	.0640	.0911	.1250
4	0	.8145	.6561	.5220	.4096	.3164	.2401	.1785	.1296	.0915	.0625
	1	.1715	.2916	.3685	.4096	.4219	.4116	.3845	.3456	.2995	.2500
	2	.0135	.0486	.0975	.1536	.2109	.2646	.3105	.3456	.3675	.3750
	3	.0005	.0036	.0115	.0256	.0469	.0756	.1115	.1536	.2005	.2500
	4	.0000	.0001	.0005	.0016	.0039	.0081	.0150	.0256	.0410	.0625
5	0	.7738	.5905	.4437	.3277	.2373	.1681	.1160	.0778	.0503	.0312
	1	.2036	.3280	.3915	.4096	.3955	.3602	.3124	.2592	.2059	.1562
	2	.0214	.0729	.1382	.2048	.2637	.3087	.3364	.3456	.3369	.3125
	3	.0011	.0081	.0244	.0512	.0879	.1323	.1811	.2304	.2757	.3125
	4	.0000	.0004	.0022	.0064	.0146	.0284	.0488	.0768	.1128	.1562
	5	.0000	.0000	.0001	.0003	.0010	.0024	.0053	.0102	.0185	.0312
6	0	.7351	.5314	.3771	.2621	.1780	.1176	.0754	.0467	.0277	.0156
	1	.2321	.3543	.3993	.3932	.3560	.3025	.2437	.1866	.1359	.0938
	2	.0305	.0984	.1762	.2458	.2966	.3241	.3280	.3110	.2780	.2344
	3	.0021	.0146	.0415	.0819	.1318	.1852	.2355	.2765	.3032	.3125
	4	.0001	.0012	.0055	.0154	.0330	.0595	.0951	.1382	.1861	.2344
	5	.0000	.0001	.0004	.0015	.0044	.0102	.0205	.0369	.0609	.0938
	6	.0000	.0000	.0000	.0001	.0002	.0007	.0018	.0041	.0083	.0156
7	0	.6983	.4783	.3206	.2097	.1335	.0824	.0490	.0280	.0152	.0078
	1	.2573	.3720	.3960	.3670	.3115	.2471	.1848	.1306	.0872	.0547
	2	.0406	.1240	.2097	.2753	.3115	.3177	.2985	.2613	.2140	.1641
	3	.0036	.0230	.0617	.1147	.1730	.2269	.2679	.2903	.2918	.2734
	4	.0002	.0026	.0109	.0287	.0577	.0972	.1442	.1935	.2388	.2734
	5	.0000	.0002	.0012	.0043	.0115	.0250	.0466	.0774	.1172	.1641
	6	.0000	.0000	.0001	.0004	.0013	.0036	.0084	.0172	.0320	.0547
	7	.0000	.0000	.0000	.0000	.0001	.0002	.0006	.0016	.0037	.0078

TABLE 5 *(Continued)*

n	x	.05	.10	.15	.20	P .25	.30	.35	.40	.45	.50
8	0	.6634	.4305	.2725	.1678	.1001	.0576	.0319	.0168	.0084	.0039
	1	.2793	.3826	.3847	.3355	.2670	.1977	.1373	.0896	.0548	.0312
	2	.0515	.1488	.2376	.2936	.3115	.2965	.2587	.2090	.1569	.1094
	3	.0054	.0331	.0839	.1468	.2076	.2541	.2786	.2787	.2568	.2188
	4	.0004	.0046	.0185	.0459	.0865	.1361	.1875	.2322	.2627	.2734
	5	.0000	.0004	.0026	.0092	.0231	.0467	.0808	.1239	.1719	.2188
	6	.0000	.0000	.0002	.0011	.0038	.0100	.0217	.0413	.0703	.1094
	7	.0000	.0000	.0000	.0001	.0004	.0012	.0033	.0079	.0164	.0312
	8	.0000	.0000	.0000	.0000	.0000	.0001	.0002	.0007	.0017	.0039
9	0	.6302	.3874	.2316	.1342	.0751	.0404	.0207	.0101	.0046	.0020
	1	.2985	.3874	.3679	.3020	.2253	.1556	.1004	.0605	.0339	.0176
	2	.0629	.1722	.2597	.3020	.3003	.2668	.2162	.1612	.1110	.0703
	3	.0077	.0446	.1069	.1762	.2336	.2668	.2716	.2508	.2119	.1641
	4	.0006	.0074	.0283	.0661	.1168	.1715	.2194	.2508	.2600	.2461
	5	.0000	.0008	.0050	.0165	.0389	.0735	.1181	.1672	.2128	.2461
	6	.0000	.0001	.0006	.0028	.0087	.0210	.0424	.0743	.1160	.1641
	7	.0000	.0000	.0000	.0003	.0012	.0039	.0098	.0212	.0407	.0703
	8	.0000	.0000	.0000	.0000	.0001	.0004	.0013	.0035	.0083	.0176
	9	.0000	.0000	.0000	.0000	.0000	.0000	.0001	.0003	.0008	.0020
10	0	.5987	.3487	.1969	.1074	.0563	.0282	.0135	.0060	.0025	.0010
	1	.3151	.3874	.3474	.2684	.1877	.1211	.0725	.0403	.0207	.0098
	2	.0746	.1937	.2759	.3020	.2816	.2335	.1757	.1209	.0763	.0439
	3	.0105	.0574	.1298	.2013	.2503	.2668	.2522	.2150	.1665	.1172
	4	.0010	.0112	.0401	.0881	.1460	.2001	.2377	.2508	.2384	.2051
	5	.0001	.0015	.0085	.0264	.0584	.1029	.1536	.2007	.2340	.2461
	6	.0000	.0001	.0012	.0055	.0162	.0368	.0689	.1115	.1596	.2051
	7	.0000	.0000	.0001	.0008	.0031	.0090	.0212	.0425	.0746	.1172
	8	.0000	.0000	.0000	.0001	.0004	.0014	.0043	.0106	.0229	.0439
	9	.0000	.0000	.0000	.0000	.0000	.0001	.0005	.0016	.0042	.0098
	10	.0000	.0000	.0000	.0000	.0000	.0000	.0000	.0001	.0003	.0010
11	0	.5688	.3138	.1673	.0859	.0422	.0198	.0088	.0036	.0014	.0005
	1	.3293	.3835	.3248	.2362	.1549	.0932	.0518	.0266	.0125	.0054
	2	.0867	.2131	.2866	.2953	.2581	.1998	.1395	.0887	.0513	.0269
	3	.0137	.0710	.1517	.2215	.2581	.2568	.2254	.1774	.1259	.0806
	4	.0014	.0158	.0536	.1107	.1721	.2201	.2428	.2365	.2060	.1611
	5	.0001	.0025	.0132	.0388	.0803	.1321	.1830	.2207	.2360	.2256
	6	.0000	.0003	.0023	.0097	.0268	.0566	.0985	.1471	.1931	.2256
	7	.0000	.0000	.0003	.0017	.0064	.0173	.0379	.0701	.1128	.1611
	8	.0000	.0000	.0000	.0002	.0011	.0037	.0102	.0234	.0462	.0806
	9	.0000	.0000	.0000	.0000	.0001	.0005	.0018	.0052	.0126	.0269
	10	.0000	.0000	.0000	.0000	.0000	.0000	.0002	.0007	.0021	.0054
	11	.0000	.0000	.0000	.0000	.0000	.0000	.0000	.0000	.0002	.0005

TABLE 5 (Continued)

n	x	.05	.10	.15	.20	p .25	.30	.35	.40	.45	.50
12	0	.5404	.2824	.1422	.0687	.0317	.0138	.0057	.0022	.0008	.0002
	1	.3413	.3766	.3012	.2062	.1267	.0712	.0368	.0174	.0075	.0029
	2	.0988	.2301	.2924	.2835	.2323	.1678	.1088	.0639	.0339	.0161
	3	.0173	.0853	.1720	.2362	.2581	.2397	.1954	.1419	.0923	.0537
	4	.0021	.0213	.0683	.1329	.1936	.2311	.2367	.2128	.1700	.1208
	5	.0002	.0038	.0193	.0532	.1032	.1585	.2039	.2270	.2225	.1934
	6	.0000	.0005	.0040	.0155	.0401	.0792	.1281	.1766	.2124	.2256
	7	.0000	.0000	.0006	.0033	.0115	.0291	.0591	.1009	.1489	.1934
	8	.0000	.0000	.0001	.0005	.0024	.0078	.0199	.0420	.0762	.1208
	9	.0000	.0000	.0000	.0001	.0004	.0015	.0048	.0125	.0277	.0537
	10	.0000	.0000	.0000	.0000	.0000	.0002	.0008	.0025	.0068	.0161
	11	.0000	.0000	.0000	.0000	.0000	.0000	.0001	.0003	.0010	.0029
	12	.0000	.0000	.0000	.0000	.0000	.0000	.0000	.0000	.0001	.0002
13	0	.5133	.2542	.1209	.0550	.0238	.0097	.0037	.0013	.0004	.0001
	1	.3512	.3672	.2774	.1787	.1029	.0540	.0259	.0113	.0045	.0016
	2	.1109	.2448	.2937	.2680	.2059	.1388	.0836	.0453	.0220	.0095
	3	.0214	.0997	.1900	.2457	.2517	.2181	.1651	.1107	.0660	.0349
	4	.0028	.0277	.0838	.1535	.2097	.2337	.2222	.1845	.1350	.0873
	5	.0003	.0055	.0266	.0691	.1258	.1803	.2154	.2214	.1989	.1571
	6	.0000	.0008	.0063	.0230	.0559	.1030	.1546	.1968	.2169	.2095
	7	.0000	.0001	.0011	.0058	.0186	.0442	.0833	.1312	.1775	.2095
	8	.0000	.0000	.0001	.0011	.0047	.0142	.0336	.0656	.1089	.1571
	9	.0000	.0000	.0000	.0001	.0009	.0034	.0101	.0243	.0495	.0873
	10	.0000	.0000	.0000	.0000	.0001	.0006	.0022	.0065	.0162	.0349
	11	.0000	.0000	.0000	.0000	.0000	.0001	.0003	.0012	.0036	.0095
	12	.0000	.0000	.0000	.0000	.0000	.0000	.0000	.0001	.0005	.0016
	13	.0000	.0000	.0000	.0000	.0000	.0000	.0000	.0000	.0000	.0001
14	0	.4877	.2288	.1028	.0440	.0178	.0068	.0024	.0008	.0002	.0001
	1	.3593	.3559	.2539	.1539	.0832	.0407	.0181	.0073	.0027	.0009
	2	.1229	.2570	.2912	.2501	.1802	.1134	.0634	.0317	.0141	.0056
	3	.0259	.1142	.2056	.2501	.2402	.1943	.1366	.0845	.0462	.0222
	4	.0037	.0349	.0998	.1720	.2202	.2290	.2022	.1549	.1040	.0611
	5	.0004	.0078	.0352	.0860	.1468	.1963	.2178	.2066	.1701	.1222
	6	.0000	.0013	.0093	.0322	.0734	.1262	.1759	.2066	.2088	.1833
	7	.0000	.0002	.0019	.0092	.0280	.0618	.1082	.1574	.1952	.2095
	8	.0000	.0000	.0003	.0020	.0082	.0232	.0510	.0918	.1398	.1833
	9	.0000	.0000	.0000	.0003	.0018	.0066	.0183	.0408	.0762	.1222
	10	.0000	.0000	.0000	.0000	.0003	.0014	.0049	.0136	.0312	.0611
	11	.0000	.0000	.0000	.0000	.0000	.0002	.0010	.0033	.0093	.0222
	12	.0000	.0000	.0000	.0000	.0000	.0000	.0001	.0005	.0019	.0056
	13	.0000	.0000	.0000	.0000	.0000	.0000	.0000	.0001	.0002	.0009
	14	.0000	.0000	.0000	.0000	.0000	.0000	.0000	.0000	.0000	.0001

TABLE 5 *(Continued)*

n	x	.05	.10	.15	.20	p .25	.30	.35	.40	.45	.50
15	0	.4633	.2059	.0874	.0352	.0134	.0047	.0016	.0005	.0001	.0000
	1	.3658	.3432	.2312	.1319	.0668	.0305	.0126	.0047	.0016	.0005
	2	.1348	.2669	.2856	.2309	.1559	.0916	.0476	.0219	.0090	.0032
	3	.0307	.1285	.2184	.2501	.2252	.1700	.1110	.0634	.0318	.0139
	4	.0049	.0428	.1156	.1876	.2252	.2186	.1792	.1268	.0780	.0417
	5	.0006	.0105	.0449	.1032	.1651	.2061	.2123	.1859	.1404	.0916
	6	.0000	.0019	.0132	.0430	.0917	.1472	.1906	.2066	.1914	.1527
	7	.0000	.0003	.0030	.0138	.0393	.0811	.1319	.1771	.2013	.1964
	8	.0000	.0000	.0005	.0035	.0131	.0348	.0710	.1181	.1647	.1964
	9	.0000	.0000	.0001	.0007	.0034	.0116	.0298	.0612	.1048	.1527
	10	.0000	.0000	.0000	.0001	.0007	.0030	.0096	.0245	.0515	.0916
	11	.0000	.0000	.0000	.0000	.0001	.0006	.0024	.0074	.0191	.0417
	12	.0000	.0000	.0000	.0000	.0000	.0001	.0004	.0016	.0052	.0139
	13	.0000	.0000	.0000	.0000	.0000	.0000	.0001	.0003	.0010	.0032
	14	.0000	.0000	.0000	.0000	.0000	.0000	.0000	.0000	.0001	.0005
	15	.0000	.0000	.0000	.0000	.0000	.0000	.0000	.0000	.0000	.0000
16	0	.4401	.1853	.0743	.0281	.0100	.0033	.0010	.0003	.0001	.0000
	1	.3706	.3294	.2097	.1126	.0535	.0228	.0087	.0030	.0009	.0002
	2	.1463	.2745	.2775	.2111	.1336	.0732	.0353	.0150	.0056	.0018
	3	.0359	.1423	.2285	.2463	.2079	.1465	.0888	.0468	.0215	.0085
	4	.0061	.0514	.1311	.2001	.2252	.2040	.1553	.1014	.0572	.0278
	5	.0008	.0137	.0555	.1201	.1802	.2099	.2008	.1623	.1123	.0667
	6	.0001	.0028	.0180	.0550	.1101	.1649	.1982	.1983	.1684	.1222
	7	.0000	.0004	.0045	.0197	.0524	.1010	.1524	.1889	.1969	.1746
	8	.0000	.0001	.0009	.0055	.0197	.0487	.0923	.1417	.1812	.1964
	9	.0000	.0000	.0001	.0012	.0058	.0185	.0442	.0840	.1318	.1746
	10	.0000	.0000	.0000	.0002	.0014	.0056	.0167	.0392	.0755	.1222
	11	.0000	.0000	.0000	.0000	.0002	.0013	.0049	.0142	.0337	.0667
	12	.0000	.0000	.0000	.0000	.0000	.0002	.0011	.0040	.0115	.0278
	13	.0000	.0000	.0000	.0000	.0000	.0000	.0002	.0008	.0029	.0085
	14	.0000	.0000	.0000	.0000	.0000	.0000	.0000	.0001	.0005	.0018
	15	.0000	.0000	.0000	.0000	.0000	.0000	.0000	.0000	.0001	.0002
	16	.0000	.0000	.0000	.0000	.0000	.0000	.0000	.0000	.0000	.0000
17	0	.4181	.1668	.0631	.0225	.0075	.0023	.0007	.0002	.0000	.0000
	1	.3741	.3150	.1893	.0957	.0426	.0169	.0060	.0019	.0005	.0001
	2	.1575	.2800	.2673	.1914	.1136	.0581	.0260	.0102	.0035	.0010
	3	.0415	.1556	.2359	.2393	.1893	.1245	.0701	.0341	.0144	.0052
	4	.0076	.0605	.1457	.2093	.2209	.1868	.1320	.0796	.0411	.0182
	5	.0010	.0175	.0668	.1361	.1914	.2081	.1849	.1379	.0875	.0472
	6	.0001	.0039	.0236	.0680	.1276	.1784	.1991	.1839	.1432	.0944
	7	.0000	.0007	.0065	.0267	.0668	.1201	.1685	.1927	.1841	.1484
	8	.0000	.0001	.0014	.0084	.0279	.0644	.1134	.1006	.1883	.1855
	9	.0000	.0000	.0003	.0021	.0093	.0276	.0611	.1070	.1540	.1855

TABLE 5 (Continued)

n	x	.05	.10	.15	.20	p .25	.30	.35	.40	.45	.50
17	10	.0000	.0000	.0000	.0004	.0025	.0095	.0263	.0571	.1008	.1484
	11	.0000	.0000	.0000	.0001	.0005	.0026	.0090	.0242	.0525	.0944
	12	.0000	.0000	.0000	.0000	.0001	.0006	.0024	.0081	.0215	.0472
	13	.0000	.0000	.0000	.0000	.0000	.0001	.0005	.0021	.0068	.0182
	14	.0000	.0000	.0000	.0000	.0000	.0000	.0001	.0004	.0016	.0052
	15	.0000	.0000	.0000	.0000	.0000	.0000	.0000	.0001	.0003	.0010
	16	.0000	.0000	.0000	.0000	.0000	.0000	.0000	.0000	.0000	.0001
	17	.0000	.0000	.0000	.0000	.0000	.0000	.0000	.0000	.0000	.0000
18	0	.3972	.1501	.0536	.0180	.0056	.0016	.0004	.0001	.0000	.0000
	1	.3763	.3002	.1704	.0811	.0338	.0126	.0042	.0012	.0003	.0001
	2	.1683	.2835	.2556	.1723	.0958	.0458	.0190	.0069	.0022	.0006
	3	.0473	.1680	.2406	.2297	.1704	.1046	.0547	.0246	.0095	.0031
	4	.0093	.0700	.1592	.2153	.2130	.1681	.1104	.0614	.0291	.0117
	5	.0014	.0218	.0787	.1507	.1988	.2017	.1664	.1146	.0666	.0327
	6	.0002	.0052	.0301	.0816	.1436	.1873	.1941	.1655	.1181	.0708
	7	.0000	.0010	.0091	.0350	.0820	.1376	.1792	.1892	.1657	.1214
	8	.0000	.0002	.0022	.0120	.0376	.0811	.1327	.1734	.1864	.1669
	9	.0000	.0000	.0004	.0033	.0139	.0386	.0794	.1284	.1694	.1855
	10	.0000	.0000	.0001	.0008	.0042	.0149	.0385	.0771	.1248	.1669
	11	.0000	.0000	.0000	.0001	.0010	.0046	.0151	.0374	.0742	.1214
	12	.0000	.0000	.0000	.0000	.0002	.0012	.0047	.0145	.0354	.0708
	13	.0000	.0000	.0000	.0000	.0000	.0002	.0012	.0045	.0134	.0327
	14	.0000	.0000	.0000	.0000	.0000	.0000	.0002	.0011	.0039	.0117
	15	.0000	.0000	.0000	.0000	.0000	.0000	.0000	.0002	.0009	.0031
	16	.0000	.0000	.0000	.0000	.0000	.0000	.0000	.0000	.0001	.0006
	17	.0000	.0000	.0000	.0000	.0000	.0000	.0000	.0000	.0000	.0001
	18	.0000	.0000	.0000	.0000	.0000	.0000	.0000	.0000	.0000	.0000
19	0	.3774	.1351	.0456	.0144	.0042	.0011	.0003	.0001	.0000	.0000
	1	.3774	.2852	.1529	.0685	.0268	.0093	.0029	.0008	.0002	.0000
	2	.1787	.2852	.2428	.1540	.0803	.0358	.0138	.0046	.0013	.0003
	3	.0533	.1796	.2428	.2182	.1517	.0869	.0422	.0175	.0062	.0018
	4	.0112	.0798	.1714	.2182	.2033	.1491	.0909	.0467	.0203	.0074
	5	.0018	.0266	.0907	.1636	.2023	.1916	.1468	.0933	.0497	.0222
	6	.0002	.0069	.0374	.0955	.1574	.1916	.1844	.1451	.0949	.0518
	7	.0000	.0014	.0122	.0443	.0974	.1525	.1844	.1797	.1443	.0961
	8	.0000	.0002	.0032	.0166	.0487	.0981	.1489	.1797	.1771	.1442
	9	.0000	.0000	.0007	.0051	.0198	.0514	.0980	.1464	.1771	.1762
	10	.0000	.0000	.0001	.0013	.0066	.0220	.0528	.0976	.1449	.1762
	11	.0000	.0000	.0000	.0003	.0018	.0077	.0233	.0532	.0970	.1442
	12	.0000	.0000	.0000	.0000	.0004	.0022	.0083	.0237	.0529	.0961
	13	.0000	.0000	.0000	.0000	.0001	.0005	.0024	.0085	.0233	.0518
	14	.0000	.0000	.0000	.0000	.0000	.0001	.0006	.0024	.0082	.0222

TABLE 5 *(Continued)*

n	x	.05	.10	.15	.20	p .25	.30	.35	.40	.45	.50
19	15	.0000	.0000	.0000	.0000	.0000	.0000	.0001	.0005	.0022	.0074
	16	.0000	.0000	.0000	.0000	.0000	.0000	.0000	.0001	.0005	.0018
	17	.0000	.0000	.0000	.0000	.0000	.0000	.0000	.0000	.0001	.0003
	18	.0000	.0000	.0000	.0000	.0000	.0000	.0000	.0000	.0000	.0000
	19	.0000	.0000	.0000	.0000	.0000	.0000	.0000	.0000	.0000	.0000
20	0	.3585	.1216	.0388	.0115	.0032	.0008	.0002	.0000	.0000	.0000
	1	.3774	.2702	.1368	.0576	.0211	.0068	.0020	.0005	.0001	.0000
	2	.1887	.2852	.2293	.1369	.0669	.0278	.0100	.0031	.0008	.0002
	3	.0596	.1901	.2428	.2054	.1339	.0716	.0323	.0123	.0040	.0011
	4	.0133	.0898	.1821	.2182	.1897	.1304	.0738	.0350	.0139	.0046
	5	.0022	.0319	.1028	.1746	.2023	.1789	.1272	.0746	.0365	.0148
	6	.0003	.0089	.0454	.1091	.1686	.1916	.1712	.1244	.0746	.0370
	7	.0000	.0020	.0160	.0545	.1124	.1643	.1844	.1659	.1221	.0739
	8	.0000	.0004	.0046	.0222	.0609	.1144	.1614	.1797	.1623	.1201
	9	.0000	.0001	.0011	.0074	.0271	.0654	.1158	.1597	.1771	.1602
	10	.0000	.0000	.0002	.0020	.0099	.0308	.0686	.1171	.1593	.1762
	11	.0000	.0000	.0000	.0005	.0030	.0120	.0336	.0710	.1185	.1602
	12	.0000	.0000	.0000	.0001	.0008	.0039	.0136	.0355	.0727	.1201
	13	.0000	.0000	.0000	.0000	.0002	.0010	.0045	.0146	.0366	.0739
	14	.0000	.0000	.0000	.0000	.0000	.0002	.0012	.0049	.0150	.0370
	15	.0000	.0000	.0000	.0000	.0000	.0000	.0003	.0013	.0049	.0148
	16	.0000	.0000	.0000	.0000	.0000	.0000	.0000	.0003	.0013	.0046
	17	.0000	.0000	.0000	.0000	.0000	.0000	.0000	.0000	.0002	.0011
	18	.0000	.0000	.0000	.0000	.0000	.0000	.0000	.0000	.0000	.0002
	19	.0000	.0000	.0000	.0000	.0000	.0000	.0000	.0000	.0000	.0000
	20	.0000	.0000	.0000	.0000	.0000	.0000	.0000	.0000	.0000	.0000

TABLE 6 Values of $e^{-\mu}$

μ	$e^{-\mu}$	μ	$e^{-\mu}$	μ	$e^{-\mu}$
.0	1.0000	3.1	.0450	8.0	.000335
.1	.9048	3.2	.0408	9.0	.000123
.2	.8187	3.3	.0369	10.0	.000045
.3	.7408	3.4	.0334		
.4	.6703	3.5	.0302		
.5	.6065	3.6	.0273		
.6	.5488	3.7	.0247		
.7	.4966	3.8	.0224		
.8	.4493	3.9	.0202		
.9	.4066	4.0	.0183		
1.0	.3679	4.1	.0166		
1.1	.3379	4.2	.0150		
1.2	.3012	4.3	.0136		
1.3	.2725	4.4	.0123		
1.4	.2466	4.5	.0111		
1.5	.2231	4.6	.0101		
1.6	.2019	4.7	.0091		
1.7	.1827	4.8	.0082		
1.8	.1653	4.9	.0074		
1.9	.1496	5.0	.0067		
2.0	.1353	5.1	.0061		
2.1	.1225	5.2	.0055		
2.2	.1108	5.3	.0050		
2.3	.1003	5.4	.0045		
2.4	.0907	5.5	.0041		
2.5	.0821	5.6	.0037		
2.6	.0743	5.7	.0033		
2.7	.0672	5.8	.0030		
2.8	.0608	5.9	.0027		
2.9	.0550	6.0	.0025		
3.0	.0498	7.0	.0009		

TABLE 7   Poisson Probabilities

Entries in the table give the probability of x occurrences for a Poisson process with a mean μ. For example, when $\mu = 2.5$, the probability of four occurrences is .1336.

x	0.1	0.2	0.3	0.4	μ 0.5	0.6	0.7	0.8	0.9	1.0
0	.9048	.8187	.7408	.6703	.6065	.5488	.4966	.4493	.4066	.3679
1	.0905	.1637	.2222	.2681	.3033	.3293	.3476	.3595	.3659	.3679
2	.0045	.0164	.0333	.0536	.0758	.0988	.1217	.1438	.1647	.1839
3	.0002	.0011	.0033	.0072	.0126	.0198	.0284	.0383	.0494	.0613
4	.0000	.0001	.0002	.0007	.0016	.0030	.0050	.0077	.0111	.0153
5	.0000	.0000	.0000	.0001	.0002	.0004	.0007	.0012	.0020	.0031
6	.0000	.0000	.0000	.0000	.0000	.0000	.0001	.0002	.0003	.0005
7	.0000	.0000	.0000	.0000	.0000	.0000	.0000	.0000	.0000	.0001

x	1.1	1.2	1.3	1.4	μ 1.5	1.6	1.7	1.8	1.9	2.0
0	.3329	.3012	.2725	.2466	.2231	.2019	.1827	.1653	.1496	.1353
1	.3662	.3614	.3543	.3452	.3347	.3230	.3106	.2975	.2842	.2707
2	.2014	.2169	.2303	.2417	.2510	.2584	.2640	.2678	.2700	.2707
3	.0738	.0867	.0998	.1128	.1255	.1378	.1496	.1607	.1710	.1804
4	.0203	.0260	.0324	.0395	.0471	.0551	.0636	.0723	.0812	.0902
5	.0045	.0062	.0084	.0111	.0141	.0176	.0216	.0260	.0309	.0361
6	.0008	.0012	.0018	.0026	.0035	.0047	.0061	.0078	.0098	.0120
7	.0001	.0002	.0003	.0005	.0008	.0011	.0015	.0020	.0027	.0034
8	.0000	.0000	.0001	.0001	.0001	.0002	.0003	.0005	.0006	.0009
9	.0000	.0000	.0000	.0000	.0000	.0000	.0001	.0001	.0001	.0002

x	2.1	2.2	2.3	2.4	μ 2.5	2.6	2.7	2.8	2.9	3.0
0	.1225	.1108	.1003	.0907	.0821	.0743	.0672	.0608	.0550	.0498
1	.2572	.2438	.2306	.2177	.2052	.1931	.1815	.1703	.1596	.1494
2	.2700	.2681	.2652	.2613	.2565	.2510	.2450	.2384	.2314	.2240
3	.1890	.1966	.2033	.2090	.2138	.2176	.2205	.2225	.2237	.2240
4	.0992	.1082	.1169	.1254	.1336	.1414	.1488	.1557	.1622	.1680

TABLE 7 (*Continued*)

x										
5	.0417	.0476	.0538	.0602	.0668	.0735	.0804	.0872	.0940	.1008
6	.0146	.0174	.0206	.0241	.0278	.0319	.0362	.0407	.0455	.0540
7	.0044	.0055	.0068	.0083	.0099	.0118	.0139	.0163	.0188	.0216
8	.0011	.0015	.0019	.0025	.0031	.0038	.0047	.0057	.0068	.0081
9	.0003	.0004	.0005	.0007	.0009	.0011	.0014	.0018	.0022	.0027
10	.0001	.0001	.0001	.0002	.0002	.0003	.0004	.0005	.0006	.0008
11	.0000	.0000	.0000	.0000	.0000	.0001	.0001	.0001	.0002	.0002
12	.0000	.0000	.0000	.0000	.0000	.0000	.0000	.0000	.0000	.0001

					μ					
x	3.1	3.2	3.3	3.4	3.5	3.6	3.7	3.8	3.9	4.0
0	.0450	.0408	.0369	.0344	.0302	.0273	.0247	.0224	.0202	.0183
1	.1397	.1304	.1217	.1135	.1057	.0984	.0915	.0850	.0789	.0733
2	.2165	.2087	.2008	.1929	.1850	.1771	.1692	.1615	.1539	.1465
3	.2237	.2226	.2209	.2186	.2158	.2125	.2087	.2046	.2001	.1954
4	.1734	.1781	.1823	.1858	.1888	.1912	.1931	.1944	.1951	.1954
5	.1075	.1140	.1203	.1264	.1322	.1377	.1429	.1477	.1522	.1563
6	.0555	.0608	.0662	.0716	.0771	.0826	.0881	.0936	.0989	.1042
7	.0246	.0278	.0312	.0348	.0385	.0425	.0466	.0508	.0551	.0595
8	.0095	.0111	.0129	.0148	.0169	.0191	.0215	.0241	.0269	.0298
9	.0033	.0040	.0047	.0056	.0066	.0076	.0089	.0102	.0116	.0132
10	.0010	.0013	.0016	.0019	.0023	.0028	.0033	.0039	.0045	.0053
11	.0003	.0004	.0005	.0006	.0007	.0009	.0011	.0013	.0016	.0019
12	.0001	.0001	.0001	.0002	.0002	.0003	.0003	.0004	.0005	.0006
13	.0000	.0000	.0000	.0000	.0001	.0001	.0001	.0001	.0002	.0002
14	.0000	.0000	.0000	.0000	.0000	.0000	.0000	.0000	.0000	.0001

					μ					
x	4.1	4.2	4.3	4.4	4.5	4.6	4.7	4.8	4.9	5.0
0	.0166	.0150	.0136	.0123	.0111	.0101	.0091	.0082	.0074	.0067
1	.0679	.0630	.0583	.0540	.0500	.0462	.0427	.0395	.0365	.0337
2	.1393	.1323	.1254	.1188	.1125	.1063	.1005	.0948	.0894	.0842
3	.1904	.1852	.1798	.1743	.1687	.1631	.1574	.1517	.1460	.1404
4	.1951	.1944	.1933	.1917	.1898	.1875	.1849	.1820	.1789	.1755
5	.1600	.1633	.1662	.1687	.1708	.1725	.1738	.1747	.1753	.1755
6	.1093	.1143	.1191	.1237	.1281	.1323	.1362	.1398	.1432	.1462
7	.0640	.0686	.0732	.0778	.0824	.0869	.0914	.0959	.1002	.1044
8	.0328	.0360	.0393	.0428	.0463	.0500	.0537	.0575	.0614	.0653
9	.0150	.0168	.0188	.0209	.0232	.0255	.0280	.0307	.0334	.0363

TABLE 7 (Continued)

x										
10	.0061	.0071	.0081	.0092	.0104	.0118	.0132	.0147	.0164	.0181
11	.0023	.0027	.0032	.0037	.0043	.0049	.0056	.0064	.0073	.0082
12	.0008	.0009	.0011	.0014	.0016	.0019	.0022	.0026	.0030	.0034
13	.0002	.0003	.0004	.0005	.0006	.0007	.0008	.0009	.0011	.0013
14	.0001	.0001	.0001	.0001	.0002	.0002	.0003	.0003	.0004	.0005
15	.0000	.0000	.0000	.0000	.0001	.0001	.0001	.0001	.0001	.0002

μ

x	5.1	5.2	5.3	5.4	5.5	5.6	5.7	5.8	5.9	6.0
0	.0061	.0055	.0050	.0045	.0041	.0037	.0033	.0030	.0027	.0025
1	.0311	.0287	.0265	.0244	.0225	.0207	.0191	.0176	.0162	.0149
2	.0793	.0746	.0701	.0659	.0618	.0580	.0544	.0509	.0477	.0446
3	.1348	.1293	.1239	.1185	.1133	.1082	.1033	.0982	.0938	.0892
4	.1719	.1681	.1641	.1600	.1558	.1515	.1472	.1428	.1383	.1339
5	.1753	.1748	.1740	.1728	.1714	.1697	.1678	.1656	.1632	.1606
6	.1490	.1515	.1537	.1555	.1571	.1584	.1594	.1601	.1605	.1606
7	.1086	.1125	.1163	.1200	.1234	.1267	.1298	.1326	.1353	.1377
8	.0692	.0731	.0771	.0810	.0849	.0887	.0925	.0962	.0998	.1033
9	.0392	.0423	.0454	.0486	.0519	.0552	.0586	.0620	.0654	.0688
10	.0200	.0220	.0241	.0262	.0285	.0309	.0334	.0359	.0386	.0413
11	.0093	.0104	.0116	.0129	.0143	.0157	.0173	.0190	.0207	.0225
12	.0039	.0045	.0051	.0058	.0065	.0073	.0082	.0092	.0102	.0113
13	.0015	.0018	.0021	.0024	.0028	.0032	.0036	.0041	.0046	.0052
14	.0006	.0007	.0008	.0009	.0011	.0013	.0015	.0017	.0019	.0022
15	.0002	.0002	.0003	.0003	.0004	.0005	.0006	.0007	.0008	.0009
16	.0001	.0001	.0001	.0001	.0001	.0002	.0002	.0002	.0003	.0003
17	.0000	.0000	.0000	.0000	.0000	.0001	.0001	.0001	.0001	.0001

μ

x	6.1	6.2	6.3	6.4	6.5	6.6	6.7	6.8	6.9	7.0
0	.0022	.0020	.0018	.0017	.0015	.0014	.0012	.0011	.0010	.0009
1	.0137	.0126	.0116	.0106	.0098	.0090	.0082	.0076	.0070	.0064
2	.0417	.0390	.0364	.0340	.0318	.0296	.0276	.0258	.0240	.0223
3	.0848	.0806	.0765	.0726	.0688	.0652	.0617	.0584	.0552	.0521
4	.1294	.1249	.1205	.1162	.1118	.1076	.1034	.0992	.0952	.0912
5	.1579	.1549	.1519	.1487	.1454	.1420	.1385	.1349	.1314	.1277
6	.1605	.1601	.1595	.1586	.1575	.1562	.1546	.1529	.1511	.1490
7	.1399	.1418	.1435	.1450	.1462	.1472	.1480	.1486	.1489	.1490
8	.1066	.1099	.1130	.1160	.1188	.1215	.1240	.1263	.1284	.1304
9	.0723	.0757	.0791	.0825	.0858	.0891	.0923	.0954	.0985	.1014

TABLE 7 *(Continued)*

10	.0441	.0469	.0498	.0528	.0558	.0588	.0618	.0649	.0679	.0710
11	.0245	.0265	.0285	.0307	.0330	.0353	.0377	.0401	.0426	.0452
12	.0124	.0137	.0150	.0164	.0179	.0194	.0210	.0227	.0245	.0264
13	.0058	.0065	.0073	.0081	.0089	.0098	.0108	.0119	.0130	.0142
14	.0025	.0029	.0033	.0037	.0041	.0046	.0052	.0058	.0064	.0071
15	.0010	.0012	.0014	.0016	.0018	.0020	.0023	.0026	.0029	.0033
16	.0004	.0005	.0005	.0006	.0007	.0008	.0010	.0011	.0013	.0014
17	.0001	.0002	.0002	.0002	.0003	.0003	.0004	.0004	.0005	.0006
18	.0000	.0001	.0001	.0001	.0001	.0001	.0001	.0002	.0002	.0002
19	.0000	.0000	.0000	.0000	.0000	.0000	.0000	.0001	.0001	.0001

					μ					
x	7.1	7.2	7.3	7.4	7.5	7.6	7.7	7.8	7.9	8.0
0	.0008	.0007	.0007	.0006	.0006	.0005	.0005	.0004	.0004	.0003
1	.0059	.0054	.0049	.0045	.0041	.0038	.0035	.0032	.0029	.0027
2	.0208	.0194	.0180	.0167	.0156	.0145	.0134	.0125	.0116	.0107
3	.0492	.0464	.0438	.0413	.0389	.0366	.0345	.0324	.0305	.0286
4	.0874	.0836	.0799	.0764	.0729	.0696	.0663	.0632	.0602	.0573
5	.1241	.1204	.1167	.1130	.1094	.1057	.1021	.0986	.0951	.0916
6	.1468	.1445	.1420	.1394	.1367	.1339	.1311	.1282	.1252	.1221
7	.1489	.1486	.1481	.1474	.1465	.1454	.1442	.1428	.1413	.1396
8	.1321	.1337	.1351	.1363	.1373	.1382	.1388	.1392	.1395	.1396
9	.1042	.1070	.1096	.1121	.1144	.1167	.1187	.1207	.1224	.1241
10	.0740	.0770	.0800	.0829	.0858	.0887	.0914	.0941	.0967	.0993
11	.0478	.0504	.0531	.0558	.0585	.0613	.0640	.0667	.0695	.0722
12	.0283	.0303	.0323	.0344	.0366	.0388	.0411	.0434	.0457	.0481
13	.0154	.0168	.0181	.0196	.0211	.0227	.0243	.0260	.0278	.0296
14	.0078	.0086	.0095	.0104	.0113	.0123	.0134	.0145	.0157	.0169
15	.0037	.0041	.0046	.0051	.0057	.0062	.0069	.0075	.0083	.0090
16	.0016	.0019	.0021	.0024	.0026	.0030	.0033	.0037	.0041	.0045
17	.0007	.0008	.0009	.0010	.0012	.0013	.0015	.0017	.0019	.0021
18	.0003	.0003	.0004	.0004	.0005	.0006	.0006	.0007	.0008	.0009
19	.0001	.0001	.0001	.0002	.0002	.0002	.0003	.0003	.0003	.0004
20	.0000	.0000	.0001	.0001	.0001	.0001	.0001	.0001	.0001	.0002
21	.0000	.0000	.0000	.0000	.0000	.0000	.0000	.0000	.0001	.0001

TABLE 7 (Continued)

x	8.1	8.2	8.3	8.4	μ 8.5	8.6	8.7	8.8	8.9	9.0
0	.0003	.0003	.0002	.0002	.0002	.0002	.0002	.0002	.0001	.0001
1	.0025	.0023	.0021	.0019	.0017	.0016	.0014	.0013	.0012	.0011
2	.0100	.0092	.0086	.0079	.0074	.0068	.0063	.0058	.0054	.0050
3	.0269	.0252	.0237	.0222	.0208	.0195	.0183	.0171	.0160	.0150
4	.0544	.0517	.0491	.0466	.0443	.0420	.0398	.0377	.0357	.0337
5	.0882	.0849	.0816	.0784	.0752	.0722	.0692	.0663	.0635	.0607
6	.1191	.1160	.1128	.1097	.1066	.1034	.1003	.0972	.0941	.0911
7	.1378	.1358	.1338	.1317	.1294	.1271	.1247	.1222	.1197	.1171
8	.1395	.1392	.1388	.1382	.1375	.1366	.1356	.1344	.1332	.1318
9	.1256	.1269	.1280	.1290	.1299	.1306	.1311	.1315	.1317	.1318
10	.1017	.1040	.1063	.1084	.1104	.1123	.1140	.1157	.1172	.1186
11	.0749	.0776	.0802	.0828	.0853	.0878	.0902	.0925	.0948	.0970
12	.0505	.0530	.0555	.0579	.0604	.0629	.0654	.0679	.0703	.0728
13	.0315	.0334	.0354	.0374	.0395	.0416	.0438	.0459	.0481	.0504
14	.0182	.0196	.0210	.0225	.0240	.0256	.0272	.0289	.0306	.0324
15	.0098	.0107	.0116	.0126	.0136	.0147	.0158	.0169	.0182	.0194
16	.0050	.0055	.0060	.0066	.0072	.0079	.0086	.0093	.0101	.0109
17	.0024	.0026	.0029	.0033	.0036	.0040	.0044	.0048	.0053	.0058
18	.0011	.0012	.0014	.0015	.0017	.0019	.0021	.0024	.0026	.0029
19	.0005	.0005	.0006	.0007	.0008	.0009	.0010	.0011	.0012	.0014
20	.0002	.0002	.0002	.0003	.0003	.0004	.0004	.0005	.0005	.0006
21	.0001	.0001	.0001	.0001	.0001	.0002	.0002	.0002	.0002	.0003
22	.0000	.0000	.0000	.0000	.0001	.0001	.0001	.0001	.0001	.0001

x	9.1	9.2	9.3	9.4	μ 9.5	9.6	9.7	9.8	9.9	10
0	.0001	.0001	.0001	.0001	.0001	.0001	.0001	.0001	.0001	.0000
1	.0010	.0009	.0009	.0008	.0007	.0007	.0006	.0005	.0005	.0005
2	.0046	.0043	.0040	.0037	.0034	.0031	.0029	.0027	.0025	.0023
3	.0140	.0131	.0123	.0115	.0107	.0100	.0093	.0087	.0081	.0076
4	.0319	.0302	.0285	.0269	.0254	.0240	.0226	.0213	.0201	.0189
5	.0581	.0555	.0530	.0506	.0483	.0460	.0439	.0418	.0398	.0378
6	.0881	.0851	.0822	.0793	.0764	.0736	.0709	.0682	.0656	.0631
7	.1145	.1118	.1091	.1064	.1037	.1010	.0982	.0955	.0928	.0901
8	.1302	.1286	.1269	.1251	.1232	.1212	.1191	.1170	.1148	.1126
9	.1317	.1315	.1311	.1306	.1300	.1293	.1284	.1274	.1263	.1251

TABLE 7 (Continued)

10	.1198	.1210	.1219	.1228	.1235	.1241	.1245	.1249	.1250	.1251
11	.0991	.1012	.1031	.1049	.1067	.1083	.1098	.1112	.1125	.1137
12	.0752	.0776	.0799	.0822	.0844	.0866	.0888	.0908	.0928	.0948
13	.0526	.0549	.0572	.0594	.0617	.0640	.0662	.0685	.0707	.0729
14	.0342	.0361	.0380	.0399	.0419	.0439	.0459	.0479	.0500	.0521
15	.0208	.0221	.0235	.0250	.0265	.0281	.0297	.0313	.0330	.0347
16	.0118	.0127	.0137	.0147	.0157	.0168	.0180	.0192	.0204	.0217
17	.0063	.0069	.0075	.0081	.0088	.0095	.0103	.0111	.0119	.0128
18	.0032	.0035	.0039	.0042	.0046	.0051	.0055	.0060	.0065	.0071
19	.0015	.0017	.0019	.0021	.0023	.0026	.0028	.0031	.0034	.0037
20	.0007	.0008	.0009	.0010	.0011	.0012	.0014	.0015	.0017	.0019
21	.0003	.0003	.0004	.0004	.0005	.0006	.0006	.0007	.0008	.0009
22	.0001	.0001	.0002	.0002	.0002	.0002	.0003	.0003	.0004	.0004
23	.0000	.0001	.0001	.0001	.0001	.0001	.0001	.0001	.0002	.0002
24	.0000	.0000	.0000	.0000	.0000	.0000	.0000	.0001	.0001	.0001

					μ					
x	11	12	13	14	15	16	17	18	19	20
0	.0000	.0000	.0000	.0000	.0000	.0000	.0000	.0000	.0000	.0000
1	.0002	.0001	.0000	.0000	.0000	.0000	.0000	.0000	.0000	.0000
2	.0010	.0004	.0002	.0001	.0000	.0000	.0000	.0000	.0000	.0000
3	.0037	.0018	.0008	.0004	.0002	.0001	.0000	.0000	.0000	.0000
4	.0102	.0053	.0027	.0013	.0006	.0003	.0001	.0001	.0000	.0000
5	.0224	.0127	.0070	.0037	.0019	.0010	.0005	.0002	.0001	.0001
6	.0411	.0255	.0152	.0087	.0048	.0026	.0014	.0007	.0004	.0002
7	.0646	.0437	.0281	.0174	.0104	.0060	.0034	.0018	.0010	.0005
8	.0888	.0655	.0457	.0304	.0194	.0120	.0072	.0042	.0024	.0013
9	.1085	.0874	.0661	.0473	.0324	.0213	.0135	.0083	.0050	.0029
10	.1194	.1048	.0859	.0663	.0486	.0341	.0230	.0150	.0095	.0058
11	.1194	.1144	.1015	.0844	.0663	.0496	.0355	.0245	.0164	.0106
12	.1094	.1144	.1099	.0984	.0829	.0661	.0504	.0368	.0259	.0176
13	.0926	.1056	.1099	.1060	.0956	.0814	.0658	.0509	.0378	.0271
14	.0728	.0905	.1021	.1060	.1024	.0930	.0800	.0655	.0514	.0387
15	.0534	.0724	.0885	.0989	.1024	.0992	.0906	.0786	.0650	.0516
16	.0367	.0543	.0719	.0866	.0960	.0992	.0963	.0884	.0772	.0646
17	.0237	.0383	.0550	.0713	.0847	.0934	.0963	.0936	.0863	.0760
18	.0145	.0256	.0397	.0554	.0706	.0830	.0909	.0936	.0911	.0844
19	.0084	.0161	.0272	.0409	.0557	.0699	.0814	.0887	.0911	.0888

TABLE 7 *(Continued)*

20	.0046	.0097	.0177	.0286	.0418	.0559	.0692	.0798	.0866	.0888
21	.0024	.0055	.0109	.0191	.0299	.0426	.0560	.0684	.0783	.0846
22	.0012	.0030	.0065	.0121	.0204	.0310	.0433	.0560	.0676	.0769
23	.0006	.0016	.0037	.0074	.0133	.0216	.0320	.0438	.0559	.0669
24	.0003	.0008	.0020	.0043	.0083	.0144	.0226	.0328	.0442	.0557
25	.0001	.0004	.0010	.0024	.0050	.0092	.0154	.0237	.0336	.0446
26	.0000	.0002	.0005	.0013	.0029	.0057	.0101	.0164	.0246	.0343
27	.0000	.0001	.0002	.0007	.0016	.0034	.0063	.0109	.0173	.0254
28	.0000	.0000	.0001	.0003	.0009	.0019	.0038	.0070	.0117	.0181
29	.0000	.0000	.0001	.0002	.0004	.0011	.0023	.0044	.0077	.0125
30	.0000	.0000	.0000	.0001	.0002	.0006	.0013	.0026	.0049	.0083
31	.0000	.0000	.0000	.0000	.0001	.0003	.0007	.0015	.0030	.0054
32	.0000	.0000	.0000	.0000	.0001	.0001	.0004	.0009	.0018	.0034
33	.0000	.0000	.0000	.0000	.0000	.0001	.0002	.0005	.0010	.0020
34	.0000	.0000	.0000	.0000	.0000	.0000	.0001	.0002	.0006	.0012
35	.0000	.0000	.0000	.0000	.0000	.0000	.0000	.0001	.0003	.0007
36	.0000	.0000	.0000	.0000	.0000	.0000	.0000	.0001	.0002	.0004
37	.0000	.0000	.0000	.0000	.0000	.0000	.0000	.0000	.0001	.0002
38	.0000	.0000	.0000	.0000	.0000	.0000	.0000	.0000	.0000	.0001
39	.0000	.0000	.0000	.0000	.0000	.0000	.0000	.0000	.0000	.0001

TABLE 8   Random Digits

63271	59986	71744	51102	15141	80714	58683	93108	13554	79945
88547	09896	95436	79115	08303	01041	20030	63754	08459	28364
55957	57243	83865	09911	19761	66535	40102	26646	60147	15702
46576	87453	44790	67122	45573	84358	21625	16999	13385	22782
55363	07449	34835	15290	76616	67191	12777	21861	68689	03263
69393	92785	49902	58447	42048	30378	87618	26933	40640	16281
13186	29431	88190	04588	38733	81290	89541	70290	40113	08243
17726	28652	56836	78351	47327	18518	92222	55201	27340	10493
36520	64465	05550	30157	82242	29520	69753	72602	23756	54935
81628	36100	39254	56835	37636	02421	98063	89641	64953	99337
84649	48968	75215	75498	49539	74240	03466	49292	36401	45525
63291	11618	12613	75055	43915	26488	41116	64531	56827	30825
70502	53225	03655	05915	37140	57051	48393	91322	25654	06543
06426	24771	59935	49801	11082	66762	94477	02494	88215	27191
20711	55609	29430	70165	45406	78484	31639	52009	18873	96927
41990	70538	77191	25860	55204	73417	83920	69468	74972	38712
72452	36618	76298	26678	89334	33938	95567	29380	75906	91807
37042	40318	57099	10528	09925	89773	41335	96244	29002	46453
53766	52875	15987	46962	67342	77592	57651	95508	80033	69828
90585	58955	53122	16025	84299	53310	67380	84249	25348	04332
32001	96293	37203	64516	51530	37069	40261	61374	05815	06714
62606	64324	46354	72157	67248	20153	49804	09226	64419	29457
10078	28073	85389	50324	14500	15562	64165	06125	71353	77669
91561	46145	24177	15294	10061	98124	75732	00815	83452	97355
13091	98112	53959	79607	52244	63303	10413	63839	74762	50289
73864	83014	72457	22682	03033	61714	88173	90835	00634	85169
66668	25467	48894	51043	02365	91726	09365	63167	95264	45643
84745	41042	29493	01836	09044	51926	43630	63470	76508	14194
48068	26805	94595	47907	13357	38412	33318	26098	82782	42851
54310	96175	97594	88616	42035	38093	36745	56702	40644	83514
14877	33095	10924	58013	61439	21882	42059	24177	58739	60170
78295	23179	02771	43464	59061	71411	05697	67194	30495	21157
67524	02865	39593	54278	04237	92441	26602	63835	38032	94770
58268	57219	68124	73455	83236	08710	04284	55005	84171	42596
97158	28672	50685	01181	24262	19427	52106	34308	73685	74246
04230	16831	69085	30802	65559	09205	71829	06489	85650	38707
94879	56606	30401	02602	57658	70091	54986	41394	60437	03195
71446	15232	66715	26385	91518	70566	02888	79941	39684	54315
32886	05644	79316	09819	00813	88407	17461	73925	53037	91904
62048	33711	25290	21526	02223	75947	66466	06232	10913	75336
84534	42351	21628	53669	81352	95152	08107	98814	72743	12849
84707	15885	84710	35866	06446	86311	32648	88141	73902	69981
19409	40868	64220	80861	13860	68493	52908	26374	63297	45052
57978	48015	25973	66777	45924	56144	24742	96702	88200	66162
57295	98298	11199	96510	75228	41600	47192	43267	35973	23152
94044	83785	93388	07833	38216	31413	70555	03023	54147	06647
30014	25879	71763	96679	90603	99396	74557	74224	18211	91637
07265	69563	64268	88802	72264	66540	01782	08396	19251	83613
84404	88642	30263	80310	11522	57810	27627	78376	36240	48952
21778	02085	27762	46097	43324	34354	09369	14966	10158	76089

TABLE 9

T_L Values for the Mann-Whitney-Wilcoxon Test

Reject the hypothesis of identical populations if the sum of the ranks for the n_1 items is *less than* the value T_L shown in the following table or if the sum of the ranks for the n_1 items is *greater than* the value T_U where

$$T_U = n_1(n_1 + n_2 + 1) - T_L$$

$\alpha = .05$		2	3	4	n_2 5	6	7	8	9	10
	2	3	3	3	3	3	3	4	4	4
	3	6	6	6	7	8	8	9	9	10
	4	10	10	11	12	13	14	15	15	16
	5	15	16	17	18	19	21	22	23	24
n_1	6	21	23	24	25	27	28	30	32	33
	7	28	30	32	34	35	37	39	41	43
	8	37	39	41	43	45	47	50	52	54
	9	46	48	50	53	56	58	61	63	66
	10	56	59	61	64	67	70	73	76	79

$\alpha = .10$		2	3	4	n_2 5	6	7	8	9	10
	2	3	3	3	4	4	4	5	5	5
	3	6	7	7	8	9	9	10	11	11
	4	10	11	12	13	14	15	16	17	18
	5	16	17	18	20	21	22	24	25	27
n_1	6	22	24	25	27	29	30	32	34	36
	7	29	31	33	35	37	40	42	44	46
	8	38	40	42	45	47	50	52	55	57
	9	47	50	52	55	58	61	64	67	70
	10	57	60	63	67	70	73	76	80	83

C Summation Notation

Summations

DEFINITION

$$\sum_{i=1}^{n} x_i = x_1 + x_2 + \cdots + x_n \tag{C.1}$$

Example: $x_1 = 5$, $x_2 = 8$, $x_3 = 14$:

$$\sum_{i=1}^{3} x_i = x_1 + x_2 + x_3$$
$$= 5 + 8 + 14$$
$$= 27.$$

RESULT 1

For a constant c:

$$\sum_{i=1}^{n} c = (c + c + \cdots + c) = nc \tag{C.2}$$

Example: $c = 5$, $n = 10$:

$$\sum_{i=1}^{10} 5 = 10(5) = 50$$

Example: $c = \bar{x}$:

$$\sum_{i=1}^{n} \bar{x} = n\bar{x}$$

RESULT 2

$$\sum_{i=1}^{n} cx_i = cx_1 + cx_2 + \cdots + cx_n \tag{C.3}$$
$$= c(x_1 + x_2 + \cdots + x_n) = c \sum_{i=1}^{n} x_i$$

Example: $x_1 = 5$, $x_2 = 8$, $x_3 = 14$, $c = 2$:

$$\sum_{i=1}^{3} 2x_i = 2 \sum_{i=1}^{3} x_i = 2(27) = 54$$

RESULT 3

$$\sum_{i=1}^{n}(ax_i + by_i) = a\sum_{i=1}^{n}x_i + b\sum_{i=1}^{n}y_i \tag{C.4}$$

Example: $x_1 = 5$, $x_2 = 8$, $x_3 = 14$, $a = 2$, $y_1 = 7$, $y_2 = 3$, $y_3 = 8$, $b = 4$:

$$\sum_{i=1}^{3}(2x_i + 4y_i) = 2\sum_{i=1}^{3}x_i + 4\sum_{i=1}^{3}y_i$$

$$= 2(27) + 4(18)$$

$$= 54 + 72$$

$$= 126$$

Double Summations

Consider the following data involving the variable x_{ij}, where i is the subscript denoting the row position and j is the subscript denoting the column position:

		Column		
		1	**2**	**3**
Row	**1**	$x_{11} = 10$	$x_{12} = 8$	$x_{13} = 6$
	2	$x_{21} = 7$	$x_{22} = 4$	$x_{23} = 12$

DEFINITION

$$\sum_{i=1}^{n}\sum_{j=1}^{m}x_{ij} = (x_{11} + x_{12} + \cdots + x_{1m})$$
$$+ (x_{21} + x_{22} + \cdots + x_{2m})$$
$$+ (x_{31} + x_{32} + \cdots + x_{3m}) + \cdots \tag{C.5}$$
$$+ (x_{n1} + x_{n2} + \cdots + x_{nm})$$

Example:

$$\sum_{i=1}^{2}\sum_{j=1}^{3}x_{ij} = x_{11} + x_{12} + x_{13} + x_{21} + x_{22} + x_{23}$$

$$= 10 + 8 + 6 + 7 + 4 + 12$$

$$= 47$$

DEFINITION

$$\sum_{i=1}^{n}x_{ij} = x_{1j} + x_{2j} + \cdots + x_{nj} \tag{C.6}$$

Example:

$$\sum_{i=1}^{2}x_{i2} = x_{12} + x_{22}$$

$$= 8 + 4$$

$$= 12$$

Shorthand Notation

Sometimes when a summation is for all values of the subscript, we use the following shorthand notations:

$$\sum_{i=1}^{n} x_i = \sum_i x_i \tag{C.7}$$

$$\sum_{i=1}^{n} x_i = \sum x_i \tag{C.8}$$

$$\sum_{i=1}^{n} \sum_{j=1}^{m} x_{ij} = \sum_i \sum_j x_{ij} \tag{C.9}$$

$$\sum_{i=1}^{n} x_{ij} = \sum_i x_{ij} \tag{C.10}$$

Answers to Review Quizzes

Chapter 1

1. T	3. F	5. b	7. a	9. b
2. F	4. T	6. b	8. b	10. b

Chapter 2

1. F	5. T	9. F	13. T	17. d
2. T	6. F	10. T	14. b	18. a
3. F	7. F	11. T	15. b	19. c
4. T	8. F	12. F	16. c	20. a

Chapter 3

1. F	5. F	8. F	11. b	14. d
2. F	6. F	9. T	12. b	15. b
3. T	7. F	10. T	13. a	16. b
4. T				

Chapter 4

1. T	5. F	9. T	13. a	17. d
2. T	6. T	10. F	14. c	18. a
3. F	7. T	11. b	15. b	19. d
4. F	8. F	12. a	16. d	

Chapter 5

1. T	5. F	9. T	13. c	17. d
2. F	6. T	10. F	14. a	18. c
3. T	7. T	11. F	15. b	19. d
4. F	8. T	12. F	16. a	20. c

Chapter 6

1. F	5. F	9. T	12. a	15. b
2. F	6. T	10. T	13. a	16. d
3. T	7. T	11. d	14. c	17. c
4. F	8. F			

Chapter 7

1. T	5. T	9. F	13. a	17. b
2. T	6. F	10. F	14. b	18. c
3. F	7. F	11. c	15. b	19. d
4. T	8. F	12. c	16. d	20. b

Chapter 8

1. F	6. F	11. T	15. b	19. a
2. T	7. T	12. F	16. b	20. b
3. T	8. F	13. F	17. a	21. b
4. T	9. F	14. c		22. c
5. T	10. T			

Chapter 9

1. F	3. T	5. F	7. d	9. d
2. F	4. T	6. T	8. c	10. b

Chapter 10

1. T	5. F	8. T	11. d	14. b
2. F	6. F	9. F	12. d	15. a
3. T	7. T	10. T	13. d	16. a
4. F				

Chapter 11

1. F	4. F	7. F	10. c	13. c
2. F	5. T	8. T	11. b	14. d
3. T	6. T	9. a	12. b	15. b

Chapter 12

1. T	6. F	11. F	16. c	21. d
2. T	7. T	12. F	17. d	22. b
3. T	8. F	13. F	18. a	23. c
4. F	9. T	14. F	19. b	24. d
5. F	10. F	15. b	20. d	

Chapter 13

1. T	6. T	11. T	15. c	19. d
2. F	7. T	12. T	16. a	20. b
3. F	8. T	13. a	17. a	21. c
4. T	9. F	14. b	18. a	22. a
5. F	10. F			

Chapter 14

1. F	4. T	7. F	9. b	11. a
2. F	5. T	8. T	10. d	12. b
3. F	6. T			

Chapter 15

1. F	6. T	11. T	16. c	20. d
2. T	7. F	12. T	17. b	21. b
3. F	8. F	13. T	18. b	22. c
4. F	9. F	14. c	19. d	23. d
5. F	10. F	15. d		

Chapter 16

1. T	5. T	9. T	12. c	15. b
2. F	6. T	10. F	13. c	16. b
3. F	7. F	11. a	14. a	17. d
4. F	8. T			

Chapter 17

1. T	5. F	8. T	11. F	14. a
2. T	6. T	9. F	12. c	15. b
3. F	7. T	10. F	13. c	16. a
4. F				

Answers to Even-Numbered Problems

Chapter 1

2. a. all adults (perhaps at least 18 years old)
   b. 1500
   c. 70%
   d. inferring 70% of the population make the same choice
4. a. sample of 391 supermarkets
   b. cost per item $1 to $5; 56% men; 51% under 30
   c. Sample only from Southern California. Cannot make supported statistical inference to nation.
6. a. women whose mothers took DES; women whose mothers did not take DES
   b. 15.8 per thousand women
   c. 7.9 per thousand women
8. a. percent defective, average filling weight, average part size, and so on
   b. 100% inspection too time consuming and expensive
10. a. correct descriptive statistic
   b. incorrect
   c. correct due to the word "estimate"
   d. incorrect
   e. incorrect
12. a. all individuals with first marriage in U.S.
   b. Sample; the population is too large for a census.
   c. median ages: 25.9 for men and 23.6 for women
   d. using sample to make inferences about the population
   e. 3.4 years for men
   f. waiting the same; 3.5 years for women.
14. a. all students taking the SAT
   b. sample of 2800 seniors
   c. 83.5%
   d. A variety of opinions are possible.

Chapter 2

2. nominal
6. a. $+3$, the number of strokes over par
   b. interval
8. c. *Who Framed Roger Rabbit?*

10. a. percentages: 49, 14, 28, 9
12. a. percentages: 5, 10, 24, 48, 13
   d. Service appears to be very good.
14. b. frequencies: 2, 23, 12, 3
   c. percent frequencies: 5.0, 57.5, 30.0, 7.5
   d. cumulative frequencies: 2, 25, 37, 40
20. b. frequencies: 14, 6, 6, 7, 14, 3
   c. percent frequencies: 28, 12, 12, 14, 28, 6
24. midpoints: 1.45, 4.45, 7.45, 10.45, 13.45, 16.45
26. cell frequencies: 25, 5, 30, 20, 5, and 15
28. a. frequencies: 8, 13, 8, 11
   b. percent frequencies: 20, 32.5, 20, 27.5
30. a. frequencies: 2, 8, 12, 5, 3
   b. percent frequencies: 6.7, 26.7, 40, 16.6, 10
32. a. frequencies: 7, 11, 4, 4, 4, 1, 1, 2, 5, 1
   b. percent frequencies: 17.5, 27.5, 10, 10, 10, 2.5, 2.5, 5, 12.5, 2.5
   c. cumulative frequencies: 7, 18, 22, 26, 30, 31, 32, 34, 39, 40
   d. cumulative percent frequencies: 17.5, 45, 55, 65, 75, 77.5, 80, 85, 97.5, 100
34. Start at 10,000 and use class width of 5,000.
   a. frequencies: 2, 0, 2, 6, 8, 5, 4, 1, 2
   b. percent frequencies: 6.7, 0, 6.7, 20, 26.6, 16.7, 13.3, 3.3, 6.7

40.

4	2	4				
5	3	6	8			
6	1	2	2			
7	5	6	7	8	9	
8	2	4	4	5	8	9	9
9	3	5	6	7	8	

42.

4	0	5		
5	1	7	9	
6	5			
7	1	4		
8	0	2	5	
9	0	4	7	7
10	2			
11	1	7		
12	5	7		

Note: Yardage recorded in ten yards.

Chapter 3

2. a. 171.25, 175, 145
   b. 145, 202.5
4. 26.3, 26.5, 18; 70th percentile = 18
6. city: 15.58, 15.9, 15.3
   country: 18.92, 18.7, 18.6 and 19.4
8. a. 48.33, 49, no mode
   b. 45, 55
10. 33, 155.76, 182.087
12. a. 146,419
   b. 6,768
   c. 44,477.658
14. Quarter milers: 5.8; milers: 2.9
16. a. 75%
   b. 89%
   c. 95%, almost all
18. a. 77.5, 9.8573
   b. yes
   c. 16%, 2.5%
20. a. 17.32
   b. 11.74, 3.43
22. a. 34.35
   b. 268.76, 16.39
24. a. 16,517; 20,648; 22,888.5; 27,416; 162,936
   b. inner: 10,496   37,568
      outer: 344   47,720
26. a. 484   1061   2472   4514   32,249
   b. inner: 0   9693.5
      outer: 0   14,873
   c. 15,980 and 32,249
28. a. 12.33
   b. 12
   c. 18
   d. 10
   e. 17
   f. 30.75
   g. 5.55
30. 98,489,368; 9,924.18; 39,500
32. 4553
34. a. 95%
   b. almost all
   c. 125 to 175
36. 51.5   227.37   15.08
38. a. 42, 62, 79, 89, 98
   b. inner: 21.5, 100
      outer: 0, 100
   c. no outliers
40. a. 211, 646, 836, 999, 1451
   b. inner: 116.5, 1528.5
      outer: 0, 2058
   c. no outliers

Chapter 4

2. a. ¼
   b. ¾
4. a. 1024
   b. $\frac{1}{1024}$
6. a. 1,000,000
   b. 5,760,000
   c. more letters
8. No, there are 4 outcomes for the two flips.
12. a. 6
   b. relative frequency
   c. .12, .24, .30, .20, .10, .04
14. a. ¼
   b. ½
16. a. 36
   b. (1,1), (1,2), · · ·, (6,5), (6,6)
   c. ⁶⁄₃₆
   d. ¹⁰⁄₃₆
   e. No, they both have the same probability.
18. a. .31
   b. .69
20. Revise probabilities so they sum to 1.
22. a. .54
   b. .34
24. a. .75
   b. .80
   c. .55
26. a. no
   b. .73
   c. .82
   d. 0
   e. .80
28. .72
30. b. .9
   c. .1
32. a. .75
   b. .667
   c. .50
   d. no
   e. no
34. a. 0
   b. 0
   c. not true
   d. They are dependent.
36. d. .95
   e. .75
   f. no
38. a. .12
   b. .24
   c. no
40. a. .1925
   b. .7075
   c. .2925

42. a. .10, .20, .09
   b. .51
   c. .26, .51, .23
44. a. .21
   b. yes
46. a. $\{E_1, E_2, E_3, E_4\}$
   b. $\{E_1, E_5, E_6, E_7, E_8\}$
   c. $\{E_2\}$
   d. $\{E_7, E_8\}$
   e. ϕ
   f. $\{E_4, E_5, E_6, E_7, E_8\}$
   g. $\{E_1, E_2, E_3, E_4\}$
   h. $\{E_1, E_2, E_3, E_4\}$
   i. $\{E_1, E_2, E_3\}$
   j. no
   k. yes
48. a. .35, .80, .67
   b. no
50. a. .9, 0
   b. no
52. a. .5, .3, .2
   b. no
   c. .40
   d. no
54. a. .30, yes it increases the probability
   b. 20%
   c. .33
56. a. .25
   b. .125
   c. .0125
   d. .10
   e. no
58. a. .20
   b. .35
   c. 65%
60. .21
62. 3.44%

Chapter 5

2. a. discrete
   b. discrete
   c. discrete
   d. continuous
   e. continuous
4. values are 0, 1, 2
6. a. yes
   b. .65
10. b. $^2/_{10}$
   c. $^7/_{10}$
12. a. .05
   b. .70
   c. .40

14. a. 2.45
   b. 2.0475, 1.43
16. b. $-.26$
   c. 24.94
   d. 4.99
18. c. 3.5
   d. 2.9166, 1.708
20. 9475, 97.34
22. a. .5220
   b. .3685
   c. .1095
24. a. .2348
   b. .0936
   c. .0000
26. b. 2
   c. .5, yes
   d. independence
   e. number of girls
30. c. 2
   d. .9409, .0582, .0009
32. .4116, .2646, .2401
34. a. .0324, .2952, .6724
   b. .1936, .4982, .3136
36. a. 2.4
   b. 1.92
38. 212.5, 31.88
40. a. .0183
   b. .0733
   c. .1465
42. a. .0821
   b. .0653
   c. .3840
44. a. .0000
   b. .0104
   c. .0821
   d. .9179
46. a. .50
   b. .0667
   c. .4667
   d. .30
48. .0756
50. a. .0112
   b. .0725
   c. .9163
   d. .0725
52. a. .2013
   b. .1244
   c. .1323
   d. .2541
54. a. no
   b. yes
   c. no
56. a. 2.4
   b. 1.94

58. **b.** 1.65, 1.0275
60. **b.** 10.65
    c. 2.1275
64. **a.** .0150
    b. .9606
    c. .5630
    d. .0388
66. **a.** .1328
    b. .1413
68. **a.** .4354
    b. .4316
    c. .1053
70. .1912
72. **a.** .2241
    b. .5767

Chapter 6

2. **b.** .50
   c. .60
   d. 15
   e. 8.33
4. **b.** .25
   c. .40
   d. 80, 5.7735
   e. zero
6. **b.** 2:17½
   c. .67
   d. .47
10. **b.** .6826
    c. .9544
    d. .5000
    e. .4772
    f. .0228
12. **a.** 3.0
    b. .5 hours
    c. .5000
    d. .4772
    e. .0228
    f. .9772
14. **a.** .3413
    b. .4332
    c. .4772
    d. .4938
    e. .4986
16. **a.** .6640
    b. .1903
    c. .1091
18. **a.** − .80
    b. 1.66
    c. 0.26
    d. 2.56
    e. − .50

20. **a.** .1587
    b. .0228
    c. .6294
    d. .0606
22. **a.** .3830
    b. .1056
    c. .0062
    d. .1603
24. **a.** .7745
    b. 36.32
    c. 19%
26. **a.** 2.865 years
    b. .6247
28. **a.** .0142
    b. .0346
    c. .5910
    d. .1271
30. **a.** 11
    b. .5780
    c. .1170
    d. .6950
32. **a.** 1
    b. .0228
    c. 0
34. **a.** .6321
    b. .3935
    c. .2386
    d. .4109
36. **a.** .0588
    b. .5200
38. **a.** 50 hours
    b. .3935
    c. .1353
40. **a.** .341
    b. .372
    c. .189
42. **a.** .2642
    b. .2736
    c. .2207
    d. .1539
44. **a.** 1.645
    b. .84
    c. − .53
    d. − .25
    e. .72
46. **b.** .20
    c. 37 minutes
    d. 35, 2.89
48. .0062
50. **a.** 17.17
    b. 87.7%
    c. 66.35 to 133.65
52. **a.** 13,020
    b. no
    c. no

54. a. 47.06%
   b. .0475
   c. 42,480
56. a. .0228
   b. Do not use die.
   c. 3.36733
58. a. 54
   b. 4.98
   c. .1841
60. a. 5.16%
   b. 57.87%
   c. 99.55%
   d. 0
62. a. .2865
   b. .3935
   c. .2493
64. a. .2775
   b. .2262
   c. .0784 years

Chapter 7

2. a. all college students receiving federal aid
   b. students whose records were studied
   c. less time and lower cost
4. a. all United flights arriving at O'Hare
   b. the 104 flights studied
   c. 2.88%
   d. prefer FAA
6. a. finite
   b. infinite
8. 20
10. a. 10
   b. 1/10
12. easier and quicker than listing all possible samples
14. a. B and A
   b. E and C
16. a. Select a random page from 1 to 853 and then select a random phone number from 1 to 400.
   b. Discard the selected number.
18.

20. a. 32
   b. 0.976
22. a. $\mu = 14$, $\sigma = 4$
   b. 17, 13, 11, 15, 16, 14, 18, 10, 14, 12

c.

   d. 14, 2.45
   e. 14, 2.45
   f. approach in e is easier
24. 72, 2.34
26. 3.45, 2.37, 1.88, 1.58
   The standard error decreases and n increases.
28. a. 2.175
   b. 2.191
   c. Numerical values are approximately the same.
   d. 0.9927
30. a. 100
   b. 2
   c. normal
   d. normal curve with mean 100 and standard deviation 2
32. a. 105
   b. 0.55
   c. normal
   d. normal curve with mean 105 and standard deviation .55
34. a. only $n = 30$ and $n = 40$
   b. normal with $E(\bar{x}) = 400$ and $\sigma_{\bar{x}} = 9.13$
       normal with $E(\bar{x}) = 400$ and $\sigma_{\bar{x}} = 7.91$
36. normal with $E(\bar{x}) = 68$ and $\sigma_{\bar{x}} = 1.58$
38. 0.4772
40. a. normal with $E(\bar{x}) = 15.9$ and $\sigma_{\bar{x}} = 0.079$
   b. 0.1020
42. a. normal with $E(\bar{x}) = 120$ and $\sigma_{\bar{x}} = 7.3$
   b. 0.0031, 0.2482
44. a. Assume population is normally distributed.
   b. 0.9266
   c. larger sample with $n \geq 30$
46. a. 20.11, 20.26, 20.31
   b. 0.7850, 0.7814, 0.7814
   c. yes; probabilities are approximately the same
48. a. 0.5036
   b. 246
50. a. $\mu = 4.5$, $\sigma = 0.9574$
   b. 2 at 3.5, 3 at 4.0, 5 at 4.5, 3 at 5.0 and 2 at 5.5
   d. 4.5, 0.606
   e. 4.5, 0.606
52. a. 67
   b. 1.5
   c. normal curve with mean 67 and standard deviation 1.5
   d. 0.9082
   e. 0.4972

54. a. no; $n/N = .01$
b. 1.290, 1.297; results approximately the same
56. $\mu = 2$, $\sigma = 0.333$
58. 1102, 5290, 4588, 157, 5498, 5055, 165, 528, 625 and 4516

Chapter 8

2. a. 0.6826
b. 4
4. No, because μ is unknown.
6. 80 to 85
8. 246.42 to 253.58
10. 61.14 to 64.86
12. 12.86 to 14.34
14. 229.53 to 250.47
16. 201.37 to 248.63
18. 33
20. a. 62
b. 385
c. 1537
22. a. 49
b. 30 to 34
24. a. can never reject H_0 with $\bar{x} < 125$.
b. type II error
26. no
28. H_0: $\mu \leq 4$
H_a: $\mu > 4$
30. $z = 2.32$; reject H_0
32. $z = -1.06$; do not reject H_0
34. $z = -2.32$; reject H_0
36. $z = -2.00$; reject H_0
38. Do not reject H_0
40. a. H_0: $\mu = 91,000$
H_a: $\mu \neq 91,000$
b. 0.0340
c. Reject H_0
42. 0.0384; do not reject H_0
44. 0.0734; do not reject H_0
46. a. 1.734
b. -1.321
c. 3.365
d. -1.761 and $+1.761$
e. -2.048 and $+2.048$
48. 7.26 to 9.74
50. 75.90 to 84.10
75.03 to 84.97
52. a. 5.55
b. 4.51 to 6.59
54. a. $t = 2.94$; reject H_0
b. 0.011
56. $t = 3.20$; reject H_0
58. 37 to 37.8

60. a. 7.74 to 10.06
b. 7.08 to 10.72
62. 316
64. 150
66. $z = -1.80$; do not reject H_0
68. $z = -5.06$; reject H_0
p-value approximately 0
70. a. $z = 1.58$; do not reject H_0
b. 0.0571; do not reject H_0
72. 9.19 to 14.81
74. 9.83 to 12.17

Chapter 9

2. a. 0.7750
b. 0.5625
c. 0.4375
4. a. 3 with $\bar{p} = .25$, q with $\bar{p} = .50$ and 3 with $\bar{p} = .75$.
b.

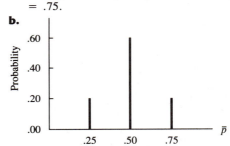

6. a. normal; $E(\bar{p}) = .33$; $\sigma_{\bar{p}} = 0.0665$
b. normal; $E(\bar{p}) = .33$; $\sigma_{\bar{p}} = 0.0470$
8. a. normal; $E(\bar{p}) = .30$; $\sigma_{\bar{p}} = 0.0458$
b. 0.9708
c. 0.7242
10. a. n is too small for normal distribution to be used.
b. 0.9476
12. 0.4714
14. 0.7825 to 0.8175
16. 0.2384 to 0.3616
18. 0.0279 to 0.0621
20. 0.4811 to 0.6189
22. a. 494
b. 543
24. a. $x = 1.31$; do not reject H_0
b. 0.1626 to 0.2574
26. $x = -1.20$; do not reject H_0
p-value $= 0.1151$
28. a. H_0: $p \geq .20$
H_a: $p < .20$
b. $x = -0.56$; do not reject H_0
30. $x = 1.37$; do not reject H_0
32. a. 1 with $\bar{p} = 1$; 8 with $\bar{p} = .75$; 6 with $\bar{p} = .50$

b.

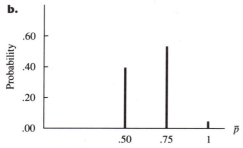

34. a. normal; $E(\bar{p}) = .15$; $\sigma_{\bar{p}} = 0.0505$
    b. 0.4448
36. 0.9525
38. 0.0438 to 0.2362
40. a. 1057
    b. 265
42. $x = -0.72$; do not reject H_0
    p-value $= 0.4716$

Chapter 10

2. a. normal distribution with 5, 2.92
    b. 0.5036
    c. 0.0436
4 a. 125
    b. 36.83 to 213.17
6. a. 1.63
    b. 1.48 to 3.52
8. $x = 2.66$; reject H_0
10. $t = -1.69$; do not reject H_0
    p-value $= 0.10$
12. $t = -1.16$; do not reject H_0
    6.45 to 11.27
14. a. $t = 2.94$; reject H_0
    b. .48 to 7.18
16. -0.0042 to 0.0856
18. 0.0902 to 0.2394
20. a. $x = -3.47$; reject H_0
    p-value close to 0
    b. -0.2357 to -0.0643
22. $x = -2.20$; reject H_0
24. 18.22 to 21.48
    17.55 to 22.15
26. $x = 1.32$; do not reject H_0
    p-value $= 0.1868$
28. 45 each
30. a. normal with equal variances
    b. 1,514,286
    c. $t = 1.11$; do not reject H_0
    d. at least .20
32. a. $t = 5.86$; reject H_0
    b. 1.4 to 3.0
34. 0.0156 to 0.1444
36. $x = -2.23$; reject H_0

38. a. $x = 3.42$; reject H_0
    b. $x = 1.94$; reject H_0
    c. $x = 2.53$; reject H_0
    d. Gum helps abstain from smoking.

Chapter 11

2. 3.95 to 7.59
4. a. 59.68 to 194.42
    b. 7.73 to 13.94
6. $\chi^2 = 36.25$; do not reject H_0
8. a. $\chi^2 = 27.20$; do not reject H_0
    b. 0.2465 to 0.7842
10. a. F distribution with 5 and 7 degrees of freedom
    b. 0.95
12. a. $F = 4$; reject H_0
    b. variance is greater on wet pavement.
14. $F = 1.56$; do not reject H_0
16. $F = 1.36$; do not reject H_0
18. a. 900
    b. 567.29 to 1,690.22
    c. 23.82 to 41.11
20. a. $\chi^2 = 47.25$; reject H_0
    b. 0.3329 to 1.1487; 0.5770 to 1.0718
22. $\chi^2 = 51.2$; do not reject H_0
24. $F = 2.20$; reject H_0
26. $F = .43$; do not reject H_0
28. a. $F = .31$; do not reject H_0
    b. $t = 0.51$; do not reject H_0
    c. $F = 0.40$; do not reject H_0
    d. $t = -2.03$; do not reject H_0
    e. no significant difference found

Chapter 12

2. no significant difference; $F = 2.54 < F_{.05} = 3.24$
4. significant difference; $F = 17.5 > F_{.05} = 3.89$
6. no significant difference; $F = 1.78 < 3.89$
8. significant difference; $F = 4.50 > F_{.05} = 3.68$
10. significant difference; $F = 9.87 > F_{.05} = 3.35$
12. significant difference; $F = 19.99 > F_{.05} = 3.10$
14. significant difference; $F = 35.34 > F_{.05}$
16. cannot reject the hypothesis that the means are equal
18. significant difference between Brand Y and Brand Z
20. significant difference; $F = 56.36 > F_{.05} = 4.46$
22. significant difference; $F = 7.11 > F_{.05} = 3.26$
24. significant difference; $F = 6.578 > F_{.05} = 4.46$
26. no significant effect due to the loading and unloading method ($F = 1.2$), the type of ride ($F = .4$), or interaction ($F = 2.8$)

28. The ANOVA procedure is used to test whether the means of k populations are equal.

30. To work correctly, the ANOVA procedure must have the population variances the same.

32. $MSTR = n \sum_{i} (\bar{x}_i - \bar{\bar{x}})^2/(k - 1)$ is an unbiased estimate of σ^2 when all the population means are the same. When they are not, the squared deviations will be larger causing MSTR to be larger.

34. significant difference; $F = 4.45 > F_{.05} = 4.26$

36. significant difference; $F = 12.90 > F_{.05}$

38. no significant difference; $F = 1.48 < F_{.05} = 3.35$

40. no significant difference; $F = 1.42 < F_{.05} = 4.26$

42. significant difference; $F = 4.02 > F_{.05} = 3.35$

44. a. significant difference; $F = 3.15 > F_{.05} = 2.99$
    b. Evaluations for instructor 2 are significantly higher than those for instructor 3.

46. No, since there was no significant difference using the ANOVA procedure.

48. no significant difference; $F = 0.61 < F_{.05} = 3.89$

50. no significant difference; $F = 1.67 < F_{.01} = 10.92$

52. significant difference due to the language translated ($F = 6.57$); type of system and interaction are not significant.

Chapter 13

2. b. $\hat{y} = 30.33 - 1.88x$

4. There does not appear to be a linear relationship. It looks curvilinear.

6. c. There does appear to be a linear relationship.
    d. $\hat{y} = 6.38 + 45.68x$
    e. $129,716

8. b. $\hat{y} = 47.61 + .44x$
    c. 85.01

10. a. $SSR = 108.47$ $SST = 114.8$
    b. $r^2 = 0.94$

12. a. $SSR = 691.723$ $SST = 1002$
    b. 69.03%

14. a. $\bar{x} = 3.8$ $\bar{y} = 23.2$
    b. Values do satisfy the equation.
    c. yes

16. a. $\hat{y} = 0.75 + .51x$
    b. $t = 1.83$; no significant relationship
    c. $F = 3.33$; we cannot reject H_0.

18. $F = 48.17$, $F_{.01} = 34.12$; selling price and square footage are related

20. $F = 106.92$, $F_{.05} = 5.32$; years of experience and annual sales are related

22. 32.42 to 41.24

24. 112.48 to 143.52

26. There is more variance associated with the estimate of an individual valve.

28. b. significant relationship
    c. $r^2 = .997$; excellent fit
    d. 89.88 to 94.93

30. a. $\hat{y} = 6.1092 + 0.8951x$
    b. $t = 6.01$; $t_{.025} = 2.306$; maintenance expense is related to usage.

32. b. yes
    c. $\hat{y} = -33.88 + .09253x$
    d. significant relationship; $F = 10.85 > F_{.05} = 4.84$
    e. $\hat{y} = 104.915$; 95% prediction interval is 24.90 to 184.93

34. a. $\hat{y} = 2.32 + 0.64x$
    b. Variance appears to increase for larger values of x.

36. c. Assumptions concerning ϵ should be questioned.

38. b. appears to be a negative linear relationship
    c. $s_{xy} = -60$
    d. $r_{xy} = -0.97$

40. $r_{xy} = 0.91$

42. a. $r_{xy} = -0.91$
    b. conclude $\rho \neq 0$

44. In regression analysis we are interested in developing a model of the relationship between a dependent variable and one or more independent variables to help in estimating values of the dependent variable. In correlation analysis we are only concerned with determining if variables are related and not in specifying an equation relating the variables.

46. The estimate of a mean value is an estimate of the average of all y values associated with the same x. The estimate of an individual y value is an estimate of only one of the y values associated with a particular x.

48. b. There appears to be a linear relationship between x and y.
    c. $\hat{y} = -55.84 + 1.67x$
    d. 44.36%
    e. 36.01%; almost identical to the observed value.

50. a. $\hat{y} = 22.173 - 0.1478x$
    b. $F = 11.33$, $F_{.05} = 7.71$; significant relationship
    c. $r^2 = 0.739$; reasonably good fit
    d. 12.295 to 17.271

52. a. There appears to be a negative linear relationship between distance to work and days absent.
    b. $\hat{y} = 8.10 - 0.34x$
    c. $F = 19.70$, $F_{.05} = 5.32$; significant relationship
    d. $r^2 = 0.71$; acceptable fit
    e. 5.22 to 7.58 days

54. a. $\hat{y} = 220 + 131.67x$
    b. $F = 54.75$, $F_{.05} = 5.32$; significant relationship

c. $r^2 = 0.87$; good fit

d. $595.72 to $897.64

56. a. $\hat{y} = 5.85 + 0.83x$

 b. $F = 57.42$, $F_{.05} = 5.32$; significant relationship

 c. 84.70 points

 d. 69.13 to 100.27

Chapter 14

2. The expected increase in the final grade point average corresponding to a one-point increase in the high school grade point average is 0.0235 when the SAT mathematics score does not change. Similar interpretation for b_2.

4. Yield can be expected to increase by 1.17 units when the temperature increases by one degree and the pressure does not change. Similar interpretation for b_2.

6. a. $\hat{y} = 2.1 + 0.0138x_1 + 0.00584x_2$

 b. advertising expenditure, per capita income in the city, store size

8. a. SSE = 507.75

 b. $R^2 = 0.92$

 c. $R_a^2 = 0.90$

 d. excellent fit

10. a. $R^2 = 0.75$

 b. $R_a^2 = 0.68$

 c. reasonably good fit

12. a. SST = 19,065.93

 b. $R^2 = 0.947$, $R_a^2 = 0.93$

14. a. MSR = 3108.188, MSE = 72.536

 b. Since $F = 42.85 > F_{.05} = 4.74$ the overall model is significant.

 c. $t = 7.26$; significant

 d. $t = 8.78$; significant

16. a. SSE = 4000, MSE = 571.43, MSR = 6,000

 b. Since $F = 10.5 > F_{.05} = 4.74$, there is a significant relationship.

18. a. $t = 3.87$; significant

 b. $t = 1.84$; not significant

 c. $t = -5.07$; significant

20. b. Yes. If x_1 and x_2 are correlated one would expect a change in x_1 to be accompanied by a change in x_2.

22. 3.1884

24. 99.08

26. d. No unusual patterns. One point does have a very large residual and would be considered an outlier.

28. b. Residual plot supports the assumptions regarding ϵ.

30. b. $F = 22.79$, $F_{.05} = 5.79$; significant relationship

c. $R_a^2 = .861$; good fit

d. β_1: $t = -5.59$; reject H_0

 β_2: $t = 6.48$; reject H_0

32. a. $2,120,000

 b. $F = 67.70$, $F_{.01} = 6.93$; significant relationship

34. a. $\hat{y} = 88.64 + 1.60x$

 b. $94,240

36. b. $F = 6.58$, $F_{.05} = 4.74$; significant relationship

 c. $R_a^2 = .554$; the fit is only fair

 d. β_1: $t = 1.59$; do not reject H_0

 β_2: $t = -3.33$; reject H_0

 e. 93,680

Chapter 15

2. a. $y = 433 + 37.4x - 0.383x^2$

 b. Significant relationship since the relationship between x and y was significant (Exercise 1).

 c. Confidence interval: 1270.41 to 1333.61

 Prediction interval: 1242.55 to 1361.47

4. b. No; curvilinear.

 c. Several possible models; e.g. $\hat{y} = 2.90 - 0.185x + .00351x^2$ ($R_a^2 = .91$) or $\ln y = 2.60 - 0.998\ln x$ ($R_a^2 = .98$)

6. a. $\hat{y} = 2.15 + .304x_1$ where $x_1 = $ months since previous service call

 b. No; $R^2 = .53$

8. a.

D1	D2	Industry
0	0	1
1	0	2
0	1	3

 $\hat{y} = $ P/E $= 7.54 + 0.183$%PROFIT $+ .213$%GROWTH $+ 2.98$D1 $- .84$D2

 b. Consider the interaction term %PROFXD1 = %PROF times D1

 $\hat{y} = $ P/E $= 10.3 + 0.373$%PROFXD1 ($R_a^2 = .636$)

10. a. $R^2 = .975$

 b. $R_a^2 = .971$

 c. Significant relationship; $F = 244.44 > F_{.05} = 2.76$.

12. a. $\hat{y} = $ %COLLEGE $= -26.6 + 0.0970$SATSCORE

 c. $\hat{y} = $ %COLLEGE $= -26.93 + 0.084$SATSCORE $+ 0.204$%TAKETEST

 d. Same as part (c).

14. a. $\hat{y} = $ WINS $= 14.3 - 0.373$TEAMINT

 b. $\hat{y} = 11.199 - 0.28$TEAMINT $+ 0.00288$PASSING $- 0.026$POINTS

 c. Same as part (b).

16. a. $\hat{y} = 10.30 + 0.373\%\text{PROFXD1}$
   b. Same as part (a).
18. a. $\hat{y} = \text{WINS} = 2.95 + 0.222\text{OPPONINT}$
   b. No unusual patterns
20. a. Observation 7
   b. Observation 7
   c. Observation 15
22. $d = 1.60$; test is unconclusive
24. a.

D1	D2	D3	Paint
0	0	0	1
1	0	0	2
0	1	0	3
0	0	1	4

   $\hat{y} = \text{TIME} = 133 + 6\text{D1} + 3\text{D2} + 11\text{D3}$
   Not significant
   b. 139
26. a. $\hat{y} = \text{AUDELAY} = 80.4 + 11.9\text{INDUS} - 4.82\text{PUBLIC} - 2.62\text{ICQUAL} - 4.07\text{INTFIN}$
   b. $R_a^2 = .312$; not evidence of a good fit.
   c. Possible curvilinear relationship between AUDELAY and INTFIN
   d. Let $\text{INTFINSQ} = (\text{INTFIN})^2$
   $\hat{y} = 112.79 + 11.6\text{INDUS} - 2.49\text{ICQUAL} - 36.6\text{INTFIN} + 6.6\text{INTFINSQ}$
   $R^2 = 59.05$
28. a. $\hat{y} = \text{AUDELAY} = 63.0 + 11.1\text{INDUS}$
   b. $\hat{y} = 112.79 + 11.6\text{INDUS} - 2.49\text{ICQUAL} - 36.6\text{INTFIN} + 6.6\text{INTFINSQ}$
   c. Same as part (b).
30. b. No outliers
   c. No influential observations
32. a. $\text{POINTS} = 394 - 5.10\text{OPPONINT}$
   b. Two usual points
   c. Outlier: observation 2
   d. Large Influence: Observation 7
34. a. $\text{AUDELAY} = 70.6 + 12.7\text{INDUS} - 2.92\text{ICQUAL}$
   b. No obvious pattern indicative of positive autocorrelation
   c. $d = 1.43$; test is inconclusive.
36.

D1	D2	Type of Browser
0	0	non browser
1	0	light browser
0	1	heavy browser

   $\text{SCORE} = 4.25 + 1\text{D1} + 1.5\text{D2}$
   Significant relationship

Chapter 16

2. $\chi^2 = 6.24$, $\chi^2_{.10} = 4.60517$; reject H_0
4. $\chi^2 = 13.23$, $\chi^2_{.05} = 7.81$; historical percentages have changed
6. $\chi^2 = 6.31$, $\chi^2_{.05} = 9.48773$; assumption of independence cannot be rejected
8. $\chi^2 = 37.17$, $\chi^2_{.01} = 9.21034$; classifications are not independent
10. $\chi^2 = 1.54$, $\chi^2_{.05} = 5.99147$; assumption of independence cannot be rejected
12. $\chi^2 = 2.44$, $\chi^2_{.05} = 5.99$; consumer preferences do not appear to vary
14. $\chi^2 = 8.11$, $\chi^2_{.05} = 5.99147$; shift and quality are not independent
16. a. 55%
   b. Political affiliation: $\chi^2 = 1.29$, $\chi^2_{.05} = 5.99$; assumption of independence cannot be rejected. Age: $\chi^2 = 6.26$, $\chi^2_{.05} = 5.99$; we can reject the null hypothesis that being for or against the sales tax is independent of voter age.
18. $\chi^2 = 5.91$; $\chi^2_{.10} = 4.61$; grade point average and the company decision are not independent.
20. $\chi^2 = 51.57$, $\chi^2_{.05} = 12.59$; perception and occupation are not independent.

Chapter 17

2. a. H_0: $p = .50$
     H_a: $p \neq .50$
   b. If less than 5 or greater than 13
   c. Reject H_0
4. $z = 1.90$; do not reject H_0.
6. $z = 3.76$; reject H_0.
8. $z = 1.27$; do not reject H_0.
10. $z = 1.89$; reject H_0.
12. $z = -2.05$; reject H_0.
14. $T = 28.5$; reject H_0.
16. $z = 3.33$; reject H_0.
18. $z = -.25$; do not reject H_0.
20. $W = 9.06$; do not reject H_0.
22. $W = 8.03$; reject H_0
24. a. $+.60$
   b. $z = 1.82$; do not reject H_0.
26. $r_s = .04$; $z = .12$; do not reject H_0.
28. $z = .82$; do not reject H_0.
30. $z = 1.18$; reject H_0.
32. $z = 2.96$; reject H_0.
   $z = 0.68$; do not reject H_0.
34. $z = -2.59$; reject H_0.
36. $z = -2.97$; reject H_0.
38. $W = 12.61$; reject H_0.
40. $r_s = -.92$; $z = -3.05$; reject H_0.
42. $r_s = .76$; $z = 2.84$; reject H_0.

Index

events, 106–107
  complement of, 112–113
  independent or dependent, 122
  intersection of, 113–114
  mutually exclusive, 124
  related, 119
  union of, 113
expected frequencies
  for contingency tables, 613
  and observed frequencies, 607
expected value
  for binomial probability distribution, 162–163
  of random variable, 147–148
  of sample mean, 230
  of sample proportion, 306
experimental units, 394
experiments, 100, 392
  counting rule for multi-step, 101
  design of, 393–394
  factorial, 421–426
  outcomes for, 101
exploratory data analysis, 83–85
exponential probability distribution, 203–206
  computing probabilities for, 205
  and Poisson distribution, 206

F

F distribution, 377, 378–379
  inverse relationship of, 379
  tables for, 380–381
F statistic, 571, 572
  for difference among treatment means using
    randomized block design, 417–418
F test, 401–404
  in multiple regression, 525–526
  in regression analysis, 469–470
factor, 394
factorial experiments, 421–426
failure, in binomial experiment, 152
fences, 84–85
finite population
  sampling from, 219–222
  sampling without replacement, 169
finite population correction factor, 232
first-order model with one predictor variable, 555
forward-selection procedure, 571, 573–574
frequencies, observed and expected, 607
frequency distribution, 25–26, 33–34
  cumulative, 35
frequency polygon, 41

G

general linear model, 551–557
goodness of fit, 521–523, 579
  of estimated regression line, 455
  test for, 606–609
Gosset, W.S., 285
graphic methods

bar graphs and pie charts, 26
scatter diagrams, 447–448, 485, 552
for summarizing quantitative data, 39–42
grouped data, measures of location and dispersion
  for, 79–82

H

hinge, upper or lower, 65
histograms, 7, 39–40
home ownership, statistics on, 57
hypergeometric probability distribution, 169–171
hypothesis testing, 253
  on difference in means of two populations,
    339–343
  on difference in proportions of two populations,
    353–355
  and interval estimates, 276–279
  level of significance of, 270
  about median, sign test for, 629–630
  Minitab for small sample procedure for, 293
  one-tailed, 274, 383
  about population mean, 267–279
  about population proportion, 319–322
  on population variance, 374–375
  on regression analysis, 480
  of significance of regression relationship,
    467–469
  for small samples, 288–289
  summary of forms for, 269
  summary of rejection rules for, 276–277
  two-tailed, 272–273, 276–278
  type I and type II errors in, 269–270
  on variances of two populations, 379, 382–383

I

independence assumption, 612
independence test, 610–614
independent events, 122
independent samples, 345
independent variables, 447
  in multiple regression, 512
  range of, 471
  in regression analysis, 528–529
  for regression models, 571–575
  residual plots against, 481–483
indicator variable, 560
inference, statistical, 7–8
infinite population, sampling from, 222
influential observations, 580–583
inner fences, 84
intelligence quotient, 181
interaction, 422, 426, 556
interactive mode, 87
interquartile range, 69
intersection of events, 113–114
  multiplication rule for computing probability,
    122–123
interval estimates

MSE (mean square due to error), 400, 412, 466–467
MSR (mean square due to regression), 469
MSTR (mean square between treatments), 399
multicollinearity, 528–529
multimodal data set, 62
multinomial probability distribution, 606–609
multiple coefficient of determination, 521–523
multiple regression, 447, 511–548
  and analysis of variance, 590–593
  computer solution for, 518
  estimated regression equation for, 516–518
  least squares method in, 514, 516
  model and assumptions for, 513–516
  residual analysis in, 533–537
  testing for significant relationship in, 524–529
multiplication rule, 122–123
multi-step experiments, counting rule for, 101
mutually exclusive events, 124
  addition rule applied to, 116

N

n factorial, 155
negative autocorrelation, 583
Newman-Kuels test, 414
nicotine chewing gum study, 329
nominal scale of measurement, 20, 624
nonparametric methods, 622–650
  conditions for, 624
  Kruskal-Wallis test, 644–647
  Mann-Whitney-Wilcoxon test, 636–642
  rank correlation, 648–650
  sign test, 625–630
  Wilcoxon signed-rank test, 631–634
nonprobability sampling technique, 245
normal equations, 517
  solving, 547–548
normal probability distribution, 186–197
  characteristics of, 186–187
  computing probabilities for, 193–197
  sample size and, 237
  standard, 187–193
null hypothesis, 267
  summary of forms for, 269

O

observations, 4, 23
  influential, 580–583
observed frequencies, and expected frequencies, 607
observed level of significance, 284
ogive, 41–42
one-tail hypothesis tests, 274
  about population proportion, 321
  for two population variances, 383
ordinal scale of measurement, 20–21, 624
  nonparametric methods for, 624
outer fences, 84

outlier detection, 77–78, 84–85
  in regression analysis, 577–579

P

p-values, 281–284
  in hypothesis testing of difference in population proportions, 355
  for small samples, 289–290
pth percentile, calculating, 63–64
paired samples, 345. *See also* matched samples
parameters, 254
  of population, 58
  of regression model, 463
  significance testing of, in multiple regression, 527–528
parametric methods, 623
partitioning the sum of squares, 407
Pearson Product Moment Correlation Coefficient, 490
percent frequency distribution, 25–26, 35
  cumulative, 35
percentiles, 62–64
pie charts, 26
planning value, 264
point estimation
  of population mean, 254
  of population proportion, 305
  properties of, 254–255
Poisson probability distribution, 165–168
  and exponential distribution, 206
  tables for, 166, 167
polygon, frequency, 41
pooled variance estimate, 334–337
pooling, 334
population, 3, 216–217
  finite or infinite, 219, 223
  parameters of, 58, 225
population covariance, 487
population distribution, and sampling distribution characteristics, 235
population mean, 60, 253–297
  determining difference among. *See* ANOVA (analysis of variance) procedure
  hypothesis testing about, 267–279
  interval estimation (large sample case), 258–264
  point estimation of, 254
  and sample mean, 225
  t distribution and, 286–288
population proportion, 304
  estimation of, 313–317
  hypothesis tests about, 319–322
  interval estimate of, 313
  sample size for estimating, 315–317
population variance, 70–71, 366–383
  between-treatments estimate of, 397–400
  hypothesis testing on, 374–375
  interval estimates of, 372–374